Conference of the South African Advanced Materials Initiative (CoSAAMI 2019)

IOP Conference Series: Materials Science and Engineering
Volume 655

Vanderbijlpark, South Africa
22 - 25 October 2019

ISBN: 978-1-7138-0912-8
ISSN: 1757-8981

Printed from e-media with permission by:

Curran Associates, Inc.
57 Morehouse Lane
Red Hook, NY 12571

Some format issues inherent in the e-media version may also appear in this print version.

Institute of Physics
Dirac House, Temple Back
Bristol BS1 6BE UK

Phone: 44 1 17 929 7481
Fax: 44 1 17 920 0979

techtracking@iop.org

Additional copies of this publication are available from:

Curran Associates, Inc.
57 Morehouse Lane
Red Hook, NY 12571 USA
Phone: 845-758-0400
Fax: 845-758-2633
Email: curran@proceedings.com
Web: www.proceedings.com

TABLE OF CONTENTS

ADVANCED ALLOYS

ADDITIVE MANUFACTURING

CATALYSIS

ami
advanced materials initiative

FOREWORD

The Advanced Materials Initiative (AMI) programme was established in 2003 by the Department of Science and Innovation (DSI) (*formerly the Department of Science and Technology (DST)*) and Science Councils that contribute towards research and development on materials science, namely, the Council for Scientific and Industrial Research (CSIR), Council for Mineral Technology (MINTEK) and South African Nuclear Energy Corporation (NECSA).

The objective of establishing this initiative was to research, develop and innovate across the advanced materials value chain. In this context, the advanced materials are defined as all materials that provide opportunities for creating new products and utilising advanced properties through the application of new technologies. From the DSI's strategic consideration, the AMI's key objectives are: to foster the creation of new industries, to improve the competitiveness of old and existing industries, and to promote technology localisation.

The significance of the AMI programme through high-end beneficiation of South Africa's mineral wealth is to position the science councils and universities to contribute towards the provision of solutions to the priority needs of our country, especially South Africa's urgent need to create jobs and assist in the alleviation of our beloved country's Triple Challenge of Poverty, Unemployment and Inequality. This is also part of the government's **nine-point plan** - adding value to our mineral wealth by advancing beneficiation and support to the engineering and metals value chain.

The Conference of the South African Advanced Materials Initiative (CoSAAMI), is an evolution of the Advanced Materials Initiative (AMI) series of conferences which offered a platform for local and international researchers and students, to present their work on topics related to the themes of the conference. The thematic areas of the conference are: powder metallurgy, materials for extreme environment, smart materials, corrosion, light weighting, computational materials science (modelling and simulation), political economy of materials science, advanced manufacturing, alloy development, polymers, ceramics, composites, nanomaterials and novel energy materials.

All papers submitted to the CoSAAMI 2019 were peer-reviewed by a panel of expert referees in accordance with the IOP Peer Review Guidelines. A total of 95 manuscripts were received and through a rigorous review process where each paper received three reviews, only 50 were selected for publishing in a conference proceedings. The papers that were not accepted for publication were offered an option for presentation as a poster.

The support of our panel of reviewers is tremendously appreciated and this conference proceedings would not be possible if it were not for their tireless and committed effort. The CoSAAMI 2019 team thanks you.

science & innovation
Department:
Science and Innovation
REPUBLIC OF SOUTH AFRICA

MINTEK
A global leader in mineral and metallurgical innovation

CSIR
our future through science

necsa
We're in your world
South African Nuclear Energy Corporation SOC Limited

IOP Conf. Series: Materials Science and Engineering **655** (2019) 011002 doi:10.1088/1757-899X/655/1/011002

CHAIRS OF THE CONFERENCE

Conference Chair

Joseph Moema – josephm@mintek.co.za

Conference Co-Chair

Dr Jones Papo – jonesp@mintek.co.za

SECRETARY OF THE CONFERENCE

Kelly Matthews – kmatthew@csir.co.za
Evah Motsego – evahm@mintek.co.za

ADVANCED MATERIALS INITIATIVE NETWORK MEMBERS

The Ferrous Materials Development Network (FMDN), Mintek
The Precious Materials Development Network (PMDN), Mintek
The Light Materials Development Network (LMDN), CSIR
Titanium Centre of Competence (TiCOC), CSIR
The New Materials Development Network (NMDN), NECSA

AMI STEERING COMMITTEE

Llanley Simpson	AMI	DSI
Wilna du Plessis	AMI	DSI
Joseph Moema	FMDN	Mintek
Dr Jones Papo	FMDN	Mintek
Dr Alain Mwamba	PMDN	Mintek
Dr Keneiloe Sikhwivhilu	PMDN	Mintek
Dr Thabang Ntho	PMDN	Mintek
Dr Robert Tshikhudo	TiCoC	CSIR
Dr Shahed Fazluddin	TiCoC	CSIR
Dr Sylvester Bolokang	LMDN	CSIR
Dr Sagren Govender	LMDN	CSIR
Jakkie Postma	NMDN	NECSA
Njabulo Siyakatshana	NMDN	NECSA

LOCAL ORGANISING COMMITTEE

Llanley Simpson	DSI	
Wilna du Plessis	DSI	
KellyAnne Matthews	Conference secretary	CSIR
Evah Motsego	Conference secretary	Mintek
Joseph Moema	Conference chair	Mintek
Dr Jones Papo	Conference co-chair	Mintek
Dr Asimenye Kapito	Technical chair	Mintek
Dr Alain Mwamba	Technical co-chair	Mintek
Mbavhalelo Maumela	Technical co-chair	Mintek
Melanie Smit	Committee member	Mintek
Veronica Morudu	Committee member	Mintek
Thembisile Khumalo	Committee member	Mintek
Hendriette Van der Walt	Committee member	Mintek
Dr Thabang Ntho	Committee member	Mintek
Dr Keneiloe Sikhwivhilu	Committee member	Mintek

IOP Conf. Series: Materials Science and Engineering **655** (2019) 011002 doi:10.1088/1757-899X/655/1/011002

TECHNICAL COMMITTEE

Editors and Chairs of the Technical Committee

Dr Asimenye Kapito	Technical chair
Mbavhalelo Maumela	Technical co-chair
Dr Alain Mwamba	Technical co-chair
Dr Ronald Machaka	Committee member

PEER-REVIEW COMMITTEE

Absalom Mabeba	Mintek
Dr Adewale Adegbenjo	University of Johannesburg
Dr Alain Mwamba	Mintek
Dr Amogelang Bolokang	CSIR
Prof. Andrew Venter	NECSA
Dr Annelize Botes	CSIR
Dr Asimenye Kapito	Mintek
Dr Babatunde Obadele	University of Johannesburg
Dr Bill Moshe	University of Johannesburg
Bongani Ngobe	Mintek
Prof. Charles Siyasiya	University of Pretoria
Dr Cornie van Sittert	CSIR
Dr Daniel Madyira	University of Johannesburg
Dr Daniel Jansen van Vuuren	North-West University
Deon Slabbert	Sasol
Dr Desmond Klenam	Self Employed
Dr Didier Nyembwe	University of Johannesburg
Dr Dreyer Bernard	Nelson Mandela University
Duduzile Nkomo	Mintek
Prof. Elma van der Lingen	University of Pretoria
Emma Molobi	Transnet Engineering
Francis Nturanabo	Vaal University of Technology
Gabisile Mahlase	Tshwane University of Technology
Dr Gary Pattrick	Mintek
Dr Gebhu Ndlovu	Mintek
Prof. Hasani Chauke	University of Limpopo
Dr Hertzog Bissett	NECSA
Jabu Skosana	CSIR
Jabulani Gama	NECSA
Dr Jaco van der Walt	NECSA
Prof. Jianqiang Zhang	The University of New South Wales
Dr Jeffrey Baloyi	Mintek
Dr John Barry	NECSA
Joseph Moema	Mintek
Prof. Kapil Gupta	University of Johannesburg
Dr Keneiloe Sikhwivhilu	Mintek
Kiyaasha Ukabhai	University of the Witwatersrand
Dr Kofi Annan	University of Pretoria
Lebedike Mampuru	Mintek
Lerato Raganya	CSIR
Prof. Lesley Cornish	University of the Witwatersrand
Dr Levy Chauke	CSIR
Dr Lerato Hlekelele	CSIR
Dr Lusubilo Mukucha	Dentist, Yusuf Dadoo
Mandy Seerane	CSIR
Marlise Jordaan	CSIR
Marandela Mulaudzi	Mintek

Mbavhalelo Maumela	Mintek
Melanie Smit	Mintek
Dr Michel Kalenga	University of Johannesburg
Milton Makhofane	NECSA
Dr Mordecai Mashamaite	University of Limpopo
Munyadziwa Ramakokovhu	Tshwane University of Technology
Dr Mxolisi Shongwe	Tshwane University of Technology
Mzuziwezulu Kekana	CSIR
Nana Arthur	CSIR
Prof. Natasha Sacks	University of the Witwatersrand
Narishe Mohlala	Nelson Mandela University
Dr Nicolus Malatji	Tshwane University of technology
Nomsombuluko Masia	Mintek
Dr Nthabiseng Maledi	University of the Witwatersrand
Dr Ntombi Mathe	CSIR
Dr Oladeji Ige	University of Johannesburg
Patrick Ngwenya	University of Pretoria
Prof. Peter Olubambi	University of Johannesburg
Pfarelo Daswa	CSIR
Philhelene Chikamhi	Nelson Mandela University
Prashant Singh	Texas A&M University
Prof. Regina Maphanga	CSIR
Dr Richard Nkhoma	Rheinmetal
Richard Couperthwaite	Mintek, Texas A&M University
Prof. Robert Knutsen	University of Cape Town
Prof. Roelf Mostert	University of Pretoria
Dr Romeo S Fono-Tamo	University of Johannesburg
Dr Ronald Machaka	CSIR
Dr Rose Modiba	CSIR
Sello Tsebe	Mintek
Senzeni Lephuthing	University of Johannesburg
Dr Shahed Fazluddin	CSIR
Dr Silethelwe Chikosha	CSIR
Stephen Masete	CSIR
Dr Steven Lötter	NECSA
Tebogo Motsai	CSIR
Dr Thabang Ntho	Mintek
Thembisile Khumalo	Mintek
Tshepo Patric Nkuna	CSIR
Tshikele Mukongo Daudet	
Tshilombo Ghislain	CSIR
Tulani Mukarati	University of Pretoria
Dr Tumelo Seadira	Mintek
Dr Ulyate Curle	CSIR
Veronica Morudu	Mintek
Vinod Kurup	University of Pretoria
Vongani Chauke	Mintek
Vuyo Motsweni	Mintek
William Matshe	CSIR
Prof. Willie du Preez	Central University of Technology, Free State
Zizo Gxowa	CSIR

EDITING AND FORMATTING MANUSCRIPT

Absalom Mabeba	Mintek
Mbavhalelo Maumela	Mintek

Nomsombuluko Masia	Mintek
Lebedike Mampuru	Mintek

The Technical Committee of the CoSAAMI 2019 Conference would like to express special gratitude to all committee members and referees who volunteered and contributed to the review process.

IOP Conf. Series: Materials Science and Engineering **655** (2019) 011003 doi:10.1088/1757-899X/655/1/011003

Peer review statement

All papers published in this volume of *IOP Conference Series: Materials Science and Engineering* have been peer reviewed through processes administered by the proceedings Editors. Reviews were conducted by expert referees to the professional and scientific standards expected of a proceedings journal published by IOP Publishing.

IOP Conf. Series: Materials Science and Engineering **655** (2019) 012001 doi:10.1088/1757-899X/655/1/012001

The effect of cast structures on texture and mechanical properties of AISI 433 ferritic stainless steel

[1,2] M Maumela, [1]W E Stumpf, [1] C W Siyasiya

[1] The University of Pretoria, Department of Material Science & Metallurgical Engineering, Private Bag X20, Hatfield Campus, Pretoria 0028, South Africa, Tel. +2712 420 3182

[2] Mintek, Advanced Materials Division, Private Bag X3015, Randburg, 2125, South Africa, Tel. +2711 709 4462

E-mail address: mbavhalelom@mintek.co.za

Abstract: The influence of the individual equiaxed and columnar cast structure on the evolution of texture during thermo-mechanical processes (TMP) was investigated using the electron backscatter diffraction (EBSD) technique. The recrystallization texture after cold working and annealing was then related to mechanical properties of the initial cast structures. It was found that the initial equiaxed cast structure exhibited a homogenous γ −fiber texture while the columnar developed an inhomogeneous texture with the predominance of both the Goss and α −fiber textures. While texture has a significant influence on surface smoothness of the drawn product, it was found that it did not significantly affect the yield strength and ultimate tensile strength (UTS), but marginally affected the elongation i.e. it was found to be slighted higher for the equiaxed than the columnar cast structure measured parallel to the rolling direction (RD).

1. Introduction

The ferritic stainless steel (FSS) AISI 433 is a newly developed grade from a parent grade AISI 430 through micro-alloying with Al. Its demand has increased for replacing expensive austenitic stainless in some engineering applications where corrosion resistance is a primary requirement. However, it is highly prone to surface ridging defects under tensile forming operations. It also has limited drawability compared to austenitic grades.

The ridging phenomenon is associated with grain colonies or clusters which exhibit heterogeneous crystallographic texture distribution in finished rolled sheet products and is only manifested when the product is subjected to a tensile strain during forming operation [1]. When the sheet is subjected to at least 10% tensile elongation, the surface displays narrow corrugated ridges, with an average of 50 μm height peaks on one side of the sheet surface coinciding with troughs on the opposite side, aligned parallel to the RD (rolling direction) [2, 3]. This behaviour is undesirable in the use of aesthetic design where surface quality is highly considered.

It has been reported that surface ridging resistance and formability are improved through recrystallized homogeneous γ −fiber texture components on the final sheet product and this is contrary to the heterogeneous distribution of texture, which degrade these properties [3]. The heterogeneous texture distribution is developed from the initial as-cast columnar structures (ICCS) texture, $\{100\}//\mathrm{ND}$ (normal direction), during the TMP i.e. it develops into the α −fiber, γ −fibre and, Goss texture components. However, the initial as-cast equiaxed structures (IECS) exhibits random texture which leads to the homogeneous γ −fiber texture after TMP [4]. The random texture of equiaxed solidification structures and elongated columnar structure with $\{100\}//\mathrm{ND}$ were confirmed by Sudipta et al. [2]. Kim et al. [5] applied electromagnetic stirring (EMS) in continuous casting to suppress the columnar grains with < 001 >//ND orientations and instead promote the grain refinement and the equiaxed grains with random texture in order to improve ridging resistance of FSS sheet.

In this study, the effect of the individual IECS versus ICCS on texture evolution during rolling and annealing process was investigated. The observed recrystallized textures were then related to the mechanical properties, by plotting stress-strain curves.

2. Experimental procedure

Industrially produced as cast steel slab sample, 200 mm thickness and of the chemical composition in Table 1, was supplied by Columbus Stainless (SA). The slab sample was ground using 2400 grit on both opposite cross-sectioned surfaces and micro-etched with aqua regia solution to reveal solidification structures. Zones containing equiaxed and columnar solidification structures were mechanically separated into equal size blocks with a thickness of 25.5 mm.

Table 1. Chemical composition of the steel slab of AISI 433

Element	C	N	Cr	Ni	Mn	Si	Cu	V	Al	Nb	Fe
Mass %	**0.048 max**	0.037	**16.40**	**0.160**	0.510	0.290	0.090	0.120	**0.261**	0.002	Bal.

The industrial hot rolling process was simulated as shown in figure.1 overleaf. Thereafter, both the ICCS and IECS strips were cold rolled 62 % to 1.5 mm final thickness and annealed at 900 °C for 3 minutes then water quenched. Samples for metallurgical analysis were obtained after each rolling stage.

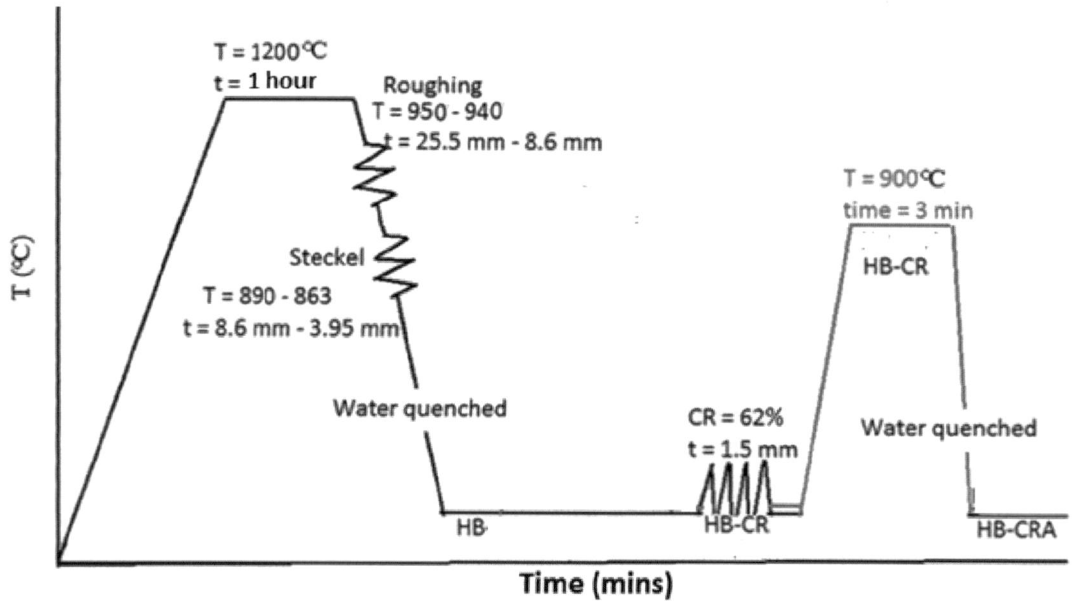

Figure 1. Thermo-mechanical schedule applied to the sample

The texture analysis was performed on the rolling plane (RD-TD) and was measured by the SEM-EBSD technique. The results were presented in Orientation Distribution Function (ODF) at section $\varphi_2 = 45°$, Bunge notation, see figure 2. Tensile testing was performed after cold working and annealing (HB-CRA) according to ASTM E 8 sub-standard (gauge length of 50 mm, thickness of 1.5 mm and width of 20 mm). Three tensile specimen were pulled in the RD per sample at a cross head speed of 2 mm/minute in an Instron Model 1175 machine. The strain was measured by using strain gauge device, attached on specimen gauge length. The specimens were pulled until fracture.

The section ($\varphi_2 = 45°$) was selected because it defines almost all bcc steel fiber texture components, that form during processing of steel, i.e. casting, rolling and annealing processing stages. The fiber texture components were mainly α –fiber, γ –fiber, Goss, Cube texture, rotated Cube texture and other texture components as are shown in figure 2.

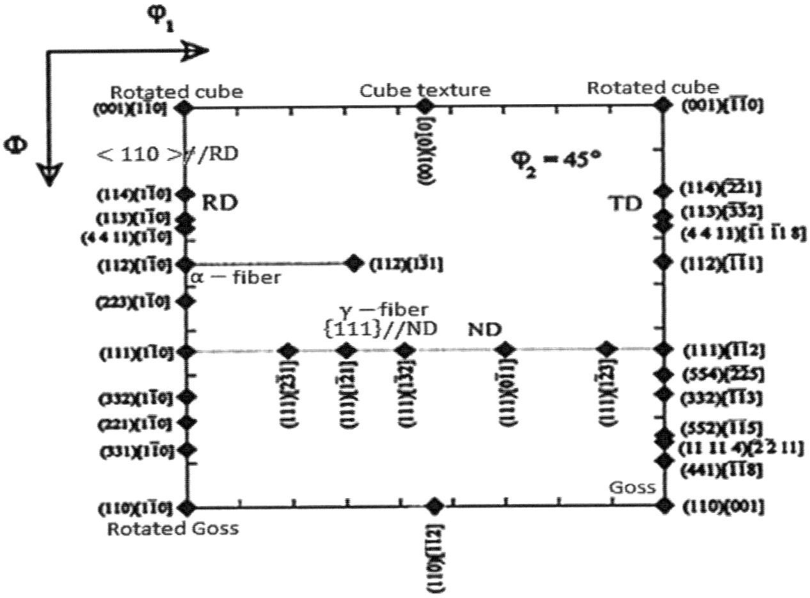

Figure 2. Ideal texture components located at ODF $\varphi_2 = 45°$ section of bcc steels

3. Results and Discussion

3.1. Cold rolling (HB-CR) texture

Figure 3 shows the ODFs for the IECS and ICCS after cold rolling (HB-CR). As may be seen, the IECS exhibited a strong γ −fibre texture components and a weak Cube texture. On the contrary, the ICCS attained two main texture components with the exception of the desired γ −fibre texture components, i.e. rotated Cube texture components with 3.574/4.011 intensity and followed by $(112)[1\bar{3}1]$ with 2.770/4.011 intensity, figure.3 (b). In other words, unlike the IECS, the ICCS promoted heterogeneous texture distribution. The IECS strongly oriented (111)-grains, which could be attributed to the pre-existing hot band (111)-grain orientations as a result of a random as-cast texture in equiaxed as-cast structures [1]. The (111)-orientation is known to store a larger amount of deformation energy than any other orientation and, therefore influences the neighbouring grains to rotate towards the same (111)-orientation during cold rolling [2].

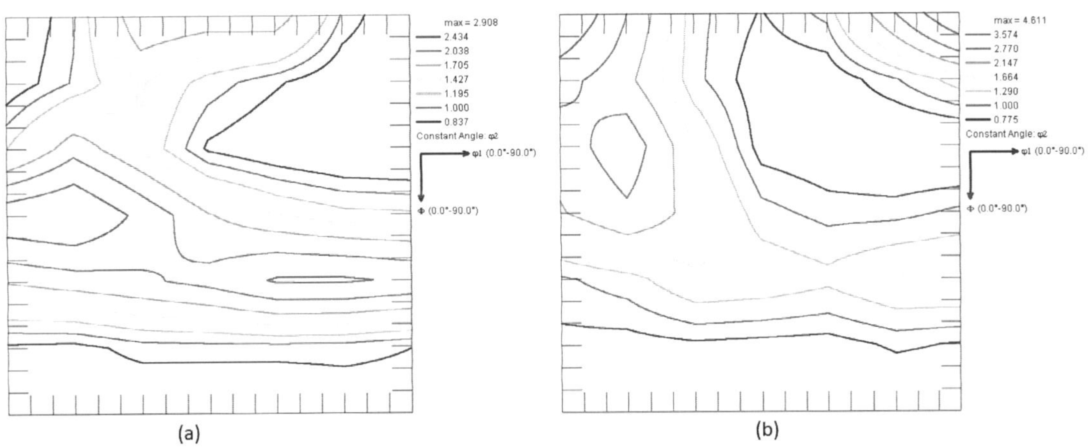

Figure 3. HB-CR rolling ODF texture at section $\varphi_2 = 45°$, (a) IECS and (b) ICCS

The ICCS (HB-CR) sample exhibited a mixture of (001) and (111)-oriented grains. The strong (001)-orientation was presumably inherited from the columnar as-cast grains with an initial (100) <uv0> as-cast texture, and which survived the hot rolling stage as (001) <110> components. Shin et al reported that the ND//<001> orientation of the columnar grains, rotates to the $\{001\} < 110 >$ family orientation after a plane strain compression [3]. The transformation of $\{001\} < 110 >$ to $\{111\} < 112 >$ family orientation is difficult because the $\{001\} < 110 >$ family component has a low Taylor factor and in turn contains low stored deformation energy, which leads to a slow transformation rate during hot working and, therefore, the $\{001\} < 110 >$ orientation persisted during the cold rolling stage. This could explain the domination of (001)-oriented grains in the ICCS specimen during subsequent cold rolling stage.

3.2. Annealing (HB-CRA) recrystallization texture

The recrystallized texture of IECS was mainly characterized by the γ −fiber texture component while the ICCS revealed clusters of γ −fiber and shifted rotated Goss texture components, figure 4. During recrystallization annealing of the cold rolled IECS, the γ −fibre texture components were strengthened and Cube texture diminished, as shown in figure 4 (a) and figure 3 (a) respectively. Conversely in IECS, the ICCS was characterized by both the γ −fibre and shifted rotated Goss texture components with strong and equal intensities, while the rotated Cube texture emerged relatively weak at 1.528 to 1.699 intensity.

Texture homogeneity was attained by IECS during the recrystallization process. As expected, the ICCS recrystallized into three discrete texture components, namely γ −fiber, rotated Cube and rotated Goss texture components at higher intensities, figure 4. Consequently, this promoted texture inhomogeneity.

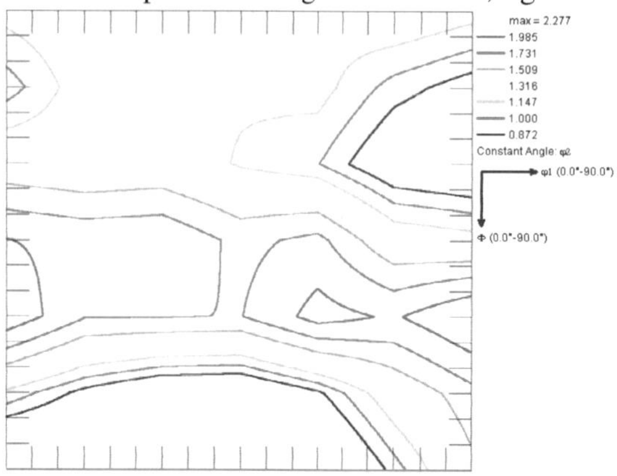

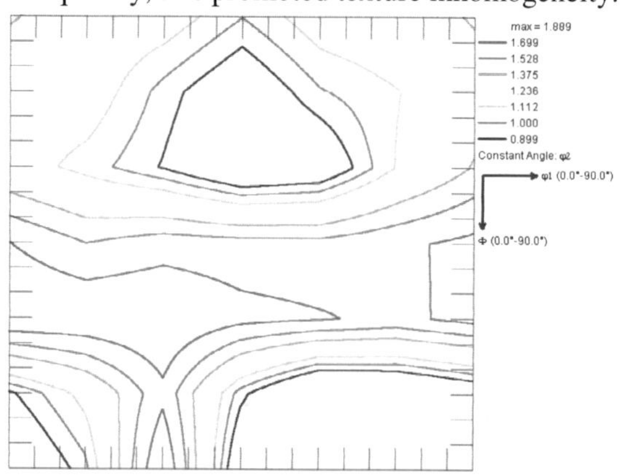

Figure 4. HB-CRA recrystallization texture at section $\varphi_2 = 45°$, (a) IECS and (b) ICCS

Recrystallization nucleation prefers high angle grain boundaries (HAGBs) of deformed grains and the new grains nucleated and grow by consuming the deformed grains while retaining their orientations [4]. In addition, grains with an orientation of high recrystallization rate, such as (111), will nucleate first by adopting that orientation and grow at an expense of other deformed grains with different orientations from (111). However, (100)-orientations are difficult to recrystallize because of their low stored deformation energy and, therefore they survive the scavenging of (111)-oriented grains while others are consumed [5]. This could be an explanation for the textures observed on both IECS versus ICCS in HB-CRA condition.

3.3. Texture effect on mechanical property

Figure 5 presents engineering stress-strain curve of both IECS and ICCS in HB-CRA condition. Both IECS and ICCS exhibited similar yield and ultimate tensile strengths of 270 and 600 MPa respectively, measured in the RD. However, the percentage elongation of the IECS was found to be slightly higher by 1.5% than the ICCS, which is an indication of better drawability.

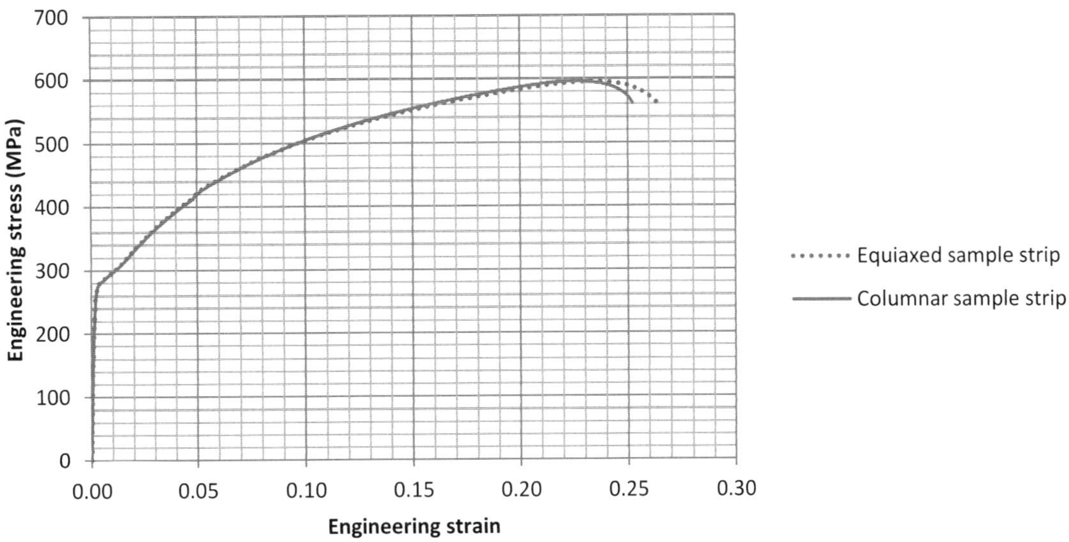

Figure 5. Typical engineering stress-strain curve in HB-CRA condition for IECS and ICCS measured in the RD.

The difference in elongation was attributed to the differences in recrystallization texture between the two samples. Nafisi et al. reported that the plastic deformation response is an anisotropic property i.e. it is influenced by the grains crystallographic texture [5]. A slightly further elongation in IECS was therefore attributed to the homogeneous distribution of recrystallized γ-fibre texture [6, 7]. Raabe et al, Knutsen et al and Zong et al observed that the annealed FSS sheets with prominently γ −fibre texture exhibited higher percentage elongation, and that texture inhomogeneity with orientations other than {111} −texture lowers percentage elongation [8, 9, and 10]. The percentage elongation of IECS material in this study supported findings of Hamada et al. [12] since the co-existence of the Cube and γ −fibre texture components in ICCS material seemed to negatively affect the percentage elongation.

The strength is also an anisotropic property, i.e. the same as elongation; therefore, texture should have an effect on it. Nafisi et al. [5] and Zong et al. [11] argued that the (001) <110> rotated cube components are responsible for the anisotropy in the strength of steel i.e. strongly anisotropic in the 45° and 90° than in the direction 0° angle to the RD. And rotated cube components provide planes that facilitate crack nucleation and propagation across the grains [5]. Therefore, this could explain the similarity in both the yield strength and the UTS measured in the direction of 0° angle to the RD in this work.

4. Conclusions

The effect of the cast structure on the evolution of texture and subsequent mechanical properties in 433 FSS was studied and the following conclusions were made:

- The IECS and ICCS cast structures influence the development of crystallographic texture differently during cold rolling and annealing processes. The IECS promoted homogeneous γ −fiber texture while ICCS promoted heterogeneous texture components with a predominance of γ −fiber, Goss and α −fiber texture components.
- The homogeneous γ −fiber texture components in IECS marginally increased the percentage elongation whereas the inhomogeneous texture in ICCS effected it negatively.

5. Acknowledgements

The authors are grateful to Mintek for the financial support and laboratory facilities, University of Pretoria for the administration, laboratory and study facilities and Columbus Stainless for the supply of material and technical support.

6. References

[1] Raabe, D. 1994. Texture of the strip cast and hot rolled ferritic and austenitic stainless steels. Material science, The institute of materials, The Institute of materials, Manuscript received 24 February 1994, final form in 16 May 1994, The institute of Metallkunde and Metallphysik, RWTH Aachen, Aachen Germany.

[2] Patra, S and Singhal, L.K. 2013. Influence of hot band annealing and cold rolling on texture and ridging of 430 stainless steel containing aluminium. Material sciences and applications, volume 4, pp 70-76.

[3] Shin. H.J, AN. J.K, Park. S.H, and Lee. D.N. 2003. The effect of texture on ridging of ferritic stainless steel. Acta Materialia 51, pp 4693-4706.

[4] Hamada, J., Matsumoto, Y, Fudanodef, and Maeda, S. 2003. Effect of initial solidification on ridging phenomenon and texture in type 430 ferritic stainless steel sheets. ISIJ International, Vol 43, no 12, pp 1989-1998.

[5] Nafisi. S, Arafin. M.A, Collins. L, AND Szpunar. J. 2012. Texture and mechanical properties of API X100 steel manufactured under various thermos-mechanical cycles. Material Science and Engineering A 531 (2012), pp 2-11.

[6] Choi. J. H, Choi. J. Y, Kim. J. C, Kim. J. J, and Kim. S. K. 2009. Control of columnar to equiaxed transition in continuous casting of 16% Cr Ferritic Stainless Steel. POSCO, Korea, Metallurgical Italian, September 2009.

[7] Blandford. P. 1989. Through thickness inhomogeneity of steel-sheet texture and its effect on material properties. A thesis submitted to the Faculty of Graduate Studies and Research in Partial Fulfilment of the requirements for the degree of Masters of Engineering, Department of Mining and Metallurgical Engineering, McGill University, Montreal Canada, November 1989.

[8] Raabe. D, Lee. J, H, Young. M, Engler. O and Park. S. 2005. Effect of through-thickness Macro- and Micro-texture gradient on ridging of 17% Cr Ferritic Stainless Steel Sheet. Division of Materials Science and Engineering, Korea University, Seoul, Korea. Steel research int 76, Vol no.11, pp 797-806.

[9] Knutsen. R.D and Wittridge. N.J. 2002. Modelling surface ridging in ferritic stainless steel. Centre for Material Engineering, Department of Mechanical Engineering, University of Cape Town, 7701 Rondebosch, RSA.

[10] Lioyd. D.J, Wu. P.D. and Huang. Y. 2006. Correlation of ridging and texture in ferritic stainless steel sheet, Department of Mechanical Engineering, McMaster University Hamilton, Ont, Canada. Material Science and Engineering A 427 (2206), pp 241-245.

[11] Zong. C, Zhu. G and Mao. W. 2013. Effect of crystallographic texture on anisotropy of yield strength in API X 100 pipeline steel. Journal of Iron and Steel research, international 2013, 20(8), pp 66-71.

[12] Hamada. J.I., Inoue. H. and Ono. N. 2011. Effect of texture on R-value of ferritic stainless steel sheets. Steel Research Laboratories, Nippon Steel Corporation, 20-1 Shintomi, Futsu, Chiba, 293-8511, Japan, Department of Material Science, Graduate School of Engineering, Osaka Prefecture University, 1-1 Gakuen-Cho, Naka-ku, Sakai, Osaka, 599-8531, Japan.

Effect of the quench temperature on the mechanical properties of a medium C Mn high Si steel during Q&P heat treatment process

V Kurup*, C W Siyasiya and R J Mostert

Department of Metallurgy and Material Science, University of Pretoria, South Africa

E-mail: vinod.kurup@up.ac.za

Abstract. A steel of composition Fe-0.2C-3Mn-2Si-0.5Al was fully austenized followed by a quench and partitioning heat treatment process (Q&P). The quench temperature was varied, which resulted in different volume fractions of retained austenite (RA) and martensite. Analysis of the phase evolution and the resulting microstructures during the Q&P process were carried out using different techniques namely, - dilatometry, FEG SEM, EBSD, and neutron diffraction. Mechanical properties were evaluated by standard tensile tests on samples quenched to different temperatures. The partitioning process was evaluated by dilatometry. The volume fraction of the RA was determined by neutron diffraction. It was found that the volume fraction of RA increased with an increase in the quench temperature contrary to the Speer model. It was also observed that the presence of bainite, which formed during the quench and partitioning temperature significantly stabilized the RA by carbon partitioning. The tensile test results indicated the optimum quench temperature for the best combination of strength and ductility and contrary to expectation, this did not occur in the specimen with the maximum amount of RA. In other words, the mechanical properties of the steel undergoing a Q&P process is influenced by the quench temperatures and is also affected the phase evolution which occurs during both the quench and partitioning process.

1. Introduction

The average use of advanced high strength steel (AHSS) has increased significantly over the past decade in the automobile sector [1]. This was mainly due to the change in material concepts to meet the requirements of OEMs in terms of performance and safety consideration. Third generation AHSS is based on multiphase microstructure targeting strength higher than 1000 MPa [2]. Speer et al. [3] proposed a quench and partitioning (Q&P) process as one of the methods to manufacture third generation AHSS. This process consists of heating steel above Ac_3 to obtain a fully austenitic structure, or between Ac_1 and Ac_3 to obtain partial austenitic structure. It is then quenched to a predetermined temperature between the M_s and M_f to obtain a certain fraction of martensite and retained austenite (RA). It is then reheated to a temperature well below Ac_1 for a certain time interval which enables the transfer of carbon from martensite to austenite, this being referred to as the partitioning stage. Finally, the sample is quenched back to room temperature. In this manner, a multiphase microstructure consisting of tempered martensite, austenite, fresh martensite and ferrite, in the case of partial austenitisation, is obtained. Some authors [4,5] have reported competing reactions, such as carbide precipitation and austenite decomposition to bainite during the partitioning stage.

In the present work, the evolution of microstructures at various quench temperatures during the Q& P process was studied. Using Speer's CCE (constraint carbon equilibrium) model [3] the optimum quench temperature at which maximum RA was obtained. The different quench temperatures were selected above and below the optimum quench temperature and the RA was measured using neutron diffraction. Dilatometry was used to identify the austenite decomposition at the partitioning temperature. Tensile testing was performed to evaluate to see the effect of microstructure on mechanical properties. The present study showed that the formation of bainite did influence the amount of retained austenite present in the microstructure. At quench temperatures greater than the optimum quench temperature the amount of RA was more than predicted by the Speer model [3]. The optimum quench temperature sample, however, exhibited better ductility than the sample with a larger amount of retained austenite.

2. Experimental Procedure

2.1 Material

A high Si content steel was produced in the lab with a composition of Fe-0.27%C -2.8%Mn -2.2% Si-0.75% Cr and 0.5% Al using induction melting. The cast ingot was hot rolled and forged to a round bar of diameter 13mm and length 30cm. Silicon was specifically added to prevent cementite formation [6]. Even though Si is a ferrite stabilizer, it also stabilizes the RA by retardation of cementite formation.

2.2 Dilatometer Experiments

A series of dilatometer experiments were conducted to obtain critical transformation temperatures before the start of the Q&P heat treatment process. Cylindrical samples 10mm x 5mm diameter were used to carry out the heat treatment in the B*ä*hr 805 D[TM] dilatometer. The specimen was placed between two quartz rods, heated by an induction coil and cooled using He gas. Both the temperature and change in length as a function of time were recorded. Using the Speer model [3], the optimum quench temperature was calculated, which gives the temperature at which the maximum RA fraction can be obtained. The optimum quench temperature was calculated as 225°C. Specimens were quenched at different temperatures above and below the predicted optimum quench temperature. The partitioning process was carried out at 400 °C before quenching to room temperature. The quench temperatures were 180, 200, 220, 225, 240 and 260 °C. The heat treatment cycle followed is show in Figure 1. Using the lever rule, the fraction of RA and martensite at each quench temperature was deduced from the change in length vs temperature plot obtained from the dilatometry.

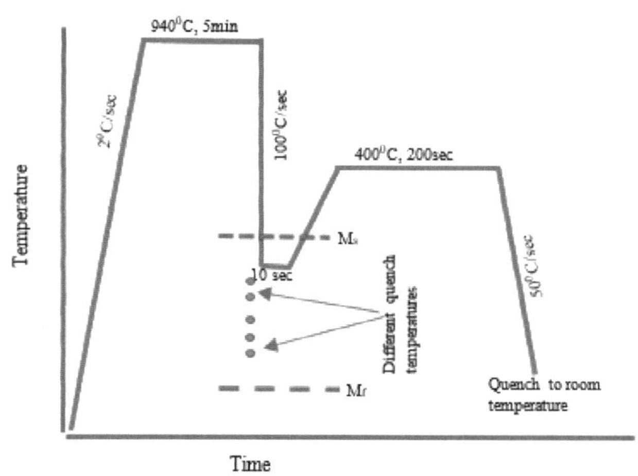

Figure 1. Schematic diagram of quench and partitioning heat treatment with different quench temperatures.

2.3 Neutron Diffraction

The volume fraction of phases after the Q&P heat treatment was determined by neutron diffraction performed at Necsa (Nuclear Energy Council of South Africa). The advantage of neutron diffraction in comparison to XRD is that the neutron beam penetrate deeper into the sample. The intensity and positions of each peak were analysed using "TOPAZ 4.2" peak fitting software. Rietveld analysis [7] was used for quantification of the volume fraction of phases. The lattice parameters of austenite and martensite were obtained from the diffraction data. In the present study, the C content of the RA was determined using the empirical relationship between lattice parameter and composition [8] as shown in equation (1):

$$a(\gamma) = 0.3556 + 0.00453w_{c} + 0.00095w_{mn} + 0.00056w_{al} + 0.0006w_{cr} - 0.0002w_{ni} \qquad \text{eq (1)}$$

where a (γ) is the lattice parameter of austenite in nm w (%) = weight % of elements.

2.4 Microstructural analysis
The standard sample preparation for metallographic analysis was followed. The microstructures were examined after each Q&P heat treatment cycle by a JEOL FEG-SEM at an accelerating voltage of 5kV. From heat treated samples 3mm diameter disc were cut out and metallographically ground to a thickness of less than 50 μm. These samples were subjected to twin-jet polishing in a "Tenopol 5" jet polishing machine. The jet polishing was carried out using 5% perchloric acid and 95% glacial acetic acid at approximately -15°C. Transmission Kikuchi Diffraction (TKD) was carried out on jet polished specimens The TDK analysis was performed at an accelerating voltage of 30 kV, tilt angle of 20° and a step size of 50 nm. The data was analysed using Aztech from Oxford Instruments.

2.5 Tensile testing:
K-Type thermocouples were spot welded to the test specimens to ensure correct temperature was attained during the heat treatment carried out in a salt bath. Sub size specimens with round cross-section of 6mm diameter and 30 mm gauge length were machined according to ASTME8/E8M-09 [9]. The tensile testing was carried out using MTS Criterion® series 45 tensile machine, on selected specimens quenched to 180, 225 and 260°C.

3. Results
3.1 Thermodynamic analysis
Thermo-Calc and JMatpro thermodynamic softwares were used to calculate the phase fractions and continuous cooling transformation (CCT) diagram of the experimental steel, see figure 2

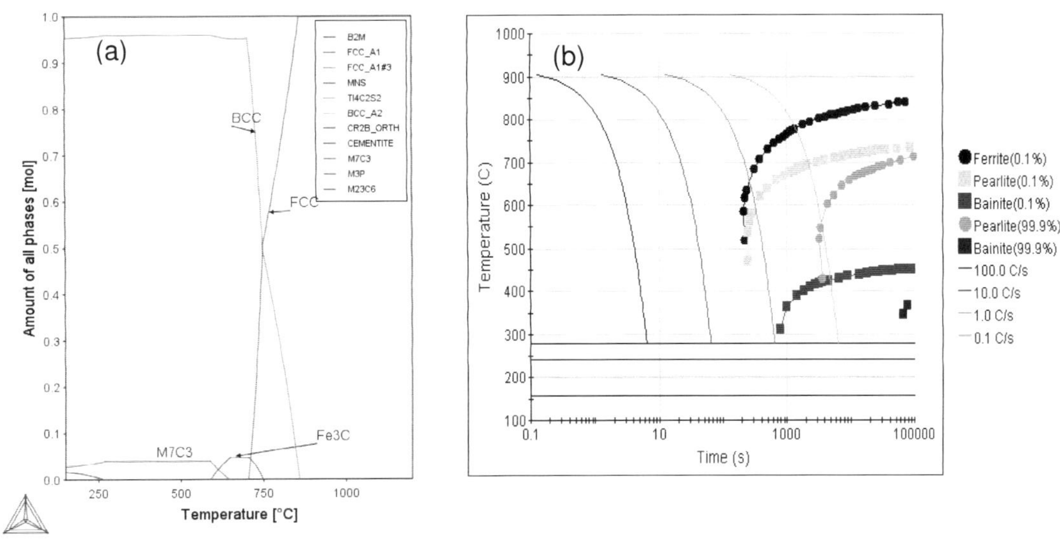

Figure 2. (a) Calculated phase fractions for a range of temperatures using Thermo-Calc (b) CCT diagram of experimental steel using JMatpro

Figure 2(a) shows the phase fraction at various temperatures. It also shows the presence of carbides at low temperature and cementite being stable above 550°C. The presence of Si in the steel shifts the formation of cementite to a higher temperature. Transition carbide stability is not affected by the presence of silicon, and its presence is shown in the temperature range of 200-250°C. Partitioning process in the Q&P heat treatment was carried out at 400°C, and from Figure 2(b), the CCT diagram predicts the formation of bainite at this temperature.

3.2 Optimum temperature calculation by constraint carbon equilibrium (CCE) model

Speer model [3] was used to estimate the optimum quench temperature which produced the maximum RA. The amount of martensite present at each temperature was determined by the K-M equation (2)

$$V^{\alpha}_{primary} = 1 - \exp(-\alpha(M_s - T)) \qquad \text{eq (2)}$$

where $V^{\alpha}_{primary}$ is the fraction of primary martensite formed at the first quench

ά = rate parameter assumed 0.011

M_s= Martensite start temperature.

M_s temperature is based on the empirical equation(3) [11]

$$M_s = 545\text{-}330*\%C+2*\%Al+7*\%Co\text{-}14*\%Cr\text{-}13\%Cu\text{-}23\%Mn\text{-}5\%Mo\text{-}4\%Nb\text{-}13\%Ni\text{-}7\%Si+3\%Ti+4\%V \qquad \text{eq (3)}$$

Using Equation 3, the M_s calculated was 366°C and this was close to the value determined by the dilatometer. During the partitioning stage, carbon is assumed to fully partition to austenite and the carbide precipitation is avoided. As a result, the carbon- enriched austenite has a lower M_s which then leads to stabilization of austenite at room temperature. If the carbon partitioning is not sufficient during the partitioning stage, the austenite after the second quench would transform to secondary martensite .The carbon content of the carbon enriched austenite was determined by CCE condition.[3] which is then substituted back in the M_s Equation 3 to obtain the new M_s for the carbon-enriched austenite. K-M equation 2 is then reapplied with new M_s to get the fraction of secondary martensite ($V^{\alpha}_{secondary}$). The optimum quench temperature was obtained when the M_s was equal to room temperature i.e 25°C. Figure. 3 shows the optimum quench temperature was 225°C with maximum retained austenite of 21.5%. The final austenite content (V_{γ}^{final}) was obtained using equation 4.

$$V_{\gamma}^{final} = 1 - V^{\alpha}_{primary} - V^{\alpha}_{secondary} \qquad \text{eq (4)}$$

where $V^{\alpha}_{secondary}$ = secondary martensite formed after second quench

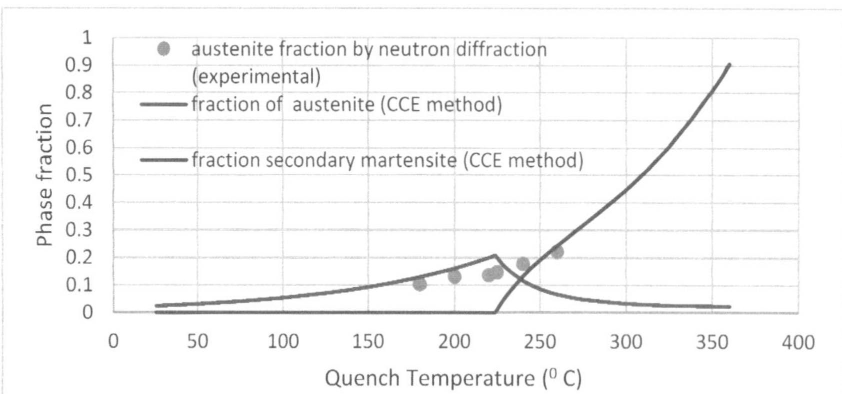

Figure 3. Volume fraction of phases (austenite and secondary martensite) at different quench temperature obtained by the CCE method and fraction of austenite determined experimentally by neutron diffraction

IOP Conf. Series: Materials Science and Engineering **655** (2019) 012002 doi:10.1088/1757-899X/655/1/012002

3.3 Dilatometry

The critical temperatures A_{c1}, A_{c3}, and M_s were determined from dilatometer experiments and were found to be 750, 860 and 370°C respectively. Figure 4 shows Q&P process carried out at different quench temperatures in dilatometer. During cooling, the first expansion took place due to martensitic transformation and then at the partitioning stage there were various degrees of expansion as indicated by a red circle in the 240 and 260°C quench case. During the second quench, there was further expansion indicating the formation of secondary martensite as seen in Figure. 4 (deviation indicated by a blue line in the case 260°C quench). The portion of the isothermal dilation during partitioning was enlarged and plotted separately in Figure 5. The 220°C quench, exhibited an initial expansion which was followed by a contraction, while the 260°C quench exhibited the highest expansion as seen in Figure 5. From Figure 5 for 200°C and 220°C quench, there was initial expansion which was related to bainite formation and followed by a contraction which was attributed to the tempering of martensite.

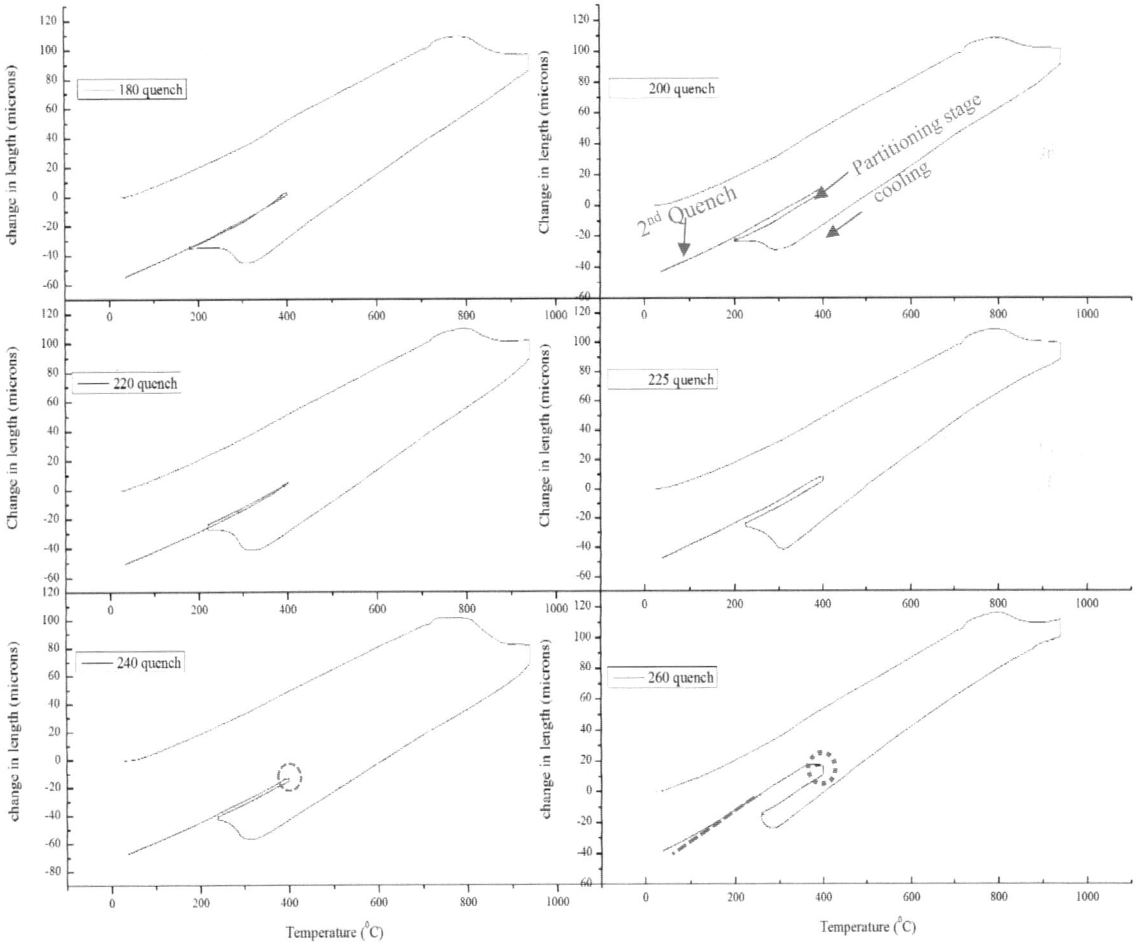

Figure 4. Change in length with the temperature at various quench temperatures during the Q&P heat treatment

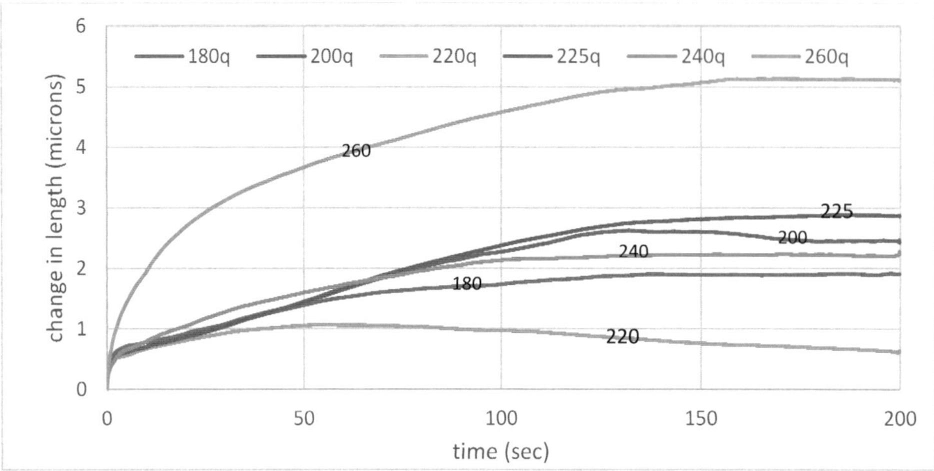

Figure 5. Change in length with time during the isothermal partitioning stage at 400°C

3.4 Neutron diffraction

The neutron diffraction results for the various quench temperatures are presented in Figure 6a. The (111) peak intensity of RA increased with increase in the quench temperatures. The volume fraction of RA obtained from the Rietveld analysis is shown in Figure 3. The results in Figure 3 indicate that the volume fraction of RA was lower than the value predicted by the Speer model up to 225°C and beyond 225°C it was higher than the predicted values. The reason for this difference can be attributed to the formation of bainite during the partitioning stage.

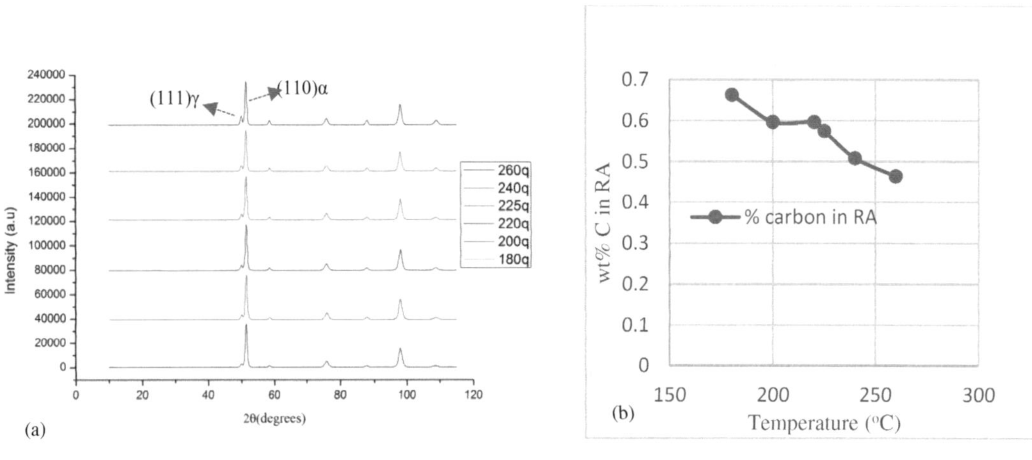

Figure 6. (a) Neutron diffraction of samples quenched at different temperatures during the Q&P heat treatment and (b) weight % C in RA at various quench temperatures

Figure 6(b) shows the weight %C in the RA after the Q&P treatment at various quench temperatures. This was determined by equation 1 using the lattice parameter obtained from the neutron diffraction results. As expected, the amount of C in the RA decreased with an increase in the quench temperature due to an increase in the amount of RA. The maximum carbon content was found to be 0.66 wt. % at 180°C quench temperature.

3.5 Microstructure analysis

The microstructures of the samples quenched at various temperatures are presented in Figure 7. In the sample quenched to room temperature, Figure 7a, the microstructure consisted of needles of martensite of different orientations with no carbides present. In Figure 7 b, c, and d, the microstructure was multiphase and complex. It consisted of following features (a) Martensite (M),elongated lath-shaped without carbides,(b) Bainite which was present in two different forms (B1)thick irregular with wavy edges also without carbides and (B2)small acicular units aligned parallel to each other.(c) Tempered martensite (TM), laths with sharp tip with carbide [12]. The RA was present in the form of thin films which was difficult to detect by SEM. However the TKD analysis of the sample quenched at 260 °C, shows the presence of RA and epsilon carbide as seen in Figure 8.

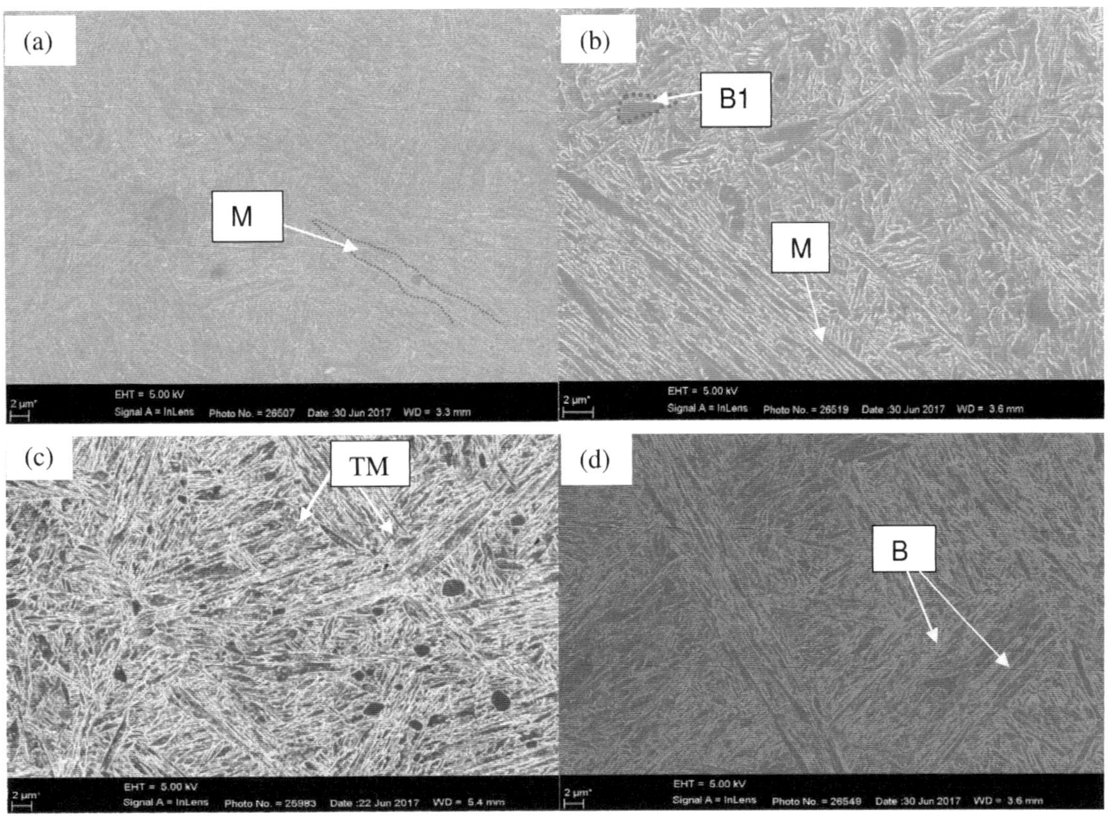

Figure 7. SEM micrographs taken in SE mode (a) direct quench to room temperature (b) 180°C quench (c) 240°C quench and (d) 260°C quench

Figures 8(a) and (b) show the band contrast image and phase colour map from TKD analysis of 260°C quench sample. Figure 8b,showed the distribution of thin films of RA occurred between the martensite laths and bainitic ferrite acicular structure in agreement with Navarro et al [12]. It also revealed the presence of epsilon carbide in the martensite lath structure. There were regions not identified by TKD analysis as shown by the dark contrast. These regions corresponded to either secondary martensite or untempered martensite formed during the second quench [13]. Marito et al. [14] reported that the high dislocation density is known to lower the quality of Kikuchi bands in EBSD patterns and deteriorates the phase identification. The presence of Si prevented the formation of Fe_3C phase but did not prevent the epsilon carbide formation [15].

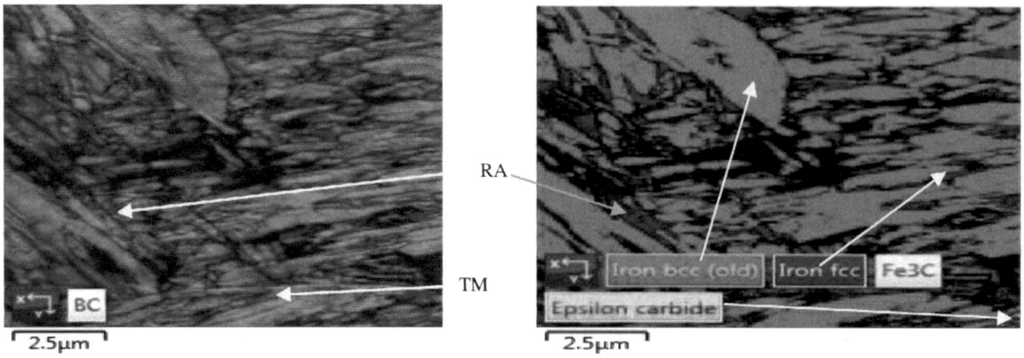

Figure 8. TKD micrograps showung (a) Band contrast and (b) Phase colour map of sample quenched at 260°C

3.6 Mechanical properties:

The tensile test results for samples subjected to Q&P heat treatment at three different quench temperature are summarised in table 1

Table 1. Tensile test results

Quench Temperature(°C)	Tensile strength (MPa)	Yield Strength (MPa)	% Elongation
180 (low)	1360± 15	1070±10	13.5±0.6
225 (optimal)	1320± 10	975±10	15.3±0.6
260 (high)	1440±20	670±10	13.3±0.8

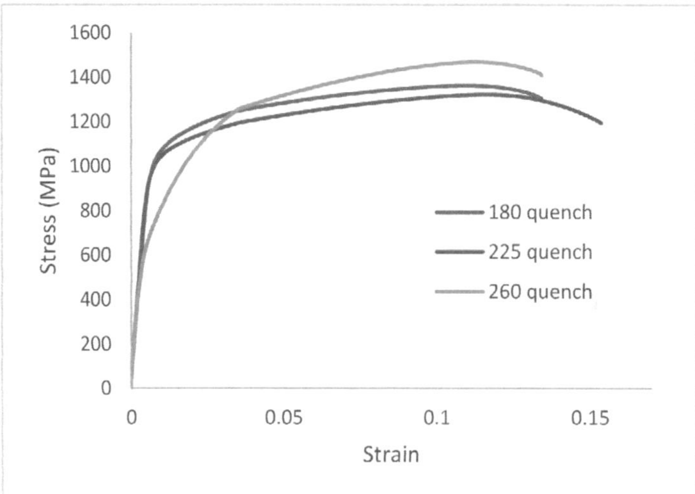

Figure 9. Stress vs strain plot for a samples subject to Q&P process at three different quench temperature

The sample quenched at low temperature (180°C quench) exhibited the highest yield strength of 1070±10. At 225°C quench, both the yield and tensile strength decreased slightly but the average total elongation slightly improved to 15.3%. At high quench temperature of 260°C, the sample had low yield to tensile ratio .This property is attractive for crash worthiness applications requiring high levels of energy absorption and work hardening.

4. Discussion

The Speer model [3] predicted the optimum quench temperature of 225°C where 21% RA would be obtained. However the neutron diffraction results showed maximum RA at 260°C. This difference was associated with the formation of bainite during isothermal partitioning process at 400°C. Figure 5 shows the expansion taking place during the partitioning process at various quench temperatures. The length change due to C partitioning has been found to be around 0.03% [16] which is quite small in comparison to what is seen in Figure 5. This larger length change was due to the isothermal decomposition of austenite to bainite along with carbon partitioning [17] . For the low-temperature quench at 200°C and 220°C, the initial expansion was due to bainite formation and was then followed by contraction due to the tempering of martensite. In the case of the low quench temperatures, the C content in the austenite was high i.e. it exceeds the T_0 value (T_0 is the temperature where free energy of parent and product phase is equal) [15]. The C enrichment in austenite during partitioning was due to bainite formation which ejects carbon to the surrounding austenite. Therefore, RA was stabilized and no further bainite transformation took place as the C in austenite exceeded the C limit at T_0 . This incomplete austenite to bainite transformation can be related to the initial expansion which is seen in the case of 200°C and 220°C quenches. In the 260°C quench, the C concentration in the austenite was low as calculated from lattice parameter values due to high volume fraction of RA. Since carbon content was below the T_0 value this lead to transformation of austenite to bainite and the rejection of carbon from bainite lead to stabilization of austenite. This led to higher RA stabilized at 260°C quench than what was calculated based on CCE [3] .

The behaviour of the stress-strain curve is a function of dislocation density evolution and this is governed by the microstructure [13]. After the Q& P process, the microstructure consisted of tempered martensite, RA and also secondary martensite for the high-temperature quench. In addition, the RA underwent strain-induced martensitic transformation (SIMT). In the 180°C quench, the stability of the RA was higher due to the higher C content (Figure 6 b), so, the extent of the RA to SIMT was expected to be low. This resulted in the higher yield strength. On the contrary, the 260°C quench had higher fraction of RA but due to its low C content in RA, its stability was low. Therefore the extent of RA to SIMT was greater leading to low yield strength to tensile strength ratio (R_p./R_m) and high work hardening observed.

The presence of secondary martensite (Figure. 4) for the 260°C quench sample led to relatively low ductility. It has been reported elsewhere that untempered martensite exhibits a low yield strength and smooth elastic-plastic transition [13] . There are several possibilities for the 260°C quench flow behaviour as proposed by different authors viz. 1) Takaki et al [18] suggested that high density of fresh dislocation are created during the phase transformation and these dislocations have high mobility eliminating yield point effects and causing a rounded stress-strain curve at lower loads; 2) Allain et al [19] suggested that inhomogeneous C distribution in as-quenched martensite can result in a austenite martensite phase mixture containing different hardness. These regions may yield at different stress on loading. In the case of the optimum quench (225°C), there is no secondary martensite formation.and lower fraction of RA compared to high temperature quench (260°C) The high C in RA lead to higher yield strength and no secondary martensite formation gave higher ductility in the optimum quench sample

5. Conclusions

The Q&P process of the medium C Mn high Si steel was studied using different techniques and the following observations were made:

- The sample quenched at 260°C yielded the highest amount RA after the Q&P process. This was attributed to the bainite formation which enriched the austenite to stabilize it.
- The discrepancy between predicted and actual amount of RA was attributed the bainite formation during Q&P process which was not considered in the prediction of the optimum quench temperature which yields maximum RA based on Speer model.
- The optimum quench (225°C) exhibited the best combination of strength and ductility, in spite of lower RA. This was due to the absence of secondary martensite and a higher stability of retained austenite.

6. Acknowledgments:
The authors would like to acknowledge Ferrous Metal Development Network (FMDN) for the financial support. The authors also want to acknowledge NMMU (Nelson Mandela Metropolitan University) for the assistance in TKD analysis. NECSA and Prof JPR de Villiers of University of Pretoria for the assistance in neutron diffraction and rietveld analysis.

7. References

1. Lesch C, Kwiaton N, Klose FB. Advanced High Strength Steels (AHSS) for Automotive Applications – Tailored Properties by Smart Microstructural Adjustments. Vol. 88, Steel Research International. Wiley-VCH Verlag; 2017.
2. Kuziak R, Kawalla R, Waengler S. Advanced high strength steels for automotive industry. Arch Civ Mech Eng [Internet]. 2008;8(2):103–17. Available from: http://www.sciencedirect.com/science/article/pii/S1644966512601976
3. Speer J, Matlock DK, De Cooman BC, Schroth JG. Carbon partitioning into austenite after martensite transformation. Acta Mater. 2003;51(9):2611–22.
4. De Moor E, Speer JG. Bainitic and quenching and partitioning steels. In: Automotive Steels: Design, Metallurgy, Processing and Applications. Elsevier Inc.; 2016. p. 289–316.
5. Toji Y, Miyamoto G, Raabe D. Carbon partitioning during quenching and partitioning heat treatment accompanied by carbide precipitation. Acta Mater. 2015;86:137–47.
6. Zhu L Juan, Wu D, Zhao X Ming. Effect of Silicon Content on Thermodynamics of Austenite Decomposition in C-Si-Mn TRIP Steels. J Iron Steel Res Int. 2006;13(3):57–61.
7. David WIF. Powder diffraction: Least-squares and beyond. J Res Natl Inst Stand Technol [Internet]. 2004;109(1):107. Available from: https://nvlpubs.nist.gov/nistpubs/jres/109/1/j91dav.pdf
8. Van Dijk NH, Butt AM, Zhao L, Sietsma J, Offerman SE, Wright JP, et al. Thermal stability of retained austenite in TRIP steels studied by synchrotron X-ray diffraction during cooling. . Acta Materialia 53 (2005) 5439–5447
9. ASTM E8/8M [Internet]. ASTM. 2016. Available from: www.astm.org
10. Koistinen D.P and Marburger R.E. A general equation presecribing the extent of austenite - martensite transformation in pure iron carbon alloys and plain carbon steels. Acta Metall Mater. 1959;7:59–60.
11. Hsu T. An approximate approach for the calculation of Ms in Fe-base alloys. J Mater Sci 1985;20:23–31.
12. Navarro-López A, Hidalgo J, Sietsma J, Santofimia MJ. Characterization of bainitic/martensitic structures formed in isothermal treatments below the Ms temperature. Mater Charact. 2017;128(March):248–56.
13. Seo EJ, Cho L, Estrin Y, De Cooman BC. Microstructure-mechanical properties relationships for quenching and partitioning (Q&P) processed steel. Acta Mater. 2016;113.
14. Morito S, Nishikawa J, Maki T. Dislocation Density within Lath Martensite in Fe-C and Fe-Ni Alloys. ISIJ Int. 2003;43(9):1475–7.
15. Ravi AM, Sietsma J, Santofimia MJ. Exploring bainite formation kinetics distinguishing grain-boundary and autocatalytic nucleation in high and low-Si steels. Acta Mater. 2016;105:155–64.
16. Santofimia MJ, Zhao L, Sietsma J. Volume Change Associated to Carbon Partitioning from Martensite to Austenite [Internet]. Vols. 706–709, Materials Science Forum. 2012. p. 2290–5. Available from: http://www.scientific.net/MSF.706-709.2290
17. Hajyakbary F, Sietsma J, Miyamoto G, Furuhara T, Santofimia M. Interaction of carbon partitioning, carbide precipitation and bainite formation during the Q&P process in a low C steel. Acta Mater. 2016 Feb 1;104:72–83.
18. Takaki S, Fujimura Y, Nakashima K, Tsuchiyama T. Effect of Dislocation Distribution on the Yielding of Highly Dislocated Iron. Mater Sci Forum [Internet]. 2007;539–543:228–33. Available from: https://www.scientific.net/MSF.539-543.228

19. Allain S, Bouaziz O, Takahashi M. Toward a New Interpretation of the Mechanical Behaviour of As-quenched Low Alloyed Martensitic Steels. ISIJ Int. 2012;52(4):717–22.

IOP Conf. Series: Materials Science and Engineering **655** (2019) 012003 doi:10.1088/1757-899X/655/1/012003

Major determinant of service life in magnesia-graphite slagline refractory lining in secondary steelmaking ladle furnace

I O Otunniyi*, Z V Theko, B L E Mokoena, B Maramba

Metallurgical Engineering, Vaal University of Technology, Vanderbijlpark, South Africa

*Email: iyiolao@vut.ac.za

Abstract. Magnesia-carbon (MgO-C) refractory bricks are used for slagline lining in secondary steelmaking to withstand the harsh basic slag and high temperature environment. In an operation using bottom purging ladle furnaces, preliminary study of residual bricks after lining failures shows less than 50 vol. % of the MgO-C bricks is typically used before the linings would fail. Slag basicity, high temperature oxidation and erosion are known as root causes of different degradation modes in this application. This work, presented in part herein, seeks to identify the dominant cause for this resource utilization inefficiency. Slag basicity excursions and thermal fluctuations versus service lives and residual thicknesses of each MgO-C brick after lining failure, were studied for five ladle linings. Optical and electron microscopy were also used in further post-mortem of the residual bricks. The operation data that showed thermal spikes above 1600 ^{0}C and acidic slags with basicity usually below the ideal of 1.67. Micrographs from the slag-refractory interfaces indicated effect of this operational extremes showing thermal shock cracking, slag penetration and evolution of phase due to the slag-refractory interactions. However, bricks evidencing these effects were not the most worn. Slag basicity attack and high temperature oxidations are diffusion controlled and did not proceed aggressively by themselves. Rather, least residual thicknesses were found in bricks from columns adjacent to the two purging gas entry points at the bottom of the ladle. Up to 86 % wear were obtained at such points, in contrast to 24 % for bricks farthest from this effect. Turbulence from the purging gas is thus responsible for localized and highest lining consumption rate that decimates lining life in these ladles. Without this purging turbulence, such as in induction stirring, it can be projected that the lining life can triple, with obvious economics and environmental advantages. While the MgO-C refractory is well able to withstand the chemical and thermal extremes of secondary steelmaking slag environment, bottom purging ladle furnace design deserves a review to improve refractory utilization.

1. Introduction

Secondary steelmaking ladle operations handling basic slags uses magnesia-carbon refractories for the slagline lining to withstand the process chemistry and temperature [1, 2]. Fluctuation in slag basicity and ladle operating temperatures are monitored and queried as underlying factors behind different wear mechanisms during slag-refractory interactions. When basicity is low, the slag behaves acidic to the lining, and magnesium oxide (MgO) can dissolve from refractory bricks into the slag. At high temperatures above 1400 ^{0}C, carbon within the refractory can reduce magnesia to magnesium [3]. Also, exposing refractory surface to the atmosphere at high temperature can cause direct oxidation of carbon in the refractory [3]. These factors can also synergise such that refractory linings continue to wear with each heat. After a set limiting residual thickness is obtained at any point in the lining, the whole lining is considered to have failed. This is to prevent slag or molten metal penetrating the backing safety lining.

However, slag basicity attack or the high temperature oxidations in this case are heterogeneous reactions. Slag viscosities is notably high, up to five poise at these temperatures, but can be orders of magnitude higher at just 100 ^{O}C lower temperature [4, 5]. The implication is that slag reaction with refractory interface will be diffusion controlled. Slag basicity and high temperature oxidation will therefore be self-limiting processes, and may not be the overriding factors responsible for refractory consumption rates seen in practice. Moreover, operations are supposed to monitor the slag basicity and operating temperatures so they do not exceed certain set values considered safe for the refractory linings. It follows that these factors may not be most decisive in determining the service lives of refractories.
In a secondary steelmaking operation, the slag line MgO-C refractory lining is projected to last a maximum of 85 heats, about 145 hours, while the alumina-magnesia-carbon barrel lining of the molten metal region serves about 165 heats though not commensurately thicker. Assessing residual bricks after breakout of the lining shows a sizeable resource waste, with less than 50 % of total lining volume consumed by the time the lining has to be replaced. In the context of sustainable and efficient resource utilization, this should not be the norm. The refractory are actually chosen to withstand these extreme conditions, while operations also monitor and control temperatures and slag basicity below safe limits. It is interesting therefore to explore what factor is overriding in determination of the service life of the MgO-C slag lining in these operations. The work reported herein is to identify the most contributing factor to the limited life of the MgO-C slag-line refractory linings in this secondary steelmaking electric arc ladle furnaces. To minimize the length of the material herein, some of the results with brief discussions are presented.

2. Methodology

Slagline linings in five ladle furnaces designated Ladle A, B, C, D and E were studied in use and after service. These are 175 tons ladle furnaces' slagline linings. Main feature of these ladle furnace are two purging plugs and a tapping hole all at the bottom. Bottom purging and tapping ladles furnaces are considered as industry standard in ladle metallurgy [6]. In these furnaces, the purging plugs for argon stirring are at 1 O'clock and 7 O'clock radial positions, while the tap hole is at 10 O'clock (Figure 1). The ladle uses combination lining concept for the working lining: the hot metal lining, for the floor and the lower side wall (the barrel), uses alumina-magnesia-carbon bricks; while the slagline (the upper side wall) uses a working lining of magnesia-carbon bricks. Behind these working linings is the safety lining. The slag line lining is projected to withstand 85 heats, while the barrel lining is projected to withstand a total of 165 heats. The second slagline campaign is for the balance of 80 heats of the barrel lining. Hence after maximum of 85 heats, the slag line is removed and relined, except the lining failed before that much heats. During use the lining is regularly scanned using thermal camera to detect hot spots indicative of the residual thicknesses. When any brick within the whole lining is worn down to 30 mm, the entire slag line is considered to have failed. It is broken out and relined. The life of lining is taken as the total number of heats endured before it failed. Of the five linings studied, B and D were first slagline campaign, while A, C and E were second.

Percentage compositions of the slag oxides CaO, SiO_2, FeO, Al_2O_3, MgO and MnO were determined in the slag for every heat, and the slag basicity was determined for each heat from the data using equation 1 [7, 8]:

$$\text{Basicity} = CaO + MgO / (SiO2 + Al2O3 + FeO + MnO) \quad (1)$$

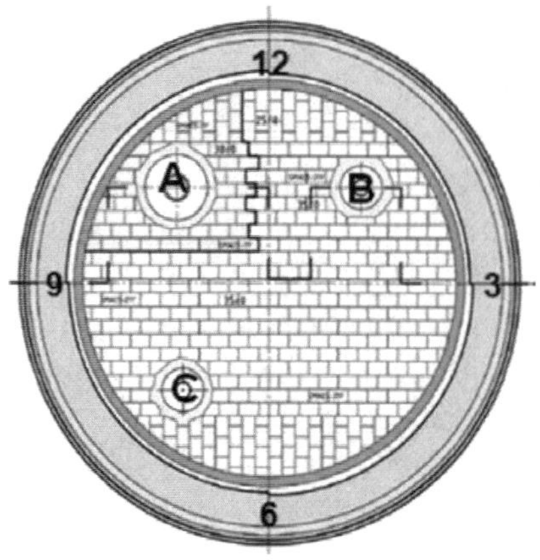

Figure 1. Representation of floor of bottom purging and pouring ladle showing locations of the purging holes, B and C, and the tapping hole A.

MgO saturation in slag is relevant in this situation to avoid MgO leaching from the MgO-C lining, and MgO saturation which varies with basicity [9] can be ensured in operation. For a reference in this operation, the ideal operating basicity is chosen to be 1.67. The resulting data are tabulated and summarized in plots of basicity variations over all the heats for each lining, reference to the ideal basicity.

To see effects of temperature extremes on the life of the lining, the ladle thermal history were studied over all the heats each lining served. The maximum process temperature recorded during each batch operation in the ladle for all the heats (85 maximum) during the entire campaign of a lining were summarised for each ladle. From these maximum temperatures during all heats of a lining, the highest and the least values were extracted and tabulated against the life of that lining. This data is finally summarised in a single plot for all the ladle linings studied.

In the course of the operation, when in-situ measurements indicate any brick within the whole lining has 30 mm residual thickness or lower, the lining life is considered over. The whole lining is broken out. The residual thickness of all bricks are taken and tabulated according to their position in the ladle. Bricks with any distinct feature such as slag penetration, crack, discoloration and least residual thickness were selected for optical and electron microscopy studies to identify overriding mechanism the brick experienced.

3. Results and Discussion

3.1 Basicity Excursions versus MgO-C Refractory Lining Life

Figures 2(a) – 2(e) show the slag basicity excursions of the five linings during operation relative to the ideal operating basicity recommended for the slag-line refractory lining.

The ideal operating basicity for these operations is set to 1.67, obtained from equation (1) based on an ideal slag composition of CaO 55 %, MgO 10 %, Al_2O_3 35 % SiO_2 < 2 % and FeO + MnO <2 % for the operation. This is within 1.52 – 1.88 which is the range of basicity allowable for MgO-C refractory. The figures show that most of the times the slags were relatively acidic, falling below the ideal basicity line for the operation, and also below the accepted range for MgO-C refractory. Lining E, for instance, experienced least basicity of 1.168 during its entire campaign (Figure 2(e)). Low basicity implies more acidic slag, with expected dissolution of MgO from the lining [10]**.** The more acidic, the more aggressive such slag is expected to be, and the shorter the lining life expected. It can be inferred that acid-base interaction and dissolution of MgO would underlie the degradation experience by the linings in all the instances, but this is not in accordance with the corresponding service lives recoded. In Figure 3, the service lives of the five linings were plotted against the maximum and the minimum basicity experienced by each lining.

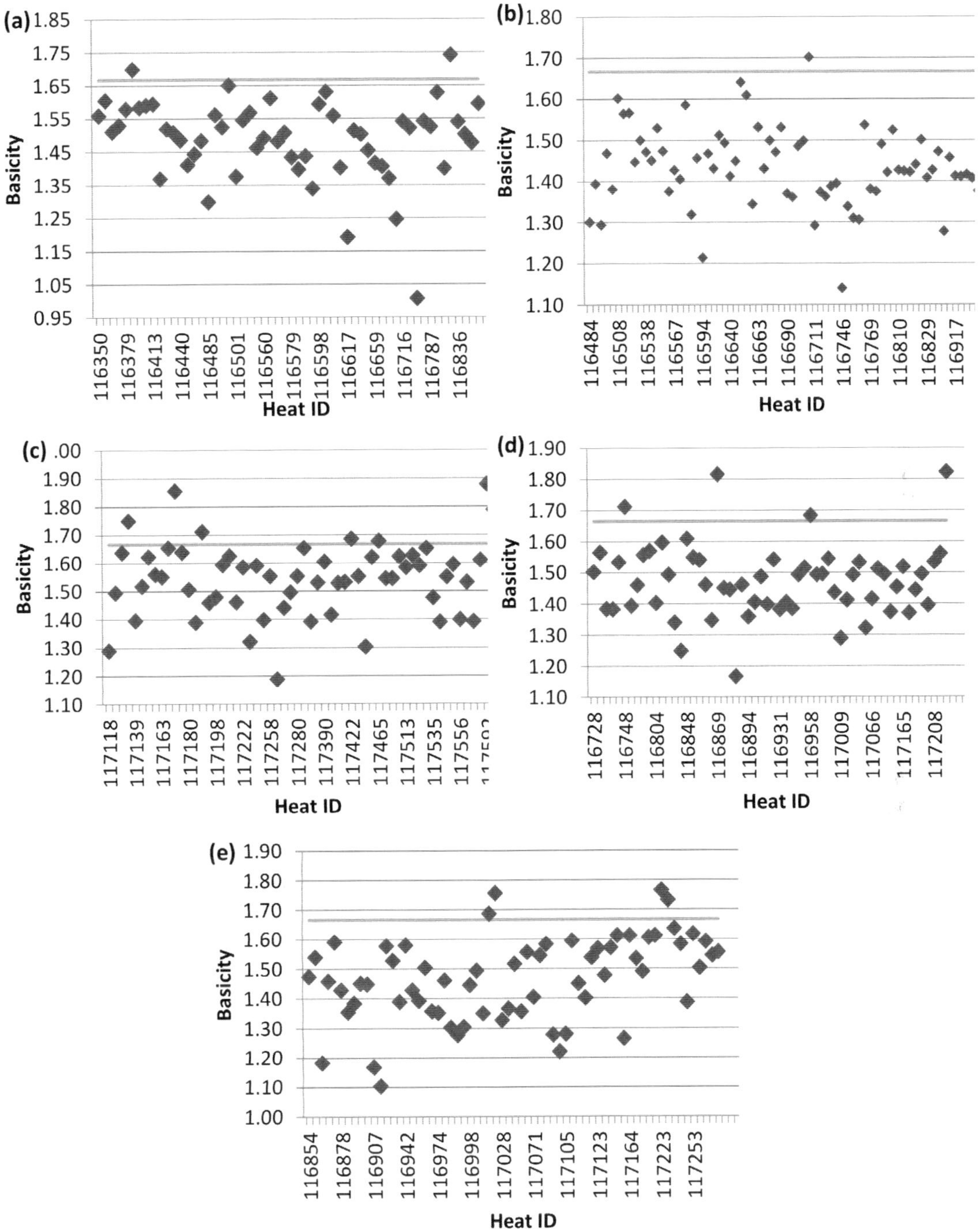

Figure 2. Basicity variations in slags over heats of the ladles furnaces studied – (a) Ladle A; (b) Ladle B; (c) Ladle C; (d) Ladle D; (e) Ladle E

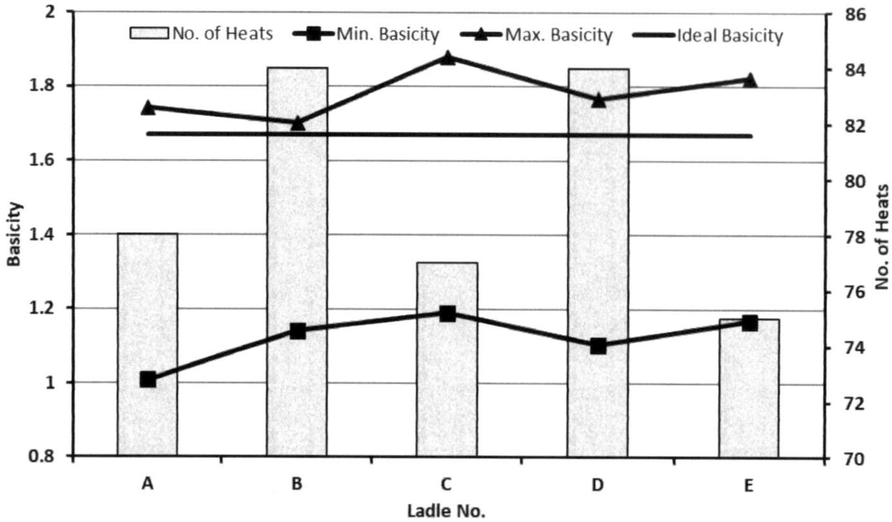

Figure 3. Service life (total number of heats endured) versus range of basicity experienced by the linings.

It can be seen that none of the ladles experienced basicity higher than allowable for MgO-C refractory; all were rather low, that is, the slags were acidic. However, Ladle A which experienced the least basicity of 1.008, compared to all others, had a service life of 78 heats. This is the highest service life of the three second campaign linings in the study (Ladles A, C and E). Ladle E, falling three full heats behind Ladle A, experienced least basicity of 1.168, which implies, Ladle A experienced slightly more acidic slag, yet with higher service live than Ladle E. The obvious inference is that another factor, as this study will show, which is contributing to degradation and wear of the refractory is having more pronounced effect on the life of the linings.

3.2 Impacts of Temperature Fluctuations on Lining Life

Figures 4(a) – 4(e) profile maximum temperatures reached during each heat over the life of the linings. From this data, the highest and the lowest of the maximum temperatures (endured by each lining during each heat over the life of the linings) were plotted against the life of each lining (Figure 5). Except on one heat where a spike of 1644 is seen in lining B (Figure 4(e)), and a dip down to 1571 during a heat in lining E (Figure 4(e)), the temperatures fluctuated mostly between 1575 and 1600 for all the heats and all the linings. This can be because some heats require a little superheat or deviations depending on the target cast item. As the operations typically seek to operate at 1590 ^{O}C, it can be said that the operating temperatures were largely under control.

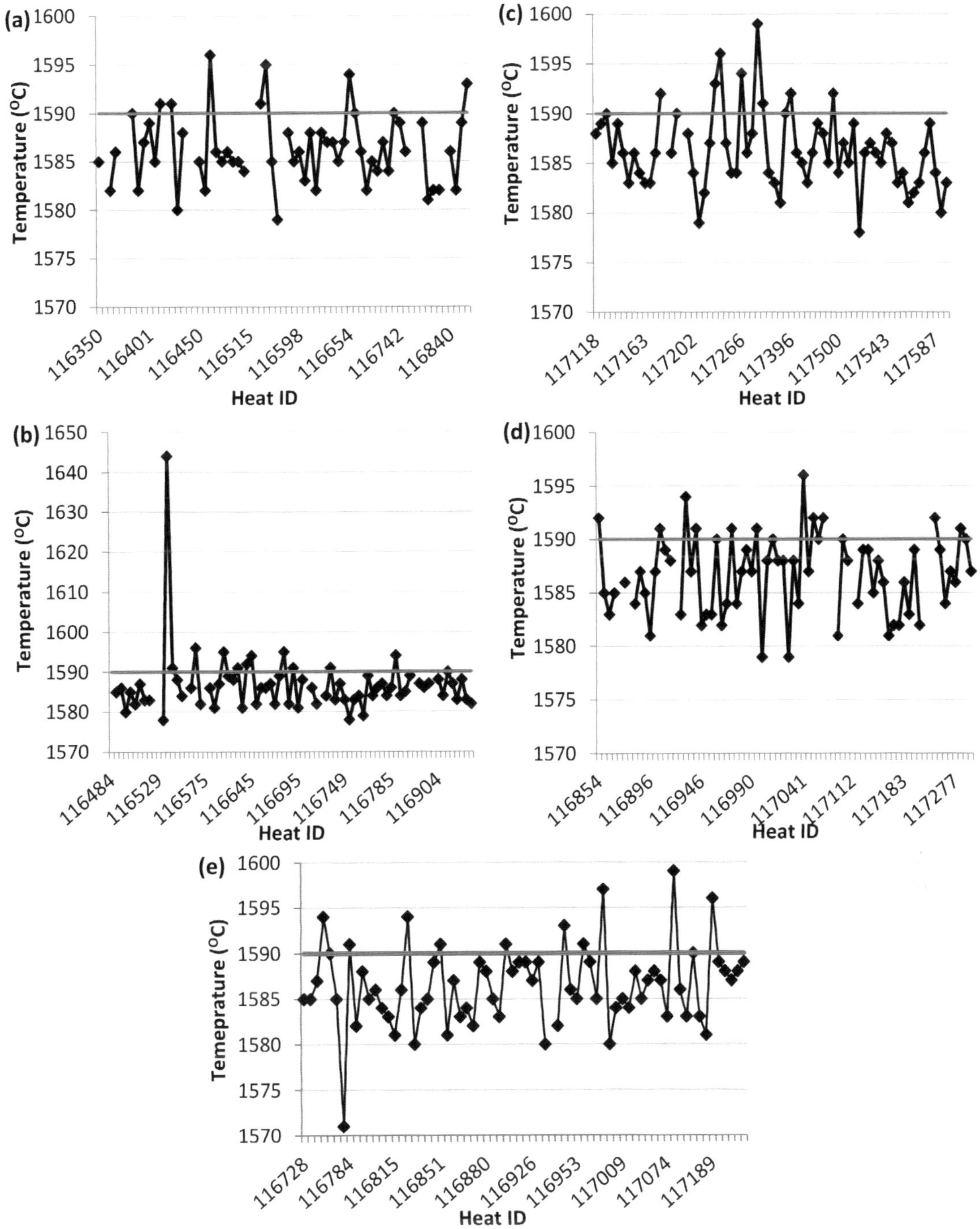

Figure 4. Maximum temperatures experienced by each lining during each heat for the different ladles (a) Ladle A, (b) Ladle B, (c) Ladle C, (d) Ladle D, and (e) Ladle E, with ideal operating temperature line indicated at 1590 ^{o}C.

However, above 1400 ^{o}C, indirect oxidation of carbon within MgO-C refractory becomes thermodynamically favourable [3], allowing:

$$MgO_{(refractory)} + C_{(refractory)} = Mg_{(g)} + CO_{(g)} \qquad \Delta G_T^0 = 646923 - 403.214T + 13.39TlnT \qquad (2)$$

At about 1590 ^{O}C (1863 K), the ΔG gives –108 kJ. Also the FeO content of the slag in principle can react with accessible carbon from the refractory as:

$$FeO_{(slag)}+C_{(refractory)}=Fe_{(l)}+CO_{(g)} \quad \Delta G_T^0 = 151125 - 151.105T \tag{3}$$

Below 1400 ^{O}C, under direct exposure to the atmosphere, direct oxidation of carbon is also possible to give carbon dioxide. Considering these possibilities, various oxidations should be happening in the lining, to bring about the refractory degradations. Contrary to what can be expected, the lining in Ladle B with highest thermal spikes of 1644 ^{O}C lasted 84 heats, as can be seen in Figure 5. The lining in Ladle D which also lasted 84 heats had the least temperature peak of 1596 ^{O}C. Comparing Ladle C and Ladle E, which both had second slagline lining campaigns, both experience same maximum temperatures of 1599 ^{O}C, but differ in service lives by two heats. From this data, temperature too may not be having the overriding effect on the service lives of the lining.

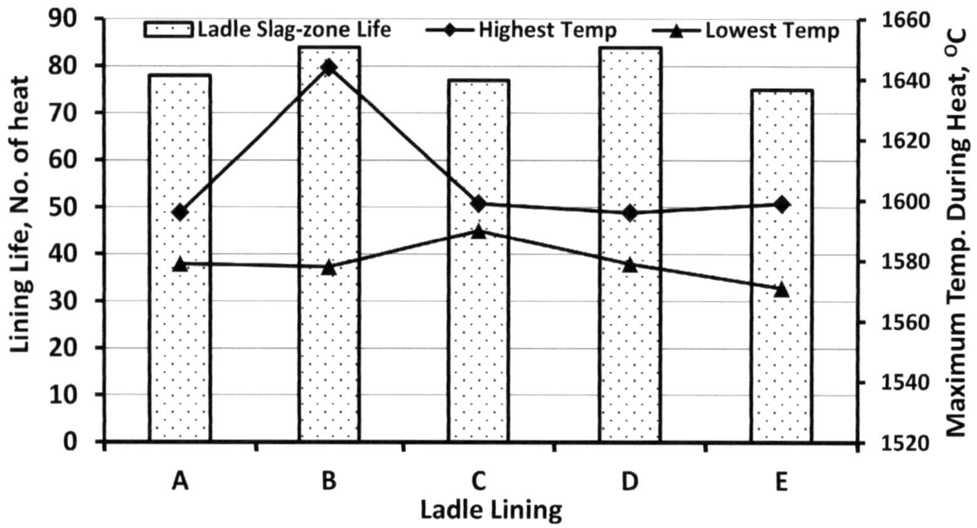

Figure 5. Highest and lowest of maximum temperatures during heats over the service life of each lining against service lives for the different linings.

Though basicity and high temperature oxidations are prominently emphasised in the slag-refractory interactions, the foregoing inferences downplay these factors for aggressiveness in refractory wear and lining service lives. Fundamentally, rates of solid-liquid reactions are not chemically controlled, but diffusion controlled [11]. The chemical reactions deriving from basicity and high temperature oxidations at the slag-refractory interfaces will not proceed aggressively between the viscous fluid such as slag and the solid refractory. Further insight from the post-mortem studies of the residual bricks address is more conclusive.

3.3 Residual Thicknesses and Non-Uniform Wear of Bricks

The MgO-C bricks newly lined are 230 mm thick. Tables 1 show residual thicknesses map of bricks after breakout from their respective locations in Ladle B at the end of the service life of the lining. The columns in the table show radial positions 1 to 12 while the rows give layers 12 to 26 that makes up the slagline lining bricks. Hence, the data at column 10 row 22 in the table, for instance, refers to residual thickness of brick from radial column position 10 and layer 22, and this can be represented as C10L22 residual thickness for this discussion.

Studying Table 1, it is striking that the bricks did not wear uniformly. Highest residual thickness of 170 mm can be see at C6L21, and some other locations, indicating least wear of 60 mm (26 %) relative to the new brick, while brick C7L19 has the least residual thickness of 35 mm, indicating wear of 195 mm (85 %). Columns 7 and 8, and columns 1 and 2, show most bricks with least residual thicknesses, while columns 5, 6, 9 – 12 has no bricks below 80 mm.

Table 1. Residual thicknesses map of bricks from Lining B at corresponding layer and radial ladle positions.

Layer	Post-production profile (mm)												
Position	6	7	8	9	10	11	12	1	2	3	4	5	6
26	150	100	95	150	160	150	140	120	145	160	150	150	150
25	150	100	100	150	170	170	150	105	145	155	150	150	150
24	130	50	70	140	130	140	160	105	135	150	150	150	130
23	120	55	60	150	150	150	150	55	90	130	130	120	120
22	120	125	130	130	135	140	140	100	100	120	120	120	120
21	170	115	115	100	110	140	150	120	120	125	120	135	170
20	130	115	100	120	120	130	130	90	80	120	125	130	130
19	130	35	60	115	120	80	120	40	70	80	110	120	130
18	130	80	90	120	120	125	130	125	110	120	120	130	130
17	135	90	90	120	120	125	140	125	110	120	125	130	135
16	140	100	100	130	130	125	130	120	110	110	110	120	140
15	140	120	130	130	130	125	130	120	115	110	110	120	140
14	125	120	100	135	130	130	135	115	115	110	110	120	125
13	125	100	95	125	125	120	130	100	105	115	125	125	125
12	120	105	105	125	125	120	130	100	100	120	115	115	120

Residual thickness map obtained for bricks from the other four linings studied show similar non uniformity. For Ladle A, least thickness of 35 mm was found at C7L15, indicating again, the 7 O'clock column as most prone to highest wear. The other three linings, C, D and E studied, show worst wear at either the 1 O'Clock or 7 O'Clock radial position.

It stands out that the 1 and 7 O' clock positions in all the linings suffer the worst wear. From the ladle design, these are the locations of the purging plugs (Figure 1). It is obvious that the purging gas caused turbulent flow of the slag against the bricks in these locations. Any weakened brick in these columns experienced aggressive wear. The slag basicity and high temperature oxidation which affects the entire lining obviously caused moderate wear, as can be seen from locations far away from these purging areas. From Table 1, some columns far away have from these zones have least residual thickness up to 120 mm, thrice that at column 7. It can be projected that without the turbulent wear, such as induction stirring, the service life of the lining can easily triple.

3.4 Micrographs of last working surfaces of residual bricks

Figures 6(a) and 6(b) show optical and electron micrographs from the last working surface of some sample residual bricks. The images capture thermal shock cracking propagating from frozen slag on the surface of the refractory inward (Figure 6(a)). Figure 6(b) shows a secondary phase that evolved from the slag refractory interaction. Residual bricks from the 1 O'clock and the 7 O'clock locations did not retain such slag penetration, or show thermal shock cracks, graphite loss or residual magnesium from oxidation effects. Rather the last working surface is largely comparable in composition to the rest of brick. Since Figures 6(a) and 6(b) show that the generally expected effects of slag-refractory interactions did actually occur in these linings, it can be inferred that the turbulent wear at the 1 O'clock and 7

O'clock locations removed any such interaction effects leaving least residual thicknesses of the bricks. It follows that wear due to the uprising purging gas from the bottom purging holes in these ladle furnaces design, not the thermal or chemical extremes, was the highest determinant of service live of the lining. This result did not admit that corrosion is the main wear mechanism in this type of operation for MgO-C degradation as some other researchers prefer to conclude [12].

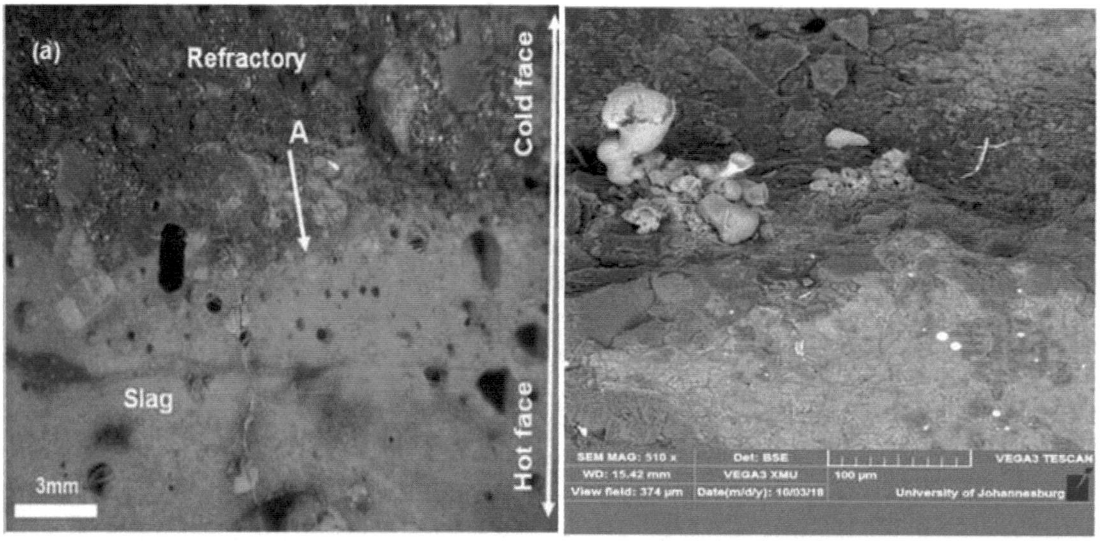

Figure 6. Slag-refractory interphase from last working surface layer of residual bricks showing (a) thermal shock crack propagating from frozen slag into the brick, and (b) phases evolved as a result of slag-refractory interaction.

4. Conclusion

Magnesia-graphite refractory slagline linings of 750 tons ladle furnaces in a secondary steel making operation were studied to determine the factor most critical to the service life of the lining. Temperature extremes and basicity fluctuations during the service life of the linings, map of residual thicknesses of bricks after breakout of the linings, and microstructures of slag-refractory interfaces from the last working surface of the residual bricks were studied. The results show that the refractories tolerated thermal spikes above 1600 OC and basicity excursions well below 1.67 ideal for the operations. Rather, the bottom purging design of the ladle, which caused localized slag turbulence, is most responsible in determining the life of the lining in service. Least residual thickness that decimated the life of the lining were obtained at lining positions closest to these purging positions. It was analysed that without this single effect, the life of the lining can triple. Magnesia-graphite refractory is well able to withstand the thermal and chemical extreme conditions of secondary steelmaking slag, but bottom purging in ladle furnaces design is the major challenge to the lining life.

5. References

[1] M.K. Kujur, I. Roy, K. Kumar, P. Chintaiah, S. Ghosh, N.K. Ghosh, Raw materials for manufacturing of Superior quality MgO-C bricks, Materials Today: Proceedings, 5 (2018) 2359-2366.
[2] T. Zhu, Y. Li, S. Sang, Z. Xie, Improved thermal shock resistance of magnesia-graphite refractories by the addition of MgO-C pellets, Materials & Design, 124 (2017) 16-23.
[3] M.-A. Faghihi-Sani, A. Yamaguchi, Oxidation kinetics of MgO–C refractory bricks, Ceramics International, 28 (2002) 835-839.
[4] J.D. Gilchrist, Extraction Metallurgy, Pergamon Press plc, 1989, (1989) 431.
[5] A. Ghosh, A. Chatterjee, Iron making and steelmaking: theory and practice, PHI Learning Pvt. Ltd.2008.

[6] M. Kirschen, A. Hanna, R. Ehrengruber, K. Zettl, Latest developments in gas purging systems for EAF, AISTech 2015 Conference Proceedings, 2015, pp. 1974-1983.
[7] A. Muller, The influence of slag basicity and composition on ConArc magnesia-carbon refractories during the blowing phase, Journal of the Southern African Institute of Mining and Metallurgy, 111 (2011) 225-230.
[8] O.P. Gupta, Elements of fuels, furnaces and refractories, Khanna Publ.1997.
[9] E. Pretorius, Slag fundamentals, An Introduction to the Theory and Practice of EF Steelmaking, Iron & Steel Society short course book, New Orleans, USA1–38, (1998).
[10] M.A. Tayeb, A.N. Assis, S. Sridhar, R.J. Fruehan, MgO solubility in steelmaking slags, Metallurgical and Materials Transactions B, 46 (2015) 1112-1114.
[11] A. Vignes, Extractive Metallurgy 1: Basic Thermodynamics and Kinetics, John Wiley & Sons2013.
[12] R.A.A. Borges, G.F.B.L. e Silva, A statistical and post-mortem study of wear and performance of MgO-C resin bonded refractories used on the slag line ladle of a basic oxygen steelmaking plant, Engineering Failure Analysis, 78 (2017) 161-168.

IOP Conf. Series: Materials Science and Engineering **655** (2019) 012004 doi:10.1088/1757-899X/655/1/012004

Effect of intercritical annealing of normalised Nb-Ti-V microalloyed plate steel on microstructural evolution

PI Kambilinya[1, a], CW Siyasiya[1, b], KM Banks[1, c], RJ Mostert[1, d]

[1] University of Pretoria, Department of Materials Science & Metallurgical Engineering, Pretoria, 0002, South Africa

E-mail: [a] pkambilinya2@gmail.com

Abstract. A homogeneous microstructure is required for consistent mechanical properties in normalised Nb-Ti-V microalloyed plate steels. Frequently, as-hot rolled microalloyed plate steels have a banded microstructure that is persistent even after normalising heat treatment (NHT), and this leads to inconsistencies and some scatter in mechanical properties. Therefore, this work focused on the influence of single-cycle normalising heat treatment (SNHT), double-cycle normalising heat treatment (DNHT) followed by intercritical annealing normalising heat treatment (INHT) on the homogenisation and mitigation of a banded microstructure. The study was conducted on a *0.13C-Nb-Ti-V* plate steel grade. The as-hot rolled microstructure was banded and had a 1.13 Anisotropy Index (AI) value. Results from the three thermal cycles revealed that the DNHT and INHT mitigated the pearlite microstructural banding and gave a more homogenized pearlite phase distribution throughout the microstructure, unlike the SNHT that retained the banding. The DNHT also exhibited the finest ferrite grain size, while the INHT exhibited the coarsest. From Vickers hardness measurements (153±5.8 HV, 157±3.6 HV and 166±4.5 HV), the UTS was approximately deduced as, 480, 490 and 530 MPa for the SNHT, DNHT and INHT respectively.

1. Introduction

A well-known phenomenon called microstructural banding usually occurs in hot-rolled microalloyed steels. This usually originates from the cast slab during dendritic solidification and/or solid-state phase transformations governed by microsegregation which then develops into a banded microstructure [1]–[3]. Microstructural banding has negative effects on the ductility and the impact toughness of microalloyed steels [4], [5]. It is reported that microstructural banding's severity and persistence due to prior microstructure memory is dependent on the rate of cooling, prior austenite grain size, and austenitising temperature [6], [7]. It was also reported that, even though high cooling rates (~ 60°C s^{-1}) during hot rolling eliminated banding, it reappeared after a separate post hot rolling intercritical treatment [3]. This behaviour of microstructure banding reversion (memory) was also investigated in a short time-high temperature normalising heat treatment (NHT), and it was reported that microsegregation was the root cause for the reappearance of the banding [5]. Therefore, this work focuses on studying the microstructural behaviour after NHT with the aim of mitigating microstructural banding and memory in a Nb-Ti-V microalloyed steel.

IOP Conf. Series: Materials Science and Engineering **655** (2019) 012004 doi:10.1088/1757-899X/655/1/012004

2. Experimental Procedure

The chemical composition for the 30 mm thick Nb-Ti-V microalloyed steel used in this study is given in Table 1. The plate was hot-rolled from 240 mm thick slabs, and the finishing rolling temperature was 860 °C followed by cooling in air.

Table 1. Chemical composition of the experimental steel, mass-%.

Steel	Chemical element											
	C	Mn	P	Si	Ni	Cr	V	Nb	Al	Ti	N	Fe
Nb-Ti-V	0.13	1.5	0.014	0.4	0.03	0.03	0.03	0.043	0.04	0.025	0.0053	Bal.

Thermo-Calc® 6.0 software, Steel and Iron Alloys Database TCFE7, was used to predict the equilibrium ferrite to austenite phase transformation temperature, Ac_e. The dilatometer specimens were wire cut from the centre of the as-hot rolled steel to a height of 10 mm and a diameter of 5 mm. An inductive Bähr dilatometer (DIL 805A/D) was used to experimentally determine the Ac_3 temperature in accordance with ASTM A1033-18. The normalising schedules in Figure 1, were also performed on the Bähr dilatometer. Macro and micro Vickers hardness tests were then conducted on the heat treated samples.

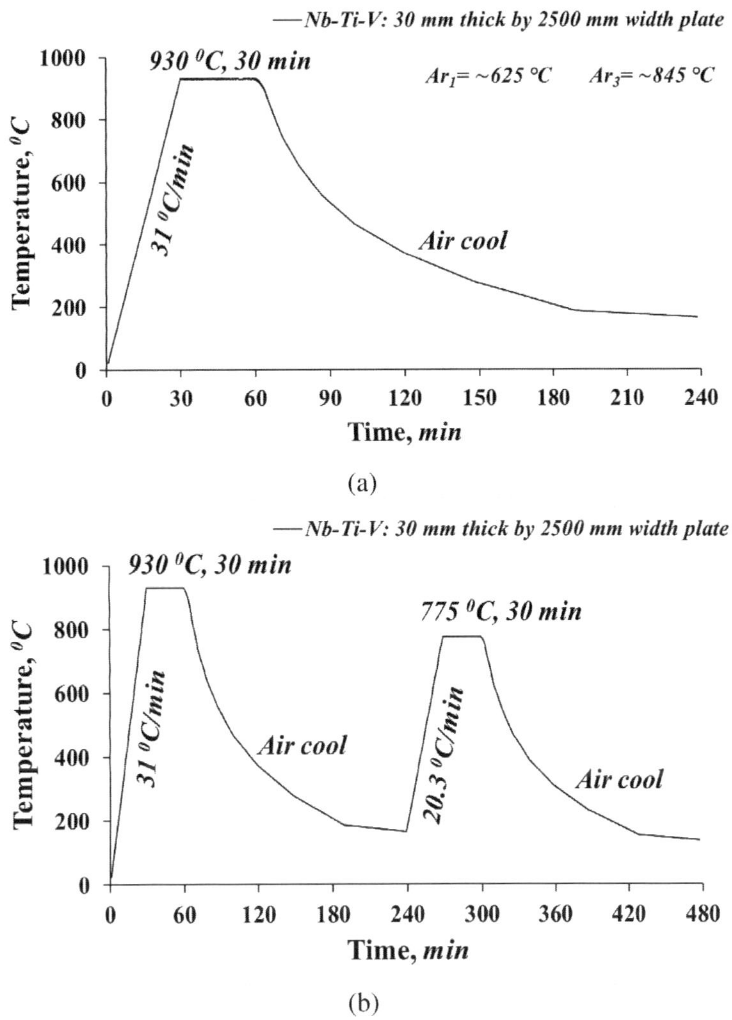

IOP Conf. Series: Materials Science and Engineering **655** (2019) 012004 doi:10.1088/1757-899X/655/1/012004

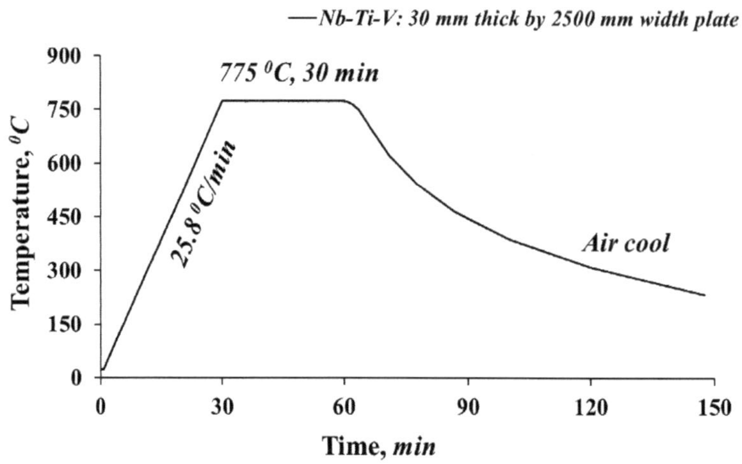

(c)

Figure 1: Applied heat treatment cycles: **(a)** single-cycle solution treatment (SNHT) in the austenite phase region at 930°C for 30 min then cooling in air; **(b)** double-cycle solution treatment in the austenite phase region at 930°C for 30 min and air cooling followed by an intercritical treatment at 775°C for 30 min and air cooled (DNHT); **(c)** intercritical treatment 775°C for 30 min and cooling in air (INHT).

Metallography was performed on both the as-hot rolled and the heat-treated samples using an *Olympus BX51M* optical microscope. With the aid of *Image J®* software, ferrite and prior austenite grain sizes (PAGS) were determined. The ferrite grain size distribution and the extent of banding were measured and quantified in accordance with respective ASTM standards: ASTM E112-13, ASTM E 1268-01, and ASTM E 562-11. The ferrite grain size distribution from the surface to the centre of the steel plate was calculated by means of the *Heyn's Lineal Intercept Procedure* [8], with test lines yielding at least 50 intercepts. The main parameters used to quantify and characterize the extent of banding in the steel's microstructure are: anisotropy index (AI); centre-to-centre mean spacing of the banded phase ($SB_{\perp}$); edge-to-edge mean free path spacing of the banded phase ($\lambda_{\perp}$); and the mean width of the banded phase given by calculating the difference between the mean spacing and the mean free spacing, Table 2. The two mean quantities give a measure of the number of bands per mm. The AI indicates how banded a matrix is, i.e., a non-banded matrix will have an AI of 1; as the degree of banding increases, so does the AI value above 1.

Through-thickness, as-rolled samples were austenitised at temperatures between 965 and 1050°C for 2 hours followed by quenching in water (at room temperature). The PAGS were determined using the lineal intercept procedure as described in ASTM E112-13.

3. Results and Discussion

3.1. As-hot rolled and PAGS microstructure analysis

Microstructural banding

According to ASTM E1268-01, Standard Practice for Assessing the Degree of Banding or Orientation of Microstructures, by observation of the optical morphology, a banded microstructure can be defined as: "the separation, of one or more phases or constituents in a two-phase or multiphase microstructure or constituent microstructure, into distinct layers parallel to the deformation axis due to elongation of microsegregation" [2]. The extent to which the as hot-rolled microstructures were banded is summarised in Table 2.

Table 2. The extent of microstructural banding in as-hot rolled 30 mm thick plate

	Nb-Ti-V Steel			
	NS[a]	QT[b]	MT[c]	AVERAGE
Mean, $\bar{X}$	17.1	15.6	15.2	**16**
SD	2.57	2.92	1.71	**2.4**
%RA	11	14	9	**11.3**
95% CI, ±	1.94	2.2	1.29	**1.81**
AI	1.07	1.21	1.1	**1.13**
$SB_{\perp}$, μm	69.3	71.5	77	**72.6**
$\lambda_{\perp}$, μm	46.4	47.2	53.9	**49.2**
($SB_{\perp}$ - $\lambda_{\perp}$), μm	22.9	24.3	23.1	**23.4**

[a] *Near-surface*
[b] *Quarter-thickness*
[c] *Mid-thickness*

From the quantitative analysis results in Table 2, the ferrite AI value (greater than 1), confirmed that the ferrite grains and the pearlite colonies were oriented in the direction of rolling. Hence, the presence of the parallel and layered ferrite-pearlite microstructure. The other banding spacing parameters ($SB_{\perp}$ and $\lambda_{\perp}$) give an insight of the distances between the bands.

Ferrite grain size distribution.

The ferrite grain size was unevenly distributed; increasing from the surface to the centre, Figures 2a-c show micrographs of ferrite (light phase) and pearlite (dark phase) at the surface and the centre respectively, together with the respective mean and standard deviations of ferrite intercept length, $\bar{l}_{\alpha}$.

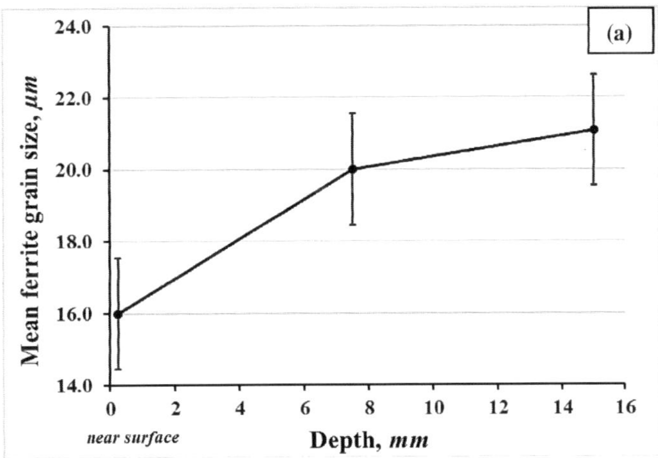

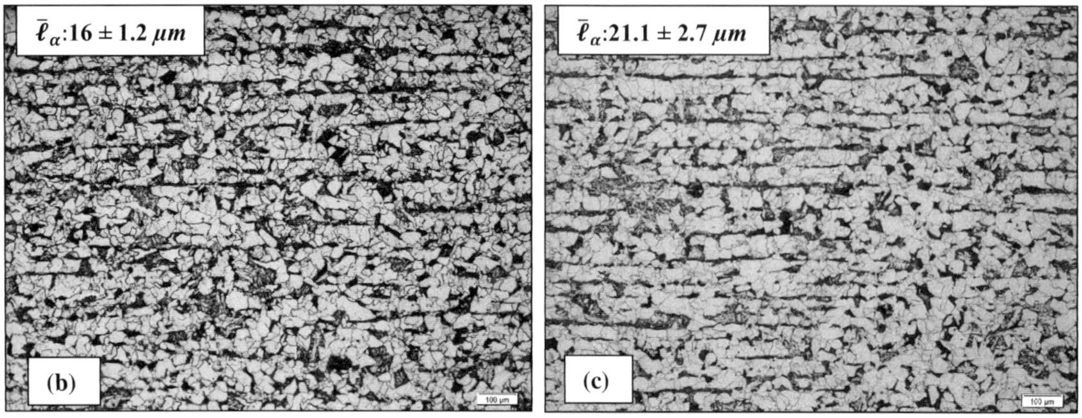

Figure 2. (a) Ferrite grain size distribution in as-hot rolled condition measured across the plate cross-section; (b) near to the surface; (c) at mid-thickness. Scale bar 100 μm, images taken in the rolling direction.

Characterisation of the prior austenite grain size (PAGS)
Figure 3 shows that the PAGS increased from the surface to the centre. Also, the PAGS difference between surface and centre becomes more pronounced at higher austenitising temperatures. Evidently, this shows some memory retention by the microstructure even after austenitisation.

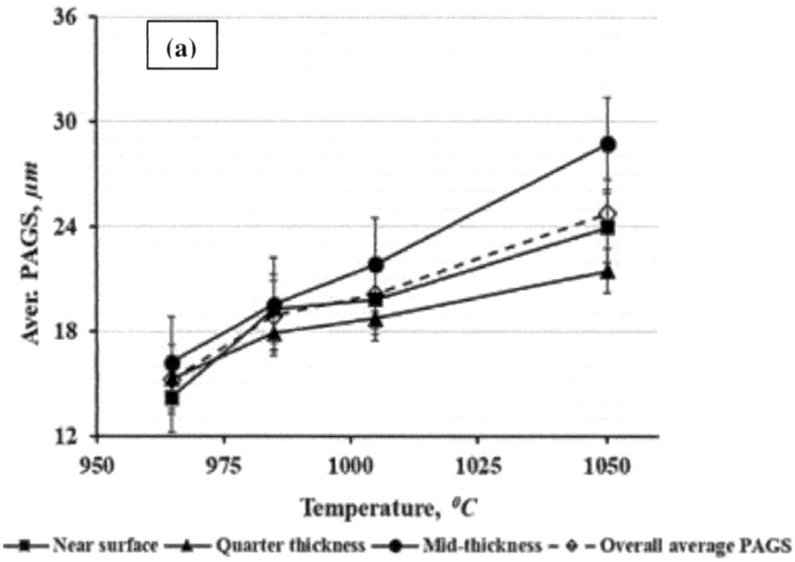

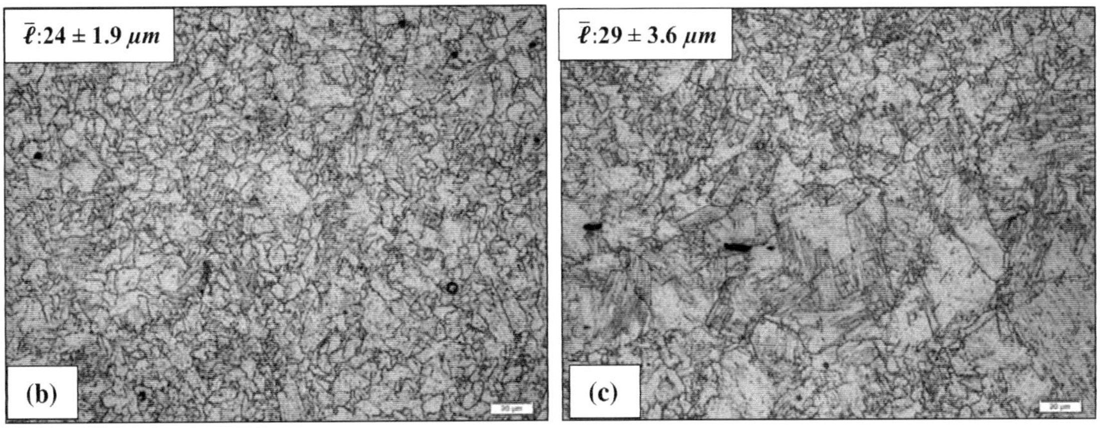

Figure 3. Micrographs of PAGS showing grain size distributions: (a) Average PAGS vs austenitising temperature. Soaked for 2hrs followed by water quenching to room temperature; (b) taken near to the surface; (c) at mid-thickness. Austenitised at 1050 °C for 2hrs followed by water-quench, then tempered at 496 °C for 26 hrs. Scale bar 20 μm.

3.2. Normalising heat treatment simulations and microstructure analysis

The hot-rolled banded structure was still present even after the 930°C for 30 min single-cycle normalising heat treatment (SNHT) simulation, Figure 4a. The microstructures after the double-cycle normalising heat treatment (DNHT), Figure 4b, and to some extent the intercritical-anneal normalising heat treatment (INHT) in Figure 4d appears to mitigate the ferrite-pearlite banded structure. Nonetheless, the INHT resulted in a coarser microstructure (ferrite) than the DNHT. The DNHT has, however, been reported to produce similar YS and UTS values compared to the SNHT [9]. During heating and austenitising in SNHT diffusion-controlled nucleation of austenite occurs predominantly at ferrite-pearlite colony interphases (i.e. sites of higher carbon concentrations). As the cementite lamellae in the pearlite dissolve, carbon and other alloying elements such as manganese [10]–[12], redistribute and attempt to homogenise in the austenite matrix. Upon air cooling from the austenite single phase region, pearlite transformation starts at the boundaries of ferrite and austenite and spreads into the austenite grains. The retention of the austenite morphological structure therefore, results in a banded pearlitic structure upon air cooling as evidenced in Figure 4a.

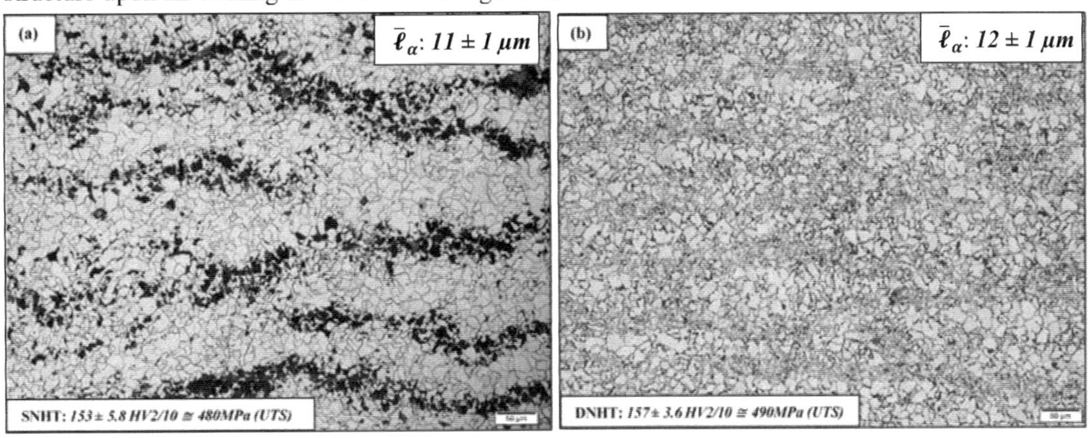

IOP Conf. Series: Materials Science and Engineering **655** (2019) 012004 doi:10.1088/1757-899X/655/1/012004

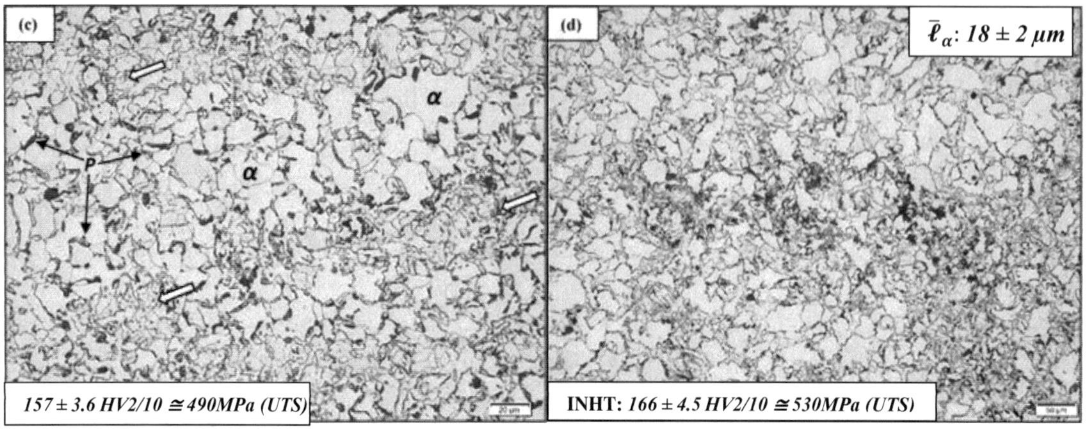

Figure 4: (a) SNHT, scale bar 50 μm; (b) DNHT, scale bar 50 μm; (c) same as in (b), taken at higher mag. (scale bar 20 μm) to show fine pearlite nodules nucleating at ferrite grain boundaries and corners. Also note, fine pearlite nucleating in the carbon-rich regions (large arrows); (d) INHT, scale bar 50 μm.

Table 3: Ferrite, pearlite micro Vickers hardness, and pearlite vol. fraction with respective HT.

	Heat Treatment Type		
	SNHT	**DNHT**	**INHT**
Average (ferrite), *HV*	176.1 ± 13.8	181.1 ± 5.1	189.6 ± 12.3
Average (pearlite), *HV*	229 ± 10.8	260 ± 23.1	272 ± 12.4
Vol. fraction (pearlite), *%*	39	30	32

During intercritical annealing upon heating and soaking, austenite formation takes place only in the carbon-rich regions that featured pearlite [13], thereby only dissolving the banded pearlite regions, this can be noted in the post intercritical annealing of the DNHT, Figure 4c pointed with large ***arrows***. Upon air cooling the carbon enriched austenite is significantly stabilised and subsequently delays the pearlite transformation and is extremely important for intercritical heat treatments [11]. Furthermore, cooling from the intercritical (α+γ) region fosters the epitaxial growth of new and fine ferrite grains at already existing ferrite interfaces [14], Figure 4c. Epitaxial ferrite, as defined by Sarwar et al. [15], is "ferrite formed by the transformation of austenite by epitaxial growth on retained ferrite (ferrite present at intercritical annealing temperature) during cooling from intercritical annealing temperature". This mechanism further enriches the surrounding austenite with carbon. Thus, cooling of austenite from the intercritical region delays the pearlite transformation compared to cooling from the single γ-phase region. Evidently, in Figure 4c the large arrows show the carbon-rich regions that had stabilised the austenite and resulted in the delayed pearlite transformation, thereby mitigating the pearlite bands after air cooling.

It was also observed in another study that while austenite enrichment with carbon delayed the austenite to pearlite transformation, it, however, increased the starting temperature of ferrite formation [16]. This explains why there is an increase in the volume fraction of ferrite, and finer pearlite as depicted in Figure 4b. Moreover, a progressive ferrite grain refinement and homogeneity in terms of pearlite distribution was observed with an increase in the number of NHT cycles.

In comparison to the SNHT and DNHT, despite being conducted at a lower temperature the INHT produced larger ferrite grains, Figure 4d. This could also have been attributed to the growth of the retained ferrite during the intercritical annealing. This is not the case with the SNHT and DNHT, where there was complete transformation to austenite during austenitisation; the ferrite grains had been refined upon air-cooling. This has also been reported by Chatterjee et al., [17]. Although the DNHT also involved intercritical annealing, the growth of retained ferrite was not as pronounced as in the INHT due to the initial cycle.

The INHT, however, produced microstructure with slightly the highest HV values despite having larger ferrite grains. This could have resulted from the refined pearlite which is slightly in higher volume fraction in the INHT than the other two heat treatments, Table 3. This can be supported by another similar study [18]. In addition, the austenite that is formed during intercritical annealing in the INHT relatively contains a higher carbon content due to the pearlite-austenite transformation as opposed to that in the DNHT. Upon transformation after air-cooling, it appears as though the INHT produced regions of ferrite with higher carbon content which contributed to the strengthening. The high austenitisation temperature in the initial cycle of the DNHT resulted in a higher carbon content partitioned to the austenite phase [11], [19]. This factor, though, is subject to further investigation.

4. Further work

This work addresses the mitigation of microstructural banding and its memory retention by investigating the effect of the DNHT and INHT on the as-hot rolled microstructure. However, more experimental work that investigates the effect of austenitisation temperature on the microstructure banding (memory retention) in the SNHT needs to be conducted. In addition, further investigation on the effect of the applied heat treatments on the mechanical properties such as; yield strength, CVN impact energy absorption and the ductile-to-brittle transition temperature (DBTT) also needs to be done.

5. Conclusions

- The double normalising heat treatment involving an intercritical austenitisation after normalising followed by air cooling mitigates the pearlite microstructural banding and homogenizes the pearlite phase distribution throughout the microstructure.
- The double normalising heat treatment further refined the ferrite grain size upon γ-decomposition during air cooling from the intercritical region. This heat treatment possibly had no impact on tensile strength as confirmed by Vickers hardness measurements.

6. References

[1] S. Cai and J. D. Boyd, "Mechanism of Microstructural Banding in Hot Rolled Microalloyed Steels," *Mater. Sci. Forum*, vol. 500–501, no. August, pp. 171–178, 2005.

[2] ASTM E 1268-01, "Standard Practice for Assessing the Degree of Banding or Orientation of Microstructures," *Am. Soc. Test. Mater. Int.*, no. Reapproved 2016, pp. 1–29, 2016.

[3] F. G. Caballero, A. Garcia-Junceda, C. Capdevila, and C. G. de Andres, "Evolution of Microstructural Banding during the Manufacturing Process of Dual Phase Steels," *Mater. Trans.*, vol. 47, no. 9, pp. 2269–2276, 2006.

[4] A. S. Bor, "Effect of Pearlite Banding on Mechanical Properties of Hot-rolled Steel Plates," *ISIJ Int.*, 2008.

[5] R. A. Grange, "Effect of microstructural banding in steel," *Metall. Trans.*, 1971.

[6] L. E. Samuels, *Light Microscopy of Carbon Steels*. 1999.

[7] T. F. Majka, D. K. Matlock, and G. Krauss, "Development of microstructural banding in low-alloy steel with simulated Mn segregation," *Metall. Mater. Trans. A Phys. Metall. Mater. Sci.*, 2002.

[8] ASTM E112 – 13, "Standard Test Methods for Determining Average Grain Size," *Am. Soc. Test. Mater. Int.*, no. Reapproved 2014, pp. 1–28, 2014.

[9] T. F. Muniz, V. B. Trindade, and G. L. Farai, "Effect of Some Heat Treatments on Microstructure and Mechanical Properties of a Microalloyed Low Carbon Steel Aiming to Attend the Oil and Gas Industry," *Proc. 23rd ABCM Int. Congr. Mech. Eng.*, no. December, pp. 6–11, 2015.

[10] G. R. Speich, V. A. Demarest, and R. L. Miller, "Formation of Austenite During Intercritical Annealing of Dual-Phase Steels," *Metall. Mater. Trans. A*, 1981.

[11] N. Fonstein, *Advanced High Strength Sheet Steels*. 2015.

[12] S. ci Li, Y. lin Kang, G. ming Zhu, and S. Kuang, "Austenite formation during intercritical annealing in C-Mn cold-rolled dual phase steel," *J. Cent. South Univ.*, 2015.

[13] R. Großterlinden, R. Kawalla, U. Lotter, and H. Pircher, "Formation of pearlitic banded structures in ferritic-pearlitic steels," *Steel Res.*, 1992.

[14] M. D. Geib, D. K. Matlock, and G. Krauss, "The effect of intercritical annealing temperature on the structure of niobium microalloyed dualphase steel," *Metall. Trans. A*, 1980.

[15] M. Sarwar, E. Ahmad, K. A. Qureshi, and T. Manzoor, "Influence of epitaxial ferrite on tensile properties of dual phase steel," *Mater. Des.*, p. 335, 2007.

[16] O. A. Girina, D. B. Bhattacharya, N. M. Fonstein, "Effect of Annealing Parameters on Austenite Decomposition in a continuously Annealed {Dual-Phase} Steel," *41st {MWSP}*, no. January 2003, pp. 403–414, 2003.

[17] A. Chatterjee, D. Chakrabarti, A. Moitra, R. Mitra, and A. K. Bhaduri, "Materials Science & Engineering A Effect of normalization temperatures on ductile – brittle transition temperature of a modi fied 9Cr – 1Mo steel," *Mater. Sci. Eng. A*, vol. 618, pp. 219–231, 2014.

[18] M. P. Engineering and N. Unive, "Inter-critical heat treatment of a low-alloy steel," vol. 40, pp. 395–405, 1994.

[19] R. G. Davies and C. L. Magee, "Physical Metallurgy of Automotive High-Strength Steels.," *Journal of Metals*. 1979.

IOP Conf. Series: Materials Science and Engineering **655** (2019) 012005 doi:10.1088/1757-899X/655/1/012005

The effect of soaking and tempering temperatures on high vanadium alloys for grinding media

AD Mabeba[1,2] and CW Siyasiya[1]

[1] University of Pretoria, Lynnwood Rd, Hatfield, Pretoria, 0002, South Africa

[2] Mintek, 200 Malibongwe Drive, Randburg 2125, South Africa

Email:absalomm@mintek.co.za

Abstract. Vanadium produces the hardest carbides and can influence wear properties positively. However, due to the minerals' price fluctuations as a result of the current economic climate, vanadium is currently expensive and seldom used in wear resistant materials despite its valuable properties. Nevertheless, it is still important to study the effects of vanadium carbides on wear properties and later compare the cost/durability ratio to the other currently used industrial alloys. Vanadium has been used as a secondary alloying element in high chromium cast iron and has produced positive results through the influence of orientation and morphology of the eutectic carbide M_7C_3. Vanadium carbides are known to be discontinuous, which is helpful with regards to impact properties. Vanadium carbides alone cannot maintain high macro-hardness values and for improved results carbon is needed to increase the strength of the matrix. In this paper, the influence of the soaking and tempering temperatures on hardness and wear properties in a high vanadium alloy, 1.91C-0.82Mn-0.96Si-0.90Ni-1.35Cr-0.25Mo-6.12V, were investigated. The soaking temperatures were 1150 and 1250°C while the tempering temperatures were varied at 100°C interval from 300 °C to 600 °C. It was found that soaking at 1250°C increased the amount of retained austenite, but at the same time also improved the wear resistance in as quenched condition as well as after tempering up to 500 °C.

1. Introduction

Abrasive wear is a combination of indentation and scratching. It is often referred to as ploughing or micro plaguing of the contact surface. Restricting indentation of abrasive particles by hardening has proved to improve wear resistance. The main factors which have considerable effect on strengthening materials are heat treatment, alloying addition, influence of carbon content, the effects of carbides and work hardening. For improved wear properties, the hardness of the surface must be between 0.5 and 1.3 times the hardness of the abrasive [1]. Most common abrasives are hard and to produce a surface hardness that matches 1.3 times their hardness values may introduce brittleness.

The use of multi-alloy phase systems where the microstructure comprises of the matrix and a carbide phase are widely used in abrasive applications. The hard secondary phases improves the macro-hardness whilst the matrix maintains the toughness of the material. From literature V produces the hardest carbides (MC). Similar to Cr, V can also ensure the stability of a carbide phase. In addition to their hardness, VC precipitates are generally discontinuous and spherical and, therefore, are less detrimental to strength and toughness. The VC precipitates are also known to have a good cohesion to the matrix owing to a cubic lattice which matches the austenitic lattice [2]. To promote their spheroisity, many authors have reported the addition of rare earths metals to the melts. The rare earth metals scavenges sulphur (S) and phosphorus (P) which segregates on the corners of the solidifying VC particles to promote dendritic growth through heterogeneous nucleation [1-3]. The dendritic carbide structures increase with C and V content as indicated Figure 1 [4]. Furthermore, the hardness increases with an increase in the volume fraction of the carbide phase to a point beyond which it declines due to increased volume fraction of the retained austenite (RA), Figure 2 [4].

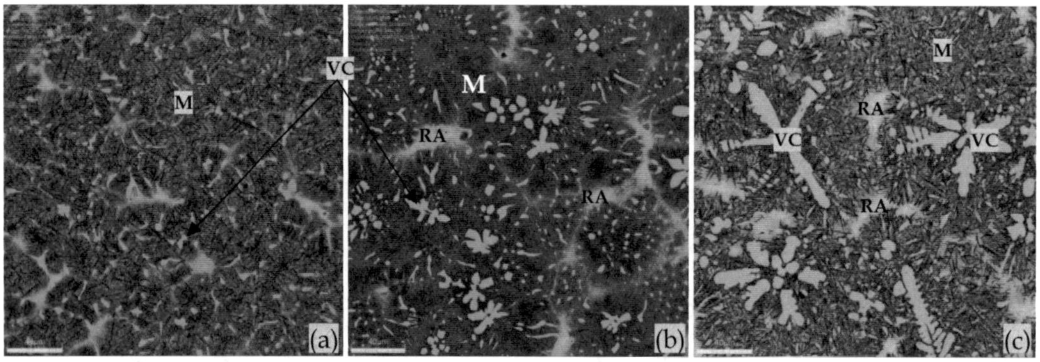

Figure 1. Micrographs of HVGM_alloys etched with Le-Pera's etchant showing martensite (M), V carbides (VC) and retained austenite (RA). (a) 1.70 mass% C, 6.74 mass% V, (b) 2.14 mass% C, 9.14 mass% V, (c) 2.91 mass% C, 11.0 mass% V [4], micrograph scale = 40um.

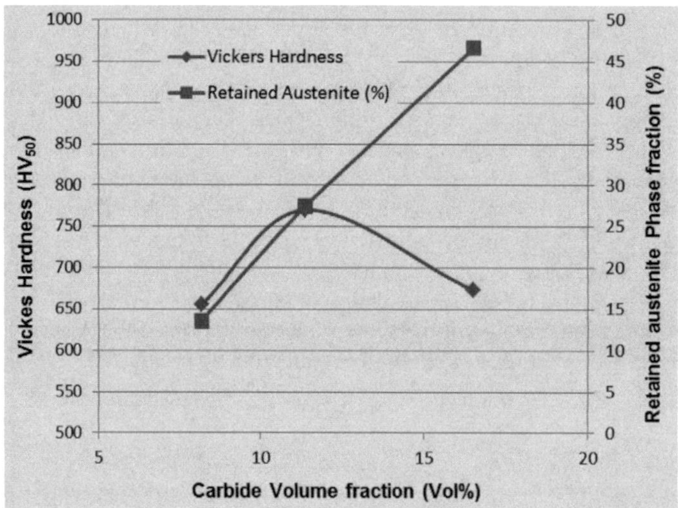

Figure 2. Photographs showing the relationship between carbide phase volume fraction, hardness and retained austenite [4].

Vanadium carbides form as VC, although a large degree of non-stoichiometry has been reported between $VC_{0.74}$ and $VC_{0.91}$ [5]. For the reaction: V + C = VC, the ratio of V to C is 4.2413:1 and for V + 0.74C = $VC_{0.74}$ and the ratio becomes 3.1412:1. For equilibrium conditions, the C needs to be in excess to strengthen the martensitic matrix. In this work the influence of soaking and tempering temperature in the high V alloy (HVGM-1) on the evolution of the microstructures and their subsequence influence on wear behaviour was investigated.

2. Experimental procedure

The experimental high HVGM-1 was produced in a 150 kg induction furnace fitted with an alumina crucible. The alloy was produced using scrap metal and ferro-alloys. The alloy was heated gradually until 1550 °C where it was fully molten and tapped into a transfer ladle. The metal was cast into sand moulds made from silica sand and sodium silicate gel dried with CO_2.

The spark analysis chemical composition was as follow: 1.91C-0.82Mn-0.96Si-0.90Ni-1.35Cr-0.25Mo-6.12V. C and high amount of V was deliberately added to form the VC carbides for abrasion resistance. Cr, Mo and Ni were added for hardenability, toughness and strength of the matrix. Cr and V were also added to suppress graphitisation by utilising most of the C for carbide formation. Si was added to aid castability by improving fluidity.

The alloy was solution treated at 1150 and 1250 °C to allow for the dissolution of VC, after which it was quenched in oil. The solution treatment temperatures were based on the simultaneous thermal analyser (STA) results which indicated that the dissolution temperatures of VC are just below 1100°C [4]. The alloy was tempered between 300 and 600 °C at 100 °C intervals to study the effects of tempering temperature of the wear behaviour of the material.

The samples cross sections were mounted, ground with silicon carbide papers and polished to 1μm. The polished surfaces were etched with Le-Pera's etcant [6] (a mixture of 4% picric acid in ethanol and 1% potassium metabisulphide in water) to reveal the microstructures. The samples were further etched with 6% potassium metabisulphide and 2% sulphamic acid [6] in water to prepare for carbide phase quantification. The carbide phase was identified, quantified and ten measurements were made from each as quenched sample. The carbides were analysed using high resolution scanning electron microscope equipped with EDX (HRSEM-EDX).

Abrasion wear was evaluated under a three-body abrasion test rig as shown in Figure 3. Prior to testing the samples were cleaned ultrasonically. The initial mass of each sample was determined using an analytical balance. After each test was completed the samples were cleaned ultrasonically and the final sample mass determined. The wear rates of the samples were calculated using material mass loss as a function of time. The specimens were pressed under a constant load of 130N against the rim of a rotating wheel for 30 minutes. The average sand flow rate between the specimen and the rubber wheel was 350 g/min according to ASTM G65.

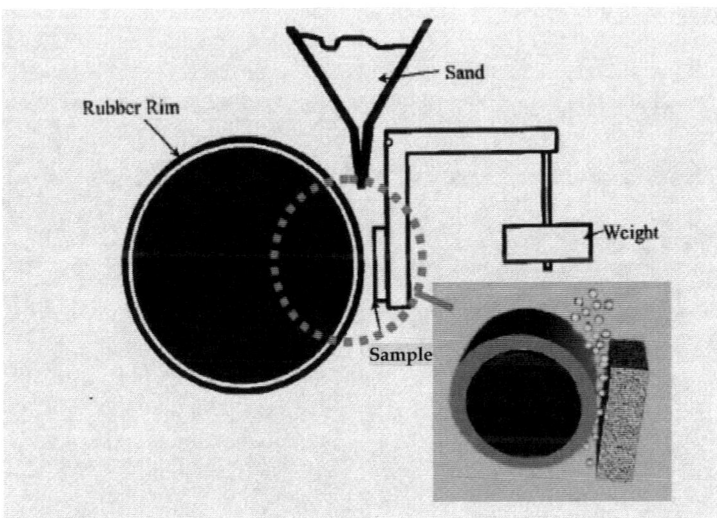

Figure-3: Schematic diagram of the dry sand/rubber wheel abrasion testing, DSRWAT, (adopted from ASTM G65)

XRD was used in order to determine the amount of the RA. The specimens were polished to 1μm and thereafter were electropolished in a 20% $HClO_4$-ethanol solution for XRD preparation to eliminate any strain induced martensite. The phase analysis was performed using a Bruker D8 equipped with a

LynxEye detector and Fe-filtered Co-Kα radiation. The phases were identified using EVA software. The relative phase amounts (wt%) was estimated using the Rietveld method (Topas program).

3. Results and Discussion

3.1. Microstructural Analysis

The as-cast microstructure is shown in Figure 4. As may be seen, the alloy, HVGM-1, consisted of pearlite and VC precipitates.

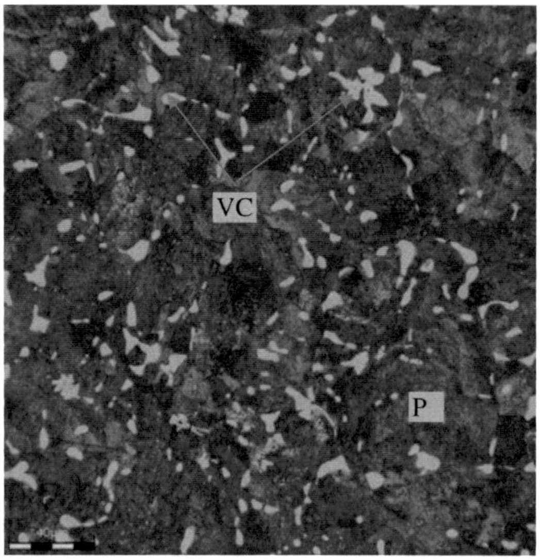

Figure-4: Micrograph of the as-cast microstructures of HVGM-1, micrograph scale = 40um.

After heat treatment, the alloy consisted of VC in a martensitic matrix. The smaller carbide structures were mostly spherical particles. The larger carbides also nucleated as spherical particles and grew heterogeneously by dendritic growth to form flower-like structures, Figure 5 (a). When the solution treatment was increased to 1250°C, the larger flower-like carbide structures dissolved partially and became spherical again as they reduced their surface energy, Figures 5 (b). The average carbide volume fractions of the as-quenched 1150°C and 1250°C was found to be 9.88 and 8.72% respectively when analysed by optical microscopy phase analysis software. Figure 6 shows the backscatter micrograph and EDX spectrum to confirm the composition of the vanadium carbides and they were found to be VC.

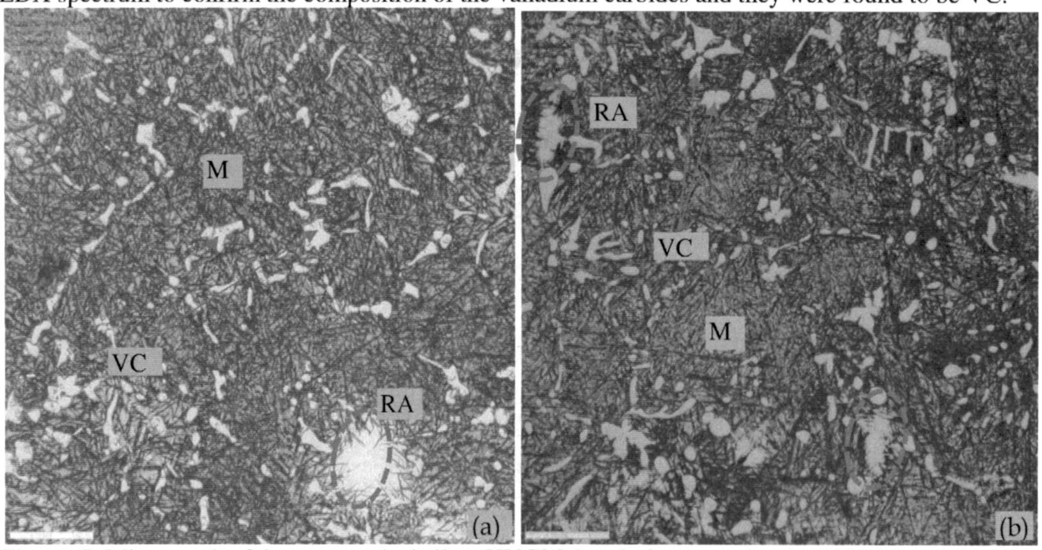

Figure-5: Micrograph of the as-quenched alloy, HVGM-1, solution treated at (a) 1150 °C and (b) 1250 °C. Micrograph scale = 40um.

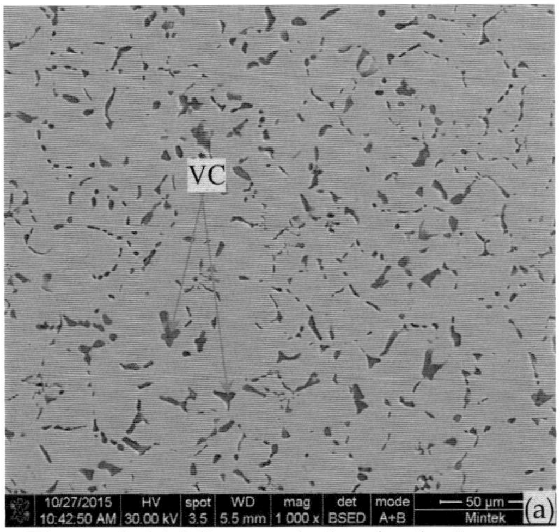

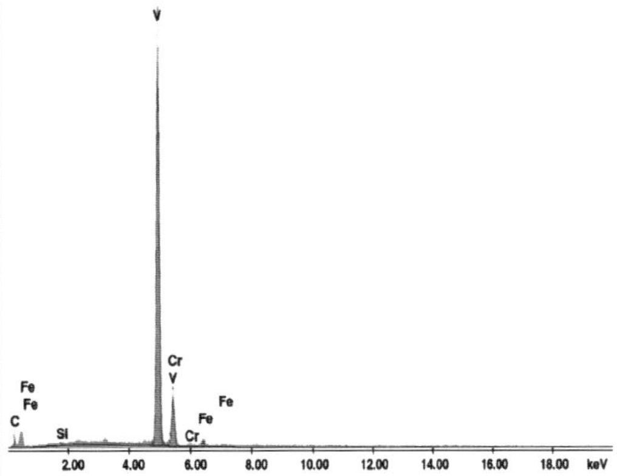

Element	C K	Si K	V K	Cr K	Fe K	Total
Wt%	18.55	0.35	79.14	0.00	1.96	100
At%	49.10	0.39	49.39	0.00	1.12	100

(b)

Figure-6: (a) Micrograph showing back scatter image of HVGM-1 1150°C and (a) EDX spectrum.

3.2. Hardness Measurements

The HRC hardness values of the alloy are graphically plotted in Figure 7. After solution treating at 1150 °C, the as-quenched hardness was higher. The relatively lower as-quenched hardness value after solution treating at 1250 °C was attributed to an increased amount of RA. After tempering at 300 °C the hardness drops significantly as a result of tempering and the hardness values were found to be similar. For both solution temperatures, the secondary peak hardness was observed at 400 °C. This was attributed to the precipitation of fine, coherent vanadium carbides. The matrix hardness dropped after tempering beyond 500 °C as the fine precipitates began to coarsen and lose their strengthening effect.

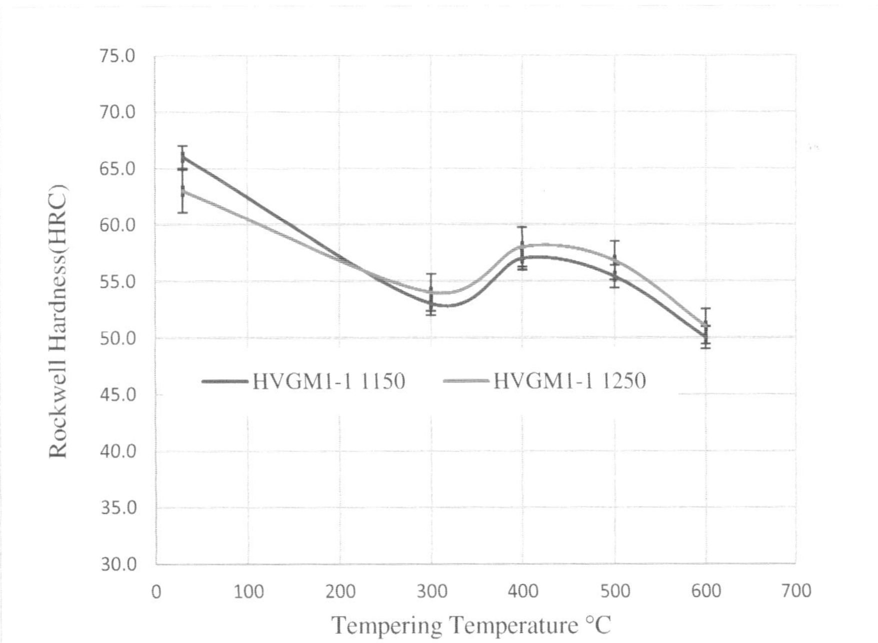

Figure-7: Rockwell hardness values as a function of the tempering temperature

3.3. Three Body Abrasion Wear Testing

The amount of RA as well as the Three Body Abrasion (TBA) wear resistance as a function of the tempering temperature are shown in Figure 8. As expected the RA increased with soaking temperature and decreased with an increase in tempering temperature. The RA only drastically reduced after tempering at 600 °C as a result of reduced stabilization of the austenite. However, as may be seen from

Fig 8, the reduction in the RA did not generally contribute to improved wear resistance because of the trade-off with coarsening of carbides. In other words, the increased wear rate at 600 °C simply meant that wear properties were best when the V precipitates were still fine. It would appear that in as-quenched condition, the combination of the hard carbide, hard martensitic and soft austenitic phases in both solution heat treatments of 1150 and 1250 °C was the optimum condition for wear resistance. Figure 9 (a) of as-quenched HVGM-1 shows less scratching from the three body abrasion test. After tempering at 600 °C, the scratches become deeper, and the softer martensitic matrix allows for the abrasion wear to plough out the carbides, Figure 9 (b).

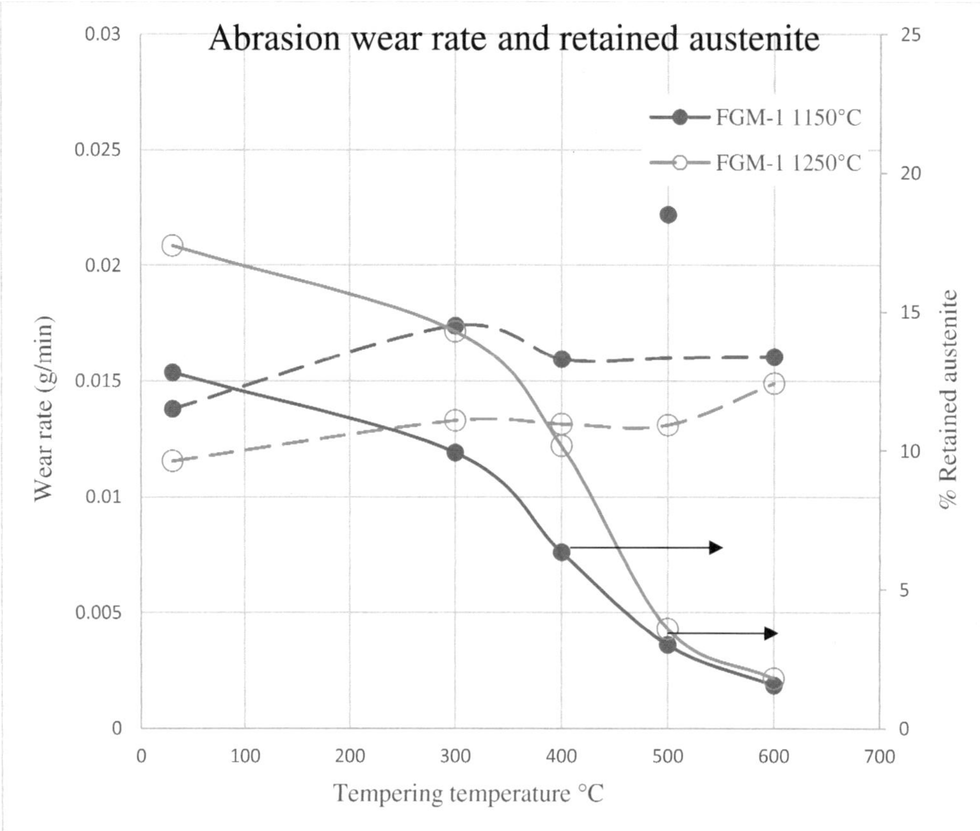

Figure-8: Three body abrasion wear rate and percentage retained austenite (RA) plotted as a function of the tempering temperature

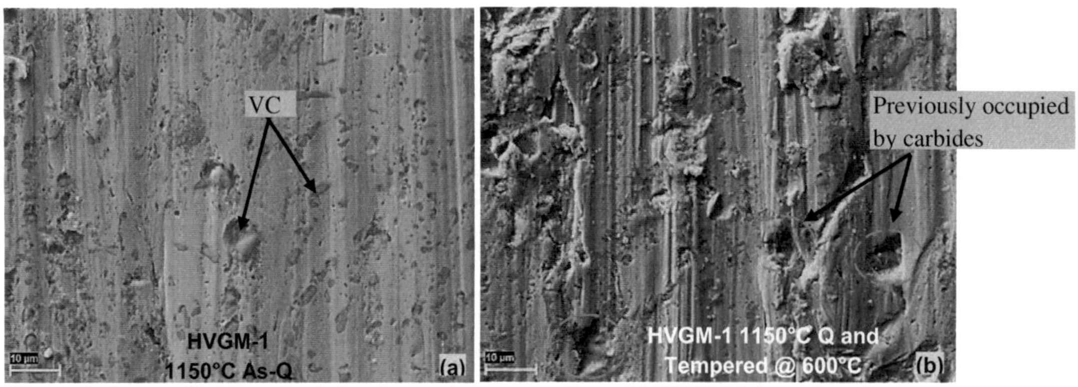

Figure-9: Dry sand/rubber wheel abrasion testing wear surfaces (a) HVGM-1 in as-quenched condition and (b) after tempering at 600 °C

4. Conclusions

- The alloy solution treated at 1250 °C with higher amount of RA exhibited better wear resistance.
- The combination of the hard carbide, hard martensitic and soft austenitic phases in both solution heat treatments of 1150 and 1250 °C was found to be optimum condition for wear resistance.

5. Acknowledgements

The financial and technical support of the Advanced Materials Division at Mintek, and the technical guidance from Department of Materials Science and Metallurgical Engineering at University of Pretoria is gratefully acknowledged.

6. References

[1] Fleishchmann A H 1995 *The manufacture and investigation of abrasion of abrasion resistant FE-V-C alloys*

[2] Efremenko, V.G., Shimizu, K., Cheiliakh, A.P., Kozarevs'ka, T.V., Chabak, Y.G., Hara H. and Kusumoto K 2013 Abrasive wear resistance of spheroidal vanadium carbide cast irons. *Friction and Wear*, 34 (6), 466-474

[3] Wang Chao Chang, Hsu Heng Tsun, Ma Quin 1990 *Formation of spheroidal carbide in vanadium white cast iron by rare earth modification*

[4] Mabeba AD and Jonck J 2016 The use of vanadium carbides to produce a wear and impact resistant alloy for grinding balls *Ferrous and Base Metals Development Network Conference 2016*, Proceedings, AMI Symposium Series S91

[5] Stephen R. Shattynski Revised May 18 1978 *The Thermochemistry of Transition Metal Carbides*

[6] G.F. Vander Voort, Metallography principles and practice *Material Science and Engineering Series*

IOP Conf. Series: Materials Science and Engineering **655** (2019) 012006 doi:10.1088/1757-899X/655/1/012006

Studying the effect of surface conditioning on the corrosion performance of titanium dental implants

N D E Masia, M Smit, A I Mwamba and J S Moema

Advanced Materials Division, Mintek, Randburg, South Africa

E-mail: nomsombulukoh@mintek.ac.za

Abstract. The biocompatibility of dental implants is dependent on good interaction between implant-human body tissue and a good osseointegration. The surface roughness plays an important role enhancing bone healing and promoting biomechanical properties at the bone/implant interface. The increase in surface roughness can simultaneously improve cell migration and attachment of an implant, and enhance the osseointegration process. However surface roughness has shown to be detrimental to corrosion, raising the need to achieve a balance between good osseointegration and corrosion performance. This study characterised the corrosion performance of commercially pure titanium grade 4 with different surface conditionings. Implants in the form of 15 mm discs and implant screws were tested in basic Ringers lactate solution of pH 5.5 at 37°C and in modified Ringers of pH 2. Corrosion tests included open circuit potential (OCP), potentiodynamic polarization test (PDP) and mass loss corrosion tests. Metallurgical characterisation showed that titanium Grade 4 consisted of equiaxed alpha (α-Ti) grains. The OCP test showed that the machined implants were nobler than the roughened implants. The corrosion rates obtained were very low, below 0.13 mm/y which is an acceptable corrosion rate for biomaterial design and application. Mass-loss tests showed that the alloys passivated regardless of the surface conditioning. The machined surface had a corrosion resistance higher than that of the roughened surface finish in all conditions.

1. Introduction

Dental implants made of titanium, especially commercially pure titanium, have become a well-established treatment modality for the replacement of missing teeth and are state of the art in the field of dental restoration [1]. The concept of osseointergration was introduced by Branemark in the mid 1960's [2]. Osseointegration is defined as a direct structural and functional connection between ordered, living bone and the surface of a load-carrying implant. It is critical for implant stability, and is considered a prerequisite for implant loading and long-term clinical success of end osseous dental implants [3].

In order to improve the speed and strength of bone formation and to allow more rapid loading of the implants, researchers have investigated the effects of modifying implant surfaces. The surface modification is considered to start from the topographical change of the surface or increasing the roughness of the titanium surface [1]. Surface roughness has an important role in enhancing bone healing and promoting biomechanical properties at the bone/implant interface through increasing the mechanical retention and allowing good stress distribution [4]. Rough surfaces are often used in clinical situations where acceleration and enhancement of osseointegration and bone interlocking is required such as in cases with poor bone quality or reduced bone volume. Based on current knowledge smooth implants are those with an average roughness of S_a less than 0.5μm, minimally rough implants have a S_a 0.5-1μm, moderately rough implants have a S_a 1-2μm and rough implants have a S_a above 2μm [2].

Alumina (Al_2O_3) is frequently used as a blasting material and produces surface roughness varying with the granulometry of the blasting media [5], [6]. However, the blasting material is often embedded into the implant surface and residue remains even after ultrasonic cleaning, acid passivation and sterilization. Alumina is insoluble in acid and is thus hard to remove from the titanium surface. In some cases, these particles have been released into the surrounding tissues and have interfered with the osseointegration of the implants. Moreover, this chemical heterogeneity of the implant surface may decrease the excellent corrosion resistance of titanium in a physiological environment [5].

The aim of this study is to characterise the corrosion performance of a machined and roughened surfaces of a commercially pure titanium grade 4.

2. Materials and Experimental Procedure

2.1 Material and Implants Preparation

Commercially pure titanium grade 4 was selected for this study because titanium has been the material of choice in several disciplines of dentistry. Titanium has excellent characteristics such as biocompatibility, osseointegration, high wear and corrosion resistance, low compatibility issues and high strength. Recent attention has been directed towards further development of this material. Table 1 below shows the chemical composition of titanium grade 4.
For this study two surface treatments were selected namely machined, and roughened. The roughened samples were prepared by Al_2O_3 blasting. Implants are usually blasted and then etched with acids such as hydrochloric, sulphuric, hydrofluoric and nitric acid. The acid attack on the surface of the dental implants results in uniform roughness with pits and craters of micrometric size, this leads to an increase in surface area. These pores enhance bioadhesion and the osteoblasts properties [2]. Each manufacturer has its own method of acid etching, by controlling the temperature, exposure time and concentration of acids.

All the samples were machined, roughened (not acid etched) and sterilised by Southern Implants (PTY) LTD. The discs were 15 mm in diameter dick and 2 mm thick; whereas the implants tested were one of their standard products, see Figure 1. The surface area of the implant was reported to be 414.15 mm^2.

Table 1. Chemical composition of titanium grade 4.

Elements (wt %)				
N	**C**	**O**	**Fe**	**Ti**
0.05% max	0.08% max	0.4% max	0.5% max	Rem/balance

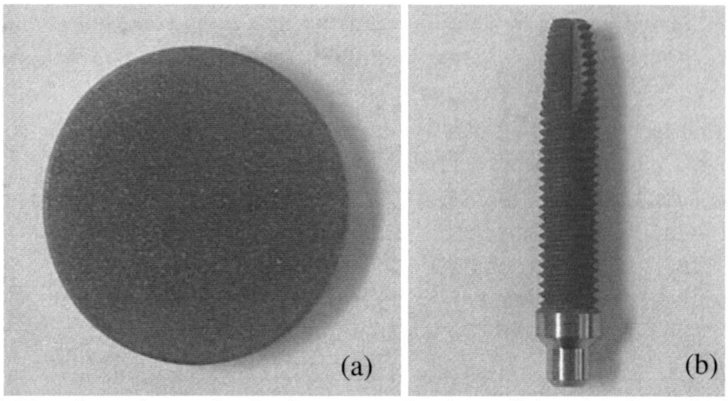

Figure 1. Roughened samples in the as-received condition (a) discs (b) implants.

2.2 *Material Characterisation*

A metallurgical characterisation was performed on each type of surface treatment to study the basic features or reference condition of each surface treatment before corrosion testing. The samples were characterized using both optical and electronic characterization techniques.

2.2.1 *Chemical Analysis*

The discs were analysed in the as-received condition using a Zeiss EVO-MA15 scanning electron microscope (SEM) fitted with Bruker energy dispersive spectroscopy (EDS) detector. EDS is a chemical microanalysis technique used in conjunction with scanning electron microscopy (SEM) for the elemental analysis or chemical characterization of a sample. Thus it gives a local elemental composition of the sample.

2.2.2 *Phase Analysis*

The discs were analysed in the as-received condition using a Bruker D8 Advance powder diffractometer, with Linxeye detector with Fe filtered Co-K alpha radiation. This method makes use of the net intensity of the main peaks of the phases, and identification is based on the crystal structure of phases that occur in amounts of greater than 3 mass %. The EVA software was used to identify the phases. Amorphous phases are not detected and the phases specified may not necessarily reflect the actual composition of phases identified.

2.2.3 *Metallography*

The discs were sectioned and mounted in an epoxy resin, ground and polished to a 1μm surface finish. The implants were then etched with Kroll's reagent (3 mL HF, 6 mL HNO_3 and 100 mL H_2O) to reveal their microstructures. This was performed to confirm the condition of the implants. Optical micrographs were taken at 2219x and 1109x magnification.

3. Corrosion Testing

The corrosion tests were performed in a basic Ringers lactate solution at 37°C, which is frequently used in *in vitro* experiments on organs or tissues. Several situations can acidify the pH of saliva such as infections, certain foods (sugary food, pickled foods, sour candies, fruits, soft drinks and juices), mouthwash products, smoking, chronic/systemic diseases and medication, may also contribute to the corrosion of dental implants [7]. In an effort to simulate this aggressive nature of the mouth it was decided to also perform tests in modified Ringers lactate solutions with respect to pH and sulphide (S^{2-}) content. Sulphide was included in the study because sulphur products are of particular concern in regards to titanium dental implants because periodontopathic bacteria secretes considerable amounts of

sulphide by products, such as hydrogen sulphide (H_2S), methylmercaptan (CH_3SH), and dimethyl sulphide ($(CH_3)_2S$), as a result of their metabolism [8].
The pH was adjusted using sulphuric acid (H_2SO_4) while the addition of sulphide was obtained by an addition of sodium sulphide (Na_2S). Table 2 below shows the chemical composition of the Ringer's lactate solutions used.

Table 2. Chemical composition of the solutions.

Solutions	**Reagents**					**pH**
	Sodium chloride (NaCl)	**Potassium chloride (KCl)**	**Calcium chloride dihydrate ($CaCl_2.2H_2O$)**	**Sodium lactate ($NaC_3H_5O_3$)**	**Sulphide (S^{2-})**	
Solution 1	30g	0.3g	0.2g	3.1g	-	5.5
Solution 2	30g	0.3g	0.2g	3.1g	-	2
Solution 3	30g	0.3g	0.2g	3.1g	0.5ppm	2

3.1 Electrochemical Testing

Two electrochemical corrosion tests were performed on the discs to assess and study the effect of different surface treatments. All the tests were performed at 37°C (body temperature, in the three Ringers lactate solutions detailed.
PC driven "ACM Gill AC" potentiostat was used to perform the electrochemical tests. The 10 mm discs were inserted in a sealed PVC holder. One graphite rod acted as a counter-electrode and a Haber-Luggin capillary made the junction with a saturated calomel reference electrode (SCE). Figure 2 shows the corrosion test set-up. All potential values are with respect to the SCE. Nitrogen was bubbled continuously through the solution during testing to create an anaerobic environment around a dental implant simulating that experience in the mouth.

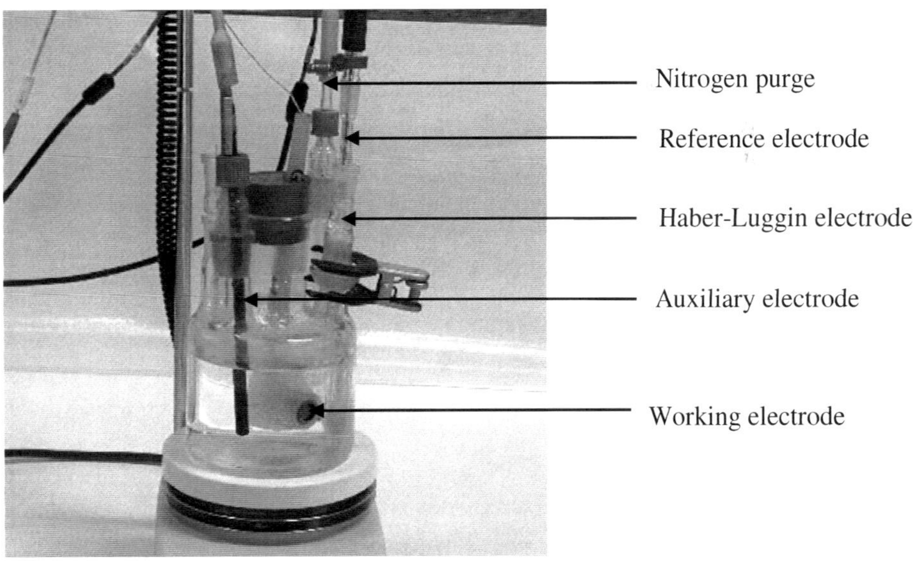

Figure 2. Corrosion cell set-up.

3.1.1 Open circuit potential (OCP)

The open circuit potential (OCP) also known as the corrosion potential was determined by measuring the voltage difference between the electrodes immersed in the solution with respect to a standard saturated calomel reference electrode. The electrodes were immersed in the solutions and the potential was recorded as a function of time. The OCP was recorded at four hours when the potential was stable (at equilibrium).

3.1.2 Potentiodynamic polarization test (PDP)

When the OCP was stable the potentiodynamic polarisation tests were recorded for each surface treatment. During these tests the potential of the treated sample is manipulated while measuring the current response. Each scan was stated after the OCP was recorded from – 250 mV to + 1500 mV versus the corrosion potential at a scanning speed of 10 mV/min. The corrosion current densities were determined by Tafel extrapolation of the rate-determining segment of the polarisation curve, and the corrosion rate was calculated by introducing the values of density and equivalent weight.

3.2 Mass-Loss Testing

Mass-loss corrosion tests were performed to assess the corrosion performance of the actual implants. To simulate the crevice conditions an implant would experience in the jaw, a crevice former was made using cast-able mounting resin, see Figure 3. This resin is non-metallic and inert, and would not chemically respond to the environment. Eight implants were exposed fully immersed in each solution in separate beakers to prevent contamination. The tests were performed for 30 days while maintaining the temperature at 37°C with the aid of a water bath.

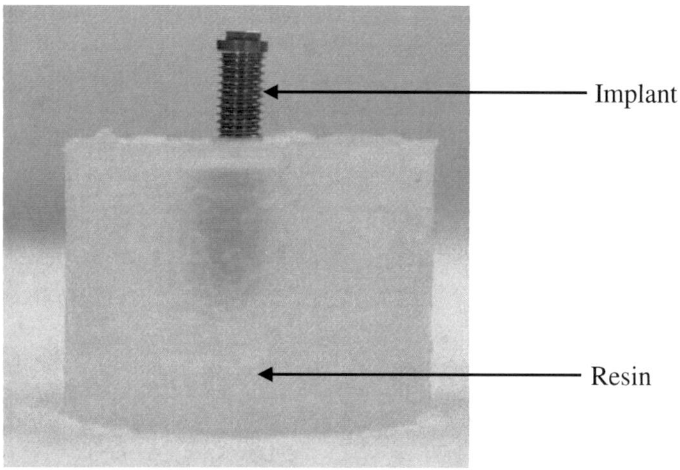

Figure 3. Implants crevice set-up.

4. Results and Discussion

4.1 Material characterisation

4.1.1 Chemical Analysis

Figure 4 shows the SEM micrographs of the machined discs. The machined implants showed striations on the surface due to the machining process while the roughened samples had a granular appearance. The EDS spectra, see Figure 5, shows that the roughened surface was contaminated with an aluminium oxide phase, where the machined sample was purely titanium.

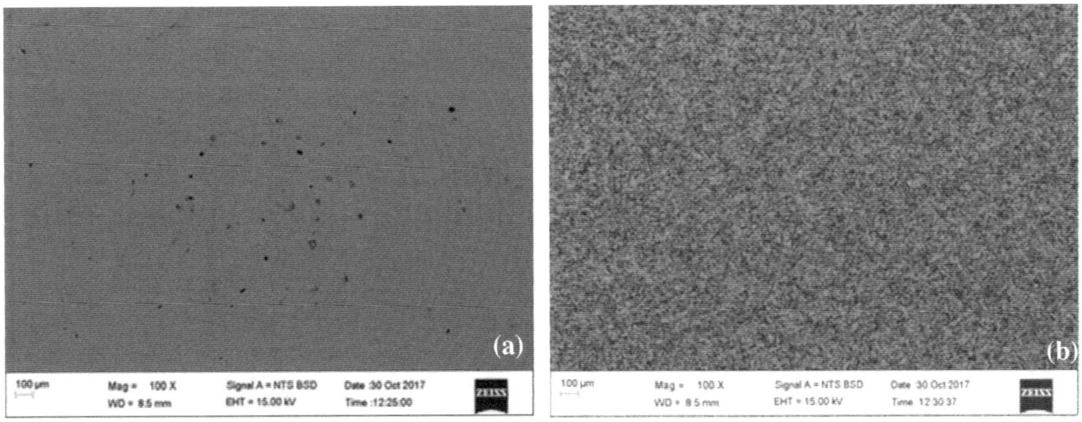

Figure 4. SEM micrographs of the discs (a) machined (b) roughened.

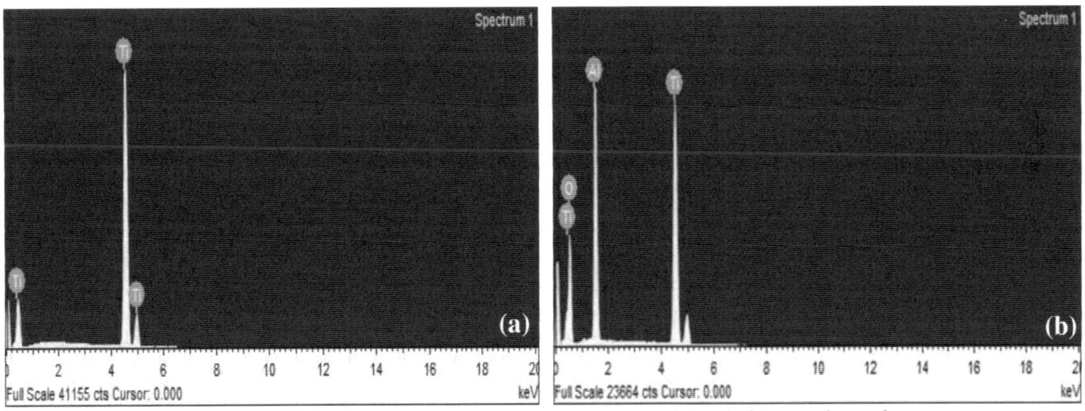

Figure 5. EDS spectra of the discs (a) machined (b) roughened.

4.1.2 *Phase Analysis*

Figure 6 shows the XRD pattern of a machined and roughened discs. The XRD pattern confirmed the EDS analysis and showed that the machined sample contained only titanium (α-Ti), where the roughened sample also showed TiO_2, and Al_2O_3. The Al_2O_3 phase probably originates from the blasting materials during roughening and the TiO_2 was probably due to a passivation response on the implants surface.

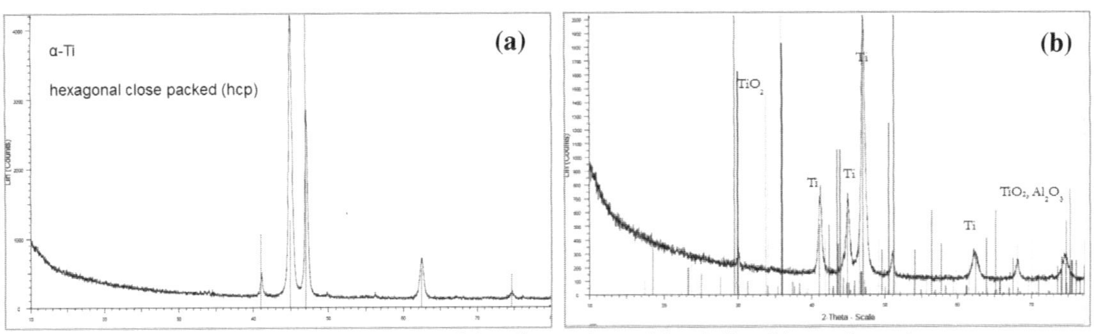

Figure 6. XRD spectra of the discs (a) machined disc (b) roughened.

4.1.3 Metallography

Figure 7 shows the cross section of the discs. The machined disc had a smooth surface while the roughened showed a rough surface due to alumina blasting. Figure 8 show the micrograph of the implants in the as-etched condition. The microstructure consisted of equiaxed alpha (α) grains in an annealed condition.

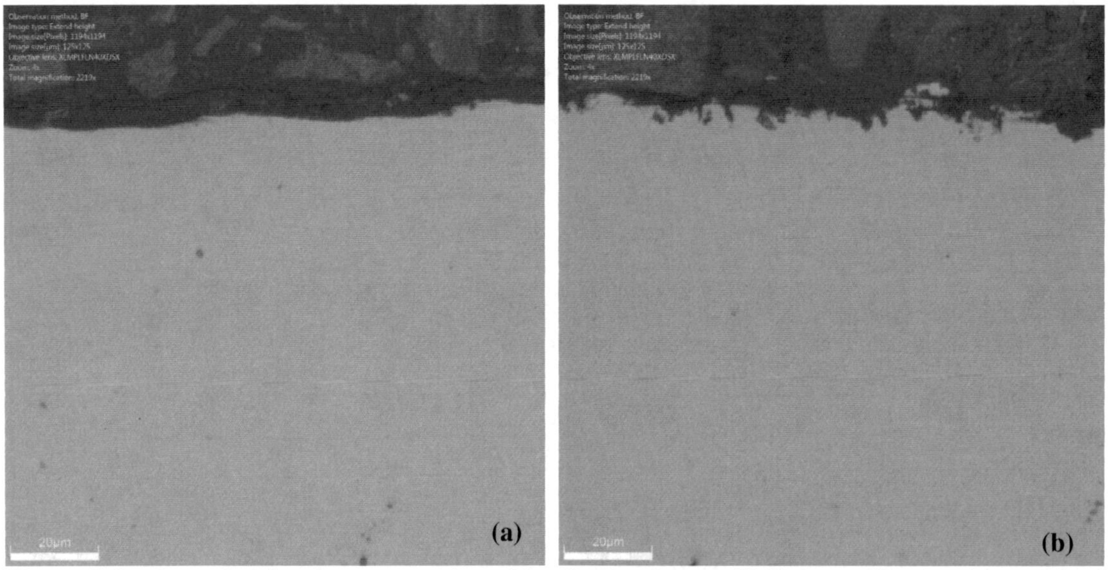

Figure 7. Implants showing the cross section of the disc implants at 2219x magnification (a) machined (b) roughened.

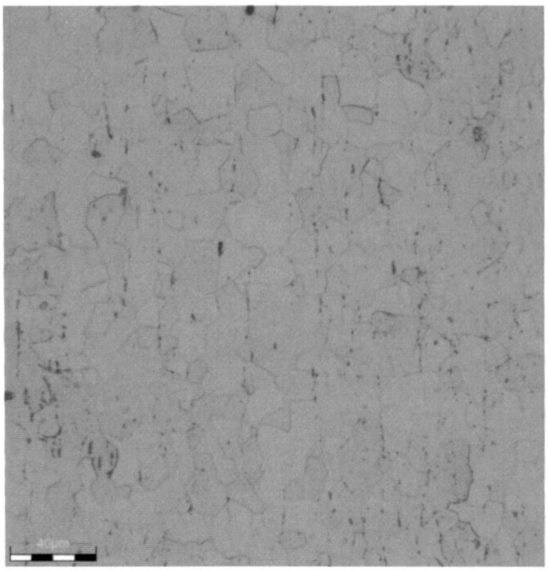

Figure 8. Titanium grade 4 micrograph showing equiaxed α grains at 1109x magnification.

4.2 Electrochemical Testing

4.2.1 Open Circuit Potential

The dynamic behaviour of the passive oxide film can be monitored by measuring the variation of open circuit potential with time, which reflects the thinning or thickening of oxide film for passive metals [9]. The OCP shift in the noble direction for the alloys suggests the formation of a passive film that reduces the corrosion rate. The OCP recorded was the last measurement after four hours. Figure 9 shows a graphical representation of the OCP results and Table 3 shows the detailed results. At four hours the OCP of the samples stabilised although the roughened showed an initial decrease in OCP. This indicates a breaking/dissolution of the air oxide film that was formed on the exposed surfaces due to the atmosphere, and could also be as a result of the aluminium contamination on the surface of the samples. The presence of the aluminium on the roughened samples also decreased the OCP as compared to the machined titanium.

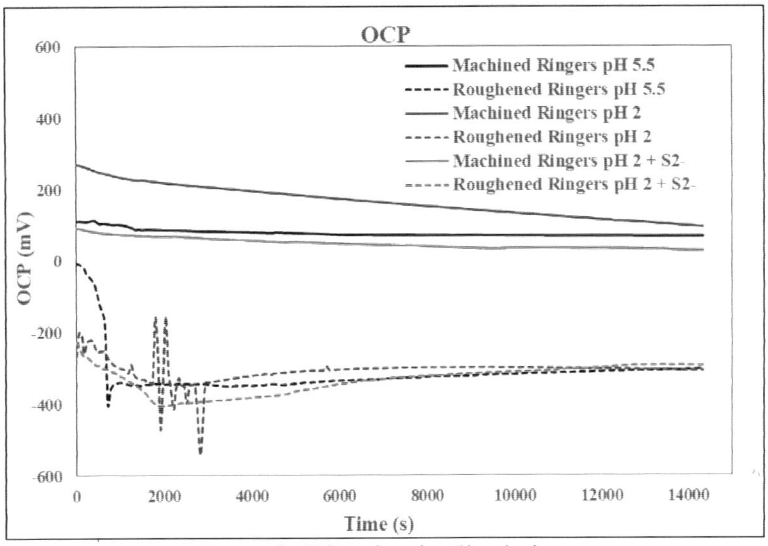

Figure 9. OCP plots in all solutions.

Table 3. OCP results

Samples	OCP (mV)		
	Ringers pH 5.5	**Ringers pH 2**	**Ringers pH 2+S**
Machined	63	85	11
	150	97	28
Roughened	-241	32	-292
	-302	-306	-106

4.2.2 Potentiodynamic Polarization Test

Figure 10 shows a graphical representation of the corrosion rates obtained while Table 4 gives the overall results. The Potentiodynamic polarisation test showed that corrosion rates obtained were very low, below 0.13mm/y which is an acceptable corrosion rate for biomaterial design and application [10]. Figure 11 shows the potentiodynamic polarisation scans.

Higher corrosion rates were observed in both the pH 2 and pH 2+S^{2-} solutions, and the effect was more pronounced on the roughened samples. The decrease in pH had little effect on the corrosion performance of the machined sample, however with the addition of sulphide S^- the corrosion rate increased significantly. The roughened samples showed significantly higher corrosion rates as compared to the machined samples. All samples showed an active to passive behaviour, and no pitting breakdown was observed.

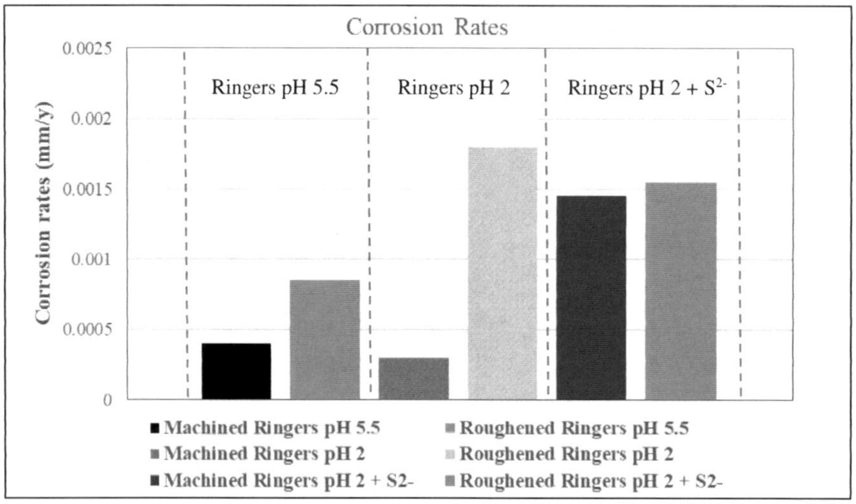

Figure 10. Graphical representation of the corrosion rates of the alloys in all solutions.

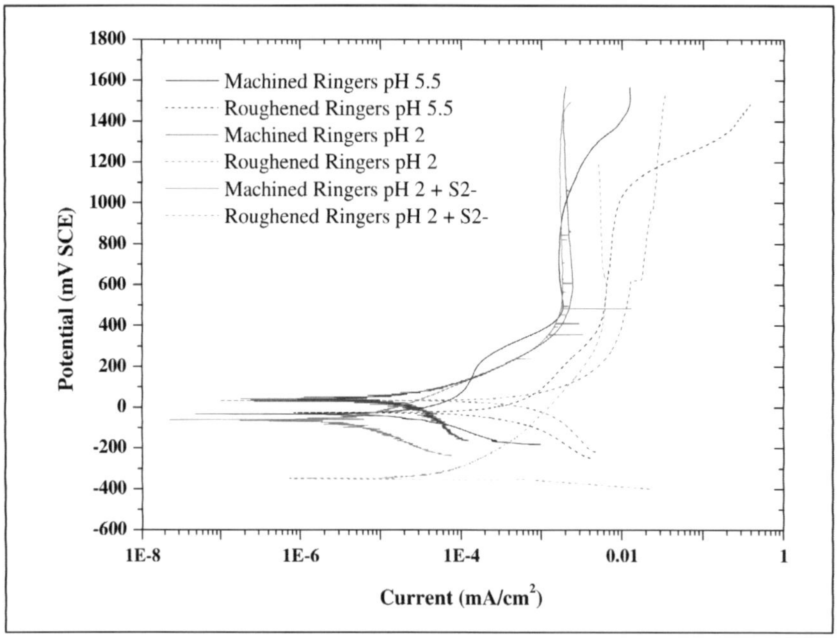

Figure 11. Plot of potential against the current.

Table 4. Potentiodynamic polarisation corrosion rates of the alloys in all solutions.

Samples		E_{corr} (mV SCE)	i_{corr} (mA/cm²)	Av. Corrosion rates (mm/y)	
				Individual	Average
Ringers pH 5.5	Machined	-296	5.1×10^{-5}	0.0005	0.0004
		-374	3.5×10^{-5}	0.0003	
	Roughened	-35	4.7×10^{-5}	0.0004	0.0008
		87	1.4×10^{-4}	0.0013	
Ringers pH 2	Machined	37	3.2×10^{-5}	0.0003	0.0003
		60	3.7×10^{-4}	0.0003	
	Roughened	27	2.6×10^{-4}	0.0023	0.0018
		-379	1.3×10^{-3}	0.0120	
Ringers pH 2+S^{2-}	Machined	-346	1.8×10^{-4}	0.0016	0.0014
		-244	1.4×10^{-4}	0.0013	
	Roughened	-62	7.2×10^{-5}	0.0006	0.0016
		2	2.8×10^{-4}	0.0025	

4.3 *Mass-loss Testing*

Mass-loss evaluations were performed after 7 and 30 days exposure. The corrosion rates were determined based on the mass-loss incurred during the exposure and the average corrosion rates were calculated, see Table 5. The corrosion rates obtained were very low in both 7 and 30 days exposure, less than 0.02mm/y indicating that the alloys tested were in the fully passive condition. Figure 12 shows the implants after testing. No change in colour/discolouration was observed on the surface of the implants after exposure.

Table 5. Mass-loss corrosion rates in all solutions.

Sample		Time (Hours)	Corrosion rates Av. (mm/y)	
			Individual	Average
Ringers	Machined	168 720	<0.02	<0.02
	Roughened	168 720	<0.02	<0.02
Ringers pH 2	Machined	168 720	<0.02	<0.02
	Roughened	168 720	<0.02	<0.02
Ringers pH 2+S^{2-}	Machined	168 720	<0.02	<0.02
	Roughened	168 720	<0.02	<0.02

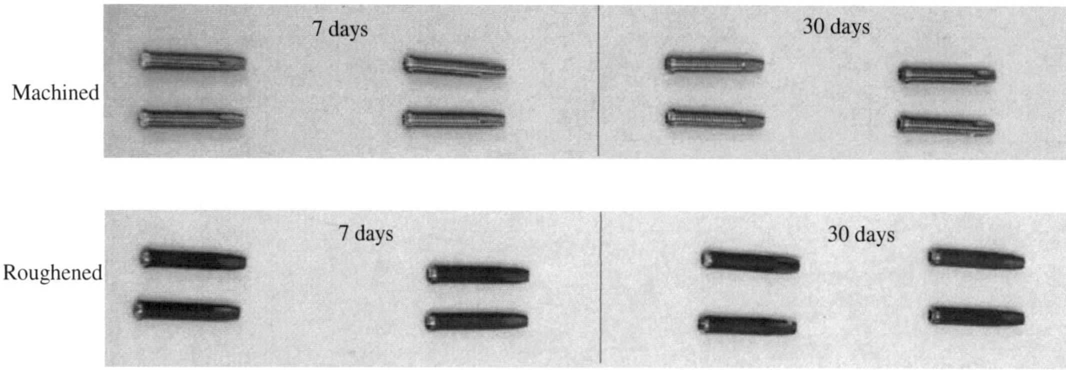

Figure 12. Photograph of the implants after testing.

5. Conclusions

The aim of this study is to characterise the corrosion performance of a machined and roughened surfaces of a commercially pure titanium grade 4. The following can be concluded:

- Metallography showed that the implants consisted of equiaxed alpha (α) grains in an annealed condition.
- The chemical composition and phase analysis showed that the machined implants consisted of titanium while the roughened implants showed a presence of aluminium particles embedded in the titanium matrix during the roughening process.
- The corrosion performance of Grade 4 titanium with machined and surface roughening were studied in standard and experimental ringers solutions. Although the implants performed quite well in the 30 day mass loss tests it is evident that longer exposure periods are required to effectively measure the effect of the surface roughness on the corrosion performance. These tests may take years, given examples from human trials.
- The electrochemical corrosion tests did however measure a significant difference in the corrosion performance of the machined and roughened samples. The increase in corrosion rate of the roughened samples can be attributed to both surface roughness and contamination of Al_2O_3 on the surface. The presence of this Al_2O_3 and TiO_2 played a significant role.
- Using electrochemistry it also determined that small changes in the environment pH and sulphide can have a significant influence on corrosion, where the sulphide addition has a larger influence on corrosion than the pH reduction, specifically related to the machined samples.

6 Acknowledgments

Authors would like to acknowledge Mintek and Southern Implants (PTY) LTD for financial assistance, sample preparation and approval to publish this work. The authors also wishes to appreciate the assistance from the Mintek colleagues.

7 References

[1] I. S. Yeo, *Surface modification of dental biomaterials for controlling bone response*. Elsevier Ltd, 2016.

[2] G. Strnad and N. Chirila, "Corrosion Rate of Sand Blasted and Acid Etched Ti6Al4V for Dental Implants," *Procedia Technol.*, vol. 19, pp. 909–915, 2015.

[3] S. Parithimarkalaignan and T. V. Padmanabhan, "Osseointegration: An update," *J. Indian Prosthodont. Soc.*, vol. 13, no. 1, pp. 2–6, 2013.

[4] A. Wennerberg and T. Albrektsson, "Suggested guidelines for the topographic evaluation of implant surfaces.," *Int. J. Oral Maxillofac. Implants*, vol. 15, no. 3, pp. 331–44, 2000.

[5] A. Sharma, G. R. Rahul, S. Poduval, and R. Sharma, "Implant Surface Micro-Design," *Glob. J. Med. Res.*, vol. 14, no. 4, 2014.

[6] M. E. Yurttutan and A. Keskin, "Evaluation of the effects of different sand particles that used in dental implant roughened for osseointegration," *BMC Oral Health*, vol. 18, no. 1, pp. 1–8, 2018.

[7] V. A. R. Barão, M. T. Mathew, W. G. Assunção, J. C. C. Yuan, M. A. Wimmer, and C. Sukotjo, "Stability of cp-Ti and Ti-6Al-4V alloy for dental implants as a function of saliva pH - an electrochemical study," *Clin. Oral Implants Res.*, vol. 23, no. 9, pp. 1055–1062, 2012.

[8] R. Harada, S. Takemoto, H. Kinoshita, M. Yoshinari, and E. Kawada, "Influence of sulfide concentration on the corrosion behavior of titanium in a simulated oral environment," *Mater. Sci. Eng. C*, vol. 62, pp. 268–273, 2016.

[9] E. X. Sun, S. Fine, and W. B. Nowak, "Electrochemical behavior of nitinol alloy in Ringer's solution," *J. Mater. Sci. Mater. Med.*, vol. 13, no. 10, pp. 959–964, 2002.

[10] R. Bhola, S. M. Bhola, B. Mishra, and D. L. Olson, "Taat1111Sjip34.Pdf," vol. 25, no. 1, 2011.

IOP Conf. Series: Materials Science and Engineering **655** (2019) 012007 doi:10.1088/1757-899X/655/1/012007

Development of lining materials for reactor vessel used in the CSIR titanium process

M P Sehoana[1,a], S Fazluddin [2,a]

[1] Light Metals, Materials Science & Manufacturing, Council for Scientific and Industrial Research, Meiring Naudé Road, Brummeria, Pretoria 0184, South Africa

E-mail: sehoanamp@gmail.com

Abstract. The CSIR Titanium powder process is a high temperature' molten salt /alkali metal titanium production process. The aim of this process is to produce titanium powder of high purity and quality that meets the specifications for commercially-pure powder. The quality of CSIR Ti-powder produced to date shows in general that there are challenges when it comes to meeting this requirement. Two of the main challenges is the reagents used employed in the process and by-products formed which are highly corrosive. These corrosive agents promote corrosion of the reactor with the corrosion product then being introduced as impurities in the titanium powder. A project to investigate lining materials to protect the vessel from corrosion and preserve product quality and purity was thus initiated. This article gives an in-depth insight into the corrosion resistance properties of different metals that can be investigated as a potential lining material in molten lithium and molten lithium chloride environment.

1. Introduction

South Africa is one of the leading producers of titanium mineral concentrates in the world [1]. Regardless of having these mineral reserves in abundance, the country does not have an established titanium production industry. The lack of local titanium production capacity and associated manufacturing capabilities were identified by the Department of Science and Technology (DST) and Department of Mineral Resources (DMR) as a key potential economic growth area.

In order to exploit this potential growth area, the DST initiated a Titanium center of competence (TiCOC) hosted by the CSIR. The main objective of DST through TiCOC was to develop a novel titanium powder production process that can be ultimately be commercialized. It is for this reason that the CSIR-Ti process, a novel titanium production process was developed and commissioned at pilot scale by the CSIR.

Titanium is currently being produced commercially via Kroll and Hunter processes[1,2]. These processes are both batch processes that employ metallothemic reduction of titanium tetrachloride to produce titanium. The Kroll process uses molten magnesium as a reducing agent while the Hunter process employs molten sodium. There are documented attempts where lithium and calcium were used as reducing agents as well[3]. The CSIR-Ti process differs from commercial processes as it aimed to produce titanium powder via a continuous process[1,3]. The aim of this process is to produce commercially pure titanium powder however the quality of CSIR Ti-powder produced to date shows in general that there are challenges when it comes to meeting this requirement[4].

Figure 1 Corrosion of molten salt pots before and after a salt leak at bench scale

One of the challenges of this process is that the molten metal that is used and the molten salt by-product that forms are very corrosive. These agents promote corrosion of the reactor vessel with a potential for corrosion product to be introduced as impurities in the titanium product. Figure 1 shows the severity of the corrosion caused by molten salt following a salt leak or a lack of proper sealing of the vessel. It was found that the use of a suitable reactor lining material can help eliminate or reduce contamination of the titanium metal product as well as reduce the risk of leakage caused by corrosion. The lining material should be able to resist corrosion by molten salt and/or molten metal at temperatures above 650°C, this being the nominal process temperature. A project to investigate lining materials to protect the vessel from corrosion and preserve product quality was thus initiated. In this project, different materials that can resist corrosion by molten lithium chloride (LiCl) and lithium (Li) will be identified and tested to identify the most suitable candidate.

Molten salts and molten alkali metals are mostly used in the heat treatment industry because of their good thermal properties. Even in these industries, the corrosion of the containment material is a major problem [5,6]. Despite the fact that the reagents are widely used in some industries, there is still limited information and understanding to serve as the basis for the selection of compatible materials [7]. The aim of this literature review is to highlight reported work on corrosion of different materials in molten lithium and lithium chloride environments, with emphasis on reaction mechanisms and corrosion trends. The review serves to identify the material with the most favorable corrosion resistant properties with the intention to test further under conditions representative of the CSIR-Ti process after which a lining material will be selected.

2. Molten salt corrosion

Molten salt corrosion generally occurs in two steps. The corrosion starts with oxidation, followed by the dissolution of the metal oxide of the containment metal [5]. Molten salts are good fluxing agents and can thus easily remove oxide scale from the metal surface [6,7]. At high temperatures molten salts also act as good solvents of precipitants, preventing passivation of the metal surface because oxides scales are not able to form[6,7]. At elevated temperatures molten salt corrosion is rarely inhibited because the salts become increasingly conductive and reactive, promoting electrochemical processes [6]. The presence of impurities in the salt, typically oxygen and water vapour can accelerate corrosion [7]. The corrosion resistance of metals to molten salt also depends on the elemental composition, i.e.,

IOP Conf. Series: Materials Science and Engineering **655** (2019) 012007 doi:10.1088/1757-899X/655/1/012007

Al, Si, Mo, etc. which affects the development of the passivating oxide [6]. Corrosion mechanisms in aqueous solutions may also be applied to that for molten salts. The main difference between the two systems is that in molten salts, the most active element is oxygen, therefore pO^{2-} is applied instead of pH [7]. Due to these similarities, the corrosion that applies in aqueous environments are also applicable in molten salt, be this in the form of uniform surface corrosion, intergranular corrosion, galvanic corrosion and pitting[5–7].

LiCl is produced as a by-product when Li is used as a reducing agent in the Ti-metal production process from $TiCl_4$. Understanding the corrosion behavior of different containment materials exposed to molten LiCl is therefore very important. Although the corrosion studies of potential containment material against molten LiCl is limited, the articles referenced below help clarify this issue and provide useful information.

Shanker at el intensively investigated high-temperature corrosion of Ni and Fe alloys in molten LiCl-KCl medium in different atmospheres[8–10]. Alloys such as 316L stainless steel, Inconel 600, Inconel 625, and Inconel 690 were investigated for corrosion in different tests with temperatures ranging from 650-850 °C for extended periods of time. The authors derived the following conclusions based on their experiments; selective leaching of Cr by molten LiCl is responsible for corrosion experienced by these alloys[8–10]. Other authors like Cho et al[11,12] and Indacochea et al [13] tried to study the corrosive properties of molten LiCl and molten LiCl-Li_2O in oxidizing and inert atmospheres.

Cho et al [11] made an observation that the corrosion rate of the Fe/ Ni alloys in molten LiCl is relatively low and the weight losses are similar but the addition of Li_2O to LiCl significantly accelerates the corrosion rate. They also noted that the corrosion rate increased with the Cr content of the alloys tested [11]. Indacochea et al [13] thus concluded that that corrosion mechanism of these materials in molten salts is different in oxidizing and inert atmospheres.

The corrosion reaction of molten salts proceeds by oxidation which is then followed by the dissolution of the metal oxide [5]. Ni and Fe alloys contain a certain percentage of Cr in their matrix, which is responsible for forming a passive Cr_2O_3 layer which acts as a protective barrier against corrosion in aqueous systems. Molten LiCl is a good fluxing agent that effectively removes and dissolves Cr_2O_3 scales from metal surfaces [5,9]. The conclusion that was made by Shanker at el [8–10] was supported by Salinas-Solano et al [14], who reported that Cr is more soluble in molten chloride salts when compared to Ni and Fe, and is therefore responsible for low resistance of some Cr-containing metals in oxidizing, molten salts environments.

The observation made by Cho et al is as a result of the introduction of oxygen ion into the system by Li_2O. It is reported that oxygen ions penetrate through the micro-channels formed in the alloys to induce internal, inter- and trans-granular corrosion[12–14]. The microchannels are formed when iron diffuses into the Cr_2O_3 layer and then dissolving in the molten salt, forming voids between the Cr_2O_3 layer and Ni-rich oxide[8]. The diffusion of Fe-ion takes place because Fe-ions have more cation defects compared to Ni ions in the Ni-rich oxides [12]. Therefore Fe also contributes to the overall corrosion of Fe and Ni alloys in molten salt environments. Shanker at el [10] reports that the better corrosion resistance properties of Ni-based alloys to molten LiCl can be attributed to the adherent and protective NiO present on the surface. Cho et al [12] made a similar observation and reported that the accumulation of Ni around the inner oxides of the alloy delays corrosion rate. In conclusion, Ni-based alloys are more resistant to molten LiCl corrosion than Fe-based alloys and the resistance to molten LiCl increases with decreasing Cr and Fe content and increasing Ni content in Fe-Ni-Cr alloys[11,10].

3. Molten metal corrosion

Molten metals are mostly used as heat transfer medium because of their excellent thermal properties such as high thermal conductivity, heat capacity and low vapour pressures [5]. Molten metals also serve as reducing agents in the production of metals[15]. Molten alkali metals like Mg, Na, and Li are used in the production of titanium with magnesium being the most common as per the Kroll process. One of the biggest challenges when working with molten metals is the corrosion of the containment metal (solid metal) by the liquid metal. Molten metal corrosion can be defined as the interaction which takes place between a solid and metal in the molten state [16]. This type of corrosion is mostly caused by the solubility of solid metal into the liquid metal. In this section of the review, the molten metal of interest is Li. In general, molten Li corrosion mechanisms can be classified as a combination of the following interactions: dissolution (uniform), impurities and interstitial reaction (inter- and trans-granular attack), alloying and compound reduction [6,15,16].

3.1. Direct dissolution

Direct dissolution is the simplest and most dominating form of molten Li corrosion. Direct dissolution takes place when solid metal atoms are released and dissolved into the molten metal in the absence of impurities [15]. Solid metals that have high solubility in molten Li will generally exhibit a higher corrosion rate [5].

Tortorelli [15] derived a mathematical expression (Equation 1) that can be used for explaining molten Li metal corrosion by dissolution. Based on the equation, corrosion by direct dissolution can be reduced by selecting a containment metal that has low solubility in molten Li or saturating molten Li with another reagent to reduce its ability to dissolve before exposure. Saturating molten Li before the exposure is however not ideal for the titanium production process due to the risk of introducing impurities into the final product.

$$J=K(C-c) \qquad (1)$$

J - Rate at which elemental species enter the solution,
C - Solubility of a particular metal in the molten metal,
C - Actual concentration of the metal in the molten metal,
K - Solution rate constant of the rate-controlling steps

The solubility of different metals in molten Li was investigated by Cleary et al [17] and the results are shown in Figure 2. Based on this study, Ni has the highest solubility in molten Li compared to Cr and Fe [17]. The susceptibility to dissolution attack in Ni-Fe-Cr alloys increases with an increase in nickel content[17]. Quain et al [18] studied the corrosion of austenitic stainless steels in molten Li and found that the corroded region of the metal had depleted Ni and Cr. In the affected region, Ni and Cr were reduced by 89-91% and 91-93% respectively. The authors then concluded that the corrosion process of the metal was due to the preferential dissolution of Ni and Cr into the molten Li [18]. It is for that reason that Ni and its alloys are normally not considered good candidate materials for containing molten Li [5]. The presence of nitrogen and carbon impurities in molten Li promotes the diffusion of Cr into the molten Li and would thus accelerate the corrosion [19,20]. Olson et al[19] report that based on the nature of dissolution attack of ferrous alloys, alloys with limited Ni and Cr content are more desirable [19]. The authors' further state that pure Fe demonstrated good corrosion properties to molten Li [19]. Figure 2, also shows that refractory metals have low solubility in molten Li when compared to Fe and Ni alloys. Investigations also indicate that refractory metals can resist the dissolution corrosion mechanism when the system contains no impurities, especially oxygen [21,22]. The effects of impurities on molten Li corrosion is discussed in the next section.

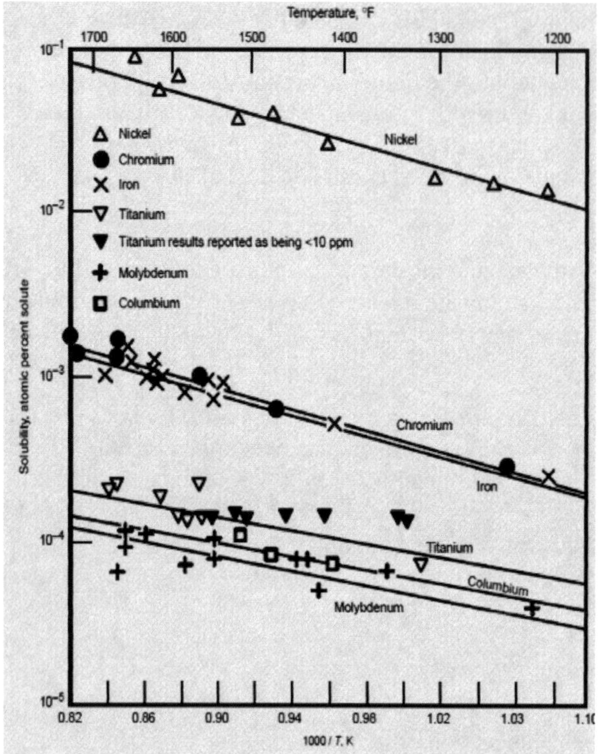

Figure 2 Solubility of different metals in molten lithium [17].

3.2. Impurity or interstitial corrosion

Impurity or interstitial reactions refers to the interactions of light elements present in containment materials (interstitial) or in the molten metal (impurities) [5,6,16,21]. The impurities active in molten metals corrosion are oxygen, nitrogen, carbon, and hydrogen but oxygen is the most important[16]. The effect of oxygen in molten-metal corrosion differs depending on whether the oxygen resides in the molten metal or solid metal [20]. When oxygen contaminants reside in the molten metal, corrosion proceeds by the dissolution of the solid metal into the molten metal. Oxygen in the molten metal melt reacts with refractory metals to form a corrosion product, which if is soluble in the melt, can accelerate corrosion [22].

Hoffman et al [23] report that the use of alloying elements such as zirconium and hafnium can improve the corrosion resistance of refractory metals[20,22]. These alloying elements form oxide layers of ZrO_2 and HfO_2 on the surface of refractory metals, are very stable and increase the threshold oxygen concentration needed for corrosion to occur[23]. Even though the solubility of refractory metals in molten Li is low, they can still be attacked by molten metal if the metals have high oxygen content. The oxygen concentration of niobium must not exceed a threshold level of 400 ppm Li in order to prevent penetration of the metal [20]. This observation of threshold concentrations was noted throughout the corrosion tests performed on refractory metals in different molten metals as reported by Hoffman et al[22]. This form of corrosion can thus be eliminated by minimizing the concentration of oxygen in the refractory metals[15]. Qian et al [18] that corrosion of Fe-Ni-Cr is also accelerated by impurities in molten Li. Nitrogen and carbon contaminants in molten Li react with Cr in the alloy metal to form Li_9CrN_5 and $Cr_{23}C_6$. The depletion of Cr on the surface of the alloy results in the diffusion of Cr out of the alloy matrix which accelerates corrosion [18,20]. Molten Li also reacts with carbon in the alloy matrix to form Li_2C_2, a process called decarburization, which leads to intergranular corrosion. It is because of this corrosion mechanism that carbon steel is not considered suitable for molten Li containment [20,21].

4. Considerations in the selection of a lining material

Information gathered from the literature discussed in previous sections can be used in selecting candidate lining material for the reactor vessel in the CSIR-Ti process. Based on the available information, Ni alloys perform better in molten LiCl environment but their efficiency decreases with increasing Fe and Cr content. However, in molten Li, Fe based alloys perform better with their performance degrading with increasing Ni content. This means that in an environment containing both molten Li and molten LiCl, Fe-Ni-Cr alloys will experience corrosion. However, with proper impurity control in the alloy as well as the molten Li and molten LiCl media, the alloys could be employed for long term service. It is important, however, that more extensive research be conducted to identify an optimum alloy composition beforehand.

The information gathered also shows that refractory metals can perform better in molten LiCl and molten Li provided there is low oxygen activity in the system. This The implies, therefore, that to achieve optimal corrosion resistance, high purity refractory metals must be employed and proper control of impurities must be maintained in the system.

Table 1 below provides useful guidelines for selecting materials to use in molten Li and molten LiCl systems. These guidelines will help in ensuring that the lining material provides optimal performance and the environment is adequately controlled so as to avoid corrosion.

Table 1 Guidelines for material selection based on molten lithium and lithium chloride reactions

Corrosion reaction	Guideline	Example
Direct dissolution	Lower activity of key elements	Reduce Ni in molten Li system Reduce Fe and Cr in molten LiCl Reduce carbon in Li and LiCl systems
Corrosion product formation	Lower reacting elements	Reduce Ni and Cr in Li system Reduce Cr and Fe in LiCl system
	In case of protective oxides, add elements to promote oxide formation	Add Zr to Nb exposed to Li Add Hf to Ta exposed to Li Add Ti to Mo exposed to Li Add Ti to carbon steel exposed to Li Add Re to W exposed to Li
Element transfer	Minimize elements being transferred	Reduce oxygen concentration in metal exposed to Li Reduce the carbon content of metals exposed to Li

5. Concluding remarks

Regardless of the growing interest in molten LiCl and molten Li, corrosion associated with these chemicals remains a significant concern. When these chemicals are used individually, their corrosion effect can be controlled more effectively but when both are present in a system, corrosion control becomes a significantly more complex task. This is due to the fact that some metals perform well in molten Li but poorly in LiCl. Some metals like refractory metals show great potential in both these environments but are sensitive to impurities. The use of refractory lining material demands therefore that both the metal and reagents have limited contaminant concentrations. Literature provides useful information and guidelines on the most suitable lining material but more research and experimental tests on different refractory metals and different grades of Fe-Ni-Cr alloys are needed to determine the optimum materials in both either case.

6. Acknowledgements

The research was supported by the CSIR and the University of Johannesburg

7. References

[1] Vuuren D S Van, Oosthuizen S J and Swanepoel J J 2017 Process for the production of crystalline titanium powder *Process Prod. Cryst. Titan. powder*

[2] Fang Z Z, Paramore J D, Sun P, Chandran K S R, Zhang Y, Xia Y, Cao F, Koopman M and Free M 2018 Powder metallurgy of titanium–past, present, and future *Int. Mater. Rev.* **63** 407–59

[3] van Vuuren D S 2015 Direct titanium powder production by metallothermic processes Titanium Powder Metallurgy (Elsevier) pp 69–93

[4] Oosthuizen S J, Swanepoel J J and van Vuuren D S 2014 Challenges experienced in scaling-up the CSIR-Ti process *Advanced Materials Research* vol 1019 (Trans Tech Publ) pp 187–94

[5] Lai G Y 2007 *High temperature corrosion and materials application* (ASM international)

[6] Patel N S, Pavlík V and Boča M 2017 High-temperature corrosion behavior of superalloys in molten salts–a review *Crit. Rev. Solid State Mater. Sci.* **42** 83–97

[7] Koger J W 1987 Fundamentals of high-temperature corrosion in molten salts *ASM Handbook.* **13** 50–5

[8] Shankar A R and Mudali U K 2008 Corrosion of type 316L stainless steel in molten LiCl–KCl salt *Mater. Corros.* **59** 878–82

[9] Shankar A R, Kanagasundar A and Mudali U K 2012 Corrosion of nickel-containing alloys in molten LiCl-KCl medium *Corrosion* **69** 48–57

[10] Shankar A R, Thyagarajan K and Mudali U K 2013 Corrosion behavior of candidate materials in molten LiCl-KCl salt under argon atmosphere *Corrosion* **69** 655–65

[11] Cho S H, Zhang J S, Shin Y J, Park S W and Park H S 2004 Corrosion behavior of Fe–Ni–Cr alloys in the molten salt of LiCl–Li2O at high temperature *J. Nucl. Mater.* **325** 13–7

[12] Cho S-H, Cho I-J, You G-S, Yoon J-S and Park S-W 2007 Corrosion behavior of Ni– Base alloys in a hot lithium molten salt under an oxidizing atmosphere *Met. Mater. Int.* **13** 303–9

[13] Indacochea J E, Smith J L, Litko K R, Karell E J and Rarez A G 2001 High-temperature oxidation and corrosion of structural materials in molten chlorides *Oxid. Met.* **55** 1–16

[14] Salinas-Solano G, Porcayo-Calderon J, Gonzalez-Rodriguez J G, Salinas-Bravo V M, Ascencio-Gutierrez J A and Martinez-Gomez L 2014 High temperature corrosion of inconel 600 in NaCl-KCl molten salts *Adv. Mater. Sci. Eng.* **2014**

[15] Tortorelli P F 1987 Fundamentals of high-temperature corrosion in liquid metals *ASM Handbook.* **13** 56–60

[16] Smith D L 1966 Effect of oxygen content on the liquid metal corrosion of molybdenum

[17] Leavenworth H W and Cleary R E 1961 The solubility of Ni, Cr, Fe, Ti, and Mo in liquid lithium *Acta.Met.* **9**

[18] Jiapu Q, Jiming C, Jinbiao C, Zengyu X, Weihao W and Chuanjie P 1991 Corrosion of austenitic stainless steel in liquid lithium *J. Nucl. Mater.* **179** 603–6

[19] Olson D L and Matlock D K 1978 The essential chemical and physical properties of ferrous alloys needed for containment of liquid lithium Lithium Needs and Resources (Elsevier) pp 315–23

[20] Klueh R L 1970 Penetration of refractory metals by alkali metals Corrosion by liquid metals (Springer) pp 177–96

[21] DeVries G 1970 The corrosion of metals by molten lithium Corrosion by Liquid Metals (Springer) pp 251–69

[22] Hoffman E E and Harrison R W 1968 The compatibility of refractory metals with liquid metals Refractory metal alloys metallurgy and technology (Springer) pp 251–87

[23] DiStefano J R and Hoffman E E 1963 Corrosion mechanisms in refractory metal - Alkali metal systems *Corros. Mech. Refract. Met. Met. Syst.* **4500** 52

IOP Conf. Series: Materials Science and Engineering **655** (2019) 012008 doi:10.1088/1757-899X/655/1/012008

Development of a mathematical equation describing the strain hardening behaviour of metastable AISI 301 austenitic stainless steel

T W Mukarati[1,*], R J Mostert[1] and C W Siyasiya[1]

[1] University of Pretoria, Private bag X20 Hatfield 0028, South Africa

*E-mail: Corresponding Author: tmcarati@gmail.com

Abstract. The strain hardening behaviour of AISI 301 metastable austenite steel was analysed by evaluating tensile data against the empirical mathematical equations of Hollomon, Ludwik and Ludwigson. It was found that these equations were inadequate to model this TRIP steel with low stacking fault energy (SFE). It was found that the fraction of strain-induced martensite could be expressed as a sigmoidal function of the applied strain. The log-log plots of true stress and true plastic strain from 5% to ε_{UTS} performed with uniaxial isothermal tests at 30 °C were thereafter adequately fitted with a sigmoidal curve. The instantaneous strain hardening exponent was determined as the slope of the above-mentioned sigmoidal curve at a specific strain value. The strain hardening exponent and the rate of strain hardening (dσ/dε) increases with deformation due to formation of strain-induced martensite to a maximum and thereafter decreases as the volume fraction of strain-induced martensite approximates saturation. The variation of the instantaneous strain hardening exponent as a function of plastic strain and the strength coefficient, K, at 30 °C was deduced. A high value of K, 1526MPa, was determined. A correlation between the extent of martensitic transformation and the value of the instantaneous strain hardening exponent was observed. This work is part of the project that seeks to develop a constitutive model describing the flow stress during plastic deformation as a function of both plastic strain and the resulting martensitic transformation at different temperatures and strain rates and which accounts for the isotropic hardening process.

1. Introduction

The work-hardening behaviour of many engineering materials have been sufficiently described by the Hollomon power law [1]:

$$\sigma = K\varepsilon^{n} \qquad (1)$$

where σ is the true stress; K is the strength coefficient, ε is true plastic strain and n is the strain hardening exponent. The Ludwik model has an additional stress factor (σ_o) for materials which show a varied yield strength with similar strain hardening [1].

$$\sigma = \sigma_o + K\varepsilon^{n} \qquad (2)$$

An equation with an additional strain term, ε_o, which accounts for pre-strain was proposed by Swift, for materials which show similar yield strength with varied strain hardening behaviour [1].

$$\sigma = K\,(\varepsilon_o + \varepsilon)^{n} \qquad (3)$$

For stable austenitic stainless steels, although the Hollomon or Ludwik equations can describe the plastic flow behaviour at high strain levels, the models were found to be insufficient in describing the plastic flow behaviour and work hardening at low strain values [2]. A modified Ludwik model was then proposed by Ludwigson to consider the deviation [3].

$$\sigma = K\varepsilon^n + \Delta \tag{4}$$

where the deviation $\Delta = \exp(K_1)\exp(n_1\varepsilon)$. K and n have the same meaning as in the Hollomon equation and $\exp(K_1) \approx$ proportional limit and n_1 is the slope of the deviation of stress from the Hollomon equation plotted against true strain, ε [3].

The Voce flow stress relationship for materials which show saturation stress at high stress/strain levels can be represented as follows [1]:

$$\sigma = \sigma_s - (\sigma_s - \sigma_1)\exp[-(\varepsilon - \varepsilon_1)/\varepsilon_c] \tag{5}$$

where σ_s, is the saturation stress, σ_1 and ε_1 are the true stress and true plastic strain values at the onset of plastic deformation, respectively, and ε_c is a constant. The Voce stress relationship is usually applicable only at high temperatures where dynamic recovery cancel out the work hardening effect during the test.

In the current work, the strain hardening behaviour of a metastable AISI 301 steel (for chemical composition see reference [4]) was investigated with a view to establish a single mathematical equation that describes the behaviour adequately.

The Ludwigson modification has earlier been found to be adequate in describing the tensile flow and work hardening behaviour of ferritic stainless steels [1]. There was no information which was found in literature about applicability of the Ludwigson equation in describing the work hardening behaviour of metastable austenitic stainless steels which undergo a phase transformation during deformation.

2. Experimental procedure

The plastic flow behaviour was studied by means of uniaxial tensile testing. Interrupted uniaxial tensile test was done at 30 °C using an Instron-type tensile machine (1175 model) fitted with an environmental chamber (3110 model) at a strain rate of 6.67 x $10^{-4}s^{-1}$. A 50 mm gauge length extensometer (Instron-type model 2630-112) was used to determine the elongation of the samples. An indication of the percentage of α′-martensite of all the specimens was initially determined using a Ferritescope (Helmut Fisher GmbH, model MP3B) at 5% engineering strain intervals. The device was calibrated using standard δ-ferrite samples supplied with it. Seven readings were taken on each sample at different points within the gauge length on the surface.

A calibration curve was later developed using X-ray diffraction to determine the percentage of α′-martensite from the Ferritescope measurements [5]:

$$\text{True } \alpha'\text{-martensite content} = 1.81 \text{ x Ferritescope reading} \tag{6}$$

3. Results

Figures 1 (a) and (b) show the engineering stress-strain curve and the true stress-true strain curves at a temperature of 30 °C, respectively. Figure 1(b) excludes tensile data points beyond the necking onset. Figure 2 shows the plastic flow log-log plot of true stress and true strain at temperature of 30 °C. Figures 3 (a) and (b) show a sigmoidal fit applied to the experimental data points of the fraction of martensite as a function of strain at 30 °C. The fraction of martensite transformed strain sensitivity obtained by differentiation of the curve in Figure 3 (a) is shown in Figure 3 (b).

Figures 4 (a) and (b) show the log-log plot of true stress-strain at temperature of 30 °C for a true strain of (a) less than 5% which shows that Hollomon fit has only been obeyed at low plastic strain with a strain hardening exponent, n of 0.11 and (b) at strains greater than 5% showing a sigmoidal flow stress relationship. Figures 5 (a) shows the variation of instantaneous strain hardening exponent as a function of increasing plastic strain for true strain values above 5%, given by the slope of Fig 4(b) utilising a number of tangent lines as a function of true strain. Figure 5(b) is a superimposition of Figure 3(a) and 4 (b) showing a correlation of the tensile flow stress and percentage of martensite transformed curves. Figure 6 (a) is a superimposition of Figure 3(b) and 5(a) showing peak instantaneous α′-martensite transformation strain sensitivity, $(dV\alpha'(d\varepsilon)^{-1})$ and peak instantaneous strain hardening exponent (n_i) as a function of strain. Figure 6 (b) shows the strength (log true stress) as a function of the percentage of martensite formed.

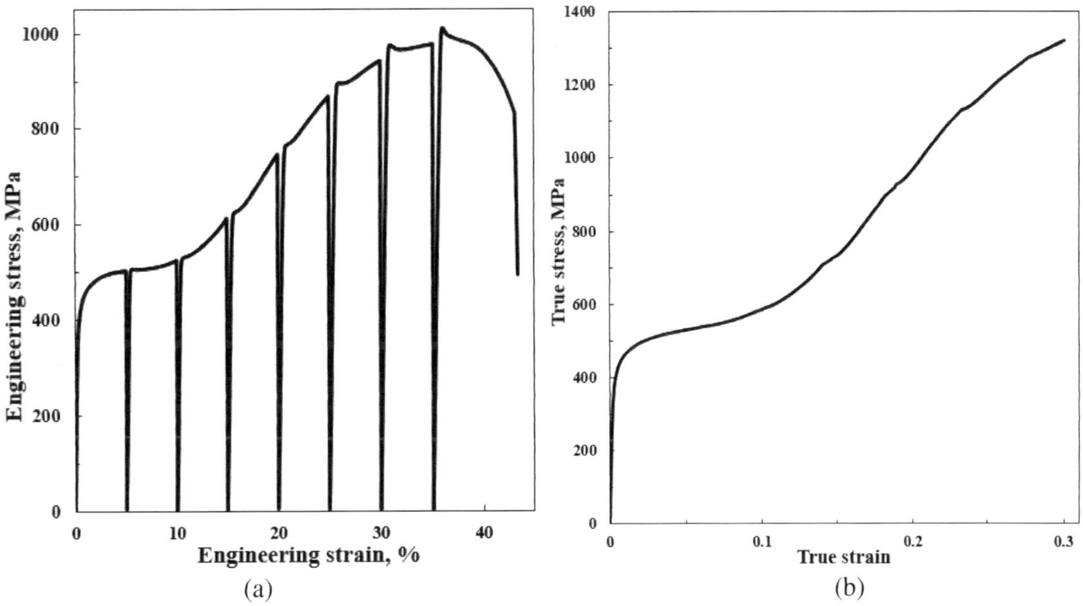

Figure 1: (a) Interrupted engineering stress-strain curve (b) True stress-true strain curve at a temperature of 30 °C.

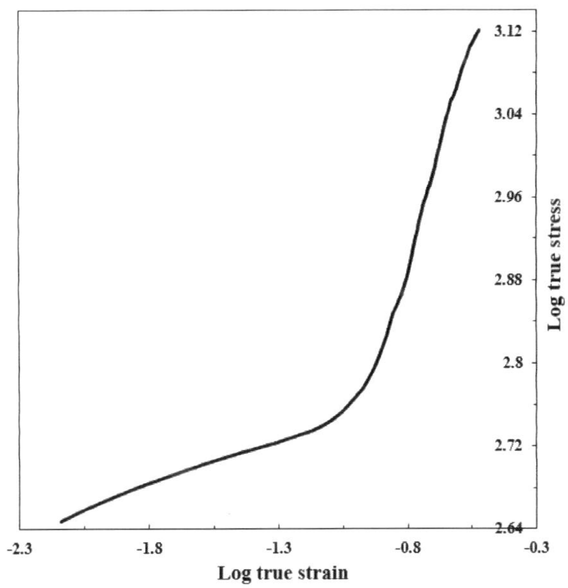

Figure 2: The plastic flow log-log plot of true stress and true strain at temperature of 30 °C from R_p to R_m

IOP Conf. Series: Materials Science and Engineering **655** (2019) 012008 doi:10.1088/1757-899X/655/1/012008

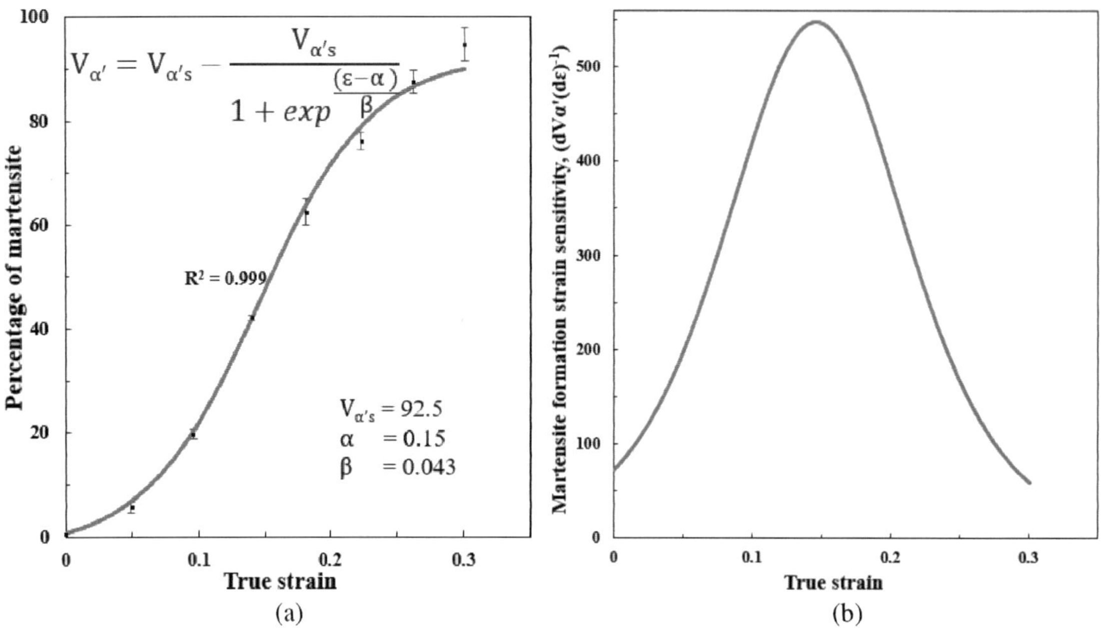

Figure 3: (a) Percentage of α′-martensite as a function of strain at 30 °C (fitted sigmoidal curve) (b) α′-martensite transformation strain sensitivity, $(dV\alpha'(d\varepsilon)^{-1})$ (dimensionless) obtained by differentiation of the curve in Figure 3 (a).

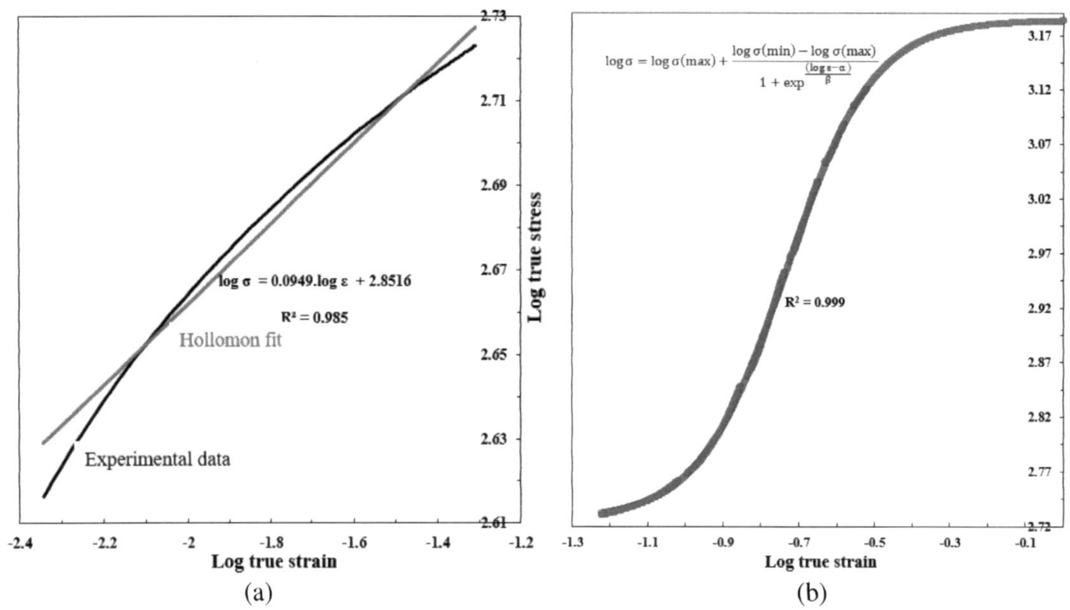

Figure 4: The log-log plot of true stress-strain at a temperature of 30 °C for true strain of (a) less than 5% (obeying the Hollomon equation) and (b) greater than 5%, showing sigmoidal flow stress behaviour (sigmoidal equation given in the figure with constants given in equation 7). [The blue line in both Figure 4 (a) and (b) represents the experimental log-log plot of true stress-strain curve whilst the red line is a Hollomon fit in Figure 4 (a) and a sigmoidal fit in Figure 4 (b) which was extrapolated to a log true strain value of 0 to determine the strength coefficient, K.]

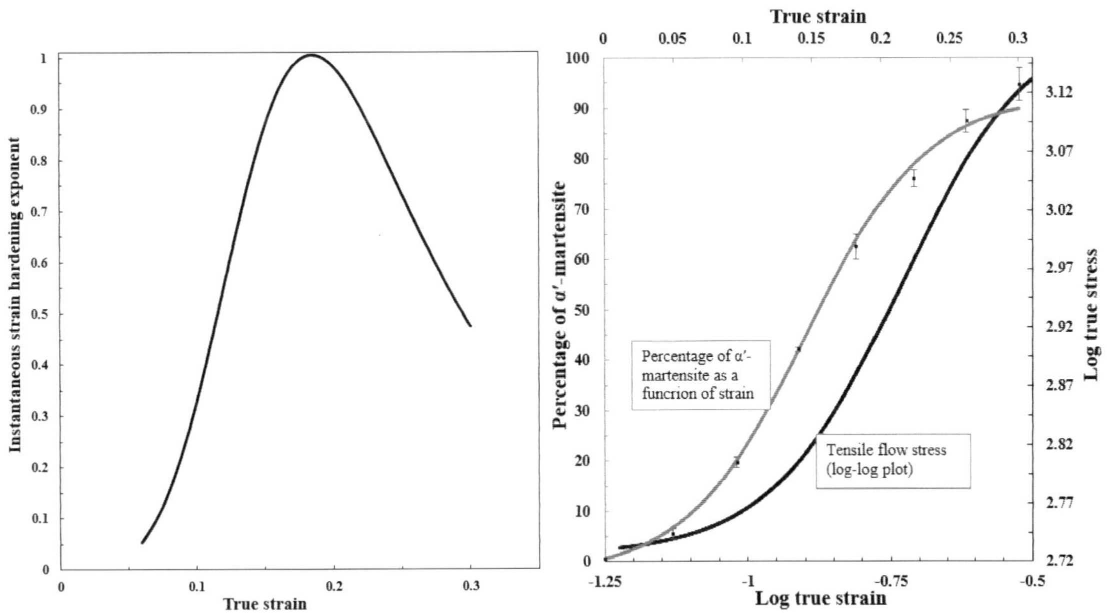

(a) (b)

Figure 5: (a) Variation of instantaneous strain hardening exponent, n_i with deformation for true strain above 5% (b) Superimposition of Figure 3(a) and 4 (b) showing a correlation of tensile flow stress with percentage of martensite transformed.

The instantaneous strain hardening exponent, (refer to Soares et al) [6] was found to generally increase with deformation due to the strain-induced martensitic transformation (SIMT). After an initial linear portion (up to $\varepsilon = 0.05$), the strain hardening behaviour becomes non-linear (Fig 4(b)), with strong strain hardening experienced, due to SIMT as shown in Figure 5(a) and 5(b).

The instantaneous strain hardening exponent was found to be variable with a peak value of 1.00 at a strain value of 0.19 and log K was found to be 3.18. The coefficient of strength, K is therefore approximately 1526 MPa as determined by an extrapolation of the sigmoidal curve to a true strain value of 1. The high value of strength is attributed to the high percentage of transformed martensite, above 90% as shown in Figure 3 (a).

It is apparent from Figure 6 (a) that the curve describing the martensite formation strain sensitivity as a function of strain has a similar shape to the instantaneous strain hardening exponent curve. However, the peak instantaneous derivative of α'-martensitic transformation was reached at lower strain than the peak instantaneous strain hardening exponent. This depicts that the strain hardening behaviour of AISI 301 steel is controlled by the formation of strain-induced α'-martensite [7]. The strength of the material is directly proportional to the percentage of martensite evolved during deformation as shown in Figure 6 (b).

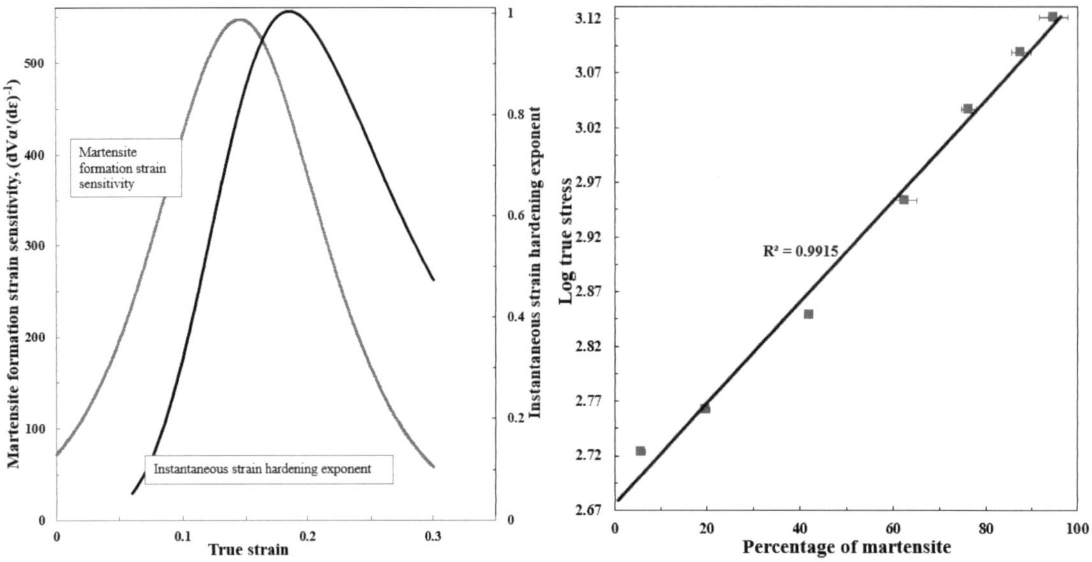

Figure 6: (a) Superimposition of Figure 3(b) and 5(a) showing peak instantaneous α′-martensite transformation strain sensitivity, $(dV\alpha'(d\varepsilon)^{-1})$ and peak instantaneous strain hardening exponent, n_i as a function of strain (b) the strength as a function of percentage of martensite.

4. Discussion

The AISI 301 alloy tested here is a lean alloyed stainless steel with high nitrogen content and very low stacking fault energy at room temperature which makes it amenable to SIMT leading to a strong increase in strain hardening with increased applied strain.

Due to the flow behaviour associated with strain-induced martensitic transformation, a sigmoidal equation was found to be the only mathematical relationship adequately describing the strain hardening of this steel, above plastic true strain levels of 0.05.

The mathematical sigmoidal function established in the true strain range from 0.05 to ε_u (strain corresponding to maximum load) is given by:

$$\log \sigma = A + (B - A)[1 + \exp(\log \varepsilon - \alpha)(\beta)^{-1})]^{-1} \quad [7]$$

with $R^2 = 0.999$, at 30 °C.

Where,

σ is the true stress in MPa,

A is the maximum log of true stress ($\log \sigma_{max}$) when the sigmoidal function levels off after the martensitic transformation reaches saturation point (value of 3.18 at 30 °C and strain rate of 6.67×10^{-4} s^{-1}).

B is the minimum log of true stress ($\log \sigma_{min}$) before any martensitic transformation (value of 2.73 at 30 °C and strain rate of 6.67×10^{-4} s^{-1}).

ε is the true strain

α is the maximum strain sensitivity which corresponds to the log true strain where there is a maximum slope of the log-log plot of true stress-true strain curve (value of -0.73 at 30 °C, $\varepsilon = 0.19$ and a strain rate of 6.67×10^{-4} s^{-1}).

β is a constant which is indicative of the extent of variation of log-log plots of true stress and true strain. The numerical value is given by, $\beta = (A - B)(4 \times n_i)^{-1}$. (Value of 0.11 at 30 °C, $\varepsilon = 0.19$ and a strain rate of 6.67×10^{-4} s^{-1}).

5. Conclusions

The detailed analysis of flow stress and strain hardening behaviour as well as the variation of the fraction of martensite induced with strain at 30 °C for AISI 301 stainless steel has demonstrated that the strain hardening behaviour at relatively high levels of plastic strain, can be adequately described by

a sigmoidal equation. This indicates that there is a correlation between the instantaneous strain hardening behaviour and the fraction of martensite induced. The strain-induced martensitic phase transformation is therefore believed to be responsible for the sigmoidal flow stress behaviour observed in this metastable austenitic stainless steel.
The instantaneous strain hardening exponent and the extent of strain hardening both increases up to a maximum as the percentage of transformed martensite increases up to the point of martensitic saturation. The high levels of instantaneous strain hardening results in a remarkable increase in strength.
The material processing in industrial practice is not performed at 30 °C only. The variation of the instantaneous strain hardening exponent, the strength coefficient, K, and strain sensitivity, α as a function of temperature will therefore be investigated in further work.

6. Acknowledgements

The authors are grateful to Columbus Stainless Steel company for financial support as well as the supply of materials, AMI-FMDN, administered by Mintek for supporting this work.

7. References

[1] D. P. R. Palaparti, B. K. Choudhary, E. I. Samuel, V. S. Srinivasan, and M. D. Mathew, "Influence of strain rate and temperature on tensile stress – strain and work hardening behaviour of 9Cr – 1Mo ferritic steel," *Mater. Sci. Eng. A*, vol. 538, pp. 110–117, 2012.

[2] G. E. Dieter and D. J. Bacon, *Mechanical metallurgy*, vol. 3. New York: McGraw-hill, 1986.

[3] J. W. Simmons, "Strain hardening and plastic flow properties of nitrogen-alloyed Fe-17Cr-(8-10)Mn-5Ni austenitic stainless steels," *Acta Mater.*, vol. 45, no. 6, pp. 2467–2475, 1997.

[4] T. W. Mukarati, R. J. Mostert, and C. W. Siyasiya, "The direct observation of surface martensite formation upon cooling to temperatures close to ambient in a heat treated AISI 301 stainless steel," *Mater. Sci. Eng. Conf. Ser.*, vol. 430, no. 1, p. 12042, 2018.

[5] T. W. Mukarati, R. J. Mostert, and C. W. Siyasiya, "Effect of temperature and strain on martensite transformation in a high-nitrogen low-carbon AISI 301 metastable austenitic stainless steel," *Unpubl. Work*, 2019.

[6] G. C. Soares, M. Carla, M. Rodrigues, and L. D. A. Santos, "Influence of Temperature on Mechanical Properties , Fracture Morphology and Strain Hardening Behavior of a 304 Stainless Steel," *Mater. Res.*, 2017.

[7] J. Talonen, P. Nenonen, G. Pape, and H. Hänninen, "Effect of strain rate on the strain-induced $\gamma \rightarrow \alpha'$-martensite transformation and mechanical properties of austenitic stainless steels," *Metall. Mater. Trans. A*, vol. 36, no. 2, pp. 421–432, 2005.

IOP Conf. Series: Materials Science and Engineering **655** (2019) 012009 doi:10.1088/1757-899X/655/1/012009

Influence of Impeller Diameter on the Performance of Centrifugal pumps.

M E Matlakala[1], D V V Kallon[2], K E Mogapi [3], I M Mabelane [4], D M Makgopa[5]

Department of Mechanical and Industrial Engineering Technology, University of Johannesburg

Corresponding Author: dkallon@uj.ac.za

Abstract: The impeller is the most important part of the centrifugal pump because it determines the flow or efficiency of the pump. The design of the impeller and its diameter are very essential when considering pump efficiency. The capacity of the pump is determined from the impeller vanes and channels. The three types of impellers include open impeller, enclosed impeller and semi-open impeller are each designed for a specific application. Impellers are made of cast iron or carbon steel and special alloys depending on their application. The research involves the study of the application of impellers in the centrifugal pumps used by Rand Water Company in Johannesburg. The paper considers how variations in the impeller diameter affect the performance of the centrifugal pump. Simulations were conducted at different diameters of the impeller. The impeller was modelled using AutoCAD and simulated using ANSYS software to determine the performance of the centrifugal pump on different diameters. Calculations were done to compare the results of simulation of the different diameters of 2000mm, 2200mm, and 2400mm. The results show that changing the diameter of a pump impeller affects the flow, head and input brake horsepower of the pump in different proportions. The centrifugal pump gives a high flow rate as the diameter of the impeller increases.

1. Introduction

An impeller is a rotor used to increase or decrease the pressure and flow of a fluid. The impeller is assembled in a centrifugal pump and it is the most important part of the pump [1]. Typical impeller of a centrifugal pump is show in figure 1. A centrifugal pump is a type of general machinery which is widely and fully utilized in the industrial and agricultural fields such as irrigation and water supply [2]. It is also used in many plants to transfer fluids from one point to the other. It normally includes four different parts: suction pipe, impeller, volute, and exit pipe. The impeller is the core part and it converts the mechanical energy into pressure energy which directly determines the transport capacity and the hydraulic performances of a centrifugal pump [3 - 5]. Optimized design of the impeller is essential and significant for the efficient operation of a centrifugal pump [6]. The fluid enters the impeller through the eye then it is pushed by the vanes/blades as the fluid passes the channel. The impeller has a bore to attach the drive shaft. The impeller is a pump component that has more influence on the pump and one can create a new pump by just modifying the impeller. It can change the flow, the differential head, and the speed. Thus the bigger the impeller the greater the flow [5]. Figure 1 shows the model of the complete impeller.

Figure 1. Centrifugal Pump Impeller

The impeller of a centrifugal pump regulates the performance of the pump [4, 5]. The design of the impeller and diameter are very important when considering pump efficiency. The capacity of the pump is determined from the impeller vanes. Impellers are made of different types of steel cast iron or carbon steel and special alloys depending on their application. The three types of impellers include forward swept impeller, radial exit impeller, and backswept impeller. Forward swept impeller is used for lower head and flow rate, the backswept impeller is used for higher flow rate and head, and radial exit impeller is used for medium head and flow rate. The pump impeller receives the liquid to be pumped and imparts velocity and pressure to it by drawing power from a motor. The speed and diameter of the impeller determines the head or pressure that the pump can generate. The rotational speed and height of the impeller blades determines the flow that the pump can accommodate [4, 5].

2. Design of 3D Solid Model of the impeller

The design of the impeller diameter is critical because it affect the performance of the centrifugal pump. The design stage of the impeller diameter in critical when coming to the performance of the centrifugal pump. The specification of the pump should be accurate in order to achieve the correct result for the flow and head required. The different impeller diameters of 2400mm, 2200mm and 2000mm were model for the purpose of determining the performance of the centrifugal pump with difference diameters of the impeller. The impeller was model using the solidworks software.

2.1. Conceptual Design

Figure 2. Conceptual Design

2.2. Detailed Design

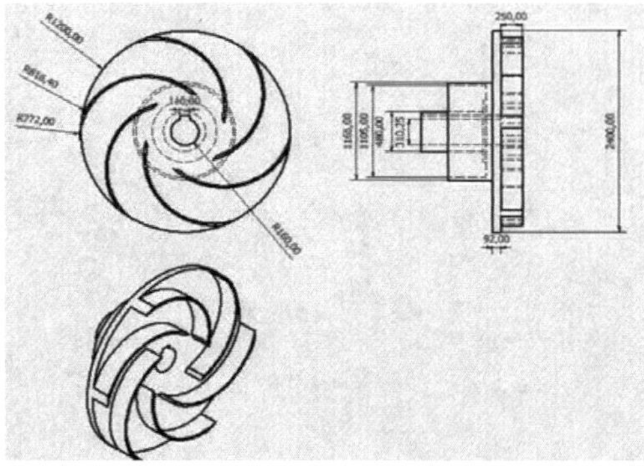

Figure 3. Design of 2400mm Impeller Diameter

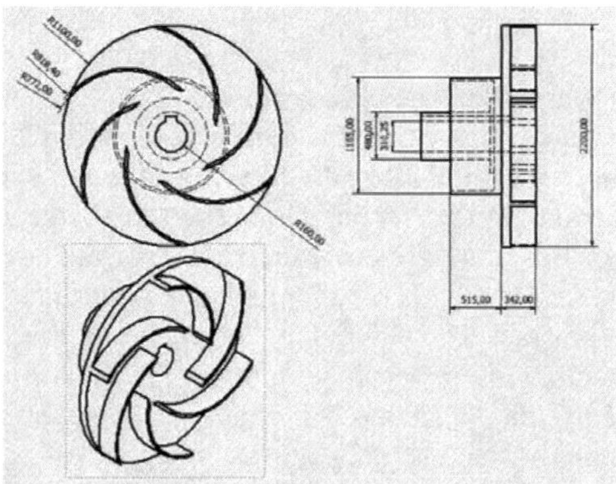

Figure 4. Design of 2200 mm Impeller Diameter

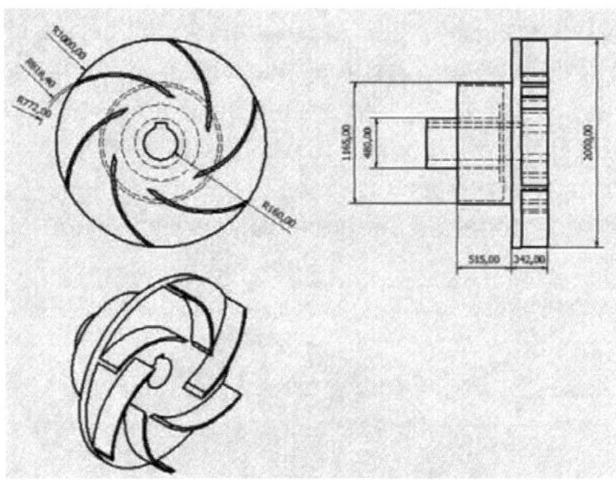

Figure 5.Design of 2200 mm Impeller Diameter

3. Calculations

These calculations are for a single-stage centrifugal pump with a Radial 6-Vanes impeller. All the calculations done were obtained with varying the diameter for an impeller and same goes with the simulation. There were assumptions made in order that the calculations can be done to test the effect of the impeller diameter on the centrifugal pump performance. The diameter was assumed based on the specification of the centrifugal pump as shown in table 1. The impeller diameters of 2000mm, 2200mm and 2400mm was used to do the calculations.

Table 1 Details of the pump

Description	SI Units	
Duty Flow Rate	(ML/d)	200
Duty Generated Head (Set)(Revised)	(m)	198
Guaranteed Demand (set)	(kW)	5203.69
Guaranteed Efficiency	(%)	86.23
Duty Generated Head (Stage 1)	(m)	50
Duty Speed (Stage 1)	(r.p.m)	730
Rated Voltage (Stage 1 and 2)	(kV)	11000
Rated Power	(kW)	1800
Rated Amps	(A)	118.00
Rated Speed 1st Stage	(r.p.m)	745
Duty Generated Head (Stage 2)	(m)	148
Duty Speed (Stage 2)	(r.p.m)	1490.00
Rated Power	(kW)	4200
Rated Amps	(A)	258
Rated Speed (Stage 2)	(r.p.m)	1490
Suction Dia.	(mm)	1400
Delivery 1st Stage Dia.	(mm)	1000
Delivery 2nd Stage Dia	(mm)	900
Total Operating Hours		149 230
Density	(kg/m^3)	1.00
Gravity	(m/sec^2)	9.79
Pump-set Utilization	(%)	52.15

3.1. *Data for the Calculation*

Suction diameter D_S = 1400 mm = 1,4 m

Discharge diameter D_D = 1000 mm = 1 m

Rotational speed N = 1490 rpm

Water flow rate at suction side Q_S = 2,315 m3/s

3.2. *Speed of Water Entry into Suction side of the Pump*

$$Q = A \times V$$

but Q= 2,315 m^3/s (1)

$$\text{and the } Area = \frac{\pi d^2}{4} = \frac{\pi(1.4)^2}{4} = 1.539m^2 \quad (2)$$

$$\text{Therefore } V = \frac{2.315}{1.539} = 1.504\ m/s$$

The following characteristics describe the performance of a centrifugal pump:

- Flow rate
- Head
- Power

Therefore, for our selected pump the following are important known variables we will use when we apply affinity laws for our calculations:

Q (Flow rate) = 200 l/day = 2,315 m^3/s
H (Head) = 198 m

P (Power) = 4200 kW
D (Impeller diameter) = 2400 mm = 2.4 m

3.3. *Affinity Laws for Centrifugal pump*

Affinity laws state that pump impellers are considered to be similar if they satisfy geometric and similarity condition. Affinity law helps to predict the performance of a centrifugal pump for speed and if the impeller is reduced in diameter within the limit of impeller design [5]. The affinity law can be shown in two ways, by keeping the rotational speed and diameter of impeller constant. Mario Savar (2009) proposed the impeller trimming method to improve the efficiency of the centrifugal pump. Impeller trimming adjusts the centrifugal pump head and flows to the actual needs. Trimming of impeller improves the performance of the centrifugal pump. The diameter of the impeller reduces from the original design can results in a low pressure, flow and power consumption. Therefore, the trimming of the impeller should not be more than75% of a pump original diameter. Different performance can be achieved and reducing impeller size allows the pump to reach specific. Affinity laws only apply to radial pumps and axial pumps.
The formulas applied for affinity laws in constant impeller diameter and constant rotational speed are as follow:

3.3.1. *Affinity Laws for centrifugal pumps:*

$$\frac{Q_1}{Q_2} = \frac{D_1}{D_2} \qquad \frac{H_1}{H_2} = \left(\frac{D_1}{D_2}\right)^2 \qquad \frac{P_1}{P_2} = \left(\frac{D_1}{D_2}\right)^3 \tag{3}$$

$$\frac{Q_1}{Q_2} = \frac{n_1}{n_2} \qquad \frac{H_1}{H_2} = \left(\frac{n_1}{n_2}\right)^2 \qquad \frac{P_1}{P_2} = \left(\frac{n_1}{n_2}\right)^3 \tag{4}$$

The Impeller diameters will be varied with $D_{original}$ as the original and two other values D_1 and D_2 as variations from $D_{original}$

D_1 = 2200 mm
D_2 = 2000 mm

3.3.2. *Applying Affinity law between $D_{original}$ and D_1:*

$$\frac{Q_{Original}}{Q_1} = \frac{D_{Original}}{D_1} \tag{5}$$

$$\frac{2{,}315}{Q_1} = \frac{2400}{2200}$$

$$Q_1 = 2{,}122 \text{ m}^3\text{/s (New flow rate)}$$

$$\frac{P_{Original}}{P_1} = \left(\frac{D_{Original}}{D_1}\right)^3 \tag{6}$$

$$\frac{4200}{P_1} = \left(\frac{2400}{2200}\right)^3$$

$$P_1 = 3235{,}069\ kW\ (New\ power)$$

$$\frac{H_{Original}}{H_1} = \left(\frac{D_{Original}}{D_1}\right)^2$$

$$\frac{198}{H_1} = \left(\frac{2400}{2200}\right)^2 \tag{7}$$

$$H_1 = 166,375\ m\ (New\ Head)$$

3.3.3. Applying Affinity law between $D_{original}$ and D_2:

$$\frac{Q_{Original}}{Q_2} = \frac{D_{Original}}{D_2} \tag{8}$$

$$\frac{2,315}{Q_2} = \frac{2400}{2000}$$

$$Q_2 = 1{,}929\ \text{m}^3/\text{s (New flow rate)}$$

$$\frac{P_{Original}}{P_2} = \left(\frac{D_{Original}}{D_2}\right)^3 \tag{9}$$

$$\frac{4200}{P_2} = \left(\frac{2400}{2000}\right)^3$$

$$P_2 = 2430,56\ kW\ (New\ power)$$

$$\frac{H_{Original}}{H_2} = \left(\frac{D_{Original}}{D_2}\right)^2 \tag{10}$$

$$\frac{198}{H_2} = \left(\frac{2400}{2000}\right)^2$$

$$H_2 = 137,5\ m\ (New\ Head)$$

The summary of the calculation results is shown in Table 2. The results indicate the performance of the centrifugal pump at the variation of the diameter of the impeller.

Table 2. Summary of the Calculation Results of the Impeller variation

	Impeller Diameter (mm)	Flow Rate (m^3/s)	Head (m)	Power (kW)
Original Impeller Diameter	2400	2,315	198	4200
New Impeller Diameter	2200	2,122	166,375	3235,069
New Impeller Diameter	2000	1,929	137,5	2430,56

4. Simulation

The simulation was done using the ANSYS software to check the effect of impeller diameter on the performance of a centrifugal pump. The simulation was done to find the Von Mises Stress, 1st Principle Stress, 3rd Principle Stress, for the diameter of 2400mm, 2200mm, and 2000mm. Figure 6, 7 and 8 shows the simulation results for 2400mm diameter. The same procedure was done to obtain the simulation results of 2200 mm and 2000mm, results obtained are summarised in table 3.

IOP Conf. Series: Materials Science and Engineering **655** (2019) 012009 doi:10.1088/1757-899X/655/1/012009

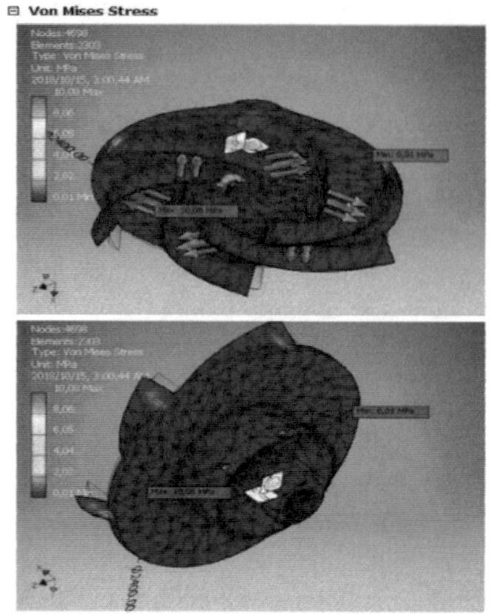

Figure 6. Von Mises Stress for 2400 mm

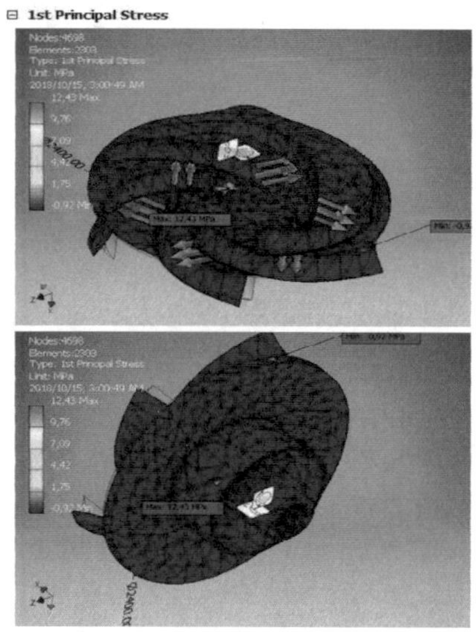

Figure 7. 1st Principle Stress for 2400 mm

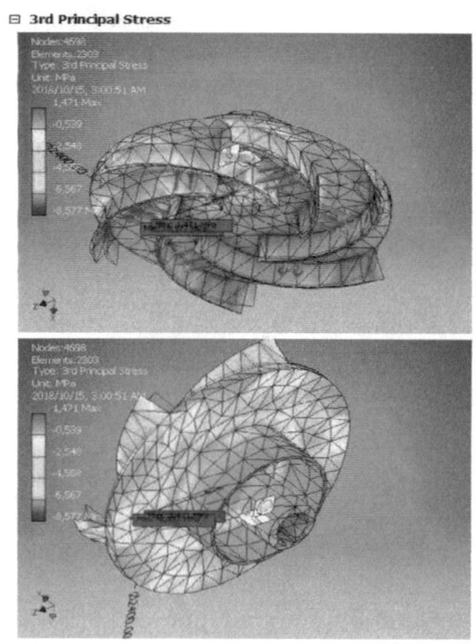

Figure 8. 3rd Principal Stress for 2000 mm

IOP Conf. Series: Materials Science and Engineering **655** (2019) 012009 doi:10.1088/1757-899X/655/1/012009

Table 3: Summary of the Simulation Results of the Impeller Variation

Stresses Experienced	**Maximum Magnitude 2400mm**	**Maximum Magnitude 2200mm**	Maximum Magnitude 2000mm
Von Mises stress	10.08 MPa	8.53 MPa	8.71MPa
1st Principal Stress	12.43 MPa	10.34MPa	8.58 MPa
3rd Principal Stress	1.47 MPa	1.045MPa	4 MPa

5. Discussion of the Results

The performance of the centrifugal pump was calculated based on varying the impeller diameter from 2400mm, 2200mm and 2000mm. The simulation was also performed in varies diameters of the impeller to test the stress of the material of the impeller. From the simulation, the results of Von Mises Stress obtained is clearly shown in figure 9. From the calculations, the characteristic graphs for the impeller diameter were obtained for the Flow rate – Impeller Diameter in figure 10, Head – Impeller Diameter in figure 11 and Impeller Diameter – Power in figure 12.

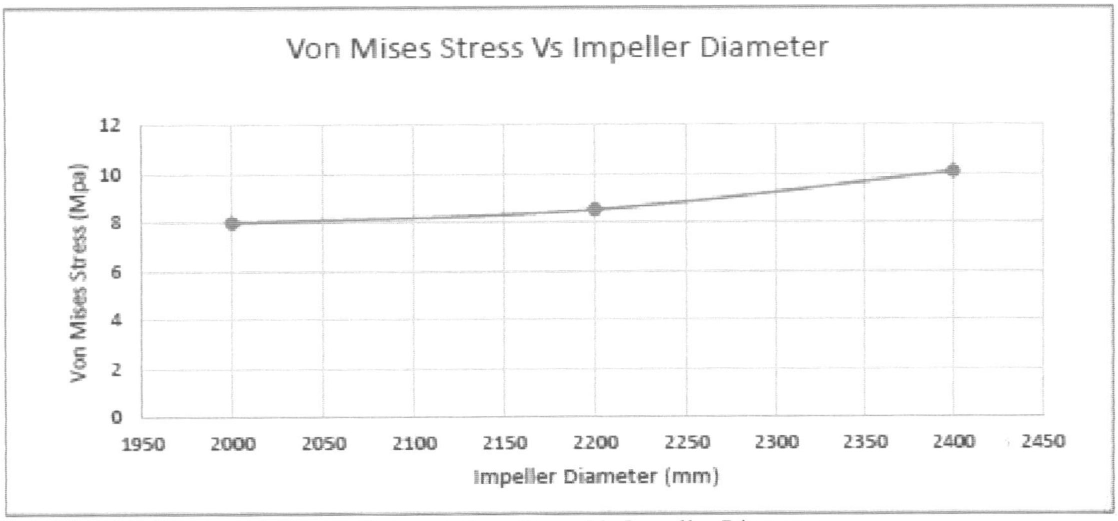

Figure 9. Von Mises Stress Vs Impeller Diameter

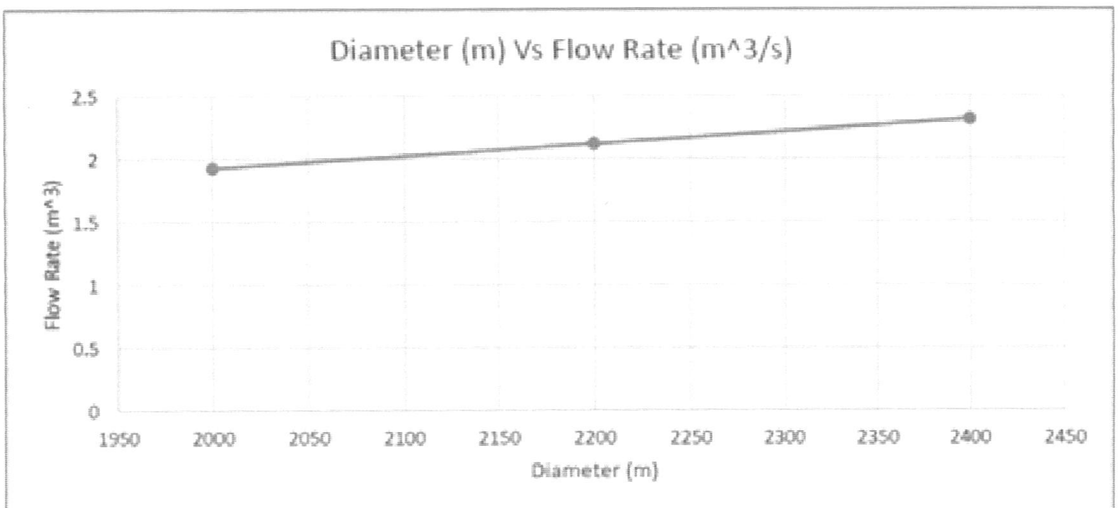

Figure 10. Flow Rate Vs Impeller Diameter

IOP Conf. Series: Materials Science and Engineering **655** (2019) 012009 doi:10.1088/1757-899X/655/1/012009

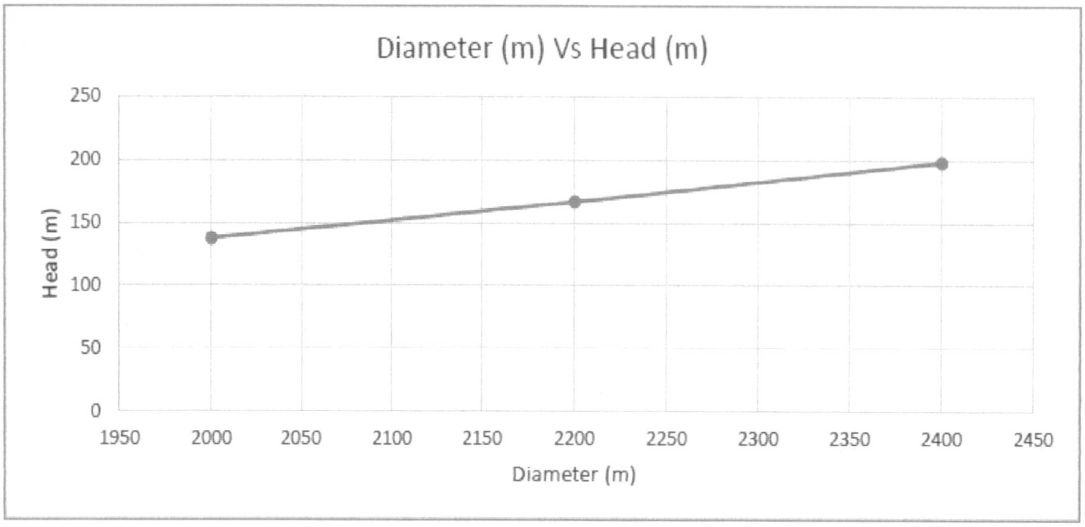

Figure 11. Head Vs Impeller Diameter

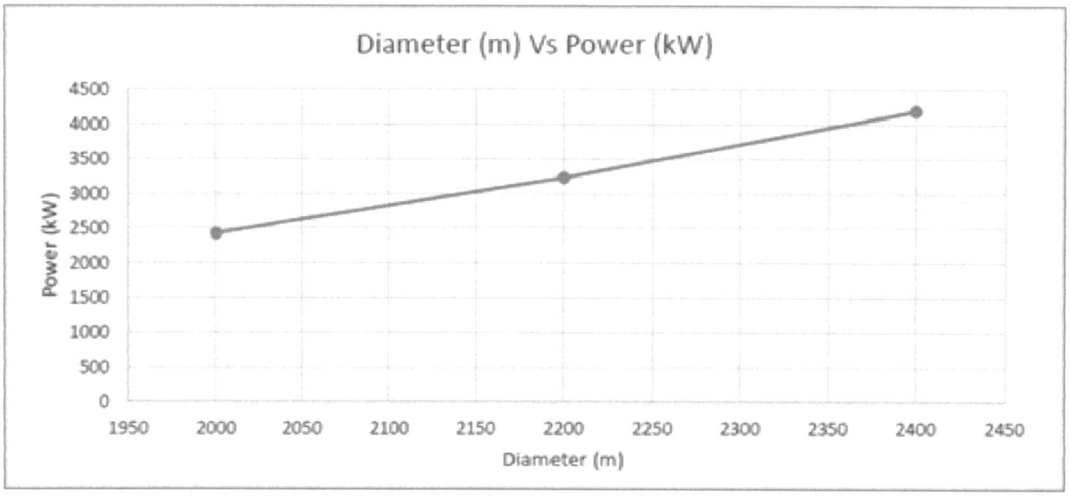

Figure 12. Power (kW) Vs Impeller Diameter (m)

The graphs in figures (9) to (12) were plotted according to the results in table 2 in order to evaluate the performance of the operating centrifugal pump with different diameter of the impeller. As indicated by the graph in figure 10 and 11, the flow rate and the head increases as the impeller diameter increases. This indicates that by increasing the impeller diameter, the performance of the pump will increase. Figure 12 shows that by increasing the impeller diameter, the power required to operate the centrifugal pump will increase, due to the weight of the impeller. The more the impeller diameter increases, more material required to build the impeller, thus more power will be required to operate the pump. For this reason the selection of the material is critical in the design and improvement of the centrifugal pump performance.

6. Conclusion

Based on the simulation results, it is observed that the stress encountered by the impeller increases with an increase in the diameter of the impeller. Both strain and displacement also become higher as we increase the diameter of the impeller. More stress is experienced at the tip of the impeller plates and in this regard, the displacement due to the applied pressure also becomes high at the tip of the plates. The mathematical analysis was done using Affinity laws. An initial original design, which included original parameters (rated power, speed, impeller diameter and the head) was used.

The objective was to see the influence of varying impeller diameter on the performance of the centrifugal pump and to do the finite element analysis of different diameters. The original impeller diameter was 2400 mm and this was then varied to 2200 mm and 2000 mm. The results obtained showed that an impeller reduction will result in a decrease in flow rate, head, and power consumption. Therefore, in order to achieve greater flow rate and head, we need a bigger impeller, but power consumption as a result of a bigger impeller is a problem as it is way too high. This greater power consumption could be reduced by decreasing the weight of the impeller. This can be achieved by selecting lighter material for our impeller, the material could be selected whilst taking the fluid we are moving into consideration and how long it will take the impeller before it corrodes, for the various diameters of the impeller as indicated on figures 6, 7 and 8. Both strain and displacement also become higher as the diameter of the impeller is increased. More stress is experienced at the tip of the impeller plates and in this regard, the displacement due to the applied pressure also becomes high at the tip of the plates.

7. References

[1] Yassi Y, Hashemloo S. Improvement of the efficiency of the Agnew micro-hydro turbine at part loads due to installing guide vanes mechanism. Energy Convers Manag 2010;51:1970–5.
[2] Saxena P, Kumar A. Hydropower development in India. In: Proceedings of the international conference on hydraulic efficiency measurement. Roorkee, India; 2010, p. 1–6.
[3] Yassi Y. The effects of improvement of the main shaft on the operating conditions of the Agnew turbine. Energy Convers Manag 2009;50:2486–94.
[4] Matlakala E.M., Kallon D.V.V., Simelane S.P., Mashinini P.M. Impact of Designh Parameters on the Performance of Centrifugal Pumps. *Procedia Manufacturing*. Volume 35. 2019. Pp 197 – 306.
[5] Matlakala E.M. A Computational Model for the Efficiency of Centrifugal Pumps. *Dissertation submitted to the University of Johannesburg*. August 2019.
[4] Prasad V, Shukla SN, Joshi SG. Performance characteristics of pump as turbine. Indian Pumps 20

IOP Conf. Series: Materials Science and Engineering **655** (2019) 012010 doi:10.1088/1757-899X/655/1/012010

Relationship between tool tilt angle, shoulder plunge depth and process energy input for pin-less friction stir welded thin Ti-6Al-4V sheets

A B Gili[1] ,D G Hattingh[1] and D Bernard[1]

[1] Nelson Mandela University, Department Mechanical Engineering, Port Elizabeth, South Africa

E-Mail: s219227535@mandela.ac.za,

Abstract. This paper reports on the initial process development towards evaluating pin-less friction stir welding as a feasible alternative joining technology for Ti-6Al-4V sheets. Initial research results were derived from "bead on plate" welds with a pin-less tool on 3 mm Ti-6Al-4V sheets to assist with decision making for joining 1 mm Ti-6Al-4V sheets. The study evaluates the shared influence of tool tilt angle and shoulder plunge depth on weld profile and process energy input for a pin-less tool. An experimental approach was followed, performing bead on plate welds with three tool tilt angles (0°, 1° and 2°) respectively at shoulder plunge depths from 0.15 mm to 0.25 mm in 0.05 mm increments. From critically evaluating the hardness profile and heat effected area below the tool shoulder, the shared influence of tool tilt angle and shoulder plunge depth could be analysed to give an indication of the best parameter combination for performing solid state welds on 1 mm Ti-6Al-4V sheet. Additionally, the thinning effect on the weld region was evaluated and compared to welds made with a pin. Future work will include quantifying surface residual stresses on the material and the relationship these stresses exhibits with the varying tool tilt and shoulder plunge depth. This work will assist in selecting weld parameter combination that will minimize tensile residual stresses in critical areas. Knowledge gained from this study forms the basis for the overall process development for joining 1 mm Ti-6Al-4V sheets by pin-less friction stir welding.

1. Introduction

The main considerations of the research projects revolve around evaluating the suitability of Friction Stir Welding (FSW), a solid-state joining technique for welding thin Ti-6Al-4V sheets. Ti-6Al-4V is a material of choice where high strength to weight ratio is desirable. FSW is a proven approach for joining, especially for Aluminium. However, the FSW of high strength, high melting point materials has proven to be more complex as tool wear becomes more prevalent [1]. Wear driven by geometrical changes of the tool has a direct influence on process energy input and material flow, effecting weld integrity. Despite these challenges the joining of Titanium, by means of this solid-state welding technique, stays attractive due to superior mechanical properties, good corrosion resistance and its ability to perform well under elevated temperatures.

The crystal structure of titanium is a close packed hexagonal crystal structure (HCP). Alloying titanium for example with aluminium, vanadium, copper, zirconium, etc. the structure can be manipulated in forming
α, β and mixed α-β Ti [2]. These structures have specific applications, however a mixed α-β structure is

regard as metallurgical more stable owing to the presences of α-β stabilizers making this type of titanium (e.g. Ti-6A-l4V) more versatile and extensively popular for engineering applications. The alpha titanium grades like pure titanium don't exhibit ductile-brittle transformation making the grades suitable for cryogenic applications. The beta titanium grades contain predominantly beta stabilisers with small amount of alpha stabilizers which increase biocompatibility [2]. Although titanium can be joint by a number of conventional fusion techniques like gas metal-, tungsten-arc, plasma and laser beam welding, these techniques present process control complications pertaining to the increase material reactivity with oxygen, hydrogen, nitrogen at the associated high welding temperatures potentially contributing to embrittlement of the welded joint.

Solid state joining (FSW), developed by Wayne Thomas [1] in the early 1990s presented an alternative solution to these inherent reactivity challenges of fusion welding, as for FSW the material is welded below the β-transus temperature of the parent material. Performing welds below the β-transus temperature with gas present in the melting pool limits effects associated with metallurgical transformation at elevated temperatures. Also, the reduction in thermal cooling cycle in FSW as a result of welding at a lower temperature reduces residual stresses in the weld region and limits the distortion [2]. Pin-less FSW is similar to conventional FSW in all ways except that the welding tool in pin-less FSW consist of a shoulder only. The tool is rotated and forced against the surface of the material to produce frictional heat at the contact area. The frictional heat plasticises (softens) the material, allowing it to be mixed by the stirring action of the pin-less tool shoulder, thereby creating a bond between the abutting sheets without reaching melting temperature of the base metal. Reported by Mashinini et al.[3] in a comparative study of friction stir and laser beam welded Ti-6Al-4V sheets, LBW experienced high tensile residual stresses in the centre of the nugget zone compared to those experienced in FSW which are more evenly distributed across the stir zone. From this same study it was reported that highest residual stress, during FSW were recorded under high transverse speed. Mashinini et al.[3] also reported hardness values for FSW which indicated higher values for the weld zone compared to the base material. However some studies indicate the minimum hardness values within the heat affected zone (HAZ) [4] and others reported the stir zone (SZ) to have the lowest values in hardness [5, 6] with hardness decreasing with an increase in rotational speed.

Tooling material and tool wear presents a challenge in the joining of alloys such as titanium by FSW. The characteristics of the tool material must include the ability to retain its strength and resist wearing while operating at elevated temperatures (+1000°C Titanium). Special tooling material such as Tungsten (cp-W), Tungsten Rhenium (W-Re), Tungsten cobalt (WC-Co), Lanthanated Tungsten ($W–La_2O_3$), Polycrystalline Cubic Boron Nitride (PCBN) etc. are needed to join high melting temperature alloys. Some of the aforementioned tool material have been reported by various authors [7-14] on FSW of Ti-6Al-4V of different thicknesses.

This study will evaluate the shared influence of tool tilt angle and shoulder plunge depth on weld heat effected profile and process energy input for a lanthanated tungsten ($W–1\%La_2O_3$) pin-less tool on a 3 mm Ti-6Al-4V sheet. This is done to inform the further develop of the process towards evaluating the feasibility of using pin-less tools for friction stir welding for joining 1 mm Ti-6Al-4V sheets. The use of a pin-less tool has originally gained consideration in pin-less friction spot welding [15-18] with few studies reporting pin-less friction stir welding as a joining technique for thin aluminium sheet. A number of studies evaluated the relationship between the tool shoulder and pin and the relationship of this combination to material thickness. The current understanding shows that tool pin is prominent for FSW of material thickness above 3 mm and as material get thinner the role of the pin becomes less prominent than the shoulder from a mixing and process energy input point of view, however the opposite is also true in that as the material thickness increase the shoulder become less prominent and the pin become important from both mixing and process energy input, this phenomenon lead to the development of "Stationary Shoulder" FSW. Additionally, with tool wear [19]being one of the associated difficulties for friction stir welding of high- strength/melting point alloys, risk of accelerated pin wear is a reality which result in tool geometry change as the weld progress, effecting process energy input and material flow which could adversely affect weld integrity

2. Experimental Procedure

Bead on plate welds in the rolling direction were performed on mill annealed 3mm thick Ti-6Al-4V square coupons with dimensions of 110 mm x 110 mm. Chemical composition (wt%) of the actual coupons are as shown in Table 1. A flat featureless shoulder pin-less tool, 14 mm diameter was manufactured from lanthanated tungsten alloy (W-1%La_2O_3) and used to implement bead on plate welds on the I-STIR Process Development System platform developed by MTS Corporation (MTS I-STIR PDS).Backing plate manufactured from Haynes 230 alloy was used during welding as it retains its strength and resists oxidation at operating temperatures up to 1149°C, making it well suited for FSW for high temperature alloys such as Ti-6Al-4V.

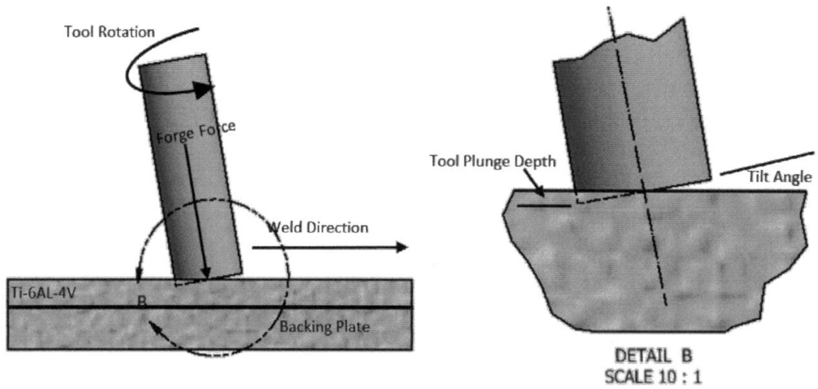

Figure 1. Pin-less friction stir welding illustration

The only two parameters varied for the purpose of this study were tool tilt angle (0°, 1° and 2°) and tool plunge depth (0.15 mm, 0.2 mm and 0.25 mm). The definition of tilt angle and plunge depth is visually illustrated in Figure 1. Bead on plate welds were produced at all plunge depths for each selected tilt angle producing a total of 9 welds (W12, W13, W14, W15, W16, W17, W19, W20 and W21 respectively) also listed in Table 4. These welds were all made at a constant rotational speed of 950 rev/min, transverse speed of 55 mm/min and a dwell time of 3 seconds in position control. Argon was used as a shielding gas for all the welds. For metallographic and micro hardness analysis specimens were extracted in the stable zone of each plate perpendicular to the tool transverse direction by water jet cutting so as to induce no heat on the specimens. The specimens were mounted, followed by mechanical polishing and were then etched with Kroll's reagent. The metallographic evaluation was performed through optical microscopy. Microhardness was performed on the Micro Vickers tester (Future Tech FM–700).

Table 1. Ti-6Al-4V alloy chemical composition

Elements	Al	V	Fe	C	N	O	Ti
wt%	6.25	4.04	0.19	0.018	0.008	0.18	Balance

3. Results and discussions

3.1 Weld surface appearance

Table 2 shows the top weld seam appearance of the bead on plate welds on Ti-6Al-4V sheet 3 mm thickness. Varying degree of discoloration was observed on the weld surface, an indication of surface oxidation as the titanium react with oxygen at elevated temperatures despite shielding with Argon. Flash formation were more prominent at welds made at 0° tilt angle and generally decrease with an increase in tool tilt angle. However, an increase in shoulder plunge depth resulted in an increase in flash formation at all tilt angles. In this scenario the material is not contained beneath the tool shoulder as is the case when a tilt angle is increased but is rather expelled as flash around the tool. With W14 and W21 ridges were formed on the weld surface resulting in a poor surface finish. Welds accomplished at 0.2 mm plunge depth exhibited a, smooth appearance over the entire weld length. Uneven profile for the first 10-30 mm of the weld length can be attributed to a number of events such as the 3 seconds dwell time which was insufficient for the tool to generate enough frictional heat before it starts to transverse, also platform stability during the weld initiation stage, resulting in discarding the first 30 mm of the weld for all the tests performed. Tool exit position presented a shiny smooth surface with no pin hole that requires run off plate as compared to tools characterized with a pin.

Table 2. Surface appearance of welds made at different tool tilt angle and tool plunge depth.

PD	Tilt Angle 0°	Tilt Angle 1°	Tilt Angle 2°
0.15 mm	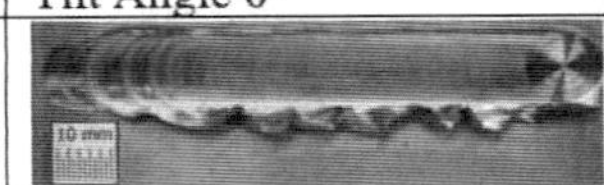 W17	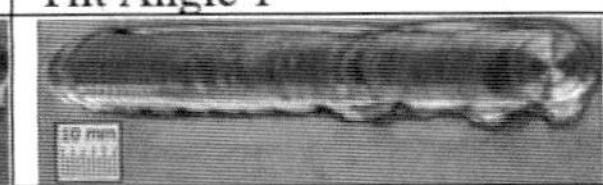 W12	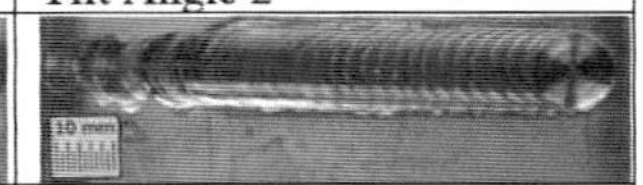 W21
0.2 mm	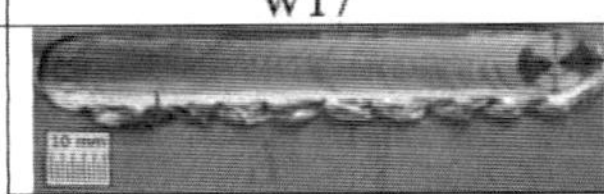 W15	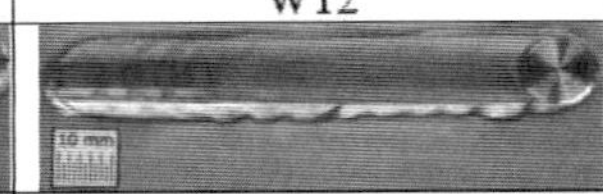 W13	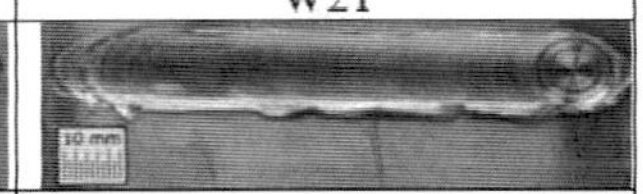 W20
0.25 mm	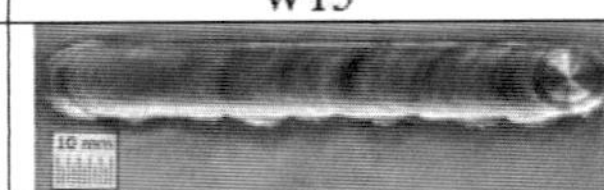 W16	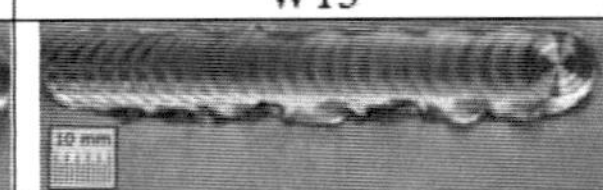 W14	 W19

3.2 Weld Macrographs

Cross-sections of the welds showed a symmetrical shape on the stir zone about the weld centerline. The macrographs are characterized by stir zone (SZ), a narrow heat affected zone (HAZ) and base material (BM) as shown by a representative sample in Figure 2. With titanium having a low thermal conductivity the formation of heat affected zone becomes very narrow and less distinct in titanium FS Welds [20]. Stir zone depth was measured at the center of the stir zone along the centerline at the interface between SZ and HAZ.

Retreating Side

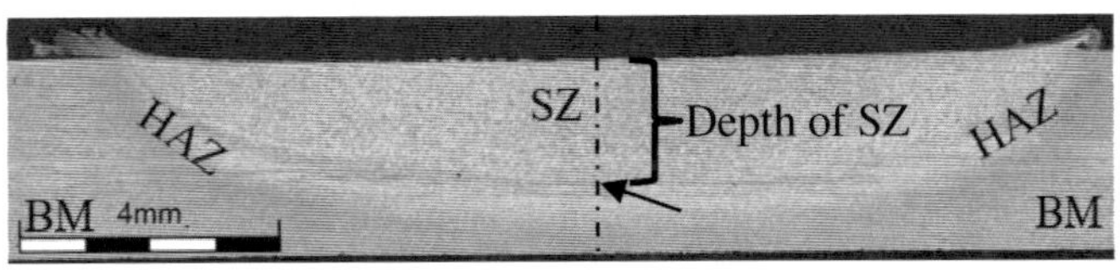

Advancing Side

Figure 2. Macrographs indicating SZ, HAZ and BM

Table 3 shows the weld cross-section perpendicular to the welding direction of Ti-6Al-4V pin-less friction stir welded sheets under different tool tilt angle and tool plunge depth. After metallographic preparation the samples revealed weld stir zone depths close to 1 mm, with weld W17 (0° tilt and 0.15 mm plunge) recording the lowest stir zone depth (0.83 mm) as indicated in Table 4. Very little overall sheet thinning was observed across the welded samples except for a parameter combination of 1° tool tilt angle and 0.25 mm shoulder plunge depth (W14) revealing irregular thinning along the advancing side(AS) and retreating side (RS)..

Table 3.Weld Macrographs at different tilt angles and plunge depths

PD	Tilt Angle 0°	Tilt Angle 1°	Tilt Angle 2°
0.15 mm	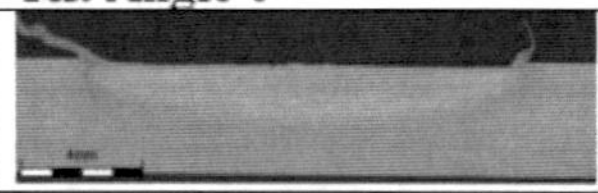	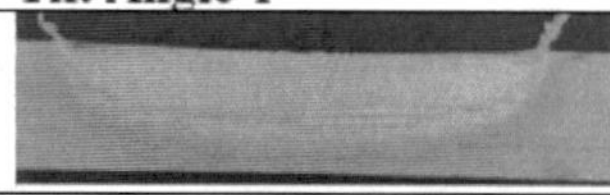	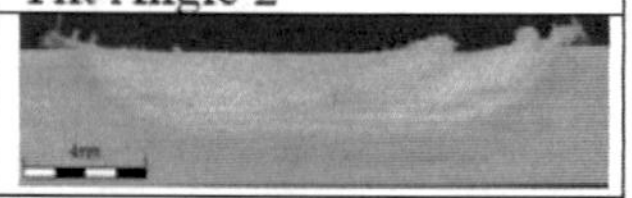
	RS W17 AS	RS W12 AS	RS W21 AS
0.2 mm	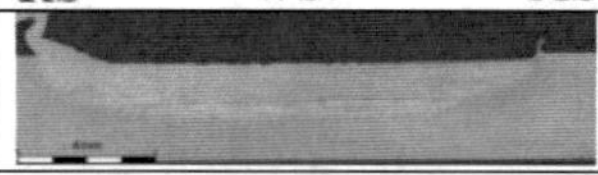	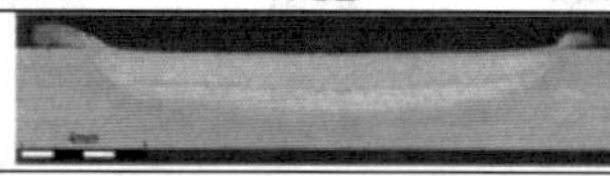	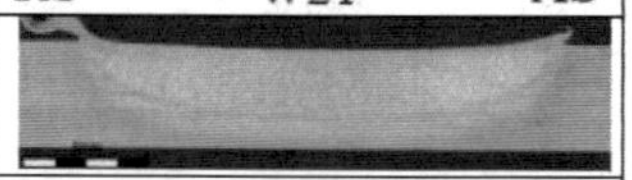
	RS W15 AS	RS W13 AS	RS W20 AS
0.25 mm	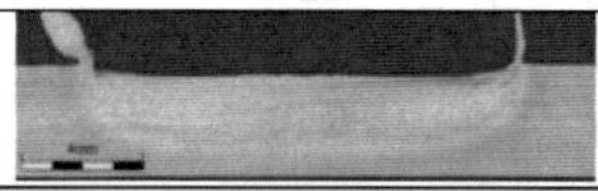	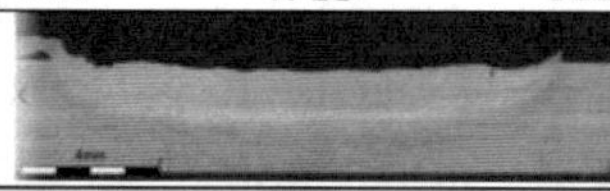	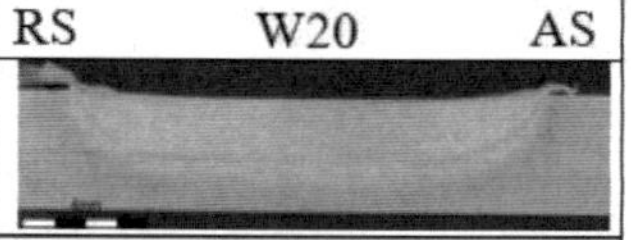
	RS W16 AS	RS W14 AS	RS W19 AS

3.3 Weld Process data

The MTS I-Stir PDS platform is equipped with transducers to record the process response during welding. The variables include spindle torque, tool downforce. The typical force and torque response are presented in
Figure 3.

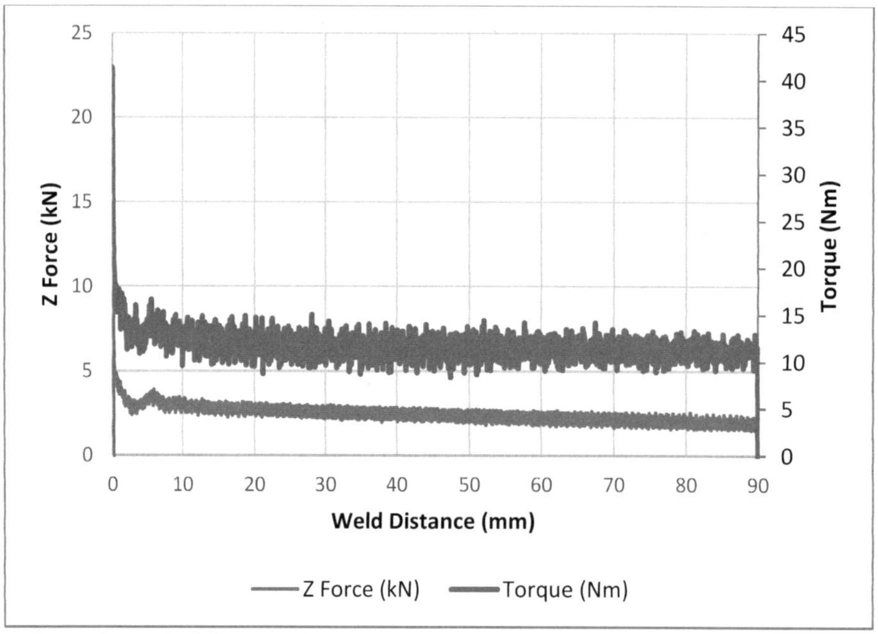

Figure 3. Typical force and torque response as measured along the weld length

The reported torque value in Table 4 was calculated from the average torque value occurring in the stable zone of the weld, coinciding with the position where the macrograph samples were removed. The torque and force response, stir zone depth, process energy input (determined from the reported torque) is shown in Table 4.

Table 4. Weld matrix, process data and average process energy input calculated

Weld	Weld Variables		Response Variables			Measured Effects
	Tilt angle(°)	Plunge depth(mm)	Torque (Nm)	Forge Force (kN)	Process Energy Input (J/mm)	Stir Zone Depth (mm)
W17	0	0.15	10.869	2.267	1116.54	0.83
W15	0	0.2	11.51	2.976	1220.707	0.97
W16	0	0.25	14.145	3.081	1284.726	1.63
W12	1	0.15	12.178	2.915	1338.98	1.42
W13	1	0.2	11.741	2.564	1299.917	1.01
W14	1	0.25	11.279	2.472	1239.153	0.92
W21	2	0.15	13.032	2.375	998.2669	1.12
W20	2	0.2	12.638	1.836	1247.834	2
W19	2	0.25	12.463	2.968	1204.431	1.51

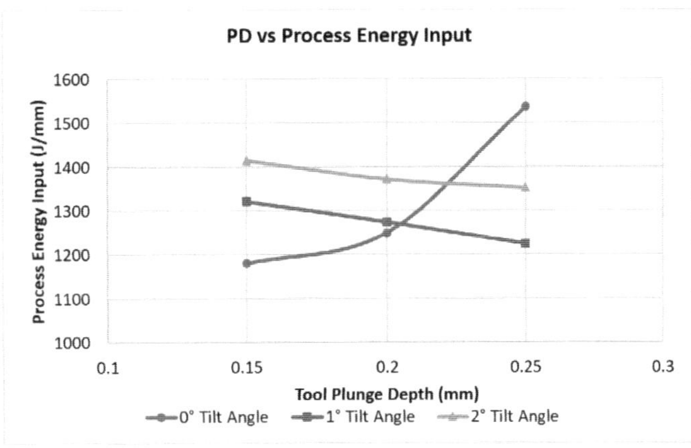

Figure 4. Effects of tool plunge depth on process energy input

The relationship between the process energy input and tool plunge depth is presented in Figure 4 for each tool tilt angle. Due to interaction between the tool shoulder and the plasticised material contact forces are developed during the process and the amount of energy required to overcome these forces is defined as process energy input. There was a general trend observed at 1° and 2° tilt angles, with the process energy decreasing with increasing plunge depth but vice versa for 0° tilt angle. The effect of not having a tilt angle (0°) on the tool results in a build-up of plasticised material in front of the tool, effecting the flow and consolidation below the tool shoulder. This resistance of material in front of the tool result in an increase in process energy input and not nessesarly in increase heat input. A negative linear relationship was observed for both the welds done at 1° and 2° tilt angles. The driving factors for a negative linear relationship could be the increased contact surface with increased plunge (increase frictional area) which led to a larger volume of material softening, thereby decreasing the required process energy input. There was no significant variation between two welds done at the same plunge depth and different tilt angles (1° and 2°) which further substantiates that tilt angle has an insignificant influence on the process energy when performing pin-less FSW at constant parameters. Further work still needs to be done which will help to explain the above observations. The highest process energy input was required with 0.25 mm plunge depth for 0° tilt.

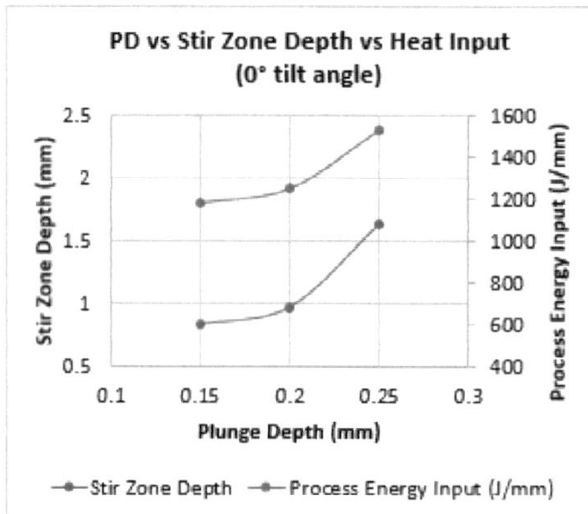

(a)

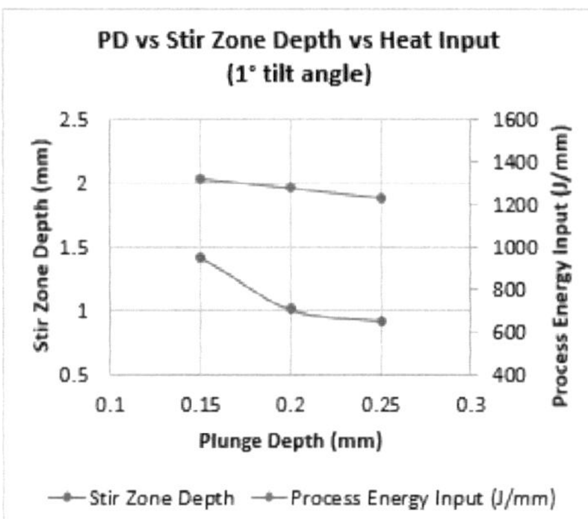

(b)

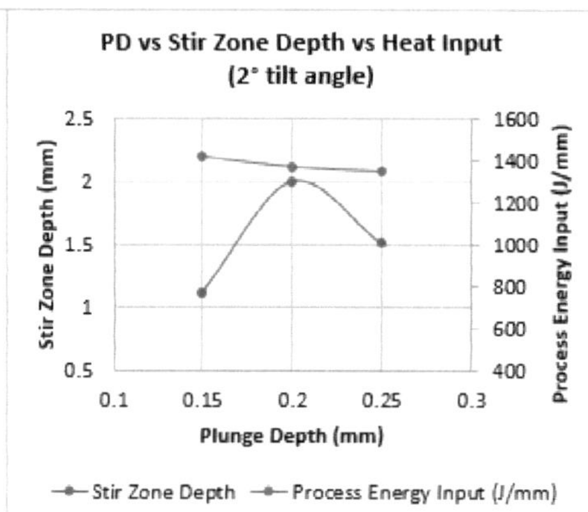

(c)

Figure 5. Tool plunge depth as a function of stir zone depth and process energy input for (a) 0°, (b) 1° and (c) 2° tool tilt angle.

The stir zone depth was measured from the weld micrographs, and the relationship between the stir zone depth and plunge depth is shown in Figure 5. To assist with interpretation and comparison the process energy input was added to Figure 5. The general observations is that 0° and to a lesser extent 2° tilt angle achieved greater stir zone depth as the plunge depth was increased. The rapid increase in stir zone depth at 2°, 0.2 mm plunge cannot be explained currently and is considered as an anomaly associated with the specific position in the weld. At 1° tilt angle we observed an associated decrease in stir zone depth with increased plunge depth. This phenomenon needs further investigation as an increase in tool plunge depth should increase the tool shoulder contact area with the material, resulting in an increase in frictional heat generated by the shoulder action. Possible explanation lies in the large volume of flash expelled at these parameters.

3.4 Hardness profile

Figure 6 shows micro Vickers hardness profile which was performed with a 500 g load for 15 s. Indents at a pitch of 0.5 mm were made along a line 0.7 mm below and parallel to the top surface of the weld.

Within the weld region the hardness profiles were fairly similar for all the welds tested, and there were no observable dependence on either tilt angle or plunge depth. The hardness in the stir zone increased from approximately 310 Hv in the parent material, to 345 Hv in the stir zone. This increase in hardness is consistent with published literature and is driven by the refined grain in the dynamically recrystallized stir zone. Variation in hardness was observed on the retreating and the advancing side of the base material, however, the variation falls within the hardness range of the base material. However a risk was identified that the samples prepared did not extend beyond the heat affected zone of the sheet on the advancing side of the weld, this could also be a contributing factor for the off-set in hardness data on the advancing side if compared to that of the retreating side. For comparison purposes the hardness was also measured 0.7 mm and 1.5 mm beneath the top surface of the parent material in an attempt to explain the large variation in hardness data. Important to note that for the parent plate the hardness at a depth of 0.7 mm ranged from 300 HV to 333 HV (average 313 HV) and at mid-thickness the ranged from 301 HV to 315 HV (average 306 HV), revealing a larger scatter in data closer to the surface.

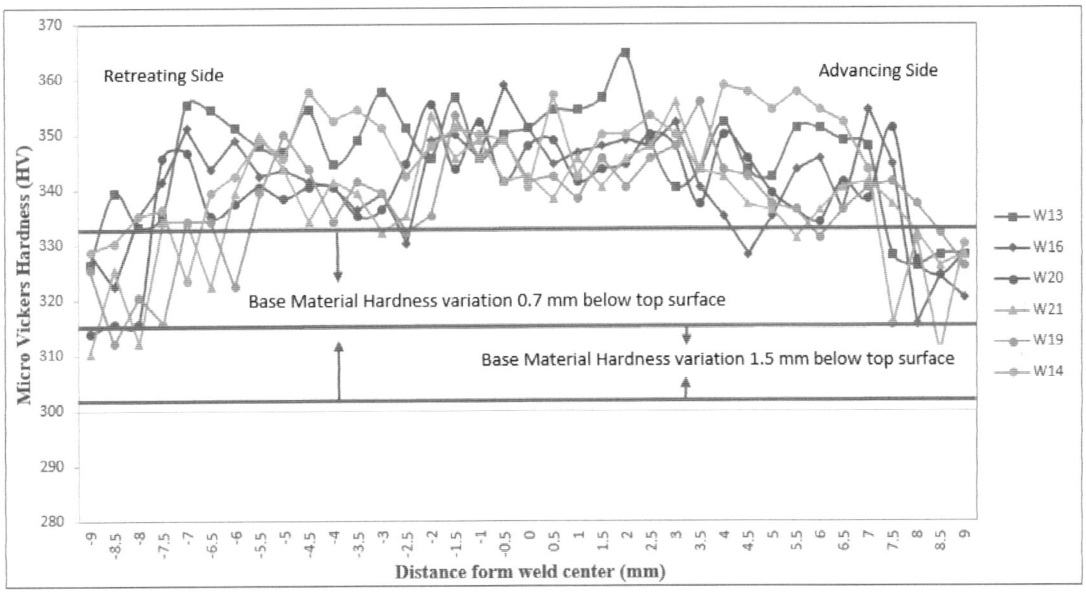

Figure 6. Effects of tool tilt angle and tool plunge depth on microhardness of FSW Ti-6Al-4V 3mm sheets

4. Conclusion

Lanthanated tungsten tool was used to perform successful bead on plate welds on 3mm Ti-6Al-4V sheets and the shared influence of various tool tilt angle and tool plunge depth on the weld profile was evaluated through macrostructure, process energy input analysis and microhardness. The conclusions to the study are summarized as follows.

- From the macrograph it could be determine that most of the selected parameter combinations resulted in a stir effected depth of more than 1 mm (except W17). This 1 mm minimum stir depth becomes an important observation for transferring this data to 1 mm thick sheet. The highest stir zone depth (2 mm) was achieve at 2° tilt and 0.2 mm plunge. Form evaluating all data and visual appearance W13 looks to be a very good weld, this will however need more evaluation.

- Process energy input and plunge depth showed a negative linear relationship (reduce process energy) for welds performed at 1° and 2° tilt angles. The driving factors for this relationship can possibly be contributed to an increase in plunge depth, which results in an increase in tool contact area. This increase in contact area generates a larger volume of plasticised material and a reduced effect of "dry friction" which will be more prominent at low plunge depth with low tilt angles.
- There was no significant variation between two welds done at the same plunge depth and different tilt angles (1° and 2°) which at this stage are indicative that for the considered tilt angles, changes process energy input will be minimal when performing pin-less FSW at constant parameters.
- The lowest hardness values were recorded in the base material and the highest hardness in the stir zone. No significant variation was observed with neither a change in tilt angle nor plunge depth.

Acknowledgments

The authors would like to extend appreciation to the DST-AMI-TiCoC program for funding provided, eNtsa staff for availing their facilities to carry out the research and National Research Foundation of South Africa (Grant Numbers: 117653) .

References

[1]. Thomas W, Threadgill P, Nicholas E. Feasibility of friction stir welding steel. Science and Technology of Welding and Joining. 1999;4(6):365-72.

[2]. Welsch G, Boyer R, Collings E. Materials properties handbook: titanium alloys: ASM international; 1993.

[3]. Mashininini PM, Hattingh DG, Lombard H. Mechanical Properties and Microstructure of Friction Stir and Laser Beam Welded 3mm Ti6Al4V Alloy. 2016.

[4]. Zhang Y, Sato YS, Kokawa H, Park SHC, Hirano S. Microstructural characteristics and mechanical properties of Ti–6Al–4V friction stir welds. Materials Science and Engineering: A. 2008;485(1-2):448-55.

[5]. Liu H, Zhou L, Liu Q. Microstructural characteristics and mechanical properties of friction stir welded joints of Ti–6Al–4V titanium alloy. Materials & Design. 2010;31(3):1650-5.

[6]. Zhou L, Liu H, Liu Q. Effect of rotation speed on microstructure and mechanical properties of Ti–6Al–4V friction stir welded joints. Materials & Design (1980-2015). 2010;31(5):2631-6.

[7]. Wang J, Su J, Mishra RS, Xu R, Baumann JA. Tool wear mechanisms in friction stir welding of Ti–6Al–4V alloy. Wear. 2014;321:25-32.

[8]. Buffa G, Ducato A, Fratini L. FEM based prediction of phase transformations during friction stir welding of Ti6Al4V titanium alloy. Materials Science and Engineering: A. 2013;581:56-65.

[9]. Wang J, Su J, Mishra RS, Xu R, Baumann JA. A preliminary study of deformation behavior of friction stir welded Ti-6Al-4V. Journal of materials engineering and performance. 2014;23(8):3027-33.

[10]. Li Z, Liu H-j, Wu L-z. Texture of friction stir welded Ti–6Al–4V alloy. Transactions of Nonferrous Metals Society of China. 2014;24(2):368-72.

[11]. Edwards PD, Ramulu M. Material flow during friction stir welding of Ti-6Al-4V. Journal of Materials Processing Technology. 2015;218:107-15.

[12]. Edwards P, Ramulu M. Fatigue performance of friction stir welded Ti–6Al–4V subjected to various post weld heat treatment temperatures. International journal of Fatigue. 2015;75:19-27.

[13]. Yoon S, Ueji R, Fujii H. Effect of rotation rate on microstructure and texture evolution during friction stir welding of Ti–6Al–4V plates. Materials Characterization. 2015;106:352-8.

[14]. Yoon S, Ueji R, Fujii H. Microstructure and texture distribution of Ti–6Al–4V alloy joints friction stir welded below β-transus temperature. Journal of Materials Processing Technology. 2016;229:390-7.

[15]. Klobčar D, Tušek J, Smolej A, Simončič S. Parametric study of FSSW of aluminium alloy 5754 using a pinless tool. Welding in the World. 2015;59(2):269-81.

[16]. Bakavos D, Chen Y, Babout L, Prangnell P. Material Interactions in a Novel Pinless Tool Approach to Friction Stir Spot Welding Thin Aluminum Sheet. Metallurgical and Materials Transactions A. 2011;42(5):1266-82.

[17]. Simoncini M, Ciccarelli D, Forcellese A, Pieralisi M. Micro-and macro-mechanical properties of pinless friction stir welded joints in AA5754 aluminium thin sheets. Procedia CIRP. 2014;18:9-14.
[18]. Chu Q, Li W, Yang X, Shen J, Li Y, Wang W. Study of process/structure/property relationships in probeless friction stir spot welded AA2198 Al-Li alloy. Welding in the World. 2017;61(2):291-8.
[19]. Dubourg L, Dacheux P, editors. Design and properties of FSW tools: a literature review.
[20]. Karna S, Cheepu M, Venkateswarulu D, Srikanth V, editors. Recent Developments and Research Progress on Friction Stir Welding of Titanium Alloys: An Overview. IOP Conference Series: Materials Science and Engineering; 2018: IOP Publishing.

IOP Conf. Series: Materials Science and Engineering **655** (2019) 012011 doi:10.1088/1757-899X/655/1/012011

The effects of Ru, Cu, Zr and Hf on mechanical properties in Ti-Pt high temperature shape memory alloys

M P Mashamaite, H R Chauke and P E Ngoepe

Materials Modelling Centre, School of Physical and Mineral Sciences, University of Limpopo, Private Bag X1106, Sovenga 0727, South Africa

Email: mordecai.mashamaite@ul.ac.za

Abstract. Shape memory alloys (SMAs) have been widely used in the fields of actuators and aerospace industry due to their pseudo-elasticity and shape memory effect which are displayed in phase transformations. The martensitic transformations (MT) of TiPt is much higher, at approximately 1273 K and this is considered to be of potential technological interest for elevated temperature SMA applications. TiPt based alloys exhibit very low shape memory effect due to low critical stress for slip deformation compared to the stress required for martensitic transformation, hence it is necessary to enhance the mechanical properties of the equiatomic alloy. The first principles approach was employed to study the effect of the third element (M = Ru, Cu, Zr, and Hf) on the TiPt shape memory alloy. The supercell approach in VASP was used to substitute Pt with Ru and Cu, Ti with Zr and Hf on the TiPt structure to evaluate their mechanical stability from elastic properties for actuators and higher temperature applications. The $Ti_{50}Pt_{50-x}Ru_x$ and $Ti_{50}Pt_{50-x}Cu_x$ decreases in density with increase in Ru and Cu concentration, whilst the $Ti_{50-x}Zr_xPt_{50}$ and $Ti_{50-x}Hf_xPt_{50}$ substitution increases with an increase in their concentration, which result in larger lattice parameters. The heats of formation suggest that $Ti_{50}Pt_{50-x}Ru_x$ substitution is more thermodynamically stable than $Ti_{50}Pt_{50-x}Cu_x$ substitution, and $Ti_{50-x}Hf_xPt_{50}$ substitution is more stable than $Ti_{50-x}Zr_xPt_{50}$. The elastic properties suggest that the ternary structures become mechanically stable with an increase of the third element. The $Ti_{50}Pt_{50-x}Ru_x$ and $Ti_{50}Pt_{50-x}Cu_x$ substitution became more ductile with the increase in concentration. Zr and Hf substitution became more ductile at higher compositions (31.75 – 43.75 at.%). The Ru and Hf substitutions have potential to be used for high-temperature applications.

1. Introduction

Shape memory amalgams (SMAs) that display shape memory impact over 373 K can possibly be utilized for higher temperature application, for example, air ship turbine motors, high-temperature actuators, thermal security, and so forth. [1]; these are known as high-temperature shape memory alloys (HTSMAs). Of all the known SMA pieces, the NiTi compound framework has been examined most broadly and is utilized in the best number of business applications. This alloy exhibits strong shape memory effect (SME), and pseudoelastic behavior under the right conditions, which makes this material ideal for a variety of applications [2]. HTSMAs are produced by adding the ternary (third elements) such as Pd, Pt, Hf, Au, and Zr to NiTi for whose temperature can be shifted anywhere between 373 and 973 K [3]. Some of these materials were reported to effectively increase the transformation temperature of NiTi [4]. Hafnium and Zr addition to NiTi has also been investigated due to their lower cost [5]. (Ti

+ Hf, Zr) - rich NiTi amalgams have weaknesses, for example, an enormous warm hysteresis, poor thermal steadiness, and fragility [6].

On the other hand, equiatomic TiPt exhibits martensitic phase transformation above 1200 K, ~400K higher than equiatomic TiAu and ~500K higher than equiatomic TiPd [7, 8]; these are categorized as HTSMAs due to their thermo-elastic B2 martensitic phase transformation to B19 above 700 K [1, 9, 10]. It was also found that TiPt alloys exhibited negligible SME (11%), due to the low critical stress for slip deformation compared to the stress required for martensitic reorientation [11]. Yoko Yamabe-Mitarai and co-workers [12] additionally detailed that Ti – 50Pt displayed a solitary yielding marvel and low quality (~450MPa) in martensite and exceptionally low quality (~20 MPa) in the B2 area.

Equiatomic Equiatomic TiPt displayed thermo-flexible martensitic phase change, and may not be utilized for high-temperature shape memory materials applications, because of irrelevant SME and low reinforcing in the austenite region. Otherwise, if the shape memory properties can be improved, the B2 phase could be utilized for higher temperature applications [13, 14]. However, research on TiPt enhancement with Zr and Ru partial substitution were discovered powerful for improving the high-temperature quality and shape memory properties, bringing about an expansion in critical stress for slip disfigurement [14]. Cobalt addition was reported to stabilize the B2 phase at lower temperatures and reduced the martensitic transformation temperature due to higher values of tetragonal shear modulus (C') [15].

Furthermore, it is important to strengthen both martensite and austenite phase against deformation, which is significant for the development of SMAs for high-temperature actuator applications [16]. In the previous work, the effect of a third element Ru, Co, Cu, Zr and Hf on the B19 $Ti_{50}Pt_{50-x}M_x$ was investigated for x=5 and Ru addition was more effective and preferred for high-temperature shape memory alloys [17]. It has been suggested that Ru and Cu are preferential for Pt substitution, while Zr and Hf showed promising shape memory trends when substituted for Ti site [18] In this study we investigated the thermodynamic and elastic properties of the B2 ternary $Ti_{50}Pt_{50-x}M_x$ M=Ru and Cu and $Ti_{50-x}M_xPt_{50}$ M=Zr and Hf shape memory alloys by employing the first principles method for understanding their martensitic transformation behavior and mechanical properties.

2. Methodology

We employed the first principles density functional theory (DFT) in the Vienna *ab initio* Simulation Package (VASP) code [19, 20] with the projector augmented wave (PAW) [21]. An energy cut–off of 500 eV was used, to achieve a good convergence of the parameters. We used the generalized gradient approximation (GGA) of the Perdew-Burke-Ernzerhof (PBE) exchange correlation functional [22]. The Brillouin zone integrations were performed with a k-spacing of 0.25 according to Monkhorst and Pack [23]. A 2×2×2 supercell of TiPt was used to substitute a portion of the Pt with Ru and Cu, and Ti with Zr and Hf.

3. Results and Discussion

3.1. Structural and Thermodynamic Properties

The calculated equilibrium lattice parameters of the B2 $Ti_{50}Pt_{50-x}M_x$ where M=Ru and Cu are shown in Figure 1 (a), B2 $Ti_{50-x}M_xPt_{50}$ M=Zr and Hf in Figure 1 (b) (x=6.25, 18.75, 25, 31.75, and 43.75), where $a = b = c$. The lattice parameters decreased with increased Ru and Cu content with $Ti_{50}Pt_{6.25}Ru_{43.75}$ having the lowest value of lattice parameter at a=3.085 Å. Conversely, $Ti_{50-x}Zr_xPt_{50}$ and $Ti_{50-x}Hf_xPt_{50}$ substitutions increased with an increase in Zr and Hf content, $Ti_{43.75}Hf_{6.25}Pt_{50}$ being the lowest at a=3.304 Å.

Figure 1 (c) and (d) show the densities of $Ti_{50}Pt_{50-x}M_x$ where M=Ru and Cu and $Ti_{50-x}M_xPt_{50}$ M=Zr and Hf against composition. The density of the $Ti_{50}Pt_{50-x}Ru_x$ and $Ti_{50}Pt_{50-x}Cu_x$ substitutions decreased with increase in Ru and Cu content, which was also verified by the lattice parameters, this might be due to their atomic radius of Ru and Cu being less than that of Pt. So partial substitution of Pt on $Ti_{50}Pt_{50-x}M_x$ with Ru was expected to decrease the density of the TiPt alloy [24]. $Ti_{50}Pt_{6.25}Cu_{43.75}$ was found to have the lowest density. On the contrary, $Ti_{50-x}Hf_xPt_{50}$ increased as 43.75>x>6.25, $Ti_{50-x}Zr_xPt_{50}$ was observed with a minimal increase with an increased Zr composition. It can also be seen that the atomic radii of Zr and Hf are much greater than both Ti and Pt, resulting in greater densities on the ternary alloys.

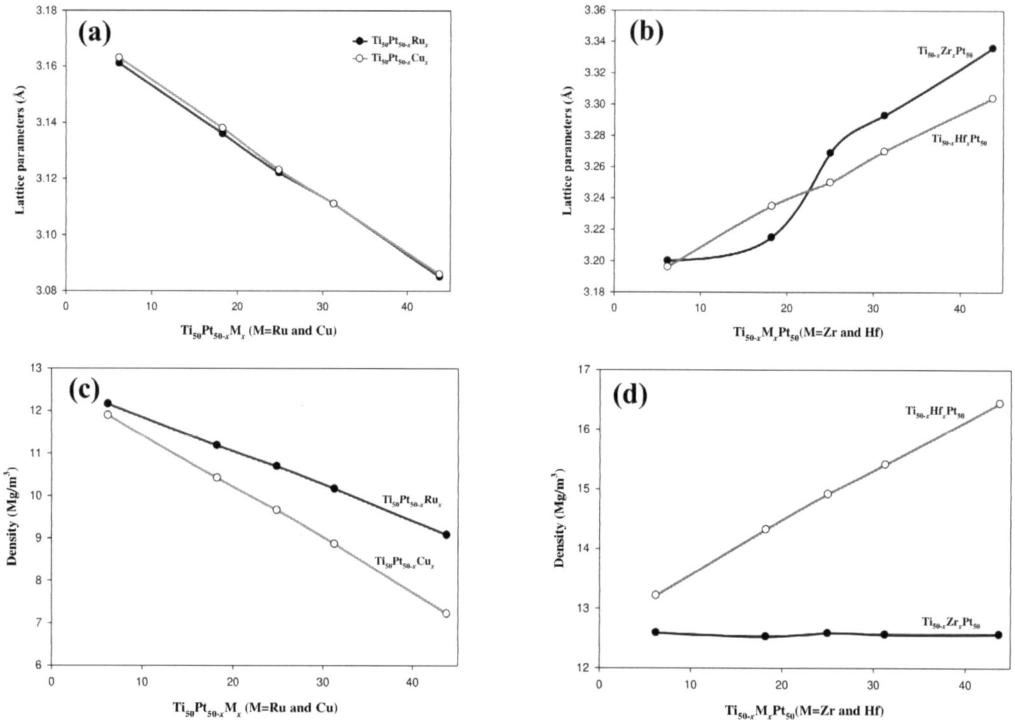

Figure 1. Equilibrium lattice parameters of (a) $Ti_{50}Pt_{50-x}M_x$ (M=Ru and Cu) (b) $Ti_{50-x}M_xPt_{50}$ (M=Zr and Hf) and densities of (c) $Ti_{50}Pt_{50-x}M_x$ (M=Ru and Cu) (d) $Ti_{50-x}M_xPt_{50}$ (M=Zr and Hf) ternaries, 43.75>x>6.25.

The heats of formation (ΔH_f), of the intermetallic phase were calculated according to the relation [25]:

$$\Delta H_f = E^{TiPt} - [(1-x)E^{Ti}_{solid} + xE^{Pt}_{solid}], \quad (1)$$

where E^{TiPt}, E^{Ti}_{solid} and E^{Pt}_{solid} are the total energies of an intermetallic compound, and basic Ti and Pt in their particular ground-state crystal structures, while x and 1-x allude to the fragmentary convergences of the constituent components.

The predicted heats of formation for $Ti_{50}Pt_{50-x}M_x$ where M=Ru and Cu and $Ti_{50-x}M_xPt_{50}$ M=Zr and Hf are shown in Figure 2. The heats of formation suggest that the small composition of $Ti_{50}Pt_{43.75}Ru_{6.25}$ (–0.748 eV/atom) and $Ti_{50}Pt_{43.75}Cu_{6.25}$ (–0.693 eV/atom) are more stable; as the Cu content is increased the $Ti_{50}Pt_{50-x}Cu_x$ became unstable. Ru is more stable than Cu substitution on $Ti_{50}Pt_{50-x}M_x$. On the contrary, $Ti_{6.25}Zr_{43.75}Pt_{50}$ (–0.908 eV/atom) and $Ti_{6.25}Hf_{43.75}Pt_{50}$ (–0.966 eV/atom) suggest that they are more stable substitutions. $Ti_{50-x}Zr_xPt_{50}$ and $Ti_{50-x}Hf_xPt_{50}$ became more stable when their concentration was increased and Hf substitution became more stable.

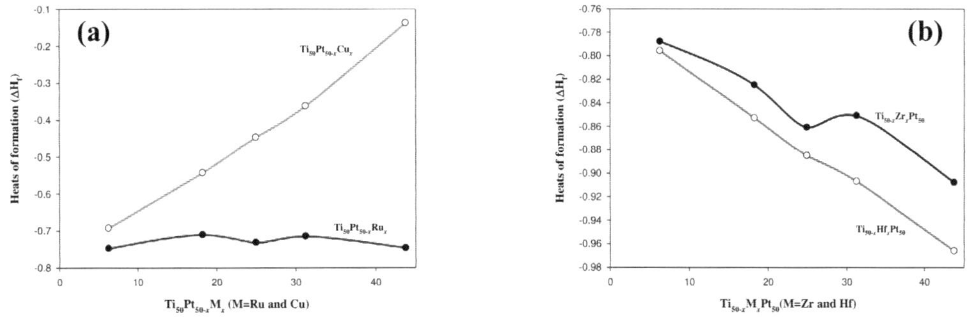

Figure 2. Heats of formation of (a) $Ti_{50}Pt_{50\text{-}x}M_x$ (M=Ru and Cu) (b) $Ti_{50\text{-}x}M_xPt_{50}$ (M=Zr and Hf), where 43.75 >x>6.25

3.2. Elastic Properties

The accurate calculation of elasticity is vital for picking up knowledge into the mechanical stability and elastic properties of compounds. The elastic constants of a cubic crystal have three (C_{11}, C_{22}, and C_{44}) independent elastic constants. Applying two sorts of strains (ε1 and ε4) on the cubic system give stresses identifying with three flexible coefficients, yielding a proficient technique for obtaining elastic constants. This technique has been effectively used to study the elastic properties of a scope of materials including metallic frameworks [26].

For a cubic crystal, the mechanical stability conditions are given by [27].

$$C_{11}>0,\ C_{44}>0,\ +\ C_{11}>|C_{12}| \text{ and } C_{11}+2C_{12}>0, \tag{2}$$

The elastic constants were evaluated to observe the impact of ternary substitution Pt with Ru and Cu, and Ti with Zr and Hf. All the independent elastic constants C_{11}, C_{12}, and C_{44} for $Ti_{50}Pt_{50\text{-}x}M_x$ where M=Ru and Cu and $Ti_{50\text{-}x}M_xPt_{50}$ M=Zr and Hf were positive which indicated mechanical stability of the structures. The $C_{11}>C_{12}$ at 25 at.% and 43.75 at.% for $Ti_{50}Pt_{50\text{-}x}Ru_x$ in Figure 3 (a), $C_{11}>C_{12}$ at 18.75–43.75 at.% for $Ti_{50}Pt_{50\text{-}x}Cu_x$ in Figure 3(b), $C_{11}>C_{12}$ at $Ti_{6.25}Zr_{43.75}Pt_{50}$ in Figure 3 (c), and $C_{11}>C_{12}$ at 25–43.75 at.% for $Ti_{50\text{-}x}Hf_xPt_{50}$ in figure 3 (d), which contributed to positive C′. However, the C_{44} increase minimally with increase in $Ti_{50}Pt_{50\text{-}x}Ru_x$ and $Ti_{50}Pt_{50\text{-}x}Cu_x$ substitutions, suggesting a good correlation between C_{44} and C′ moduli, while decreasing minimally for $Ti_{50\text{-}x}Zr_xPt_{50}$ and Hf substitution. The positive shear suggests that all the ternary compositions above satisfied all conditions of mechanical stability of the cubic crystal.

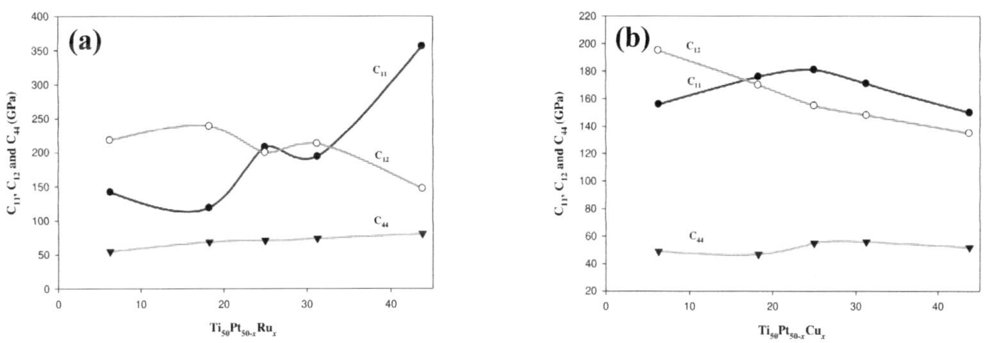

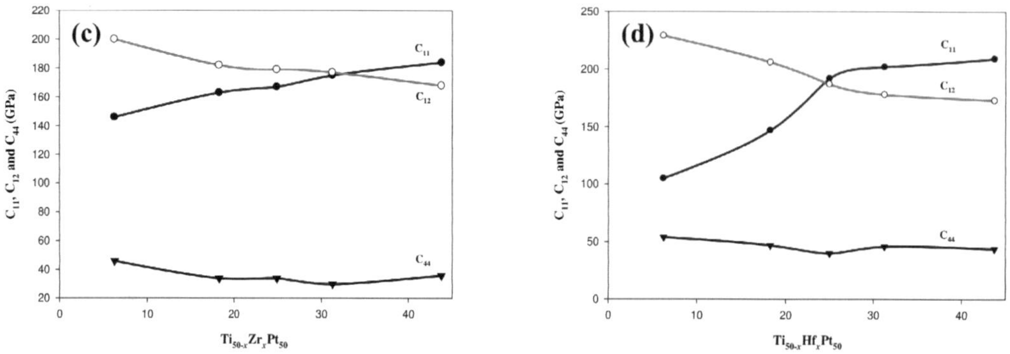

Figure 3. Independent elastic properties C_{ij} (C_{11}, C_{12}, and C_{44}) (GPa) of (a) $Ti_{50}Pt_{50\text{-}x}Ru_x$ (b) $Ti_{50}Pt_{50\text{-}x}Cu_x$ (c) $Ti_{50\text{-}x}Zr_xPt_{50}$ (d) $Ti_{50\text{-}x}Hf_xPt_{50}$, where $43.75>x>6.25$

The Ru and Cu substitution became mechanically stable with an increase in content, 25 – 43.75 at.% Ru and Cu from 18.75 at.% as shown in Figure 4 (a), Zirconium substitution stabilizes around 43.75 at.%, whilst Hf from 25 at.% in Figure 4 (b). The C′ increases with increase in Ru, Cu, Zr, and Hf content, it is also observed that smaller C′ values lead to larger anisotropy (A). These suggest that the Group 4 substitution stabilizes with an increase in Zr and Hf content, Ru belongs to Group 8, a similar trend was observed with the heats of formation; this might be due to Ru having a lower atomic radius than both Ti and Pt. $Ti_{50}Pt_{50\text{-}x}Cu_x$ in Figure 4 (a) stabilized around 18.75 at.% and it also noted that Cu has an atomic radius less than that of Pt, hence the observed curve.

Furthermore, Ru substitution reduced the martensitic transformation temperature of TiPt [14] more than $Ti_{50}Pt_{6.25}Cu_{43.75}$, indicated by higher C′, which is consistent with the experimental findings at lower concentrations for Ru substitution. The C′ moduli of $Ti_{50}Pt_{6.25}Ru_{43.75}$ correspond with A of 0.78, closer to unity ($A \approx 1$) [28]. The Hf substitution in Figure 4 (b) increased the martensitic transformation temperature of the $Ti_{50\text{-}x}Pt_{50}$ since it gave the lowest C′ at 6.25 at.%, with an A value of 0.87. The Bulk modulus (B) measures the level of firmness, or the energy required to deliver a given volume disfigurement [29]. In Figure 4 (c), Bulk moduli for $Ti_{50}Pt_{50\text{-}x}Ru_x$ increased with an increase in Ru content, suggesting that the structure became more hardened, while $Ti_{50}Pt_{50\text{-}x}Cu_x$ decreased with increased Cu content, suggesting that it became less hardened. $Ti_{50\text{-}x}Zr_xPt_{50}$ and $Ti_{50\text{-}x}Hf_xPt_{50}$ in Figure 4 (d) both decreased with increase in Zr and Hf content. These suggest that to maintain the hardness of $Ti_{50}Pt_{50\text{-}x}Ru_x$ can be used with higher substitutional content, while $Ti_{50}Pt_{50\text{-}x}Cu_x$, $Ti_{50\text{-}x}Zr_xPt_{50}$, and $Ti_{50\text{-}x}Hf_xPt_{50}$ may have to be used at smaller substitutional content.

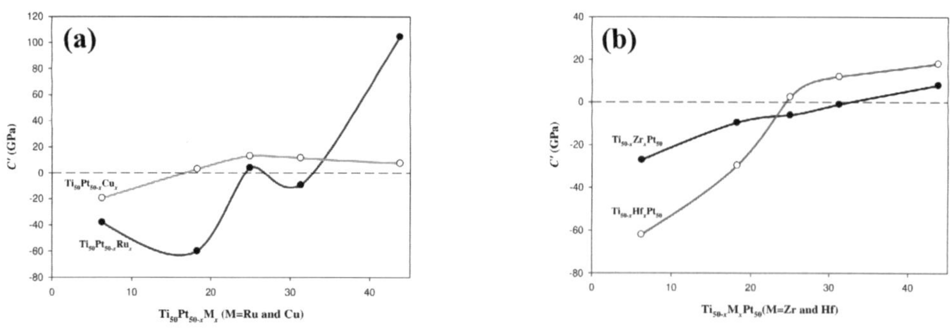

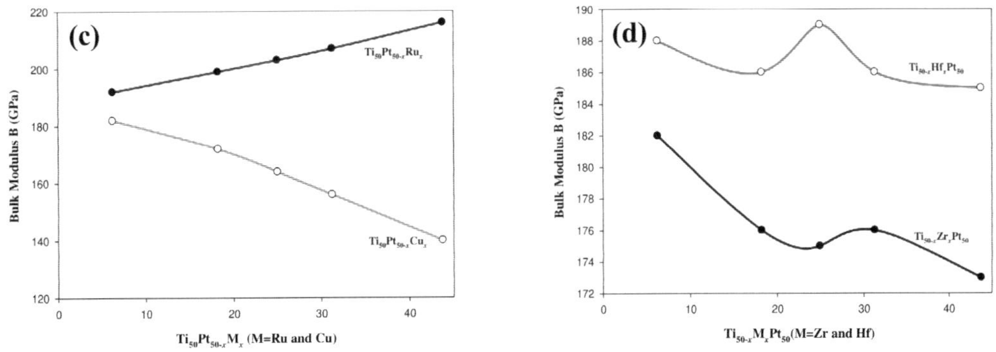

Figure 4. Shear moduli C′ (GPa) for (a) $Ti_{50}Pt_{50-x}M_x$ (M=Ru, Cu) (b) $Ti_{50-x}M_xPt_{50}$ (M=Zr and Hf) and their Bulk moduli (B) (a) $Ti_{50}Pt_{50-x}M_x$ (M=Ru and Cu) (b) $Ti_{50-x}M_xPt_{50}$ (M=Zr and Hf).

According to Pugh, for metals to be considered ductile the Bulk to shear B/G ratio must be >1.75, otherwise, the metal is brittle [30]. In Figure 5 (a), the structures displayed ductility with an increase in Ru and Cu content, since the B/G >1.75. A similar trend was observed in Ti substitution with Zr and Hf in Figure 5 (b), and the dip on Zr 25 at.% was caused by lower values of G. Lower atomic percentages (6.25 at.%) for (Pt with Ru and Cu, Ti with Zr and Hf) were attributed to negative B/G ratio, which did not satisfy ductility conditions. Therefore, Ru at 6.25–18.75 at.%, Cu at 6.25 at.%, Zr at 6.25–25 at.%, and Hf at 18.75 at.% were brittle.

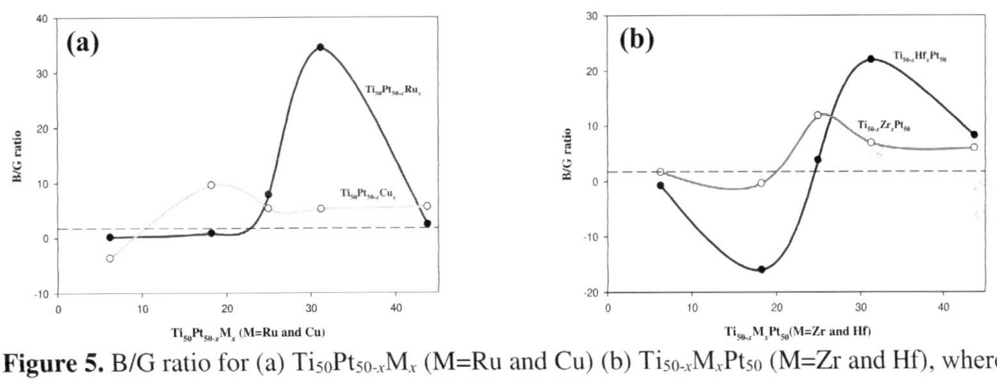

Figure 5. B/G ratio for (a) $Ti_{50}Pt_{50-x}M_x$ (M=Ru and Cu) (b) $Ti_{50-x}M_xPt_{50}$ (M=Zr and Hf), where $43.75>x>6.25$.

4. Conclusions

A computational modelling study on the thermodynamic and mechanical properties in $Ti_{50}Pt_{50-x}M_x$ (M=Ru and Cu) and $Ti_{50-x}M_xPt_{50}$ (M=Zr and Hf) where $0<x<50$ was performed. The effect of the lattice parameter depends on the type of the alloying element and sub-lattice. The Ru and Cu showed a decrease in lattice parameter, while Zr and Hf show an increase. The thermodynamic stability for the ternary framework was explored by somewhat substituting a portion of the Pt with Ru and Cu, and Ti with Zr and Hf, we observed that Ru (–0.748 eV/atom) at 6.25 at.% and Hf (–0.966 eV/atom) at 43.75 at.% substitutions were the most stable structures. The elastic constant of both for Pt substitution and Ti substitution were found to be stable with all the moduli obeying the elastic stability criterion. The addition of Ru and Cu content in Pt and Zr and Hf in Ti increased C′ moduli of the cubic phase leading to a positive anisotropy. $Ti_{50}Pt_{50-x}Ru_x$ and $Ti_{50-x}Hf_xPt_{50}$ increases the transformation temperature. This demonstrated Ru having a place with Group 8 of the Periodic Table go about as B2 phase stabilizer than Cu in Pt substitution, followed by Hf in Ti substitution.

Acknowledgements
The simulations were completed utilizing computer resources at the Materials Modelling Centre (MMC) at the University of Limpopo and the Centre for High Performance Computing (CHPC) in Cape Town. The authors recognize the South African Research Chair Initiative of the Department of Science and Technology and the National Research Foundation is profoundly perceived.

References

[1] Ma J, Karaman I and Noebe R D 2010 High temperature shape memory alloy *Int. Mater. Rev.* **55** 257

[2] Kumar P K and Lagoudas D C 2008 Introduction to shape memory alloys *Modeling and Eng. Application Lett.* 11

[3] Funakubo H (Ed.), 1987 Shape Memory Alloys, *Gordon and Breach Science Publishers*

[4] Suzuki Y, Xu Y, Morito S, Otsuka K and Mitose K 1998 Effects of boron addition on microstructure and mechanical properties of Ti–Td–Ni high-temperature shape memory alloys *Mater. Lett.* **36** 85

[5] Carl M, Smith J D, Van Doren B and Young M L 2017 Effect of Ni-content on the transformation temperatures in NiTi-20 at. % Zr high temperature shape memory alloys *Metals* **7** 511

[6] Hsieh S F, Wu S K 1998 Memory alloys a study on ternary Ti–rich TiNiZr shape memory alloys *Mater. Charact.* **41** 151

[7] Donkersloot H C and van Vucht J H N 1970 Martensitic transformations in gold-titanium, palladium-titanium and platinum-titanium alloys near the equiatomic composition *J. Less-Common Metals*. **20** 83

[8] Kawamura T, Tachi R, Inamura T, Hosoda H, Wakashima K, Hamada K and Miyazaki S 2006 Effects of ternary additions on martensitic transformation of TiAu *Mater. Sci. Eng.* **A438** 383

[9] Otsuka K, Oda K, Ueno Y, Piao M, Ueki T and Horikawa H 1993 The shape memory effect in a $Ti_{50}Pd_{50}$ alloy *Scripta Metall. Mater.* **29** 1355

[10] Yamabe-Mitarai Y, Arockiakumar R, Wadood A, Suresh K S, Kitashima T, Hara T, Shimojo M, Tasaki W, Takahashi M, Takahashi S and Hosoda H 2015 Ti(Pt, Pd, Au) based high temperature shape memory alloys *in Materials Today: Proceedings* **2** S517–S522

[11] Yamabe-Mitarai Y, Hara T, Miura S and Hosoda H 2010 Shape memory effect and pseudoelasticity of TiPt *Intermetallics* **18** 2275

[12] Yamabe-Mitarai Y, Hara T, Miura S and Hosoda H 2006 Mechanical properties of Ti–50(Pt,Ir) high-temperature shape memory alloys *Materi. Trans.* **47** 650

[13] Mahlangu R, Phasha M J, Chauke H R, Ngoepe P E 2013 Structural, elastic and electronic properties of equiatomic PtTi as potential high-temperature shape memory alloy *Intermetallics* **33** 27

[14] Wadood A, Takahashi M, Takahashi S, Hosoda H and Yamabe-Mitarai Y 2013 High-temperature mechanical and shape memory properties of TiPt–Zr and TiPt–Ru alloys *Mater. Sci. Eng.* **A564** 34

[15] Mahlangu R 2015 Computational modelling studies of $Ti_{50}Pt_{50-x}M_x$ (M: Co, Ni, Pd, Ir) shape memory alloys. *University of Limpopo: Thesis*

[16] Noebe R, Gaydosh D, Padula S, Garg A, Biles T, Nathal M. 2005 Properties and potential of two (Ni, Pt)Ti alloys for use as high-temperature actuator metals *Proc. SPIE Int. Symp. Smart Struct. Mater.* **5761** 364

[17] Mashamaite M P, Chauke H R and Ngoepe P E 2018 Mechanical properties and stability of TiPt-M (M=Ru, Co, Cu, Zr and Hf) for high temperature applications *IOP Conf. Ser.: Mater. Sci. Eng.* **430** 012019

[18] Bozzolo G, Mosca H O and Noebe R D 2007 Phase structure and site preference behavior of ternary alloying additions to PdTi and PtTi shape-memory alloys *Intermetallics* **15** 901

[19] Hohenberg P and Kohn W 1964 Inhomogeneous electron gas *Phys. Rev. B.* **136** 864

[20] Kohn W and Sham L J 1965 Self-consistent equations including exchange and correlation effects

Phys. Rev **140** A1133
[21] Blöchl P E 1994 Projector augmented-plane wave method *Phys. Rev. B.* **50** 17953
[22] Perdew J P, Burke K and Ernzerhof M 1996 Generalized gradient approximation made simple *Phys. Rev. Lett.* **77** 3865
[23] Monkhorst H J and Pack J D 1976 Special points for Brillouin-zone integrations *Phys. Rev. B.* **13** 5188
[24] Wadood A and Yamabe-Mitarai Y 2014 TiPt–Co and TiPt–Ru high temperature shape memory alloys *Mater. Sci. Eng.* **A601** 106
[25] Fernando G W, Watson R E and Weinert M 1992 Heats of formation of transition-metal alloys: Full-potential approach and the Pt-Ti system *Phys. Rev. B.* **45** 15 8233
[26] Zhang J M and Guo G Y 1997 Microscopic theory of the shape memory effect in TiNi *Phys. Rev. Letters* **78** 4789
[27] Mehl M J and Klein B M 1994 First principles calculations of elastic properties of metals *Inter. Comp.* **1** 1
[28] Goumri-Said S and Kanoun M B 2007 Theoretical investigations of structural, elastic, electronic and thermal properties of damiaoite $PtIn_2$ *Comp. Mat. Sci.* **43** 243
[29] Kittle C 1976 Introduction to solid state physics, 5th ed., *New York: John Wiley & Sons*
[30] Pugh S F 1954 Relations between elastic moduli and plastic properties of polycrystalline pure metals *Philos. Mag* **45** 843

Carbide-free bainitic steels for rail wheel applications

A Kapito[1,2], R J Mostert[2], W E Stumpf[2] and C W Siyasiya[2]

1. Advanced Materials Division, Mintek, 200 Malibongwe Drive, Randburg, 2125, South Africa

2. Department of Materials Science and Metallurgical Engineering, University of Pretoria, South Africa

E-Mail: asimenyek@mintek.co.za[a]

Abstract: In South Africa, forged wheels for rail cars are imported but cast wheels are manufactured locally, although recent developments indicate that forged wheels will in future be manufactured locally. The forged wheels are generally manufactured from AAR Class C steel, which is a high carbon, pearlitic steel used for rail application. Railway wheels are a costly component of the railway wagon as they experience wear and damage during application. Improvements in the mechanical properties are thus desired. Against this background, a project is reported on the development of durable rail steel alloys for railway wheel applications.
Carbide-free bainite is a novel microstructure comprising bainitic ferrite and retained austenite/martensite but without coarse carbides in the interlath positions. The absence of carbides is achieved through the addition of a high silicon (~2wt%) content to the steel. This carbide-free bainite can achieve high tensile strength (>1000MPa) and toughness (40J, 20°C) as well as good wear resistance. These alloys have found application in areas where high strength, toughness and wear resistance are required, such as in rail steels, and have been deemed the "next generation" of rail steels.
Carbide-free experimental alloys were produced in the laboratory and tested for mechanical properties such as hardness, tensile strength and impact toughness. The properties of the laboratory alloys were compared to those of standard Class C alloys used in South Africa to determine their potential use as railway wheel alloys.

1. Introduction

1.1. Pearlitic Steels for Rail Applications

Rail steels are generally made of medium and high-carbon steel with a pearlitic microstructure (Figure 1). The alternating plates, or lamella, of cementite and ferrite are ideal for the type of operating conditions experienced during rail transport [1],[2]. Solid wheel steel grades for heavy haul are mainly governed by the Association of American Railroad's (AAR) standards [3], [4] and the main steel grades particularly recommended for heavy haul are the medium-high carbon pearlitic steels [3], [4].

Figure 1. Microstructure of pearlite, a lamellar mixture of ferrite and cementite [1].

To extend the life cycle of rail steels, hardness has traditionally been the primary property to exploit, see Figures 2 and 3 [2]. Figure 2 shows the historical development of rail steels and it is clear that the tensile strength of the rail steels has increased continuously over the years [8]. This is mainly due to an increase in the hardness of the steel, to reduce the wear rate, and also due to the development of technologies in the steel-making process [8]. In Figure 2, R200, R260, R320Cr and R350HT are names for pearlitic rail steels grades, developed over the years, with Brinell hardness of 200BHN, 260BHN, 320BHN and 350BHN respectively. Figure 2 also shows the steel named Dobain®, a specially heat-treated bainitic high-grade steel developed in Europe in the 2000's for rail steel applications.
The hardness of a pearlitic rail steel is increased by decreasing the interlamellar spacing and increasing the content of alloying elements, such as manganese and chromium but particularly carbon [2]. Figure 3 shows that a fine pearlitic spacing results in a steel with a higher hardness and yield strength [5]. Therefore the shorter the interlamellar spacing, the better the mechanical properties of the rail steel.

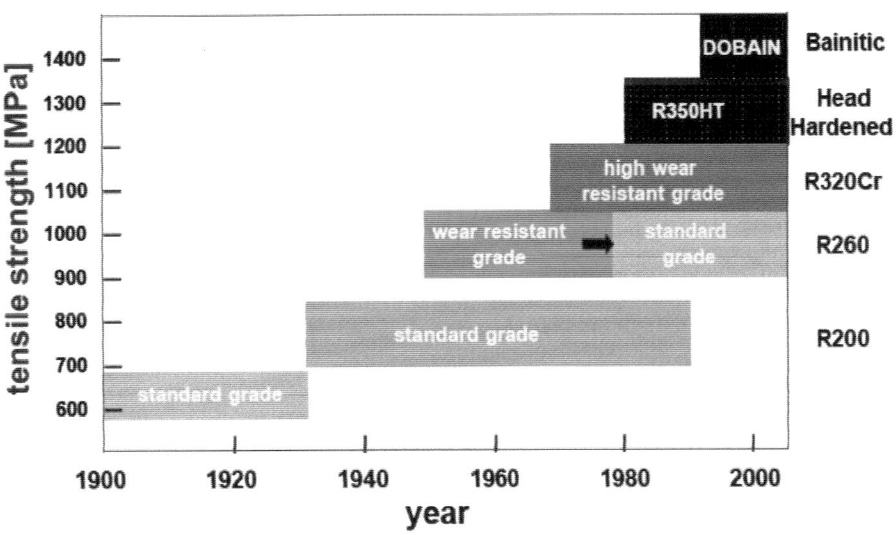

Figure 2. Trends in the development of rail steels from the 1900's to 2000's showing the different grades of steel developed with an increase in tensile strength and hardness [8].

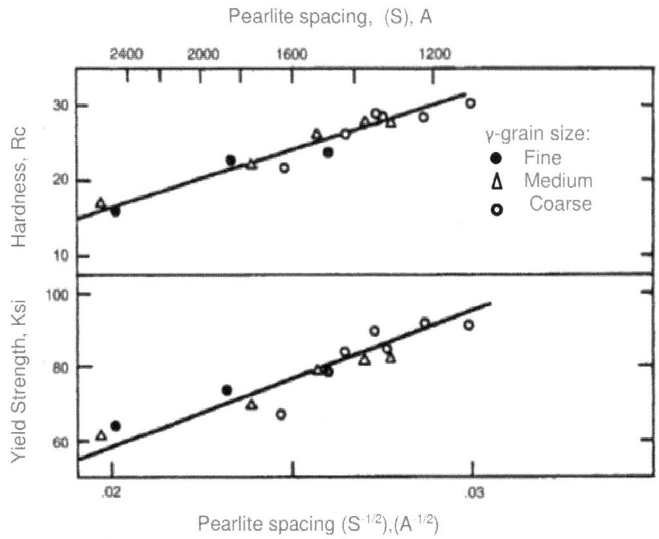

Figure 3. Hardness and yield strength as a function of pearlite interlamellar spacing in a fully pearlitic microstructures [5].

1.2. The Minimum Pearlite Interlamellar Spacing (d)

Pearlite grows in a lamellar structure with the pearlite spacing (d) defined as the perpendicular distance across two consecutive lamellae [6], [7]. Interlamellar spacing is generally measured using the scanning electron microscope (SEM) or the transmission electron microscope (TEM), except if it is very coarse and optical microscopy can be used [7]. The minimum pearlite spacing that can be achieved after transformation from austenite has a thermodynamic relationship as indicated in equation 1:

$$d_{\min} = -\frac{2\gamma_{SF}}{\Delta H_v} \times \frac{1}{\Delta T}. \quad [6] \qquad \textit{Equation 1}$$

Where γ_{SF} = surface energy of the Fe_3C/α interface and ΔH_v = change in enthalpy and $\Delta T = T - T_e$ (the undercooling below the eutectic temperature and T = transformation temperature and T_e is the eutectoid temperature) [6].

The minimum pearlite spacing decreases with a decrease in transformation temperature. The actual pearlite spacing that is achievable by the current production methods is very close to the theoretical d_{min}. Experimentally the actual spacing will always be greater than the theoretical minimum. This limits the further development of pearlitic steels with shorter pearlitic spacings. This has led to the search of alternative rail microstructures such as bainite.

2. Carbide-Free Bainitic Steels

Bainitic steels inherently have better toughness, ductility and weldability than the conventional pearlitic steels. Most modern bainitic steels are designed with much reduced carbon and other alloying elements to form lean alloys [9]. The reduced alloy concentration gives better weldability and improved toughness. The conventional bainitic steels have coarse interlath carbides and generally have poorer toughness and wear resistance than the pearlitic rail steels. This has focused attention on the development of carbide-free bainitic steels.

In the recent decades, Bhadeshia *et al* have developed and introduced high silicon, high carbon bainitic steels [10]-[16]. It is known that the precipitation of cementite during bainitic transformation can be suppressed by alloying the steel with about 1.5 wt% of silicon [10]-[15]. An interesting microstructure results when this silicon–alloyed steel is transformed into upper bainite. The carbon that is rejected into the residual austenite, instead of precipitating as interlath cementite, remains in the austenite and stabilises it down to ambient temperature. The resulting microstructure consists of fine plates of bainitic ferrite separated by carbon–enriched regions of austenite, see Figure 4 [11]. This microstructure is called carbide-free bainite.

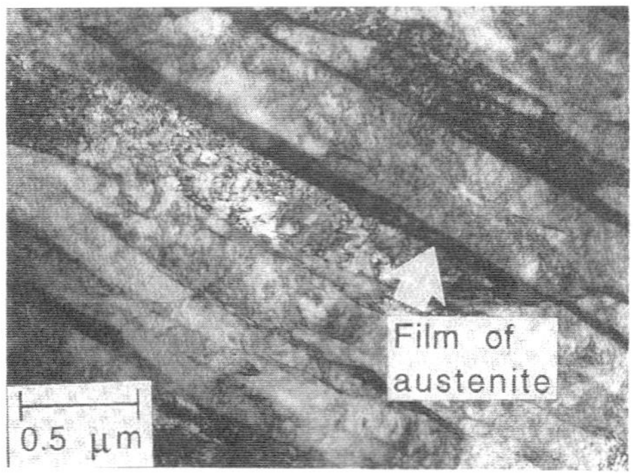

Figure 4. Transmission electron micrograph of a mixture of bainitic ferrite and stable austenite [11].

The alloying elements used to produce this microstructure are common to steel production, i.e. carbon, manganese, boron, etc. with particular addition of a high silicon content to supress cementite formation. The carbide-free bainitic steels have better fatigue properties than the pearlitic steels, accommodating longer crack lengths and growth rates before fracture. They also exhibit higher strength.
A number of carbide-free bainitic steels have thus been developed for rail applications [17], [18]-[35]. There is a government push to shift South Africa's freight transport from road to rail, cutting both high logistical costs and carbon emissions. This will require longer lasting railway components, such as wheels, to take up the higher load. In South Africa, rail AAR Class C forged wheels are used for heavy haul transportation. Durable wheel alloys would cut the maintenance costs of the railway industry significantly. In this study the carbide-free bainitic alloys were explored as possible alternatives for forged rail wheel applications.

3. Experimental Procedure

Melting was done at 1650°C in an induction melting furnace and casting, in air, using a 150 kg alumina crucible, into a designed sand mould. Aluminium (0.2 wt%) was added to the melt for deoxidation. After casting the ingots were allowed to air cool in the moulds. After cooling, the moulds were broken and the ingots (45 mm thickness) were hot forged and hot rolled at 900 °C into 15 mm thick plates, sufficient thickness for the production of the tensile and impact specimens. Chemical composition analysis was conducted using spark emission spectrometry and hardness using Vickers hardness testing (load of 200g, dwell time of 15s). For Micro-Vickers the load used for 30g for a dwell time also of 15s. Neutron diffraction was conducted to measure the retained austenite in the experimental alloys. Optical microscopy and transmission Kikuchi diffraction (TKD) was used to analyse the microstructures of the alloys. Tensile testing was conducted according to ASTM E8 and impact testing according to ASTM E23.

4. Results

4.1. Chemical Composition

Four Alloys E to H with variations in silicon content, were manufactured to study their tensile and impact properties. Table 1 gives the chemical compositions of the experimental alloys. The chemical composition of the four alloys comprised typical alloying elements used to produce rail steels, but with a high silicon content of 1-2 wt% used to retard the formation of cementite in the steel. Alloy E with a low silicon content of 0.36 wt% was produced as a reference alloy, to determine if this low silicon content will produce carbides in the microstructure. The results are an average of three readings.

Table 1. Chemical composition of experimental Alloys E to H, wt%

Alloy	C	Si	Mn	P	S	Cr	Mo	Al	B
E	0.19	0.36	1.41	0.017	0.009	0.5	0.26	0.003	0.0003
F	0.26	1.85	1.37	0.018	0.010	0.6	0.18	0.009	0.0007
G	0.29	1.47	1.38	0.016	0.011	0.5	0.17	0.018	0.0008
H	0.28	1.00	1.57	0.024	0.012	0.5	0.23	0.006	0.0003

4.2. Heat Treatment

The martensite start (M_s) and bainite start (B_s) temperatures for alloys E to H were calculated using well-known empirical equations as shown below, see Table 2 [6].

$M_s = 539 - 423C - 30.4Mn - 17.7Ni - 12.1Cr - 11.0Si - 7.0Mo$

$B_s = 630 - 45Mn - 40V - 35Si - 30Cr - 25Mo - 20Ni - 15W$.

The transformation temperatures (T_1) used to isothermally heat treat the alloys were chosen to be ~10°C above the M_s temperature, to maximise the volume fraction of bainite produced. The alloys were austenitised at temperatures T_A, which was ~50°C above the A_{C3} temperature and then quenched into a salt bath to a temperature T_1 to form bainite. The alloys were soaked for about an hour in the salt bath and then water quenched. Previous testwork was conducted to determine the time required in a salt bath for bainite formation and from these results it was found than an hour was enough to produce bainite in 15mm thick samples.

Table 2. The heat treatment temperatures for alloys E to H.

	Calculated °C		Calculated °C	Transformation °C	
Alloy	M_s	B_s	A_{C3}	T_A	T_1
E	405	594	836	900	415
F	367	583	844	900	380
G	355	577	835	900	365
H	356	555	830	900	370

4.3. The Microstructures of Alloys E to H

Figures 5 and 6 shows the microstructures attained for Alloys E to H, after heat treatment, comprising predominantly bainite (B) with some patches of martensite (M).

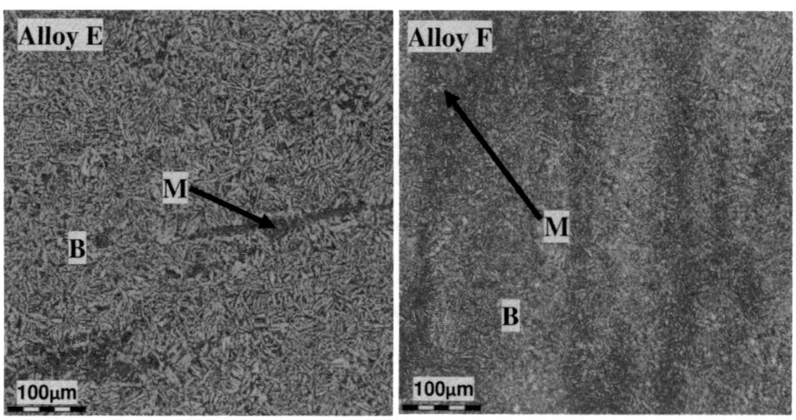

Figure 5. Microstructures of Alloys E and F showing bainite (B) and martensite (M).

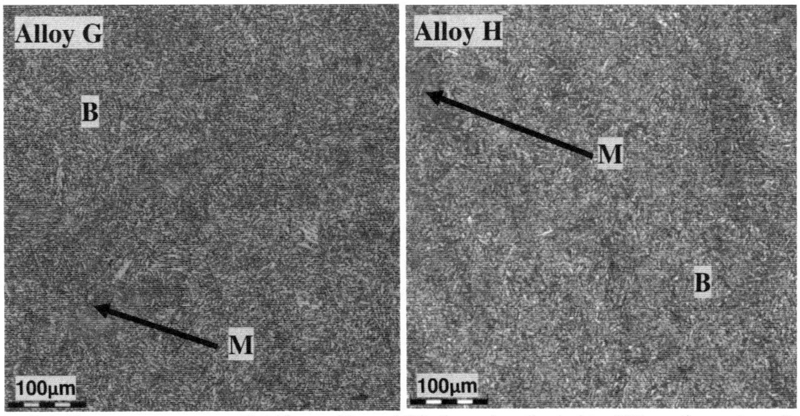

Figure 6. Microstructures of Alloys G and H showing bainite (B) and martensite (M).

To determine the volume fraction of the bainite, retained austenite and martensite in the microstructures of the alloys E to H, point counting and neutron diffraction were used, see Table 3. From the point counting it can be seen that the alloys have a high volume fraction of the bainite (B) phase, i.e. between 87%-98%. Only the martensite and bainite phases were counted using the point counting method. Neutron diffraction measured the retained austenite for alloy E (which had the lowest silicon content) to be 2.6%. Alloys F and H had high retained austenite contents of 20% and 25% respectively and alloy G had a retained austenite content of 9%. The average Vickers hardness of the alloys is also shown here together with the micro-Vickers hardness of the martensitic phase.

Table 3. Point counting, neutron diffraction and hardness measurements for alloys E to H, M = martensite and B = bainite, FCC = austenite, BCC = ferrite (a=b=c= 0.287)

	Point Counting		**Neutron Diffraction**		**Hardness (Hv)**	
Alloy	**M**	**B**	**BCC**	**FCC**	**Average**	**Micro-Vickers (Martensite Phase)**
E	12	88	97.4	2.6	284±6	498±8
F	13	87	80.0	20.0	412±41	503±3
G	2	98	90.6	9.4	352±7	296±6
H	10	90	74.5	25.5	295±2	467±9

4.4. Transmission Kikuchi Diffraction (TKD) for alloys E to H

TKD results for alloys E to H are given in Figures 7. The microstructure of alloy E, with a low silicon content of 0.3 wt%, showed evidence of cementite (yellow phase) in the microstructure along the boundaries of the bainite plates. Alloys F to H showed a microstructure of bainite and retained austenite with no cementite.

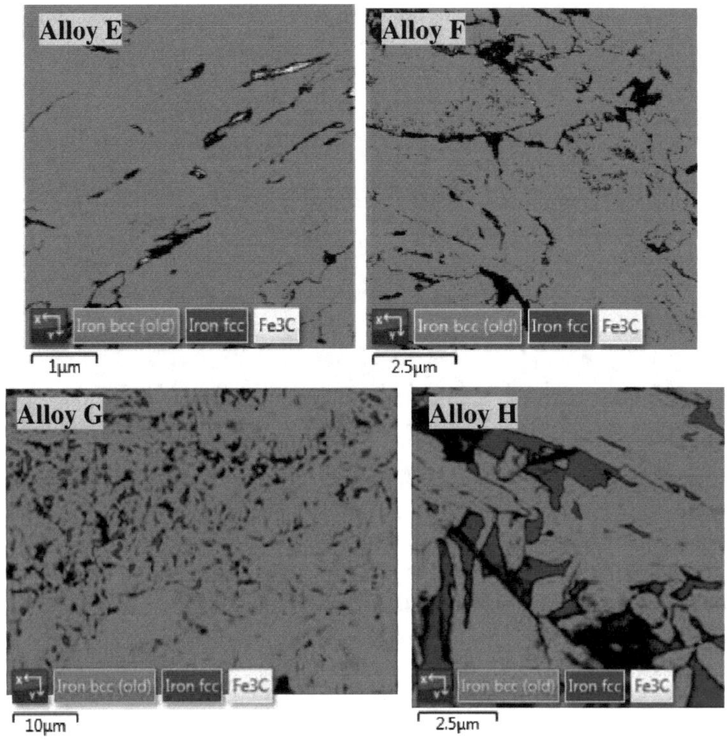

Figure 7. TKD micrographs of experimental bainitic alloys E to H showing FCC-austenite (blue) and BCC-ferrite (red phases). The black area is not a phase but regions in the sample that were not well resolved. 'old' refers to an old file name used to save the micrographs.

4.5. Tensile Tests

Rail wheels experience rolling contact fatigue and it is known that a high yield and tensile strength improves the rolling contact fatigue resistance of rail steels [8][22][31]. Table 4 shows the tensile test results conducted on alloys E to H after heat treatment. AAR Class C rail steel was used as the reference alloy. The AAR standard for Class C does not stipulate any requirements for the yield strength properties. Alloys E to F had a yield strength ranging from 714-829 MPa. The tensile strength for AAR Class C is required to be >1050 MPa (Figure 8) and alloy F achieved a tensile strength of 1224 MPa while the other alloys had a lower tensile strength, alloy F also showed the highest yield strength. The hardness of alloys F is higher than the hardness range stipulated for the AAR Class C alloy in the rim. The elongation of alloys E to F is higher than the 7% stipulated for the AAR Class C alloy (Figure 9). The Alloy F showed the lowest elongation.

Table 4. Tensile properties (yield strength (YS), ultimate tensile strength (UTS), elongation, reduction in area (RA) of alloys E to H and AAR Class C (forged) rail wheel steels.

Alloy	0.2% YS (Mpa)	UTS (Mpa)	Elongation (%)	RIA (%)	Hardness (HV) average	Neutron Diffraction		Point Counting	
						% Retained austenite	BCC	M	B
E	722±18	919±29	13.7±0.88	20.0±5	284±6	2.6	97.4	12	88
F	829±9	1224±44	9.4±0.6	9.1±0.8	412±41	20	80	13	87
G	738±58	978±53	10.5±2.1	17.3±6.5	352±7	9.4	90.6	2	98
H	714±45	960±54	11.9±1.5	18.1±11.8	295±2	25.5	74.5	10	90
AAR Class C*	-	>1050	7 (min)	-	320-376 (Rim)	-	-		

A. Kapito, J.Jonck and G. Maruma [39]

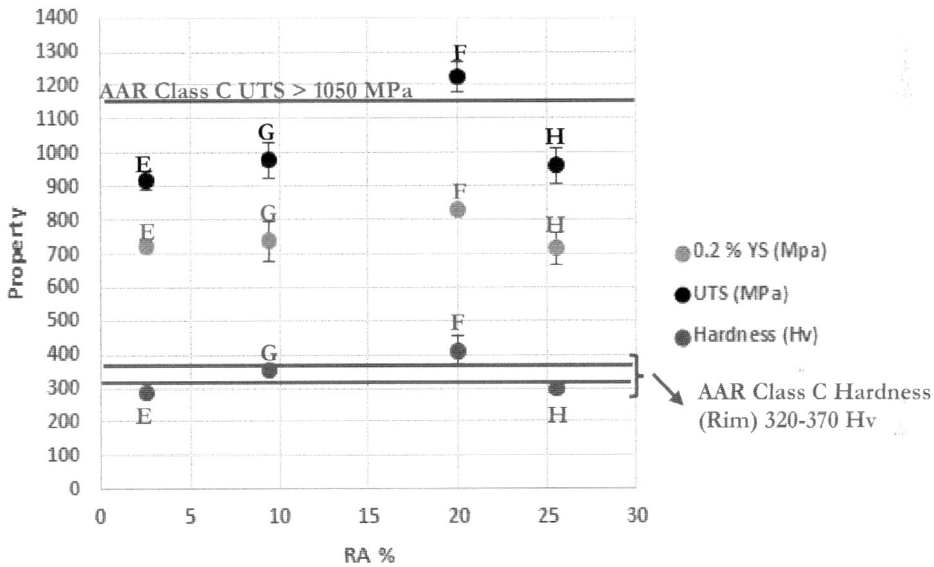

Figure 8. 0.2% yield (YS), ultimate tensile strength (UTS) and hardness versus retained austenite (%RA) results for Alloys E to G compared to AAR Class C alloy.

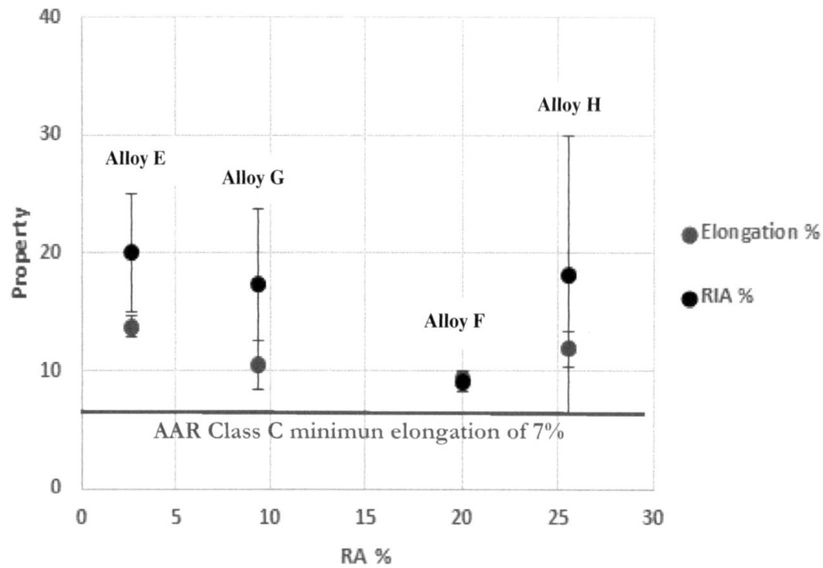

Figure 9. Elongation and reduction in area (RIA) versus retained austenite (%RA) results for alloys E to G compared to AAR Class C alloy.

Figure 10 shows that the carbide-free alloys H to G fall in the grade for advanced high strength steel (AHSS) on the elongation vs tensile strength graph for steels. Alloy E is not carbide-free as it showed carbides in the microstructure due to its low silicon content and is not included in the diagram. Alloys F-H show a combination of high strength and moderate ductility. The ductility is a result of the high amount of retained austenite. The high strength emanates from the combination of fine bainite and the transformation induced plasticity (TRIP) effect. The TRIP effect occurs due to the formation of strain induced martensite during tensile testing, increasing the tensile strength of the steel.

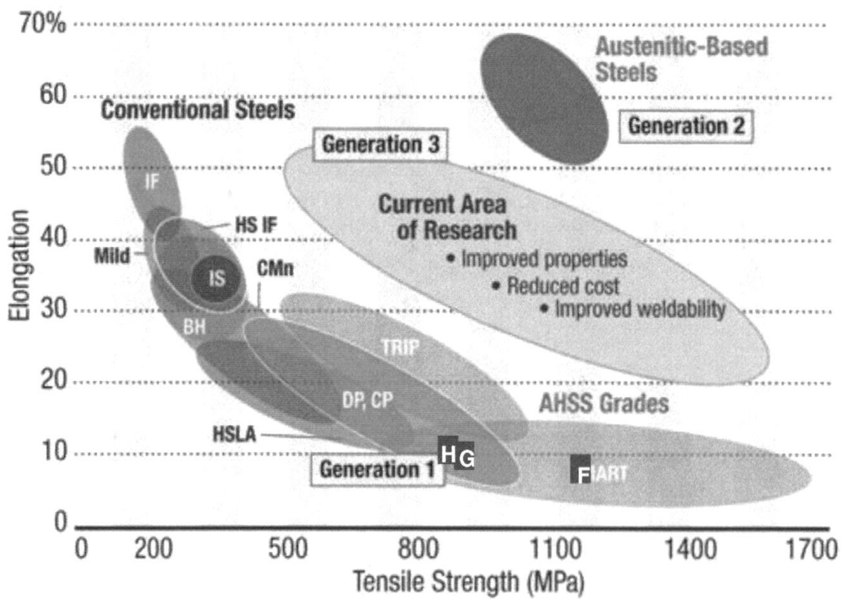

Figure 10. Elongation (%) vs tensile strength (MPa) data for conventional, advanced high strength and austenitic-based steels. The current area of research is also shown for the development of new generation steels [40].

4.6. Charpy V-Notch Impact Testing.

Research has found that high toughness results in a steel with a high resistance to crack initiation and propagation during rolling contact fatigue [8][22][31]. The results for the charpy v-notch impact testing conducted on alloys E to H are given in Figure 11. Alloys E and F show the lowest impact energy absorbed, followed by Alloy G. Alloy H shows the highest impact energy absorbed. Alloys E, F and G show a low impact energy of ≤ 5J up to 100°C, whilst Alloy H shows an impact energy of about 6J up to 50°C, after which its impact energy increases drastically. The ductile to brittle transition temperature (DBTT) for Alloys E, F and G is at high temperatures > 120°C, but Alloy H has a DBTT at about 80°C. Samples from AAR Class C wheel alloy, in the rim, were tested at room temperature and the results are shown in Table 5. Alloy H showed similar energy absorbed at room temperature as AAR Class C. Alloy H has the highest retained austenite of 25% and, from previous results, showed the highest elongation and reduction in area but lowest hardness. The bainitic alloys F and H although containing high retained austenite contents, showed low CVN values, due to the formation of brittle martensite during impact testing.

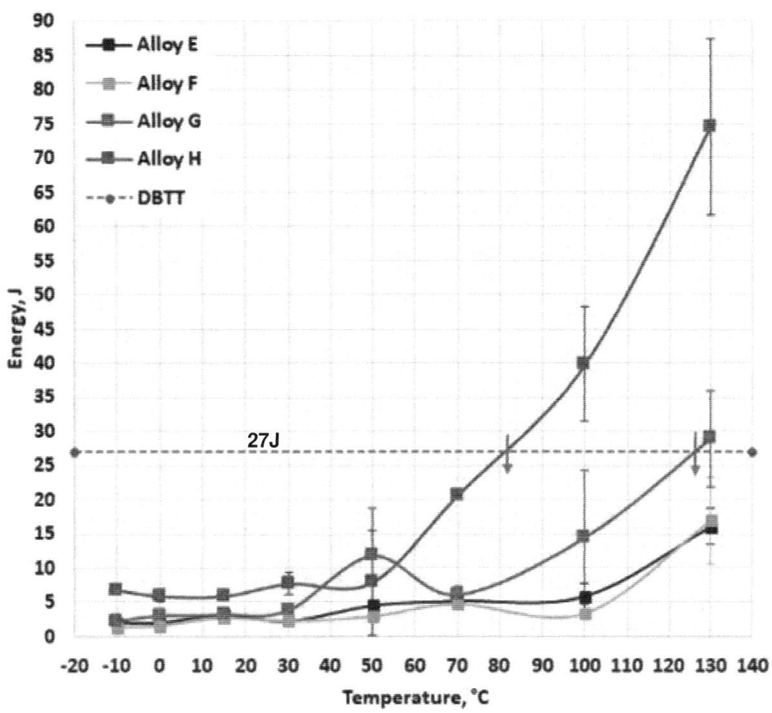

Figure 11. Impact toughness results for Alloys E to H, showing the ductile to brittle transition temperature (DBTT) calculated at 27J.

Table 5. Impact Properties of alloys E to H and AAR Class C at room temperature (RT).

Alloy	*Average Impact energy absorbed at room temperature (J)*	*Retained Austenite*
AAR Class C Forged Wheel	*6.6**	-
Alloy E	*2.3±0.4*	*2.6*
Alloy F	*2.3±0.5*	*20*
Alloy G	*3.8±0.6*	*9.4*
Alloy H	*7.7 ±0.7*	*25.5*

**A. Kapito, J.Jonck and G. Maruma [39]*

5. Discussion

The carbide-free bainitic alloys show beneficial mechanical properties of rail wheel applications. These properties however have been tested on 15mm thick laboratory samples. To transfer the beneficial properties of the carbide-free to actual wheel samples, with a thickness of about 130mm, heat treatment will need to be tailored for the alloys. The carbide-free alloys also contain martensite and the tolerable amount in thick rail wheels needs to be determined to avoid crack initiation during continuous rolling. The carbide-free alloys also have a high retained austenite which can transform to brittle martensite during operation. The amount of retained austenite also needs to be tailored.

6. Conclusions

Transmission Kikuchi diffraction analysis revealed cementite in the microstructure of alloy E with 0.36 wt% silicon, which was produced as a reference alloy. This alloy thus was not carbide-free. Carbide-free bainite was formed using common alloying elements used to produce steels but with a high silicon content >1 wt% in alloys F-G. The carbide-free alloys showed a mixed microstructure of martensite, bainite and retained austenite. The carbide-free alloys had an elongation > 7 % and Alloy F showed a UTS >1050MPa. The impact energy results showed that alloy H with the higher retained austenite of 25% showed better impact properties. The experimental alloys studied in this project generally showed poor impact properties.

7. Acknowledgement

We would like to thank the following people and organisations, without which this study would not have been possible: the University of Pretoria for providing technical support and academic supervision, Mintek, for technical and financial support, The Department of Science and Technology (DST), Advanced Materials Initiative (AMI)-Ferrous Metals Development Network (FMDN) for their financial support, the Nelson Mandela University, Department of Physics and Centre for HRTEM for transmission kikuchi diffraction (TKD) analysis and the Nuclear Energy Council for South Africa (NECSA) for their neutron diffraction analysis.

8. References

[1] N. Ridley, A Review of the Data on the Interlamellar Spacing of Pearlite, Metallurgical Transactions A Volume 15a, June 1984, pp. 1019-1036

[2] B. Bramfitt, *A Metallurgical Perspective of the Role of Rail Steel in the Growth of America*, 2011 AIST Adolf Martens Memorial Steel Lecture, A Publication of the Association for Iron & Steel Technology, January 2012, pp. 158-172

[3] https://www.scribd.com/document/355255964/ihha-guidelines-to-best-practices-for-heavy-haul-railways-operations-wheel-rail-interface-issues-pdf, cited November 2016

[4] M Diener and A Ghidini, *Materials for Heavy Haul Solid Wheels: New Experiences*, Special Issue Paper, Proc. IMechE Vol. 224 Part F: J. Rail and Rapid Transit, JRRT356, 19 April 2010, 421-428

[5] ASM International, *High-Carbon Steels: Fully Pearlitic Microstructures and Applications*, cited August 2015 2005 ASM International, Steels: Processing, Structure, and Performance (#05140G), ASM International, Materials Park, Ohio, USA, pp. 281-295

[6] W.E. Stumpf, *Phase Transformations in Metals and their Alloys*, The University of Pretoria, Materials and Metallurgical Engineering, Course NFM 700, 2010

[7] G. Vander Voort, The Interlamellar Spacing of Pearlite, April 9, 2015 https://vacaero.com/information-resources/metallography-with-george-vander-voort/1437-the-interlamellar-spacing-of-pearlite.html, cited September 2017

[8] G. Gersh and R. Heyder, *Advanced Pearlitic and Bainitic High Strength Rails Promise to Improve Rolling Contact Fatigue Resistance*, https://www.researchgate.net/publication/266035565_Advanced_pearlitic_and_bainitic_high_strength_rails_promise_to_improve_rolling_contact_fatigue_resistance, cited June 2017

[9] A. Gianni, A. Ghidini and A. Ekberg, *Bainitic Steel Grade for Solid Wheel: Metallurgical, Mechanical and In-Service Testing*, Proc, IMechE Vol. 223 Part F: J. Rail and Rapid Transit, pp 163-171, www.ndt.net, cited 28 September 2014.

[10] http://www.transnetfreightrail-tfr.net/MDS/Pages/Overview.aspx, cited November 2016

[11] H.K.D.H. Bhadeshia, *New Bainitic Steels by Design*, University of Cambridge, University of Cambridge, www.msm.cam.ac.uk/phase-trans, cited June 2013

[12] H.K.D.H. Bhadeshia and D. V. Edmonds, *The Bainite Transformation in a Silicon Steel*, Metallurgical Transactions A, Volume 10A, July 1979, pp. 895-907

[13] B.P.J. Sandvik, *The Bainite Reaction in Fe-Si-C Alloys: The Primary Stage*, Metallurgical Transaction A, Volume 13A, May 1982, pp. 777-787

[14] J. Cornide, C. Garcia-Mateo, C. Capdevila and F.G. Caballero, *An Assessment of the Contributing Factors to the Nanoscale Structural Refinement of Advanced Bainitic Steels*, Journal of Alloys and Compounds 577S (2013) S43–S47

[15] H. Amel-Farzad, H.R. Faridi, F. Rajabpour, A. Abolhasani, Sh. Kazemi and Y. Khaledzadeh, *Developing Very Hard Nanostructured Bainitic Steel*, Materials Science and Engineering A559 (2013) 68-73

[16] F.G. Caballero, M.K, Miller, S.S. Babu and C. Garcia-Mateo, *Atomic Observations of Bainite Transformation in a High Carbon High Silicon Steel*, Acta Materialia 55 (2007) 381-390

[17] C. Lonsdale and D. Stone, *Some Possible Alternatives for Longer-Life Locomotive Wheels*, Proceedings of IMECE02, 2002 ASME International mechanical Engineering Congress & Exposition, November 17-22, New Orleans, Louisiana, pp. 239-244

[18] F.C. Zhang, B. Lv, C.L. Zheng, Q. Zou, M., Zhang, M. Li, and T.S. Wang, *Microstructure of the Worn Surfaces of a Bainitic Steel Railway Crossing*, Wear, 268, 2010, 1243-1249

[19] S. Ahmed and A. Zervos, *Railway Ballast*, Engineering and the Environment, University of Southampton, http://www.southampton.ac.uk/~muvis/case_studies/06_Railway_ballast.html, cited September 2014

[20] http://www.trackguy.com/trackwork_101.htm, cited September 2014

[21] R.P.B.J. Dollevoet, *Design of an Anti-Head Check Profile Based on Stress Relief*, PhD Thesis, University of Twente, Enschede, The Netherlands

[22] *Rolling Contact Fatigue: A Comprehensive Review*, DOT/FRA/ORD-11/24, Final Report November 2011, U.S. Department of Transportation, Federal Railroad Administration, Office of Railroad Policy and Development Washington, DC 20590, http://ntl.bts.gov/lib/43000/43400/43400/TR_Rolling_Contact_Fatigue_Comprehensive_Review_final .pdf, cited September 2014

[23] *Transport Safety Board of Canada*, Railway Investigation Report No. R11D0099, Non-Main-Track Derailment, Agence Métropolitaine de Transport, Commuter Train No. 805, Mile 73.84, Saint-Hyacinthe Subdivision, Montreal, Quebec, 09 December 2011, http://www.tsb.gc.ca/eng/rapports-reports/rail/2011/r11d0099/r11d0099.pdf, cited September 2014

[24] M.R. Zhang and H.C. Gu, *Microstructure and Properties of carbide Free Bainite Railway Wheels Produced by Programmed Quenching*, Materials Science and Technology, Vol. 23, No. 8, 2007, pp. 970-974

[25] *Fracture and Fatigue Damage Tolerance of Bainitic and Pearlitic Rail Steels*, Research Results, U.S Department of Transportation, Federal Railroad Administration, RR06-02, February 2006, www.fra.dot.gov/Elib/Document/2136, cited September 2014

[26] H.A. Aglan, Z.Y. Liu, M.F. Hassa and M. Fateh, *Mechanical and Fracture Behaviour of Bainitic Rail Steel,* Journal of materials Processing Technology, 151, 2004, 268-274.

[27] V.V. Pavlov, L.A. Godik, L.V. Korneva, N.A. Kozyrev and E.P. Kuznetsov, *Railroad Rails Made of Bainitic Steel*, Metallurgist, Vol.51, Nos 3-4, 2007, pp.209-212

[28] H.K.D.H. Bhadeshia and J. Vijay, *Improvements in and Relating to carbide-Free Bainitic Steels and Methods of Producing Such Steels*, Patent No. WO 96/22396, 25 July 1996

[29] H. Kageyama, M. Ueda and K. Sugino, *Process for Manufacturing High-Strength Bainitic Steel Rails with Excellent Rolling-Contact Fatigue Resistance*, Patent No. US 5382307, 17 January 1995

[30] M. Ueda, K. Uchino, K. Iwano and A. Kobayashi, *Bainite Type Rail Excellent in Surface Fatigue Damage Resistance and Wear Resistance*, Patent No. AU 737977 B2, 6 September 2001

[31] J. Jaiswal, P. Sécordel and F. Fau, *Corus Premium Grades to Adress Wear And Rolling Contact Fatigue,* https://www.arema.org/files/library/2010_Conference_Proceedings/Corus_Premium_Grades_to_address_Wear_and_Rolling_Contact_Fatigue.pdf, cited September 2014

[32] I. Poscmann, E. Tschapowetz and H. Rinnhofer, *Heat Treatment Process and Facility for Railway Wheels, Tyres and Rings,* Advanced Forging Technologies, www.werkstoff-service.de, accessed 12 April 2014.

[33] R. Zhou, Y. Jiang, D. Lu, R. Zhou and Z. Li, *Development and Characterization of a Wear Resistant Bainite/Martensite Ductile Iron by Combination of Alloying and a Controlled Cooling Heat-Treatment*, Wear, 250, 2001, 529-534.

[34] CQU, http://www.cre.cqu.edu.au/projects/rail_grinding.htm, cited 11 May 2014.

[35] R. Lewis and U.Olofsson, *Mapping Rail Wear Regimes and Transitions*, Wear 257 (2004) 721-729

[36] H.A. Aglan, and M. Fateh, *Fracture and Fatigue Crack Growth Analysis of Rail Steels*, Mechanics of Materials and Structures, Volume 2, No. 2, February 2007.

[37] S. Khare, K. Lee and H.K.D.H. Bhadeshia, *Carbide-Free Bainite: Compromise between Rate of Transformation and Properties*, Metallurgical and Materials Transactions A, Volume 41A, April 2010, pp. 922-928

[38] H.K.D.H. Bhadeshia, *Bainite in Steels*, Second Edition, Institute of Materials, March 2001, ISBN 1 86125 112 2 (H)

[39] A. Kapito, G. Maruma and J. Jonck, Metallurgical Characterisation of 34-Inch Advanced Casting Technique (ACT) Class B Cast and Forged Rail Wheel from Transnet Engineering Koedospoort, Mintek Internal Report No. 42228, 18 December 2015

[40] S. Keeler, The Science of Forming- Automotive Steels--Online Technical Resources, 1 May 2012, https://www.metalformingmagazine.com/enterprise-zones/article/?//2012/5/1/Automotive_Steels--Online_Technical_Resources, visited January 2019

IOP Conf. Series: Materials Science and Engineering **655** (2019) 012013 doi:10.1088/1757-899X/655/1/012013

EBSD characterisation of the B2 orientation in γ-TiAl based alloy

M N Mathabathe[1, 2,*], A S Bolokang[1], G Govender[1], C W Siyasiya[2], and R J Mostert[2]

[1]Council of Scientific Industrial Research, Material Science and Manufacturing, Light Metals, Meiring Naude Road, P O Box 395, Pretoria, South Africa
[2]Department of Material Science and Metallurgical Engineering, Faculty of Engineering, Built Environment and Information Technology, University of Pretoria, South Africa

E-mail address: nmathabathe@csir.co.za

Abstract. Electron backscattered diffraction (EBSD) has been conducted to determine the orientation relationship of the remnant B2 phase in an α-solidified γ-Ti-45Al-2Nb-0.3Si (at. %) based intermetallic alloy. The morphology and microstructural characterisation was investigated using conventional methods. The results showed that the B2-grains exhibits a Blackburn orientation relationship (BOR) with the α_2 grains viz. {111}B2 // {0001}α_2 and {110}B2 // {11-20}α_2. Moreover, mis-orientation axes of B2-grains were found in the clusters between <101> and <111> in 57-63^0 angles, in the Y0 directions. In addition, it was also observed that the overall B2 grains in the alloy structure was textured and comprised of high angle boundaries of misorientations above 15^0 angles.

1. Introduction

The γ-based titanium aluminides are lightweight materials and are in continuous demand because of the higher temperature engine efficiencies in automotive, aerospace and energy industries. These intermetallic alloys offer a good combination of mechanical properties, and therefore, can withstand temperatures up to 800 ^{0}C, especially with additions of Nb and Mo, alloying elements [1]–[3]. The cast intermetallic alloys based on Ti-45Al, which solidify via the β-phase (retained as B2-phase at room temperature) consist of an isotropic equiaxed and texture-free microstructure with modest micro-segregation whereas, peritectic alloys (solidification via α-phase) show anisotropic microstructures with significant texture and segregation, respectively [1].

The β→α transition is a critical phase transformation in titanium and its alloys, because it plays a predominant role in ascertaining the microstructure and mechanical properties. This transformation reaction occurs by the Burger orientation relationship (BOR) viz. $\{110\}_\beta$ // $\{0001\}_\alpha$ and $<111>_\beta$ // $<1100>_\alpha$, which can result in crystallographic equivalent twelve α-variants [4]. Studies [5-8] have shown that the α-phase exhibits an orientation relationship between the inherited γ-phase. Through the EBSD technique, nucleation mechanism of the γ-grains originated at α/α-grain boundaries was extensively investigated. It was reported that further cooling lead to a diffusion-controlled precipitation of γ-grains to a Blackburn orientation relationship with the α-phase i.e. (0002)α // (111)γ and $[11\bar{2}0]\alpha$

[*] To whom any correspondence to be addressed.

// $[1\bar{1}0]\gamma$ with one of the neighbouring α-grain, forming α/γ colonies and few of the γ-grains did not exhibit this BOR [5-8].
Therefore, the aim of this study is to investigate the orientation relationship of the remnant B2-phase with respect to the α-peritectic grains of the heat-treated α-solidified γ-Ti-45Al-2Nb-0.3Si based intermetallic alloy.

2. Experimental work

An arc button melted Ti-45Al-2Nb-0.3Si (at. %) alloy was isothermally heat-treated in the α-transus temperature of 1335 ^{0}C and cooled slowly in the (α_2+γ) phase region in order to obtain a fully lamellar structure [2-3]. The alloy was characterised by SEM equipped with energy-dispersive X-ray spectroscopy (EDS) capabilities for chemical analysis. Phase identification was investigated by X-ray diffraction (XRD) using CuKα radiation λ = 0.154062nm and 2θ from 20-90^{0} angles. PANalytical's X'pert High Score program equipped with data organised in ASCII OR XRDML files was used to indexed lattice parameters of the obtained phases [2,9]. Orientations of the B2 and α (retained as α_2) phases and their relationships were investigated through orientation imaging microscopy (OIM) software in SEM-EBSD, using Channel 5™ system developed by HKL® of which the procedure is reported elsewhere [6].

3. Results and discussion

3.1. Microstructure analysis

The phases formed after the isothermal treating and cooling process reported in our previous study [3] were: γ-TiAl, α_2-Ti_3Al, and B2-($Ti_5Si_3Al_xNb_x$) evenly distributed as shown in **Fig. 1**. The chemical composition of the phases by SEM-EDS method is illustrated in **Table 1**:

Table 1. Chemical composition of the heat-treated Ti-45Al-2Nb-0.3Si alloy (at. %).

Element	**Composition (γ)**	**Composition (α_2)**	**Composition (B2)-Ti_5Si_3**
Ti	49.29±1.58	62.89±0.72	54.24±1.47
Al	48.15±0.46	34.41±1.89	33.38±0.81
Nb	2.32±0.89	2.12±0.63	2.62±1.19
Si	0.23±1.28	0.57±1.51	9.76±1.30

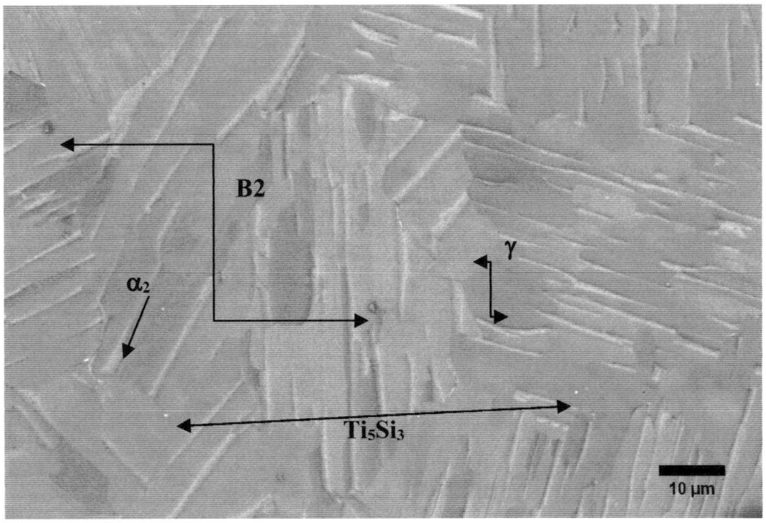

Figure 1: SEM-backscatter image (BSI) of the as-heat treated γ-Ti-45Al-2Nb-0.3Si intermetallic alloy.

3.2. EBSD analysis

The alloy was isothermally treated at temperatures of 1335 ^{0}C, for attempting to minimise the γ and β phases, followed by thermal stabilisation at 900 ^{0}C with subsequent furnace cooling. This resulted in a fully lamellar (FL) structure in **Fig. 1** mailny comprising of (α_2+γ). This resulting microstructure α_2/γ colonies, with small amount B2 phase retaining after solution annealing along with few Ti_5Si_3 particles as shown in **Fig. 1**. In addition, the XRD results in **Fig. 2** validates the stable phases present at room temperature. The lamellar structure (α_2/γ) emanated from the γ lamellae precipitating in the transformed α-grains during the cooling process ($\beta \rightarrow \alpha$ solid state transformation) [3].

Fig. 3 demonstrates that the {111}B2 // {0001}α_2 and {110}B2 // {11-20}α_2 is in agreement with the Burgers orientation relationship, see dotted circles in **Fig. 3b**. This is supplemented by the inverse pole figures in **Fig. 4a** showing the most intensity of B2 grains in the Y0 direction. According to the corresponding colour coding in **Fig. 4b** the highest intensity at a value of 10 indicates that the misorientation axes of the B2-grains is seen in the clusters between <101> and <111> directions in 57-63^0 angle **Fig. 4b**. This is substantiated by the mis-orientation angle distribution in **Fig. 4c** which, exhibits a large number of boundaries with misorientations above 15^0 when compared to the uncorrelated distribution. Furthermore, the results demonstrate that the misorientation angles 10-63^0 between the α-grains originate from the same β-grain, corresponding to 10^0 53′, 60^0, 60^0 83′, 63^0 26′ or 90^0 [6]. As may be seen, the high angle boundary is a notable feature of the microstructure.

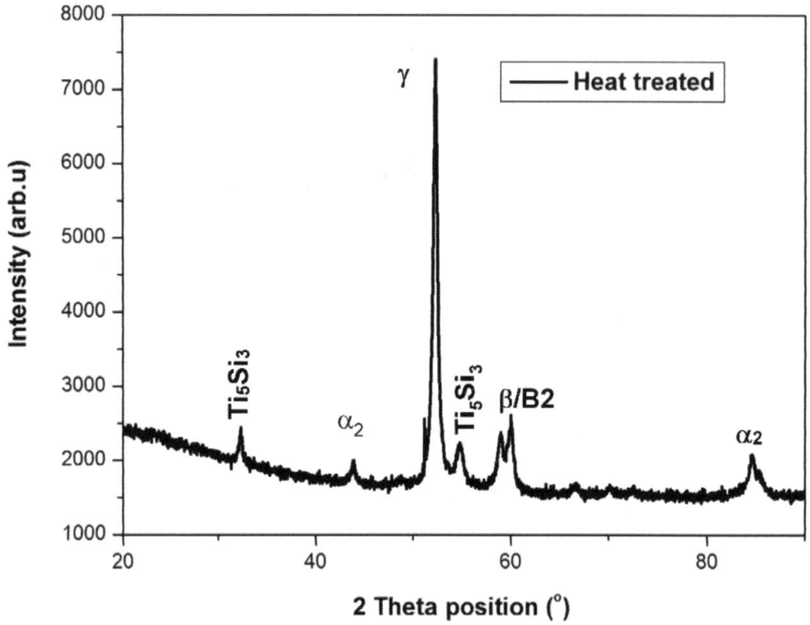

Figure 2. XRD pattern of the heat-treated γ-Ti-45Al-2Nb-0.3Si based intermetallic alloy.

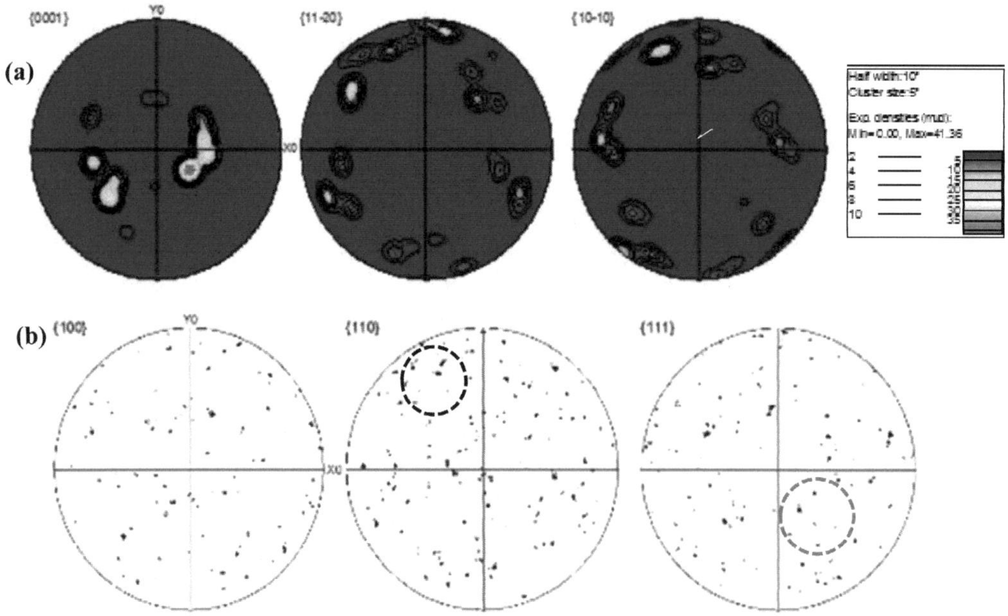

Figure 3. Pole figures of a) α_2, and b) B2 grains indicating orientation relationship.

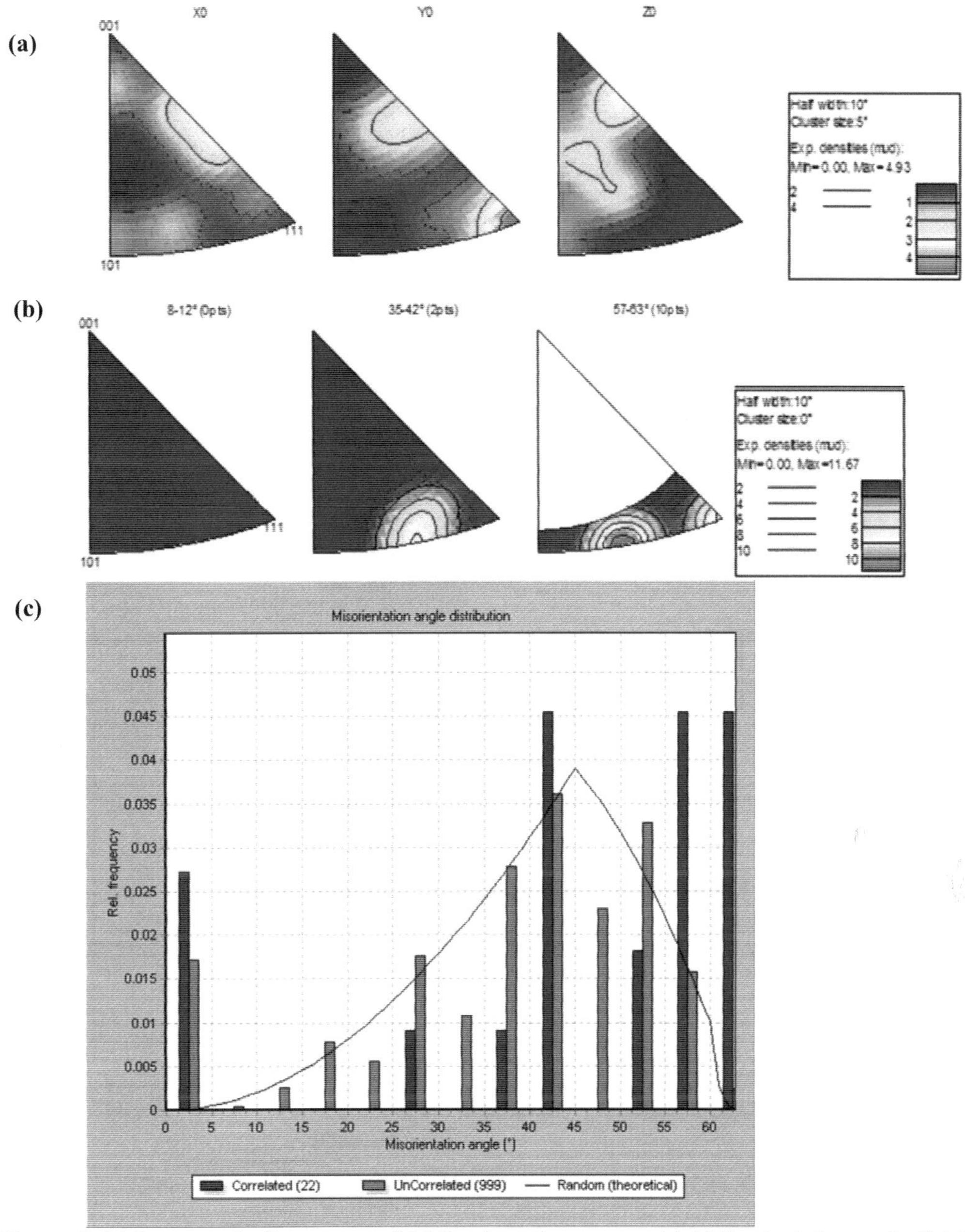

Figure 4. (a) Inverse pole figure maps, (b) Misorientation axes, and (c) Misorientation angle distribution.

4. Conclusions

The EBSD analysis was conducted in order to characterise the orientation of the residual B2 phase of the heat-treated and stabilised γ-Ti-45Al-2Nb-0.3Si alloy, and the following conclusions were drawn:

- The orientation of the B2 phases, with respect to peritectic α-grains followed the Burger OR: $(0002)\alpha \,//\, (111)\gamma$ and $[11\bar{2}0]\alpha \,//\, [1\bar{1}0]\gamma$ and $\{111\}B2 \,//\, \{0001\}\alpha_2$ and $\{110\}B2 \,//\, \{11\text{-}20\}\alpha_2$, respectively.
- It has been shown that the mis-orientation axes of B2-grains were found in the clusters between <101> and <111> in 57-63^0 angle, in the Y0 direction.
- Misorientation angle distribution showed that a large number of boundaries with misorientations above 15^0 (high angle boundary) were a significant feature of microstructure and possible product of texture.

References

[1] Clemens H, Chladil H F, Wallgram W, Zickler G A, Gerling A, Liss K D, Kremmer S, and Güther V 2008 *Intermetallics.* **16** 827-833
[2] Mathabathe M N, Bolokang A S, Govender G, Mostert R J, and Siyasiya C W 2018 *Vacuum* **154** 82–89
[3] Mathabathe M N, Govender S, Bolokang A S, Mostert R J, and Siyasiya C W 2018 *J. Alloys Compd.* **757** 8–15
[4] Chen Y, Cheng L, Yang G, Lu Y, and Han F 2018 *Metals.* **8** 605
[5] Dey S R, Bouzy E, and Hazotte A 2006 *Intermetallics.* **14** 444–449
[6] Mathabathe M N, Bolokang A S, Govender G, Mostert R J, and Siyasiya C W 2018 *J. Alloys Compd.* **765** 690–699
[7] Schwaighofer E, Clemens H, Mayer S, Lindemann J, Klose J, Smarsly W, Güther V 2014 *Intermetallics* **44** 128-140
[8] Schmoelzer T, Liss K –D, Kirchlechner C, Mayer S, Stark A, Peel M, Clemens H 2013 *Intermetallics* **39** 25-33
[9] Mathabathe M N, Govender G, Siyasiya C W, Mostert R J, Bolokang A S 2019 *Mater. Charact.* **154** 94-102

IOP Conf. Series: Materials Science and Engineering **655** (2019) 012014 doi:10.1088/1757-899X/655/1/012014

Microstructure characterisation and microhardness of P92 steel heat treated at the transformation temperatures

J Obiko [1,a], L H Chown [1,2,b], D J Whitefield [1,2,c]

[1] School of Chemical and Metallurgical Engineering, University of the Witwatersrand, 1 Jan Smuts Avenue, Johannesburg, 2000, South Africa

[2] DST-NRF Centre of Excellence in Strong Materials, hosted by the University of the Witwatersrand

[a] Japheth.Obiko@students.wits.ac.za, [b] Lesley.Chown@wits.ac.za,

[c] David.Whitefield@wits.ac.za

Abstract. Modern power plant components for the transport of superheated steam are manufactured from creep resistant steels, such as ASTM grade P92, that have microstructural stability and excellent mechanical properties at high temperature and pressure. Before application as pipes, tubes and connectors, P92 steel undergoes a heat treatment of austenitisation and tempering to achieve a tempered martensite microstructure, with carbides along the prior austenite grain boundaries and carbonitrides within the matrix. In the power plant, pipes and tubes are joined by welding, inducing residual stresses and inhomogeneous microstructures in the weld joint, which reduce creep resistance. As in most steels, the temperature gradient from the fusion line to the base metal leads to different microstructural zones in the weld joint. The fine-grained heat affected zone (FGHAZ), which forms at temperatures near the transformation region, acts as a weak zone for Type IV cracking during creep. The lower and upper critical transformation temperatures for the P92 steel in this work are 800°C-850°C (A_{e1}) and 900°C-950°C (A_{e3}) respectively. As-received samples were heated to temperatures of 830, 850, 900°C to simulate the fine grain heat affected zone, or the very high temperature of 1200°C to simulate the coarse grain heat affected zone (CGHAZ), held at that temperature for 90 minutes and then tempered at 730°C to 780°C for 120 minutes. The microstructures before and after heat treatment were observed by scanning electron microscopy and optical microscopy, and Vickers microhardness measurements were done. After tempering, coarse carbides were seen on the prior austenite grain boundaries (PAGBs) and nitrides were observed in the matrix. In general, the hardness increased with an increase in the austenitisation and tempering temperature combination.

1. Introduction

The use of fossil fuel for electricity generation has received global attention due to the shortage of coal, oil and gas and also due to environmental regulations to reduce carbon dioxide (CO_2) emissions [1]. The thermal efficiency of the power plants has been improved by increasing the steam temperatures and pressures [2,3]. Ultra-supercritical power plants are projected to operate at high temperatures and pressures of ~650°Cand 30 MPa [4,5]. Creep resistant 9-12 wt% chromium steels, such as P91, E911 and P92, have been developed to operate at high temperature and pressure [2]. P92 steel was developed from modified P91 steel and exhibits higher creep strength [5]. These steels are mainly used for structural components such as steam boiler tubes and pipes [4] .The high-temperature phase stability of these steels has been achieved by the addition of various micro-alloying elements, such as Mo, W, V,

Nb, Co and Mn [3]. These additions enhance the solid solution and precipitation strengthening, hence improving creep resistance [6,7]. P92 steel also has a lower coefficient of thermal expansion, higher resistance to corrosion, good weldability and higher thermal conductivity than older generation 9%Cr steels [8,9].

After production, P92 steels undergo heat treatments of normalisation and tempering before being supplied for industrial application [5]. The normalisation process is usually done in the temperature range of 1040°C to 1100°C for the duration of 20-30 minutes, followed by tempering at 760°C to 780°C for 30-40 minutes [9,10]. After tempering, the steel exhibits a tempered martensitic microstructure with carbides along the prior austenite grain boundaries (PAGBs) and carbonitrides finely distributed in the matrix [11,12]. The presence of precipitates along the PAGBs and martensite lath boundaries prevents subgrain growth and so improves the strength of the steel [12]. The commonly identified precipitates include $M_{23}C_6$ (containing Fe, Cr, Mo, W and C) carbides and MX (containing V, Nb, C and N) carbonitrides [6]. Nagode *et al.* [13] showed that MX precipitates finely distributed in the matrix act as a barrier to dislocation movement during creep deformation. Di-Gianfrancesco *et al.* [14] reported that both $M_{23}C_6$ and MX precipitates enhance creep resistance, as the precipitates hinder dislocation movement during creep deformation [9].

Hald [15] found that the microstructural stability of 9 wt% Cr steels largely depends on precipitate stability for long term exposure to high temperature and pressure. An optimised normalisation and tempering process of P92 steels leads to phase transformation from austenite to a refined tempered martensite structure [11]. Normalisation is usually done above the upper critical transformation temperature (Ae_3), which is ~900-950°C for P92 steel, depending on the specific composition [16].The effect of heat treating P92 between the lower and upper critical temperatures followed by tempering has not been widely reported.

This paper presents the microstructure evolution of two P92 steels with differing amounts of chromium and tungsten, which were subjected to different combinations of normalisation and tempering temperatures. The 'normalisation' temperatures were selected to be similar to those attained during welding: 850-950°C in the fine grained heat affected zone (FGHAZ) and ~1200°C in the coarse grained heat affected zone (CGHAZ). After the different heat treatment cycles, the precipitate distribution was compared and the hardness variation was also reported.

2. Materials and Methods

The two P92 steels in this study are named A and B, and their chemical compositions are given in Table 1. P92 steels are usually supplied in the as-normalised and tempered condition. Samples were cut from the as-received P92 steels and were heat treated (N) for 90 minutes and tempered (T) for 120 minutes. Four sets of N+T temperatures were used: at 830°C (N)/730°C (T), 850°C (N)/750°C (T), 900°C (N)/760°C (T), and 1200°C (N)/780°C (T), to produce various microstructures. The FGHAZ reaches peak temperatures of ~900°C

Specimens for microhardness and microstructure analysis were prepared using standard metallographic procedures. After grinding and polishing, the specimens were etched with Villella's reagent (5 ml HCl + 1 g picric acid + 100 ml ethanol) for microstructure investigation using a Leica DM 6000M optical microscope and a Zeiss Sigma field emission scanning electron (FEGSEM) microscope. Microhardness of the as-received and heat treated samples were measured using a 1 kgf load and a dwell time of 10 seconds with a FUTURE-TECH FM-700 microhardness tester.

Thermo-Calc thermodynamic simulation software with the TCFe5 database [17] was used to determine the equilibrium Ae_3 and Ae_1 transformation temperatures.

Table 1. Chemical composition of the as-received steels (wt %).

Steel	C	Mn	S	P	Si	Cr	Mo	Ni	Cu	Al	V	Nb	W	Co
A	0.10	0.39	0.011	0.014	0.20	8.29	0.65	0.19	0.08	0.012	0.16	0.093	2.07	0.015
B	0.11	0.51	0.005	0.010	0.22	9.37	0.50	0.17	0.27	0.006	0.19	0.130	1.76	0.028

3. Results and Discussion

3.1. ThermoCalc analysis

The thermodynamic calculations were done to determine the phase transformation temperatures (A_{e1} and A_{e3}) and dissolution temperatures of precipitates using ThermoCalc with the TCFe5 database. For the two P92 steels in this work, the equilibrium phase transformation and dissolution temperatures of $M_{23}C_6$ and MX precipitates are given in Table 2.

Table 2: ThermoCalc predicted transformation temperatures and precipitate dissolution temperatures (in °C)

Steel	Ae_1	Ae_3	$M_{23}C_6$	MX
A	830	945	870	1201
B	816	917	888	1250

3.2. Microstructural analysis

3.2.1. As-received base material.

Micrographs of the two P92 steel in the as-received (heat treated + tempered) condition are shown in Figure 1 a-d. Both steels had a tempered martensite microstructure with well-defined prior austenite grain boundaries. It has been reported that the creep strength of P92 steel is enhanced by the presence of precipitates such as $M_{23}C_6$ (M = Fe, Mo, W, Cr) carbides and MX (M = V, Nb; X = C, N) carbonitrides [18]. $M_{23}C_6$ carbides nucleate along the PAGBs and lath boundaries, while MX precipitates are usually randomly distributed in the matrix [8].The SEM micrographs in Figure 1c-d show precipitates on the lath b oundaries which are assu med to be $M_{23}C_6$ carbides.

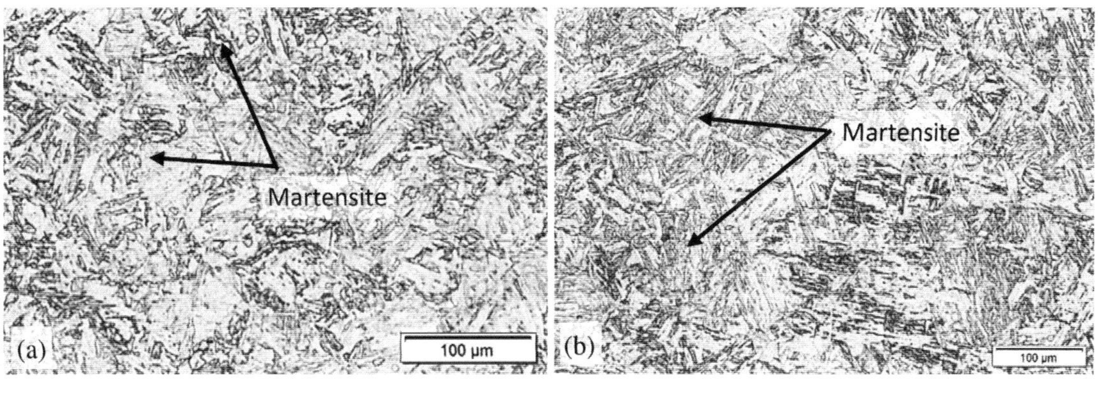

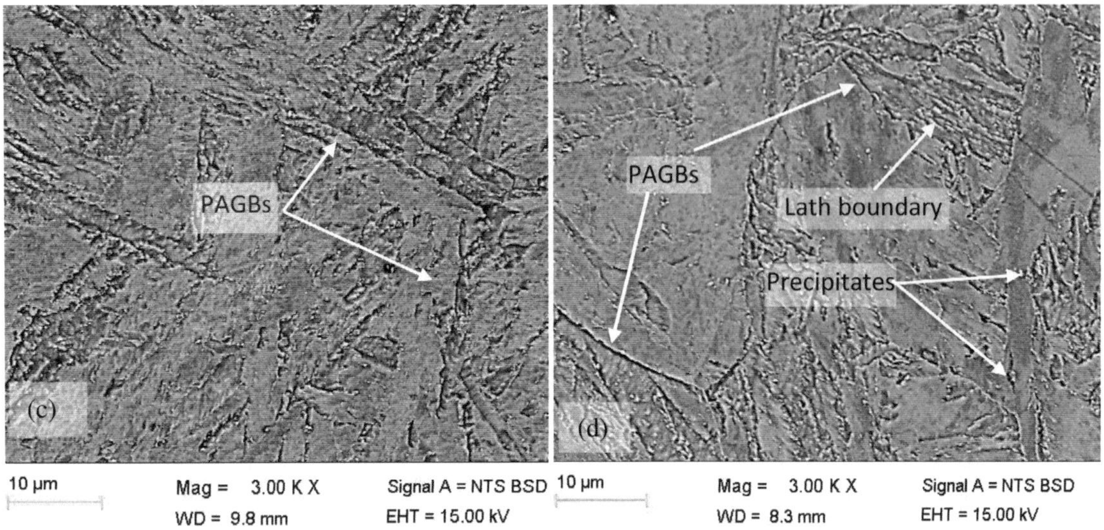

Figure 1. Optical and SEM micrographs of as-received Grade P92 steel: a) and c) steel A, b) and d) steel B.

3.2.2. Optical microstructure after heat treatment.

The optical images for steels A and B after various heat treatment procedures are shown in Figure 2. Figure 2 a and b show optical micrographs of alloy A and B after heat treatment at 830°C and tempering at 730°C. It was observed that this heat treatment had changed the initial tempered martensite lath structure of the as-received samples into finer martensitic lath structures. Similar behaviour was also observed after heat treatments of 850°C/750°C and 900°C/760°C (Figure 2 c - f). The tendency of 9-12% Cr to form fresh lath martensitic structures and fine grain sizes even at low cooling rates has been reported in literature [4]. During normalisation, there is a complete dissolution of precipitates into the matrix resulting in a homogeneous austenitic microstructure [19] . However, in this work, two of the initial heat treatment temperatures were below the precipitate dissolution temperatures of $M_{23}C_6$ carbides (870°C (steel A) and 888°C (steel B)) which would have led to only partial dissolution. The tempering stage reduces the dislocation density, breaking of laths and reduction of solid solution strengthening due to the precipitation and coarsening of the carbides [20]. Figure 2 g-h show the microstructures of steel A and steel B after normalisation at 1200°C (far above the Ae_3 temperature and the $M_{23}C_6$) and tempering at 780°C. After this treatment, both steels A and B showed tempered lath martensite microstructures with coarse grain sizes.

Figure 2 e) shows the formation of ferrite containing coarse carbides within the matrix of tempered martensite. The ferrite forms either due to the incomplete transformation of ferrite into austenite on heating and finally martensite on cooling or the contribution of alloying elements.[21] Ryu and Yu [22] reported that ferrite retention is a function of alloying elements that affect the expansion and contracting of the austenite field. Wang *et al.*[23] investigated the role of cobalt in 9-12 wt% Cr steels, and observed that δ-ferrite was suppressed by adding 1% Co and reducing the nickel content to 0.4% depending on the presence of the ferrite and austenite stabilisers. Similarly, Sireesha *et al.* [24] reported that δ-ferrite retention is likely to occur in steels containing ferrite stabilisers. They proposed using an empirical formula for the Cr equivalent and the Kaltenhauser ferrite factor (FF) to determine the tendency of ferrite retention in the matrix (in wt %):

$$Cr_{eq} = Cr + 6Si + 4Mo + 1.5W + 11V + 5Nb + 12Al + 8Ti - 40C - 2Mn - 4Ni - 2Co - 30N - Cu \quad (1)$$

$$FF = Cr + 6Si + 4Mo + 8Ti + 2Al + 4Nb - 2Mn - 4Ni - 40C + N \quad (2)$$

The Cr_{eq} is used to determine the ferrite formation in 9-12 wt% Cr since it gives the amount of chromium percentage in Fe-Cr phase at which equal amount of δ-ferrite will form at the same temperature [22]. The retention of ferrite has been reported to occur preferentially along the grain boundary and at triple points [22] when Cr_{eq} is greater than 10 and FF is greater than 8. For the current study, steel A had a Cr_{eq} of 11.1 and an FF value of 7.0, while steel B had 11.4 and 7.2. The prediction of Equations 1-2 is correct as ferrite phase was observed in the microstructure as shown in Figure 2(e) even though FF was slightly below the reported empirical value. Even though steel B is more prone to form ferrite, the presence of ferrite phase in steel A may be attributed to the cooling rate. The formation of ferrite along the grain boundary seen in Figure 2e may be attributed to the amount of ferrite and austenite stabilisers in the two steels. However, this requires further analysis.

3.2.3. SEM microstructure after heat treatment.

The SEM secondary electron images showing the microstructure of heat treated and tempered samples of steel A and B are shown in Figure 3 and Figure 4 respectively. The steels exhibited a martensitic structure with distinct prior austenite grain boundaries (PAGBs) delineated by the coarse $M_{23}C_6$ precipitates with some also on subgrain boundaries. Figure 3 a) and b) show precipitate distribution along the former PAGBs of the steel before heat treatment, while Figure 3d) shows the newly formed PAGBs decorated with precipitates after the heat treatment.

From Figure 4, steel B exhibited a tempered martensitic microstructure with fine prior austenite grain sizes. The evolution of precipitates clearly delineates the newly formed grain boundaries. Maddi *et al.* [11] reported that $M_{23}C_6$ carbides precipitate along the austenite and lath boundaries. The $M_{23}C_6$ carbides impede dislocation movement during creep deformation, hence improving creep strength [25].

Steel A Steel B

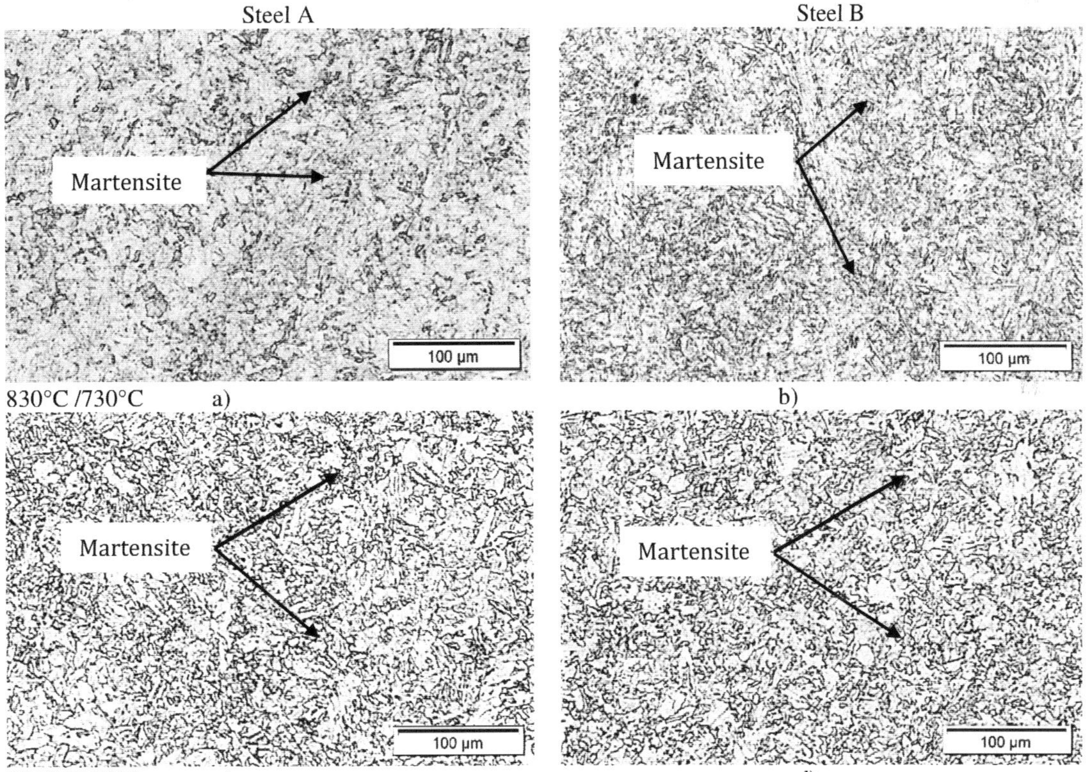

830°C /730°C a) b)

850°C/750°C c) d)

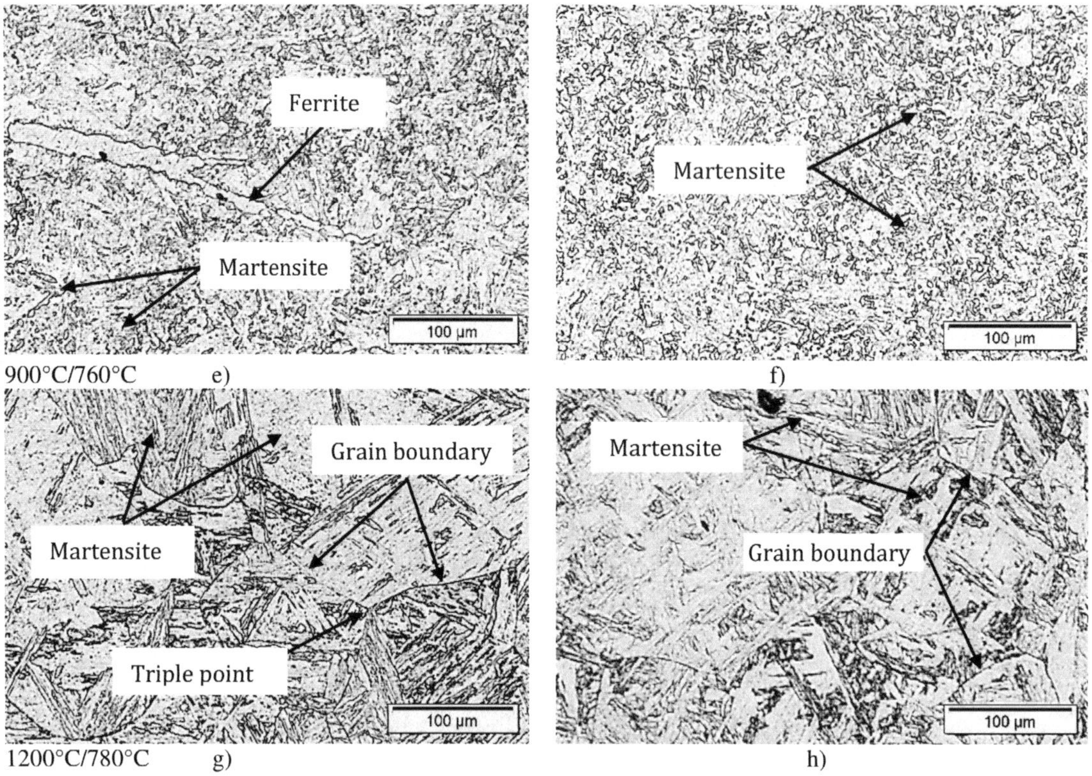

Figure 2. Optical micrographs after heat treatment at different heat-treated (N) + tempering (T) temperatures.

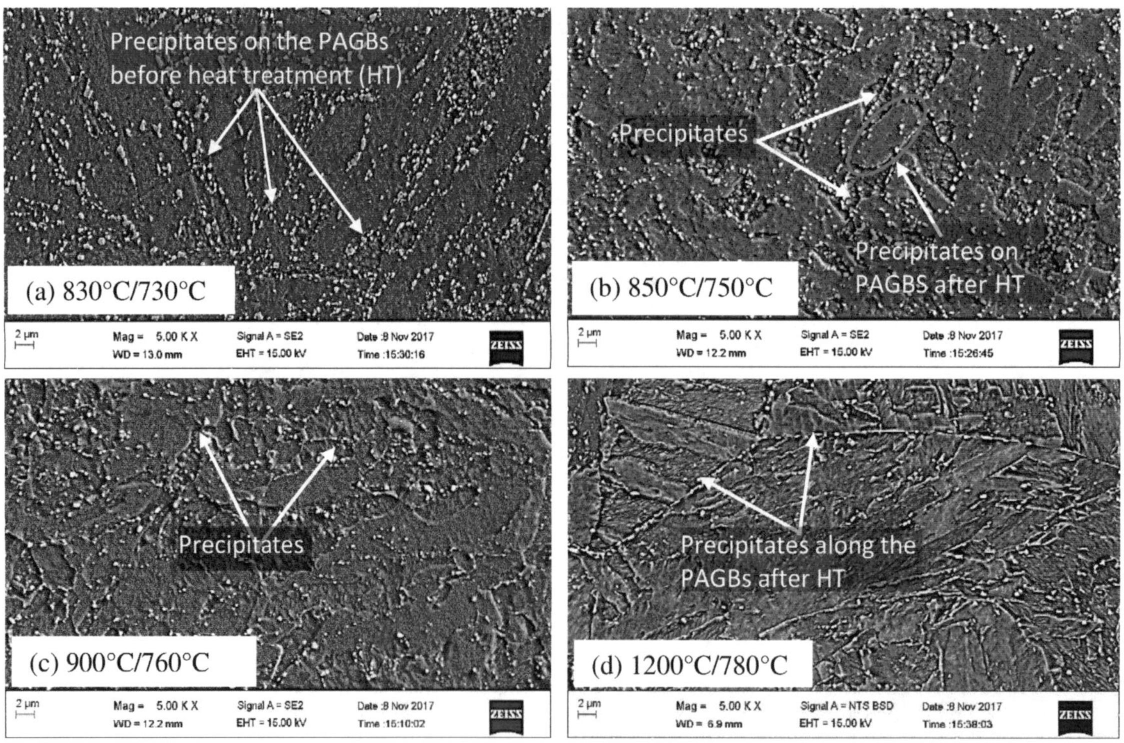

Figure 3. Secondary electron micrographs of steel A after heat treatment.

3.2.4. Quantitative image analysis.

Quantitative analysis of the larger, visible precipitates evolved during heat treatments for the two steels was done using the Image J image analysis program. The results are shown in Figure 5-7. Figure 5 shows the variation in the average particle size, expressed as particle area (μm^2), with increase in the normalising temperature (and thus tempering temperature). The particle area continuously increased for steel B as the normalising-tempering temperatures increased. While for steel A the particle area increase at slow rate and then dropped as the normalisation temperature increase to 1200°C.

The percentage area fraction of particles for steel A increased between 830°C to 900°C normalisation temperatures then drastically decreased as the normalisation-tempering temperature increased (1200°C/780°C). For steel B, the percentage area fraction of particles increased rapidly from 830°C to 900°C as shown in Figure 6. Further increases in the normalisation temperature gave only an insignificant change in percent area fraction. It was observed that the population density of precipitates decreased as the normalisation temperature increased for steel B, which can be attributed to the formation of large grain sizes as the heat treatments temperature increases hence, reducing the number of grain boundaries for the precipitation of carbides [8]. While for steel A, it was noticed that the population density of particles decreased from 830°C to 900°C and then started to increase with further increase in temperature to 1200°C. This increase in the particle population density can be attributed to the formation of smaller grain sizes which leads to more grain and lath boundaries per unit area available for the precipitation of carbides.

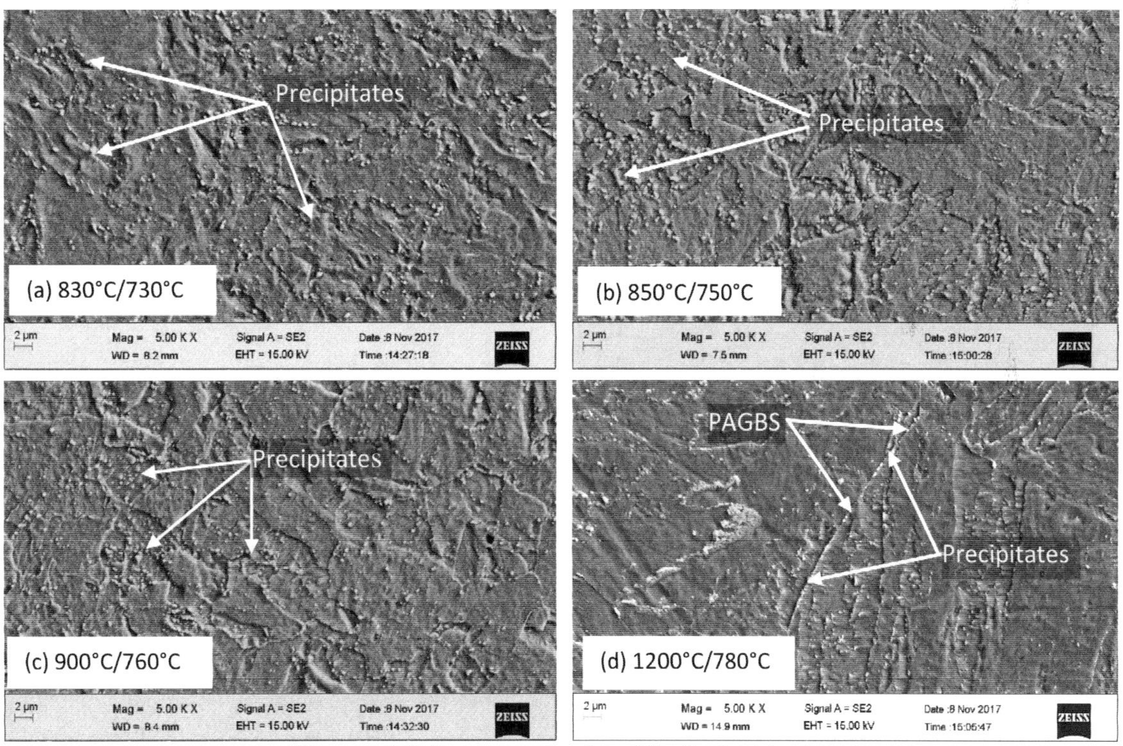

Figure 4. SEM micrographs of steel B after heat treatment.

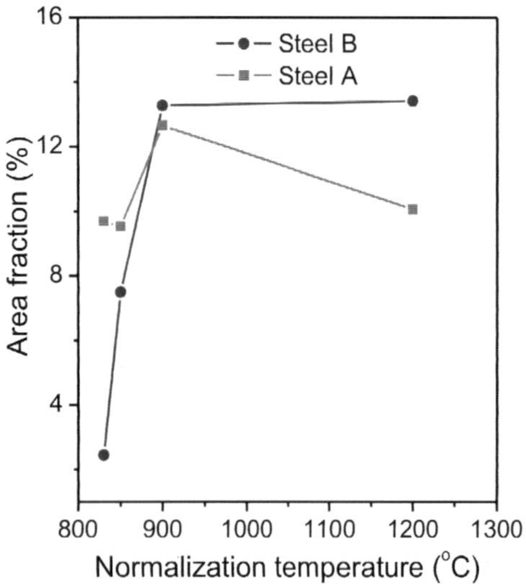

Figure 5.Variation of area fraction of precipitate s with normalising temperature

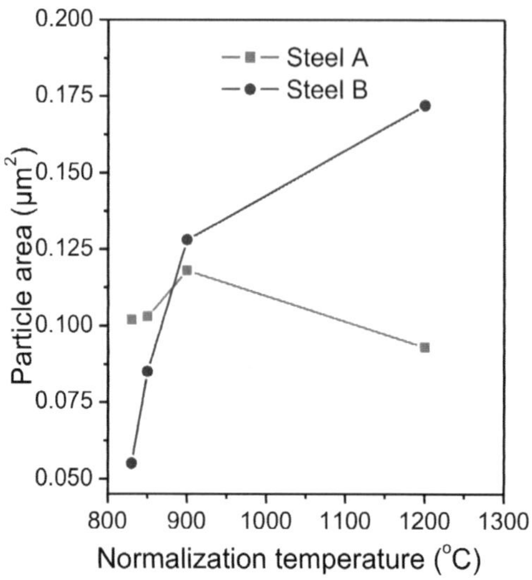

Figure 6. Variation of average particle area size with normalising temperature.

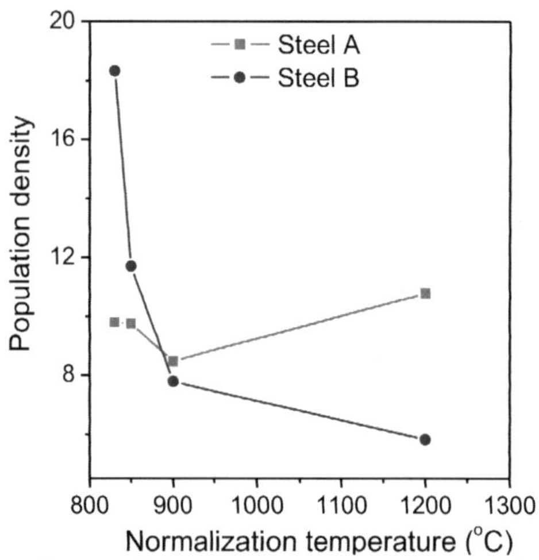

Figure 7. The effect of normalising temperature on the particle population density.

3.3. Microhardness Analysis

The microhardness values of as-received steels A and B were 239 ±7 HV_1 and 230 ±7 HV_1 respectively, which is within the design specifications range of 210-265 $HV_{0.3}$ [16]. The variation in microhardness for the as-received and heat treated steels is shown in Figure 8. It was observed that for steel A the microhardness increased as the heat treatment temperatures increased. However, steel B showed unique behaviour after the heat treatment of 830°C/730°C, with a high hardness of 241HV_1, similar to that of the as-received material. After this heat treatment (830°C/730°C), full dissolution of carbides would not have taken place. The hardness of Steel B then increased with increasing heat treatment temperature from 850-1200°C. The subsequent tempering process would have led to coarsening of the precipitates, which reduces the precipitate nucleation sites and lowers the area fraction of precipitates, thus leading to solid solution hardening [26]. At 830°C, steel B had low area fraction of precipitates (Figure 5), indicating high solid solution hardening due to incomplete precipitation of carbides hence, and high hardness. At 850°C, steel B had the lowest hardness value of 216 HV_1 compared to 227 HV_1 at 900°C and 245 HV_1 at 1200°C as shown in Figure 8.

The variation in hardness values was attributed to the microstructural changes that occur during the heat treatment processes. It has been reported that grain size, solid solution strengthening, precipitation of carbides and carbonitrides and thus the reduction of N and C from the solid solution during tempering account for the hardness variations in 9-12% Cr steels [5,19]. Pandey *et al.* [27] reported that an increase in normalisation temperature transforms the microstructure into single phase austenite which on cooling transforms to martensite. The formed martensite structure increases the microhardness due to an increase in dislocation density. In the current study, there was an increase of 11.5 ± 7 % in the microhardness with an increase in the heat treatment temperatures.
Lower hardness values were observed at heat treatment temperatures of 850°C/750°C and 900°C/760°C for steel A and B than the as-received steel samples, as the tempering process softens the formed brittle martensite structure after austenitisation. Similar low hardness values have been observed to occur in the Fine Grained Heat Affected Zone (FGHAZ) of the weld joint subjected to peak temperature near the Ac_3 transformation temperature (which is slightly higher than the Ae_3) for 9-12 wt% Cr steels [25], [28]. Wang *et al.* [19] reported that hardness lowering characteristics can be attributed to the recovery of the prior austenite grain boundaries leading to the formation of coarse sub-grains and the reduction in dislocation density. According to Abe and Tabuchi [29] lower hardness was observed when the base metal was heat treated at peak temperatures close to Ac_1 (~850°C for P92 steel). High hardness was observed at 1200°C for both steels which exhibited microstructures similar to that of the Coarse Grained Heat Affected Zone (CGHAZ) after welding. This is why steel A and B were subjected to normalisation temperatures for P92 which are experienced within the weld joint.

The hardness of steel B after heat treatment at 830/730°C was similar to the as-received steel as there would have been negligible change to the strengthening precipitates on heating up to the Ae_1. The low hardness values for both steels after the 850/750°C treatment was likely due to coarsening and dissolution of $M_{23}C_6$ precipitates, which would have been completely dissolved by 870°C (steel A) and 888°C (steel B). With further increase in the normalising temperature to 900°C, the $M_{23}C_6$ would be fully dissolved for both steels and could precipitate, increasing the hardness. The MX precipitates of steel A would have been fully dissolved at 1201°C (A), so would precipitate on cooling, further increasing the hardness. The trend in hardness with heat treatment temperature from 850 - 1200°C was similar for the two steels, with steel A having the lower values. Although the predicted Ae_1 of steel A (830°C) coincided with the heat treatment temperature of 830°C, the high hardness seen in steel B was not observed.

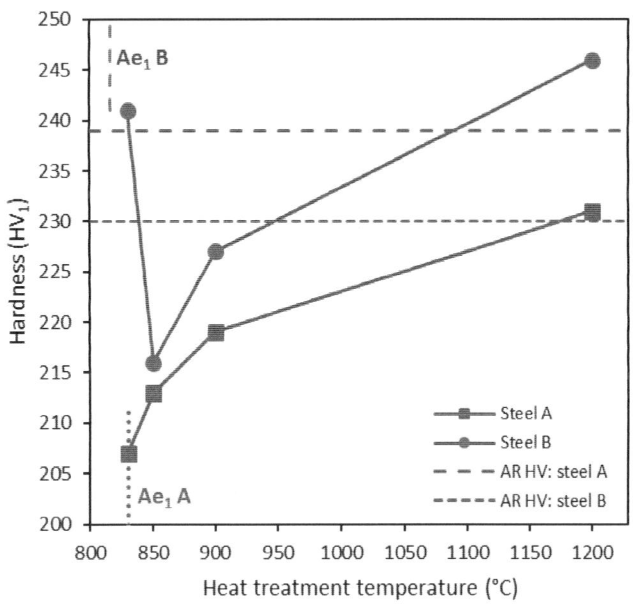

Figure 8. Hardness variation at different heat treatment conditions.

4. Conclusions

- From the ThermoCalc analysis of the two P92 steels, results showed that the lower and upper critical phase transformation temperature (Ae1) range from 800-830°C and 900-945°C. $M_{23}C_6$ carbide dissolution temperature was determined to be 870°C (steel A) and 888°C (steel B) while MX precipitates dissolved at 1201°C (steel A) and 1250°C (steel B).
- After normalising and tempering, both steels contained martensite lath structures. From the SEM images, coarse $M_{23}C_6$ precipitates were seen to occur along the prior austenite grain boundaries.
- Quantitative analysis showed that the heat treatment temperatures had an influence on the precipitate formation and distribution. As the temperatures increased, the population density of the precipitates decreased. Steel B showed a more rapid increase in the area fraction and particle size with an increase in the heat treatment temperature compared with steel A. However, area fraction and particle size of steel A decreases at 1200°C.
- The microhardness values for the two steels increased with an increase in the heat treatment temperatures. Steel B had higher hardness values than steel A in all test conditions which was attributed to the solid solution hardening effect.

5. Acknowledgements

The authors would like to acknowledge and appreciate with gratitude the DST-CSIR Interbursary Support (IBS) programme for financial support.

6. References

[1] Guo X, Gong J, Jiang Y, Wang X and Zhao Y 2015 "Microstructures and high-temperature mechanical properties in 9Cr-0.5Mo-1.8W-VNb steel after aging at 650°C," *Mater. High Temp.*, **vol. 32**, no. 6, pp. 566–574.

[2] Vuherer T, Dunder M, Milovic L, Zrilic M and Samardzic I 2013 "Microstructural Investigation of the Heat-Affected Zone of Simulated Welded Joint of P91 Steel," *Metalurgija*, **vol. 52**, no. 3, pp. 317–320, 2013.

[3] Ennis P, Zielinska-Lipiec A, Wachter O and Czyrska-Filemonowicz A 1997 "Microstructural stability and creep rupture strength of the martensitic steel P92 for advanced power plant," *Acta Mater.*, **vol. 45**, no. 12, pp. 4901–4907.

[4] Zhang Z, Holloway G and Marshall A 2008 "Properties of T/P92 Steel Weld Metals for Ultra Super Crtitical (Usc) Power Plant," *Weld. World*, **vol. 6**, no. 1, pp. 1–13.

[5] Ennis P and Czyrska-Filemonowicz A 2003 "Recent advances in creep-resistant steels for power plant applications," *Sadhana*, **vol. 28**, no. 3–4, pp. 709–730.

[6] Mandal A, Guguloth K, and Bandyopadhayay T 2016 "Characterization of Microstructure and Mechanical Properties of 9.2Cr0.45V0.7C Heat-Resistant Steel," *Metallogr. Microstruct. Anal.*, **vol. 5**, no. 2, pp. 95–107.

[7] Oñoro J 2006 "Weld metal microstructure analysis of 9-12% Cr steels," *Int. J. Press. Vessel. Pip.*, **vol. 83**, no. 7, pp. 540–545.

[8] Barbadikar D, Deshmukh G, Maddi L, Laha K, Parameswaran P, Ballal A, Peshwe D, Paretkar R, Nandagapol M and M. Mathew 2015 "Effect of normalizing and tempering temperatures on microstructure and mechanical properties of P92 steel," *Int. J. Press. Vessel. Pip.*, **vol. 132–3**, pp. 97–105.

[9] Hurtado-Noreña C, Danón C, Luppo M and Bruzzoni P 2015 "Evolution of Minor Phases in a P91 Steel Normalized and Tempered at Different Temperatures," *Procedia Mater. Sci.*, **vol. 8**, pp. 1089–98.

[10] Ohgami M, Naoi H, Kinbara S, Mimura H, Ikemoto T and Fujita T 1997 "Development of 9CrW tube, pipe and forging for ultra supercritical power plant boilers," *Nippon Steel Technical Report*, no. 72. pp. 59–64.

[11] Maddi L, Ballal A, Peshwe D, Paretkar R, Laha K and Mathew M 2015 "Effect of tempering temperature on the stress rupture properties of Grade 92 steel," *Mater. Sci. Eng. A*, **vol. 639**, pp. 431–438.

[12] Kaneko K, Matsumura S, Sadakata A, Fujita K, Moon W, Ozaki S, Nishimura N and Tomokiyo Y 2004 "Characterization of carbides at different boundaries of 9Cr-steel," *Mater. Sci. Eng. A*, **vol. 374**, no. 1–2, pp. 82–89.

[13] Nagode A, Kosec L, Ule B, and Kosec G "Review of creep resistant alloys for power plant applications," *Metalurgija*, **vol. 50**, no. 1, pp. 45–48.

[14] Di Gianfrancesco A, Vipraio S, and Vendit D 2013 "Long term microstructural evolution of 9-12%Cr steel grades for steam power generation plants," *Procedia Eng.*, **vol. 55**, pp. 27–35.

[15] Hald J 2008, "Microstructure and long-term creep properties of 9-12% Cr steels," *Int. J. Press. Vessel. Pip.*, **vol. 85**, no. 1–2, pp. 30–37.

[16] Peng N, Ahmad B, Muhamad M and Ahadlin M 2014 "Microstructure Characterization for P92 Steels Subjected to Short Term Overheating above Critical Transformation Temperatures," *Appl. Mech. Mater.*, **vol. 625**, pp. 114–117.

[17] T. Software 2002 "THERMO-CALC & DICTRA, Computational Tools For Materials Science J-O Andersson, Thomas Helander,Lars Hdghmd, Pingfang Shi, Bo Sundman," *Calphad Comput. Coupling Phase Diagrams Thermochem.*, **vol. 26**, no. 2, pp. 273–312.

[18] Saini N, Mulik R and Mahapatra M 2019 "Prior-austenite grain refinement in P92 steel using double austenitization treatment," *Mater. Res. Express*, **vol. 6**, no. 2 pp 1-12.

[19] Wang S, Peng D, Chang L, and Hui X 2013 "Enhanced mechanical properties induced by refined heat treatment for 9Cr-0.5Mo-1.8W martensitic heat resistant steel," *Mater. Des.*, **vol. 50**, pp. 174–180.

[20] Pandey C and Mahapatra M 2016 "Effect of Heat Treatment on Microstructure and Hot Impact Toughness of Various Zones of P91 Welded Pipes," *J. Mater. Eng. Perform.*, **vol. 25**, no. 6, pp. 2195–2210.

[21] Pandey C, Mahapatra M, Kumar P and Saini N 2018 "Some studies on P91 steel and their weldments," *J. Alloys Compd.*, **vol. 743**, pp. 332–364.

[22] Ryu S and Yu J 1998 "A New Equation for the Cr Equivalent in 9 to 12 Pct Cr Steels," vol. 29A, no. June, pp. 1573–74.

[23] Wang X, Zhan L, Pan Q, Liu Z, Liu H and Tao Y 2010 "Microstructure and creep properties of high Cr resisting weld metal alloyed with Co," *J. Zhejiang Univ. A*, **vol. 11**, no. 10, pp. 756–760.

[24] Sireesha M, Albert S and Sundaresan S 2001 "Microstructure and mechanical properties of weld fusion zones in modified 9Cr-1Mo steel," *J. Mater. Eng. Perform.*, **vol. 10**, no. 3, pp. 320–330.

[25] Sakthivel T, Vasudevan M, Laha K, Parameswaran P, Chandravathi K, Selvi S, Maduraimuthu V and Mathew M 2014 "Creep rupture behavior of 9Cr-1.8W-0.5Mo-VNb (ASME grade 92) ferritic steel weld joint," *Mater. Sci. Eng. A*, **vol. 591**, pp. 111–120, 2014.

[26] Pandey C, Mahapatra M, Kumar P, Prakash K, Saini N, Thakare J and Kumar S 2019 "Study on effect of double austenitization treatment on fracture morphology tensile tested nuclear grade P92 steel," *Eng. Fail. Anal.*, **vol. 96**, pp. 158–167.

[27] Pandey C, Giri A and Mahapatra M 2016 "Effect of normalizing temperature on microstructural stability and mechanical properties of creep strength enhanced ferritic P91 steel," *Mater. Sci. Eng. A*, **vol. 657**, pp. 173–184.

[28] Wang Y and Li L 2016 "Microstructure Evolution of Fine Grained Heat Affected Zone in Type IV Failure of P91 Welds," *Weld. J.*, **vol. 95**, no. Jan, pp 27-36

[29] Abe F and Tabuchi M, "Microstructure and Creep Strength of Welds in Advanced Ferritic Power Plant Steels," *Sci.and Tech. weld. and Joining*, **vol. 9**, no. 1, pp. 22–30.

Effects of Mo content on the microstructural and mechanical properties of as-cast Ti-Mo alloys

N Moshokoa[1,2,3,a*], L Raganya[1,2,3,b], B Obadele[3,4], P Olubambi[3] and R Machaka[1,2,3]

[1]Light Metals, Manufacturing Cluster, Council for Scientific and Industrial Research, Meiring Naudé Road, Brummeria, Pretoria 0185, South Africa

[2]Titanium Centre of Competence, Manufacturing Cluster, Council for Scientific and Industrial Research, Meiring Naudé Road, Brummeria, Pretoria 0185, South Africa

[3]Centre for Nano engineering and Tribocorrosion, School of Mining, Metallurgy and Chemical Engineering University of Johannesburg, Doornfontein Campus Johannesburg, South Africa

[4]Department of Chemical, Materials and Metallurgical Engineering, Botswana International University of Science and Technology, Palapye, Botswana

[a*]Abigailthabi@gmail.com, [b]LRaganya@csir.co.za

Abstract: The effects of Mo content on the mechanical and microstructural properties of as-cast Ti-10Mo wt.% and Ti-15Mo wt.% alloys were measured in order to evaluate their possible use in biomedical applications. The microstructure and phase analysis of as-cast specimens were studied using an optical microscopy and X-ray diffraction (XRD). The micro-Vickers hardness properties of the alloys were examined using Vickers equipment and the elastic modulus was determined using the tensile test. Results indicated that the Mo content influenced the structure-property relationship. The relative density increased significantly with the addition of Mo, the optical microscopy results showed that CP Ti was characterised with an α'-phase lamellar structure and with the addition of Mo, the microstructure composed of β phase with equiaxed grains of different sizes with sub-grain boundaries were observed. The measured XRD patterns showed that the CP Ti consisted of α'-phase and with the addition of Mo, the resultant alloys consisted of the β-phase. The micro-Vickers hardness and elastic modulus increased significantly after the addition of 10 wt.% Mo and then decreased significantly when the Mo content was raised to15 wt.%.

1. Introduction

Commercial pure titanium (CP Ti) and Ti6Al4V alloy are state-of-the-art implant materials with excellent specific strength, biocompatibility and low elastic modulus as compared to stainless steel and cobalt alloys [1–3]. However, more recently published reports have highlighted increasingly alarming risks associated with CP Ti and Ti6Al4V alloys implants because of implant failure and biomechanical incompatibilities [4–7].

Titanium implants are typically directly fixed to the host bone material during orthopaedic surgery. Ideally, the mechanical interlock of the bone tissue and implant surface at the bone-implant interface

zone is expected to strengthen with no other adverse. However, CP Ti and Ti6Al4V alloys suffer from biomechanical incompatibility due to their elastic modulus (~ 110 GPa) that is significantly higher than that of the cortical bone (10 ~ 30 GPa) [5,8,9]. The elastic moduli mismatch may eventually lead to severe stress shielding and ultimate implant failure due to insufficient integration [6,7,10].
In addition to mechanical failure, the reported release of V- and Al-ion from Ti6Al4V in the body environment further limits the biomedical use of Ti6Al4V and similar alloys. Reported adverse health issues associated with toxic ions release include Alzheimer and neuropathy diseases [4,5] and allergic reactions.
Broad research has been focused on the development and design of metastable β-type Ti alloys that are non-toxic and have a low young's modulus. The development and design of non-toxic β-type Ti alloys with modulus comparable to the cortical bone is also a contemporary research topic in the area of advanced biomedical materials. According to Mohammed, Khan and Siddiquee [11], and other recently published reports [3,7,12,13], β-type Ti alloys typically possess young modulus values as low as 50-90 GPa (significantly lower than that of Ti6Al4V). Consequently Recent studies have shown that β-type Ti alloys with alloying elements such as tungsten (W), niobium (Nb), zirconium (Zr), tantalum (Ta), tin (Sn) and molybdenum (Mo) have a low elastic modulus, biocompatibility, relatively good mechanical properties and corrosion resistance [12,14]. This article is limited to the study of the microstructure, mechanical properties of Ti-Mo binary alloys with Mo content of up to 15 wt.% after Ho *et al.* [15] and Oliveira *et al.* [16].
In this work, the alloy compositional design was carried out using three different methods: firstly, the Mo equivalent (Mo_{eq}) value method after Gordin *et al.* [17]; secondly, the average electron concentration (*e/a*) after Ikehata *et al.* [18]; and thirdly, the beta stabilizing index (K_β) after Moiseev and Antipov [19]. The Mo_{eq} value, calculated according to Eq. (1) below, is defined as the sum of the weight averages of the elements (wt.%) present in an alloy, according and is increasingly used to determine the stability of the β-phase in Ti alloys [17,20,21]. According to Bania, a Mo_{eq} value of greater than 10 wt.% is needed to stabilize the isomorphous β phase in Ti-Mo alloys.

$$\mathbf{(Mo_{eq})_B = 1.0\ Mo + 0.67\ V + 0.44\ W + 0.28\ Nb + 0.22\ Ta + 2.9\ Fe + 1.6\ Cr + 0.77\ Cu + 1.11\ Ni + 1.43\ Co + 1.54\ Mn - 1.0\ Al.} \quad \text{Eq. (1)}$$

K_β is yet another approach used to determine the β-phase stability in Ti alloys [19,22]. K_β is given in Eq. (2) as:

$$K_\beta = \frac{C_i}{\beta_{ci}} \quad \text{Eq. (2)}$$

where C_i is the composition of β stabilizing element and β_{Ci} is the critical beta concentration or the concentration at which β phase is expected to be stable [19].
If the value of K_β lies in the 1.0 to 1.5 range ($1.0 < K_\beta < 1.5$), then the β-phase in the alloy was found to be metastable, beyond this value the β phase found to be stable [22]. The *e/a* ratio is used to predict the formation of the athermal omega-phase in Ti alloys [18,23]. The ratio is the average number of the valence electrons in each atom of Ti-Mo binary alloys. According to Ikehata *et al.*, the formation of the athermal omega-phase is at its maximum when the *e/a* ratio is equal to 4.13 and at its minimum when the *e/a* ratio is 4.30 and above this minimum point the β-phase becomes dominant [18]. Ho *et al.* [15] and Oliveira *et al.* studied a series of binary arc melted Ti-Mo alloys using microstructural analysis and mechanical properties [15,16]. Their experimental results showed that the microstructure was strongly dependent on the molybdenum content, Ti-15Mo alloy had a lower elastic modulus than that of CP Ti, but was higher than that of Ti-7.5Mo alloy. The current study aims to investigate the effects of Mo content on the mechanical and microstructural properties of as-cast Ti-Mo alloys.

2. Experimental Methods

2.1 Material preparation

Ti-Mo ingots were fabricated by vacuum arc melting furnace technique using high purity CP Ti (99.9% and ASP of 28.07 μm) and high purity Mo (99.9% and ASP of 16.84 μm) powders as raw materials. Each ingot was re-melted at least three times and flipped between each melting to ensure chemical homogeneity. The nominal composition of the Ti-10Mo and Ti-15Mo alloys which are denoted as alloys AA and AB, respectively are shown in Table 1. The calculated values of Mo_{eq}, *e/a* ratio and K_β stabilizing of the alloys are also listed in Table 1.

Table 1: Composition of Ti-Mo binary alloys.

Alloy ID	Composition [wt.%]	Mo_{eq} [wt. %]	*e/a* ratio	K_β
CP Ti	100 CP Ti	0	4	0
AA	Ti-10.02 Mo	10.00	4.24	1.0
AB	Ti-15.05 Mo	15.00	4.3	1.51

2.2 Characterization of as-cast ingots

2.2.1 Microstructural characterization

The microstructures of the as-cast Ti-Mo alloys were observed by means of the Leica DMI500 M optical microscope (Leica GmbH, Germany). Samples for optical microscopy observation were prepared by following the metallographic preparation technique and etched using Kroll solution (85 ml distilled water, 5 ml HNO_3 and 10 ml HF). The apparent densities were determined by Archimedes principle (OHAU). XRD was used for phase identification. The diffraction patterns were measured at room temperature in the Bragg-Brenton geometry (continuous scanning at a rate of 0.01°/s) using the PW3064/60 XPERT PRO diffractometer system (PANalytical BV, The Netherlands) and Cu Kα (45 kV, 40 mA, $\lambda_1 = 0.1540598$ nm, $\lambda_2 = 1.544426$ nm and $5° \leq 2\theta \leq 90°$) radiation conditions. The diffraction patterns were analysed using X'Pert HighScore Plus computer software (PANalytical BV, The Netherlands).

2.3 Mechanical testing

Micro-hardness was measured using a Vickers micro-indenter (FM-700, Japan) with a load of 500 g for 15 s on the sample. Ten different measurements were taken and the minimum and maximum values were discarded before averaging. Specimens for tensile tests were cut using an electro-discharge machine. The dimensions of the tensile specimens were 3.0 mm in thickness and 4.0 mm in width, and a length of 35 mm; the specimens had a gauge of 3 mm × 4 mm × 30 mm. The Instron™ 1342 tensile test machine was utilised with an appropriate extensometer to record the stress-strain response of the alloys until fracture. The Young's modulus of the studied alloy was analysed therefrom.

3. Results and Discussion

3.1 Microstructural Characterization

The effect of Mo contents on the phase constitution and microstructure of the as-cast ingots is presented in Figures 1 and 2. Figure 2 (a - c) and (d - f) specifically combines the OM and SEM micrographs of the as-cast alloys, in that order.

Figure 1 illustrates the XRD spectra of the as-cast CP Ti, Ti-10Mo and Ti-15Mo alloys. The as-cast CP Ti XRD profiles are characterised by typical fully α' hexagonal-phase with peaks matching those attained by Ho *et al.* [15] and Oliveira [16] in the as-cast condition. From Figure 1, the addition of 10 wt.% (alloy AA) of Mo into CP Ti, resulted in the retention of the β-phase and nucleation of a small volume fraction of the α'' phase. Further, the addition of 15 wt.% Mo (alloy AB) results in the retention of the majority β-phase and the suppression of the formation of α'' phase. These results are in accordance with the experimental findings of Oliveira [16].

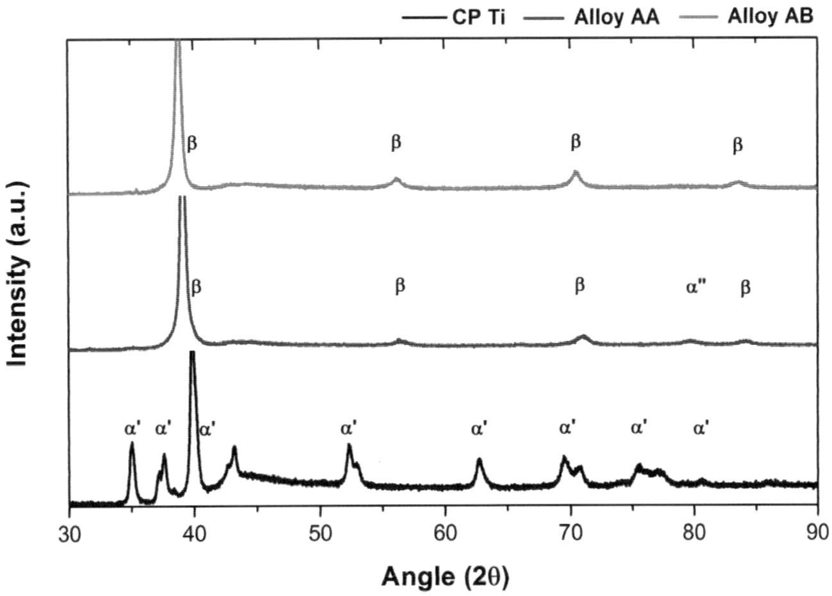

Figure 1: The XRD patterns of as-cast CP Ti, alloy AA, and alloy AB.

The OM as-cast CP Ti micrograph Figures 2(a) is characterised by a homogeneous and typical metastable 'feather-like' microstructure consisting of coarse plates with partitioned regions filled with small secondary plates and matches the one obtained by Bania [21]. The same microstructure was obtained after SEM characterization as seen in Figure 2(d). OM micrographs shown in Figures 2(b) and 2(c) confirm that the addition of 10 wt.% Mo (alloy AA) to CP Ti plays an important role in the formation of sub-grains of martensitic α" phase and equiaxed β phase microstructure with grains of various sizes. Observations of the suppression of other phases with increasing Mo content further to 15 wt.% (alloy AB) are consistent with findings presented by Chen *et al.* [24] and by Zhang *et al.* [25] which also investigated as-cast alloys. Figure 2(e) is the SEM micrographs of Ti-10Mo wt.%, also show equiaxed β-grains with sub-grains inside the grains and acicular α" structures nucleating on the β grain boundaries. Figure 2(f), SEM micrographs of Ti-15 wt.% Mo, on the other hand, revealed a microstructure with β equiaxed grains and but not the nucleation of the sub-grain microstructures.
Increasing Mo content appears to stabilize the β-Ti phase and the stability of the β-phase is in line with the predictive models (Table 1) where the K_β index of both alloys AA and AB ranges from 1.0 - 1.5.

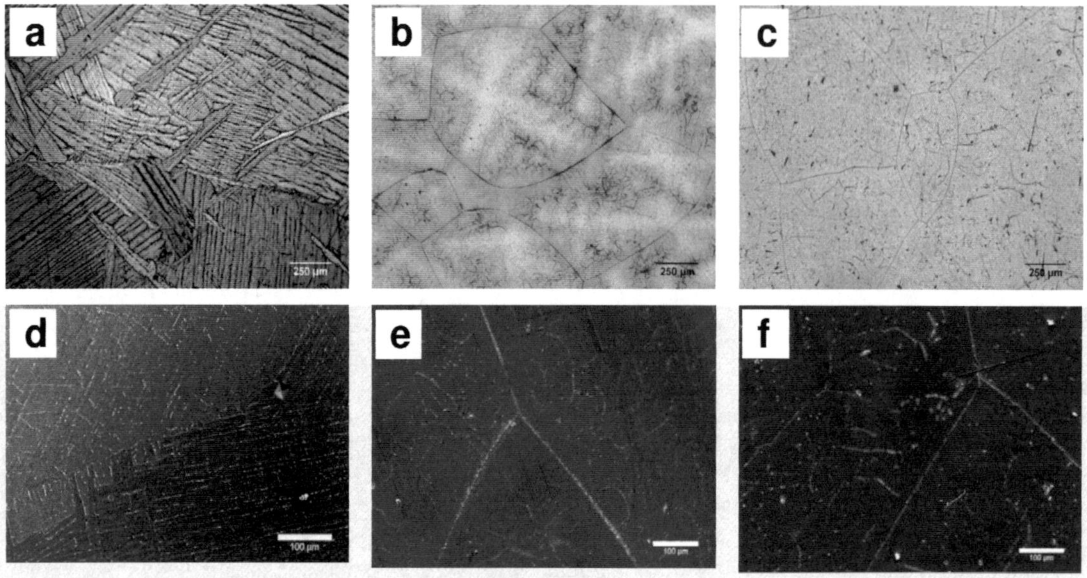

Figure 2: The optical micrographs of CP Ti (a), alloy AA (b), and alloy AB (c) respectively. The corresponding BSE SEM micrographs of CP Ti (d), alloy AA (e), and alloy AB (f).

3.2 Mechanical Properties

The relative density values of as-cast specimens are shown in Figure 3(a) below. The addition of Mo to CP Ti tends to significantly increase the relative density of alloys AA and AB. Figure 3(b) also demonstrates the micro-hardness of as-cast CP Ti, alloy AA and alloy AB. The micro-hardness properties values of the Ti-Mo alloys are comparable with findings prior reported by Ho *et al.* [15]. While the addition of Mo is expected to significantly increase in the micro-hardness of the alloys due to the solid solution effect [26], the interpretation of the observed increase is rather peculiar. According to Bagariatsikii *et al.* [27], the observed increase in the micro-hardness with increasing Mo content is due to the precipitation of the omega (ω) phase during cooling.

XRD and SEM analyses (in Figures 1 and 2) do not hint of the existence of this ω-phase (possibly hiving rise to the hardening effect) tentatively due to the phase's nanometer-sized domains or small volume fraction well below the detection limit of the conventional XRD [28]. According to Ikehata *et al.* [18], *e/a* ratio values (between 4.13 and 4.30) predict the possible existence of the ω-phase; alloy AA's *e/a* ratio predicts the presence of ω-phase while AB shows that the β phase is metastable, Moiseev and Antipov reported that the K_β stabilizing index of alloys AA and AB ranges from 1.0 - 1.5 and this shows that β is metastable at those alloys (Table 1) [19].

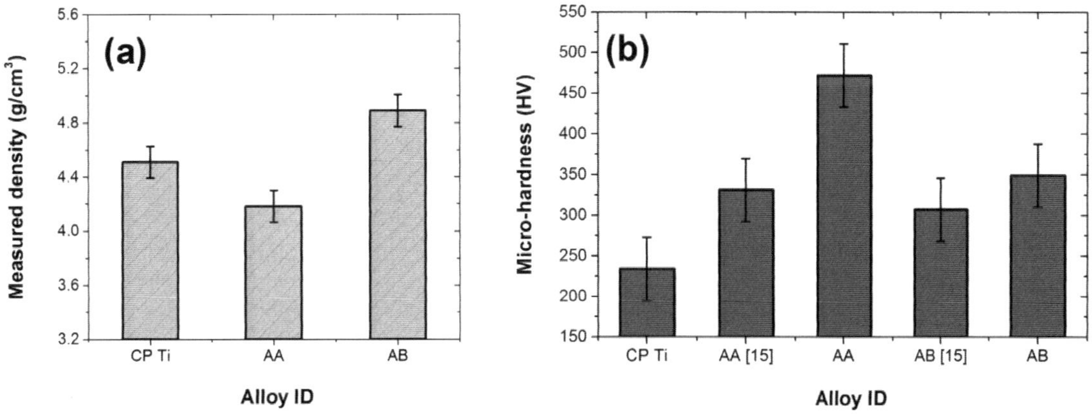

Figure 3: (a) Relative density of as-cast CP Ti, AA and AB and (b) Micro-hardness of as-cast CP Ti, AA and AB in comparison to prior reported values for alloys AA and AB [15].

The presence of the ω-phase in the Ti-Mo alloy is characterised by increase the mechanical properties [5,29]. The elastic moduli of as-cast CP Ti, alloy AA and alloy AB are illustrated in Figure 4; likewise, an increase in the Mo content is associated with a significant decrease in the measured elastic modulus that was attributed to the stabilization of the β-phase and suppression of the ω phase. The elastic modulus values obtained from this study are consistent with those obtained and reported by Gabriel, Nunes and Soares [30] and Ho [31] in the literature.

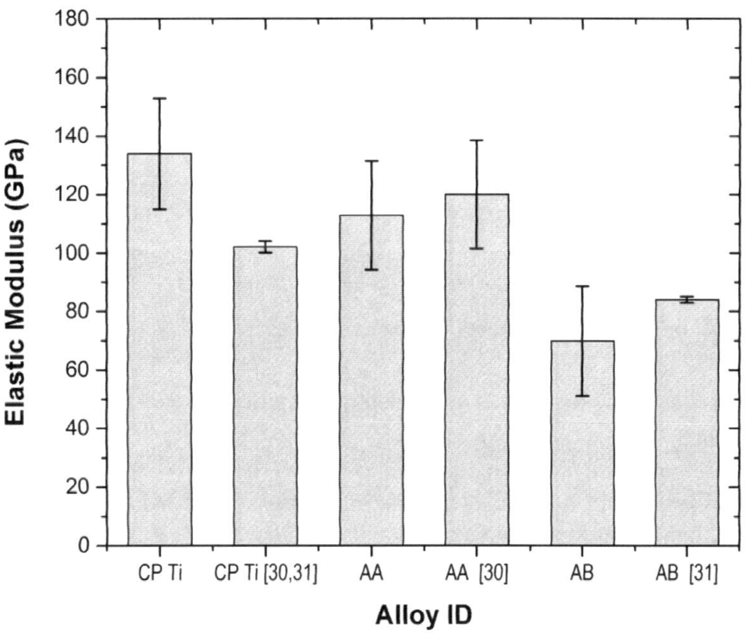

Figure 4: The Elastic modulus of as-cast CP Ti, AA and AB

4. Conclusions

Based on the density, microstructures, phase constituent and mechanical properties of the studied alloys, the following conclusions were drawn:

- The K_β stability index revealed that both alloys AA and AB had a beta stabilizing index in the range of 1.0-1.5 which indicated the formation of metastable β phase, while the *e/a* ratio predicated the ω-phase nucleation in alloy AA and its suppression in Alloy AB.
- The microstructure of CP Ti showed lamellar; with the addition of Mo, dendrites formed within the beta grains. As the Mo content increased, the dendritic structure decreased and the β equiaxed structure became the dominated phase.
- The micro-hardness of the binary Ti-Mo alloys was found to be higher than that of CP Ti. The addition of 10 wt.% Mo to CP Ti increased the hardness significantly due to the presence of the ω-phase, the hardness significantly decreased after the addition 15 wt.% Mo.
- The elastic modulus of CP Ti was higher than that of binary Ti-Mo alloys. With the addition of Mo, the elastic modulus decreased until Mo content of 10 wt.% the increase in the Mo content is due to the presence of the ω-phase as they have a tendency of increasing the modulus of elasticity and the strength. The elastic modulus then decreased as the Mo content increased to 15 wt.%.

5. Acknowledgement

I would like to acknowledge the NRF Thuthuka fund and the CSIR (Light Metals) for generously funding this project. My supervisors Ms Lerato Raganya, Prof Ronald Machaka, Prof Olubambi and Dr Babatunde Obadele for their guidance and mentorship. Mr Edison Mahuma from Mintek, Ms Pfarelo Daswa (CSIR) , Mr Steven Masete (CSIR) and Dr Kofi and the Mineral science lab (University of Pretoria)

6. Reference

[1] Donachie M J 2000 *Titanium: A Technical Guide.* (ASM International)

[2] Niinomi M 2002 Recent metallic materials for biomedical applications *Metall. Mater. Trans. A* **33** 477–86

[3] Banerjee D and Williams J C 2013 Perspectives on titanium science and technology *Acta Mater.* **61** 844–79

[4] Eisenbarth E, Velten D, Müller M, Thull R and Breme J 2004 Biocompatibility of β-stabilizing elements of titanium alloys *Biomaterials* **25** 5705–13

[5] Nag S, Banerjee R and Fraser H L 2005 Microstructural evolution and strengthening mechanisms in Ti–Nb–Zr–Ta, Ti–Mo–Zr–Fe and Ti–15Mo biocompatible alloys *Mater. Sci. Eng. C* **25** 357–62

[6] Niinomi M and Nakai M 2011 Titanium-Based Biomaterials for Preventing Stress Shielding between Implant Devices and Bone **2011**

[7] Kuroda D, Niinomi M, Morinaga M, Kato Y and Yashiro T 1998 Design and mechanical properties of new β type titanium alloys for implant materials *Mater. Sci. Eng. A* **243** 244–9

[8] Madonsela J, Matizamhuka W, Machaka R, Shongwe B and Yamamoto A 2018 Sintering Performance and Corrosion behavior of biomedical Ti-24Nb-4Zr-8Sn alloy produced using SPS in various simulated body solutions *IOP Conf. Ser. Mater. Sci. Eng.* **430** 012037

[9] Mantshiu A, Matizamhuka W, Yamamoto A, Shongwe B and Machaka R 2018 Cytocompatibility evaluation of nano-sintered Ti-15Zr-4Nb-2Ta-0.2Pd alloy produced by spark plasma sintering technique *IOP Conf. Ser. Mater. Sci. Eng.* **430** 012036

[10] T D K, Kawasaki H, Yamamoto A, Hiromoto S and Hanawa T 2005 Mechanical properties and microstructures of new Ti – Fe – Ta and Ti – Fe – Ta – Zr system alloys **25** 312–20

[11] Mohammed M T, Khan Z A and Siddiquee A N 2014 Beta Titanium Alloys: The Lowest Elastic Modulus for Biomedical Applications: A Review Surface Modifications through FSP View project MACHINING View project *Int. J. Chem. Nucl. Metall. Mater. Eng.* **8** 726–31

[12] Kolli R and Devaraj A 2018 A Review of Metastable Beta Titanium Alloys *Metals (Basel).* **8**

506
[13] Geetha M, Singh A K, Asokamani R and Gogia A K 2009 Ti based biomaterials, the ultimate choice for orthopaedic implants–a review *Prog. Mater. Sci.* **54** 397–425
[14] Welbourne T M, Neck H and Meyer G D 2012 The entrepreneurial growth ceiling: Using people and innovation to mitigate risk and break through the growth ceiling in initial public offerings *Manag. Decis.* **50** 778–96
[15] Ho W F, Ju C P and Chern Lin J H 1999 Structure and properties of cast binary Ti-Mo alloys *Biomaterials* **20** 2115–22
[16] Oliveira N T C C, Aleixo G, Caram R and Guastaldi A C 2007 Development of Ti–Mo alloys for biomedical applications: Microstructure and electrochemical characterization *Mater. Sci. Eng. A* **452–453** 727–31
[17] Gordin D M, Gloriant T, Texier G, Thibon I, Ansel D, Duval J L and Nagel M D 2004 Development of a -type Ti–12Mo–5Ta alloy for biomedical applications: cytocompatibility and metallurgical aspects *J. Mater. Sci. Mater. Med.* **15** 885–91
[18] Ikehata H, Nagasako N, Furuta T, Fukumoto A, Miwa K and Saito T 2004 First-principles calculations for development of low elastic modulus Ti alloys *Phys. Rev. B* **70** 174113
[19] Moiseev V N and Antipov A I 1995 Aluminium effect on beta-phase stability in beta-titanium alloys *Metalloved. i Termicheskaya Obrab. Met.* 30–5
[20] Ho W F 2008 Effect of omega phase on mechanical properties of Ti-Mo alloys for biomedical applications *J. Med. Biol. Eng.* **28** 47–51
[21] Bania P J 1994 Beta titanium alloys and their role in the titanium industry *JOM* **46** 16–9
[22] Nnamchi P S, Obayi C S, Todd I and Rainforth M W 2016 Mechanical and electrochemical characterisation of new Ti-Mo-Nb-Zr alloys for biomedical applications *J. Mech. Behav. Biomed. Mater.* **60** 68–77
[23] Niinomi M and Niinomi M 2003 Recent research and development in titanium alloys for biomedical applications and healthcare goods Recent research and development in titanium alloys for biomedical applications and healthcare goods **6996**
[24] Chen Y Y, Xu L J, Liu Z G, Kong F T and Chen Z Y 2006 Microstructures and properties of titanium alloys Ti-Mo for dental use *Trans. Nonferrous Met. Soc. China (English Ed.* **16** 121–36
[25] Zhang L, Wang K, Xu L, Xiao S and Chen Y 2015 Effect of Nb addition on microstructure , mechanical properties and castability of ü -type TiíMo alloys *Trans. Nonferrous Met. Soc. China* **25** 2214–20
[26] Yu H-N, Hsu H-C, Wu S-C, Hsu S-K and Ho W-F 2017 Structure and Mechanical Properties of As-Cast Ti–5Sn–xMo Alloys *Materials (Basel).* **10** 458
[27] Welding T, Review C, Iii B, Co K, Wi K, Co G A, Division C, Force A, Aeronautical W, Oh W and The I Tages Are Derived: Detail Part Cost and Weight Are Reduced (I.E., They Are Near-Net-Shape, Unitized Structures)', Overall Structural Size Is Not Limited
[28] Sabeena M, Murugesan S, Mythili R, Sinha A K, Singh M N, Vijayalakshmi M and Deb S K 2015 Studies on ω Phase Formation in Ti-Mo Alloys Using Synchrotron XRD *Trans. Indian Inst. Met.* **68** 1–6
[29] Hickman B S 1969 The formation of omega phase in titanium and zirconium alloys: A review *J. Mater. Sci.* **4** 554–63
[30] Gabriel S B, Nunes C A and Soares G de A 2008 Production, Microstructural Characterization and Mechanical Properties of As-Cast Ti-10Mo-xNb Alloys *Artif. Organs* **32** 299–304
[31] Ho W-F 2008 A comparison of tensile properties and corrosion behavior of cast Ti–7.5Mo with c.p. Ti, Ti–15Mo and Ti–6Al–4V alloys *J. Alloys Compd.* **464** 580–3

SiC and Al surface cladding of Ti-6Al-4V for improved wear properties – The binary advantage

U O Okoli, I O Otunniyi, I D Adebiyi

The Vaal University of Technology, Department of Metallurgical Engineering.Private Bag X021, Andries Potgieter Blvd, Vanderbijlpark, South Africa.

E-mail: ugomartins860@yahoo.com

ABSTRACT: Ti-6Al-4V enjoys high specific strength, toughness and excellent corrosion resistance, but suffers high friction coefficient, severe adhesive wear and sensitivity to fretting wear. Laser surface cladding is one of the surface modification techniques used for improving on these demerits. Using a 4.4Kw continuous wave Nd-YAG laser processor, the surface cladding of as received Ti-6Al-4V alloy was done with mixtures of SiC and Al powders in different ratios. The SiC powder can serve as reinforcing particles and helps improve surface wear properties in resulting surface metal matrix composite while aluminium improves on the wetting between the metal and the reinforcement SiC phase. The success of this, however, depends on the careful optimisation of the powder ratio, which dictates the resulting phases and microstructural evolutions. Study of the cladding with scanning electron microscope (energy dispersive spectroscopy), optical microscope and x-ray diffraction revealed formation of intermetallic phases such as $AlSiTi_2$, $Ti_7Al_5Si_{12}$, $TiSi_2$, $AlSi_3Ti_2$, Al_6Ti_{19}, Ti_2VAl, $AlTi_3$, $AlCTi_2$, Ti_3SiC_2 and Ti_5Si_4, which should enhance micro-hardness and tribological properties of the surface relative to the bulk as-received alloy. 90:10 weight percentage ratio of SiC and Al powder mixture gave the best microhardness with a value of 1169.95 Hv, compared to 323.67 Hv of the as-received, and mass loss under abrasive wear tests was only 40% of the as received. The morphology obtained for the abraded surfaces is also indicative of this improvement. In carefully optimised proportions and laser condition, the binary combination of Al and SiC powders suffice for well improved Ti-6Al-4V surface properties.

1. Introduction

Some properties of Ti-6Al-4V that make it an invaluable material in the aerospace industry are; high specific strength (strength to density ratio), toughness, and excellent corrosion resistance [1]. Ti-6Al-4V is a preferred titanium alloy because of its right balance of characteristics which includes strength, creep characteristics, fracture toughness, and the ability to retain its mechanical properties at elevated temperatures [2, 3]. The Ti-6Al-4V alloy would have enjoyed better patronage in more diverse applications if not for its poor tribological properties. Two main factors cause the poor tribology of Ti-6Al-4V: one, weak resistance to plastic shearing and poor work hardening; causing, the alloy to react poorly to adhesion, abrasion and delamination. Two, the oxides formed under high-temperature flashes under friction, has a low protective capacity, as this oxide is easily removed by spalling, and exposes a new surface to wear. The surface matrix also gets embrittled by dissolved oxygen from the atmosphere, which tends to reduce its mechanical and tribological properties [4]. Laser cladding is one of the means to alleviate these surface challenges of Ti-6Al-4V. Laser cladding utilises high power density available from a focused laser source to heat and melt the substrate's surface while injecting alloying elements into the melt pool. The process often results in the formation of an alloyed zone confined to a shallow depth from the surface, it also results in the formation of a unique metallurgical bond with beneficial microstructures and high concentration of critical elements, due to rapid cooling. A range of unique microstructures can be achieved by the processing method. These structures primarily depend on the laser processing parameters and powder/alloy compositions [5-7]. Silicon carbide displays exemplary configurations of physical and mechanical properties; it has a relatively low density of 3.21g/cm^3 at 25 ^{0}C making it attractive to the aerospace industries, excellent oxidation resistance and retention of strength at high temperatures with fracture stress of about 835 MPa at 1527 ^{0}C [8, 9]. Aluminium alleviates the challenge of poor interfacial bonding strength encountered in metal matrix composites (MMC), in this case between the Ti-6Al-4V matrix and the SiC reinforcement phase. This, however, requires careful optimisation of powder ratios and laser power conditions to ensure the evolution of microstructures favourable for improved properties. The focus of this work is exploring varying aluminium and silicon carbide powders ratios after first optimising the laser power and operation. The report here focuses on the powder ratio optimisation for good wetting of the laser melt pool during laser cladding. The coating depth, microstructural evolutions, microhardness and enhanced abrasive wear resistance were studied.

2. Experimental

2.1 Materials and Methods

Ti-6Al-4V is the substrate in this work and was sourced from Zhejiang Shenji Titanium Industry Co Ltd, in annealed conditions, with the dimensions of 70 mm x 70 mm x 5 mm; the nominal composition of the alloy is 6.01 wt% Al, 3.84 wt% V, 0.30 wt% Fe 0.15 wt% Si,0.10 wt% C, 0.15 wt% O, 0.15 wt% N, Ti the balance. The Ti-6Al-4V substrate was sandblasted (to increase the intensity of the laser radiation at the surface) [10] and afterwards rinsed with acetone. Silicon carbide of different aluminium additions was used for laser cladding. The powders were sourced from TLS Technik GmBH. The particle size distribution (PSD) analysis indicates that the average powder particle size of the SiC powder 106μ was while Al was 75μ. Table 1 shows the powder compositions of the four powder mixes used in the investigation.

Table 1. Power mix variation and compositions.

Sample	Mix	Cladded surface	Powder Composition
1	Mix1 (Single)	CS1	100 wt. % SiC
2	Mix2 (Binary)	CS2	90 wt. % SiC + 10 wt. % Al
3	Mix3 (Binary)	CS3	80 wt. % SiC + 20 wt. % Al
4	Mix4 (Binary)	CS4	70 wt. % SiC + 30 wt. %Al

A 4.4 kW rofin Sinar continuous-wave Nd: YAG (neodymium-doped yttrium aluminium garnet) laser equipped with an off-axis nozzle for powder feeding and a fibre-optic beam delivery system, was used for laser surface cladding of the substrate. Optimal laser parameters were established before the laser alloying. The optimisation process involved several trial runs, in which several laser parameters which include power, scan speed, beam diameter, the powder feed rate was varied to determine the optimum parameter, that is the parameter that yielded the best coating. Determination of the best coating was done on the spot after each trial run, by examining the quality of the finish. Table 2 is the optimum laser processing parameters.

Table 2. Optimum laser cladding parameters.

Powder Composition	Power (Kw)	Scan Speed(m/sec)	Beam Diameter(mm)	Powder feed rate(g/pm)	Argon gas flow rate (L/min)	Track overlap percentage
100 wt. % SiC	1.2	0.8	4	5	2	50
90 wt. % SiC + 10 wt. % Al	1.2	0.8	4	5	2	50
80 wt. % SiC + 20 wt. % Al	1.2	0.8	4	5	2	50
70 wt. % SiC + 30 wt. %Al	1.2	0.8	4	5	2	50

2.2 Characterisation of As Received and Cladded Surfaces

The as-received Ti-6Al-4V and cladded samples were cross-sectioned using a Struers Discotom-2 cutting machine which has a cut-off wheel; The samples were ground using SiC paper ranging from 200 to 1200 grit and then polished using diamond impregnated cloths ranging from 3 to 1 micron; Finally, the samples were cleaned, dried and etched with krolls reagent (6 ml HNO3 + 2 ml HF + 92 ml H20) for microscopic observation and characterisations.

Scanning electron microscope (SEM) equipped with energy-dispersive x-ray spectroscopy (EDS) was employed for the microstructural analysis, with phase constituents of the alloyed surface characterised using x-ray diffractometer (XRD). The scan was taken between 5^0 and 90^0 2 thetas (2Θ), and the phases present were identified using Deffrac Eva software with an in-built international centre for diffraction data (ICSD) database. The cladded zone, as shown in Figure 1, was subjected to microstructural analysis. Characterisation of the cladded zone was done in other to have a perfect view of the microstructural evolutions in the alloyed zone.

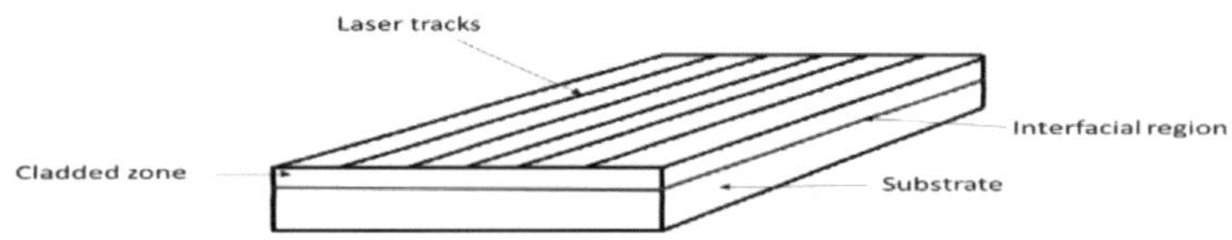

Figure 1. Schematic illustration laser

2.3 Microhardness and Tribological analysis

The specimen was tested for surface hardness using a micro Vickers hardness tester according to ASTM E384 standards: microhardness measurements along the depth direction for all the cladded samples, with a load of 1000 grams, as schematically illustrated in Figure 2. A depth direction indentation from the surface through the substrate was done, Average of 4 indentations were taken and reported as the average microhardness of the cladded surfaces.

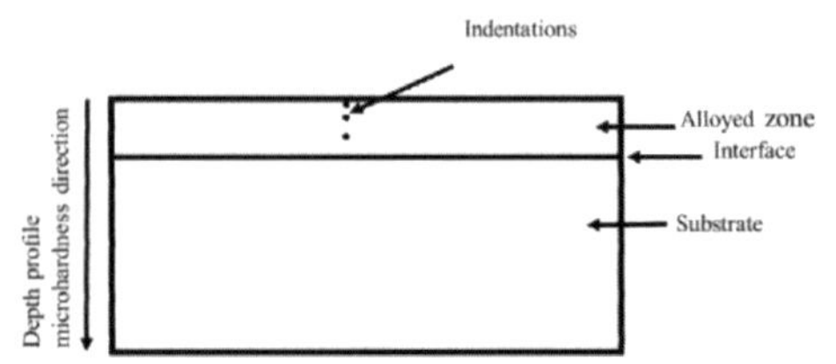

Figure 2. Schematic Illustration of microhardness along depth direction of coatings.

Three bodies abrasive wear tests were carried out to determine the wear rate of the samples under severe abrasive conditions. Silica sand with a particle size range of about 0.3–0.65mm was used as the third body, at a load of 1500N at a constant speed of 100- rpm. A relatively high load (1500N) was used to ensure severe and proper abrasion of the cladded surface and substrate with the stainless-steel wheel. Figure 3 is a schematic illustration of the operation. The wear mechanisms of the best surface (in terms of coating depth, homogeneity and microhardness) was compared with that of the substrate at the same testing conditions.

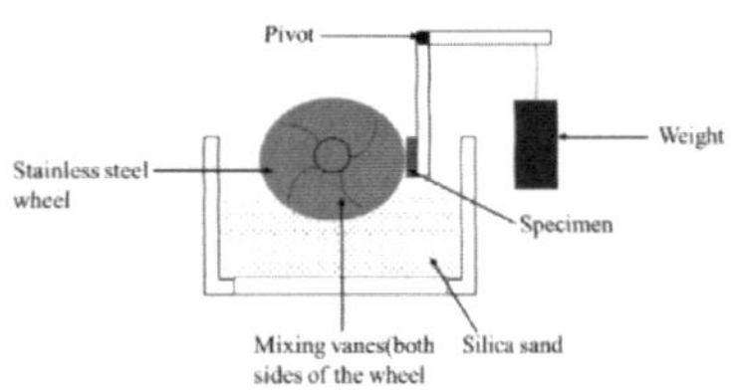

Figure 3. Schematic illustration of the three-bodies abrasive wear set-up.

The samples were weighed before and after abrading using a 1 mg precision electronic balance. The samples were abraded thrice, each time for five minutes, after which they were rinsed in acetone to ensure removal of all wear debris. Average wear rate per minute over the 15 minutes total abrasion time was calculated for the samples. Evaluation of the worn surface to determine the morphology of the wear track was done using the Scanning electron microscope.

3. Results and Discussions

3.1 Coating Thickness

Figure 4 depicts the average coating thickness of the laser cladded sample surfaces, which was about 712, 1047.81, 1087.35 and 909.42 microns under different conditions. The variations in the coating depths result from degrees of interactions in the melt pool, which also dictates the metallurgical bond between the coating and the substrate. The formation of phases in the coating layer depends on the wettability of the contact surface, nucleation and growth properties of each phase, chemical reactions between interacting phases and diffusion rates of the reacting species through the interlayer[11]. Wettability could be defined as the propensity of a fluid to spread on or stick to a solid surface amid other immiscible fluids. Wettability points to the interaction between the fluid and solid phases. The contact angle of fluid with the solid phase defines wettability. Laser scans develop a large thermal gradient between the centre and the edge of the melt pool at the surface, temperature gradient gives rise to a corresponding variation of the surface tension between the centre and the edge of the melt pool, consequently, thermocapillary forces for fluid flow are induced, known as Marangoni flow [12]. The Marangoni flow in combination with poor wetting conditions in the melt pool causes the restriction of the melt from spreading outwards on the underlying surface hence preventing proper interaction between the substrate and the pool [13]. The SiC powder has limited wettability with the Ti-6Al-4V melt, and this restricts depth of the coating [14].

The coating depth recorded in the binary powder coatings was much higher than that of the single powder cladding (CS1, 100 wt% SiC) which was 712 microns. The coating depth enhancement could be attributed to enhanced wetting of the SiC in the melt pool during the laser cladding process. The coating depth is beneficial to the underlying substrate as it provides cover and extends the lifetime of the substrates in service. Aluminium alleviates the challenge of interfacial bonding strength encountered in metal matrix composites (MMC), between the matrix and reinforcement phase, the addition of aluminium improves the poor wetting conditions between ceramic powders in the metal matrix. The addition of aluminium significantly improved the wetting of SiC. Aluminium wet SiC even at the vicinity of its melting point, the wettability the carbides are improved because of the formation of new phases [15]. Good wetting in the melt pool, enhanced the outward spreading of the melt to the underlying surface, thus expediated proper interaction between the substrate and the pool [13]. Proper interaction between the substrate and the pool led to the formation of new intermetallic phases, adequate surface bonding between the coating and the substrate and thicker coating. The evidence of the enhanced wetting of the melt pool is shown in the optical micrographs and XRD data of the single and binary powder cladded surfaces. However, a thicker coating may not immediately imply better wear behaviour and protection. The relevant characterizations for informed decisions are discussed following.

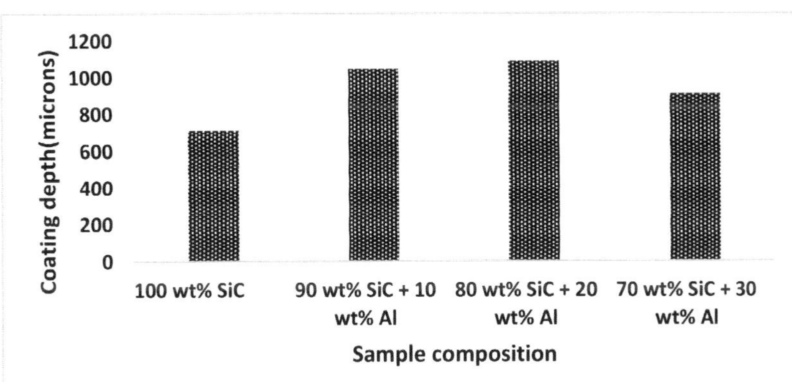

Figure 4. Average coating depths of Laser cladded Ti-6Al-4V using binary and single powders of SiC and Al.

3.2 Phase Evolution

Figures 5 and 6 are the optical and electron micrographs of the laser cladded surfaces, respectively. The optical micrographs of the laser cladded surfaces in Figure 5 show the coatings samples are composed of several distinct phases. CS1 had only the dark irregular phase noticeable, while the binary powder coatings witnessed some other remarkable transformations in the coating matrix. These further matrix transformations evidence the enhanced wetting of the binary powders in the melt pool. CS2 shows the residual SiC in a darker matrix relative to the other cladded surfaces. At higher aluminium, (CS3 and 4) the dark matrix phase became remarkably lighter with other distinct phase evolutions.CS4 had a needle-like matrix, and a threadlike matrix evolves in the alloy matrix (Figure 5d). With a distinct evolution of different phases at higher aluminium proportion, the coating properties would have changed remarkably. The SEM/EDS study gave further information on the composition of these phases. The SEM shows the presence of the residual SiC phase in a dark matrix. The irregular phases are uniformly distributed, and they can be seen to have good interfacial bonding with the Ti-6Al-4V matrix. A uniform cladding layer with no cracks was obtained for all powder mix ratios. From the EDS data (Figure 7, spectrum 1), the residual SiC phase consists of silicon and carbon, confirming that the phase is residual SiC. Dissolution of SiC and the substrate alloy gives rise to a melt pool where Si, Ti and C are the main components in different quantities based on the powder mix ratios. The resultant phase composition is heavily dependent on the prevailing thermodynamic of the elements [16]. Comparing Figure 6b, 6c and 6d, it was apparent that an increase in the percentage of aluminium powder in the matrix increased the darker matrix under the SEM. This indicates the presence of a phase with lower secondary electron intensity. In optical micrograph, this appears as a brighter matrix. The EDS of the CS4 shows elemental compositions of the needle-like matrix, which had close to 50 wt% aluminium (Figure 7, spectrum 3). It is expected that this huge quantity of aluminium will affect the service properties of this sample in some definite ways.

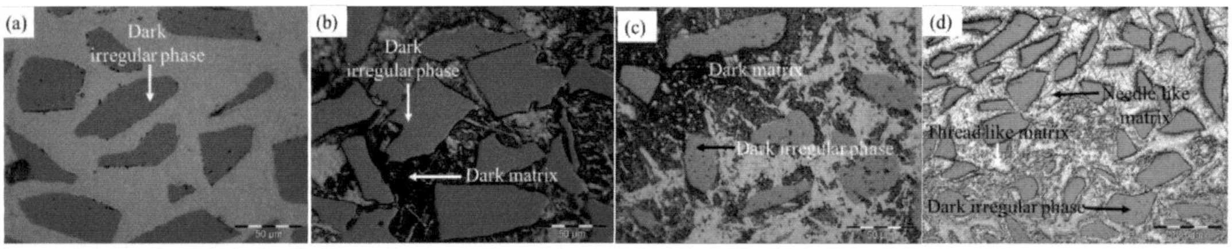

Figure 5. Optical bright fields micrographs of laser cladded Ti-6Al-4V with (a)100wt% SiC (b) 90 wt % SiC, 10 wt% Al (c) 80 wt % SiC, 20 wt% Al and (d) 70 wt % SiC, 30 wt% Al.

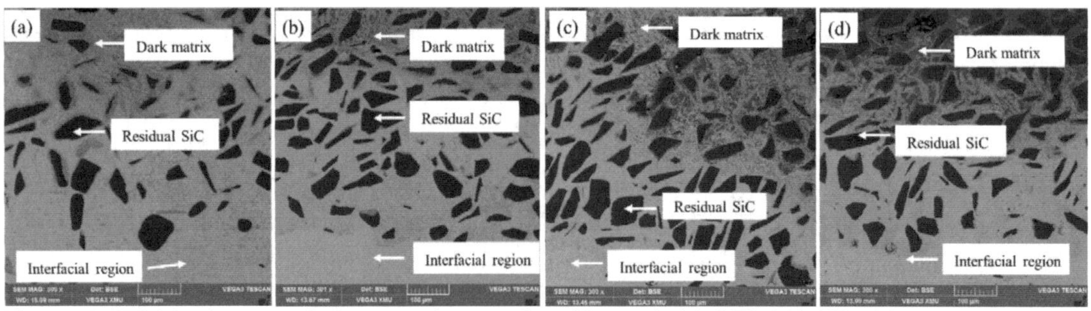

Figure 6. SEM backscattered electron micrographs of laser cladded Ti-6Al-4V with (a) 100 wt% SiC (b) 90 wt % SiC, 10 wt% Al (c) 80 wt % SiC, 20 wt% Al and (d) 70 wt % SiC, 30 wt% Al.

IOP Conf. Series: Materials Science and Engineering **655** (2019) 012016 doi:10.1088/1757-899X/655/1/012016

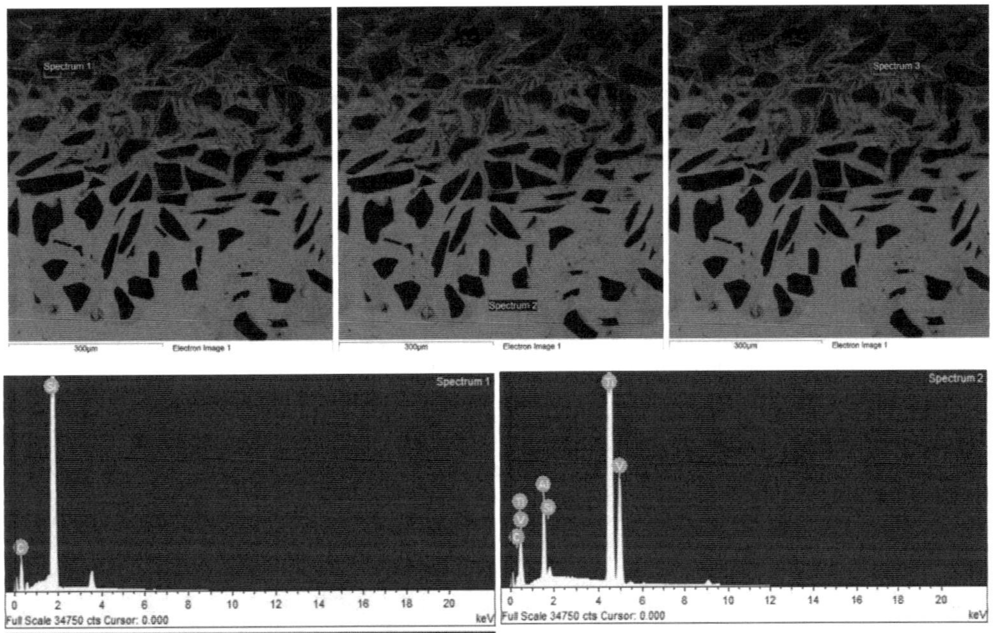

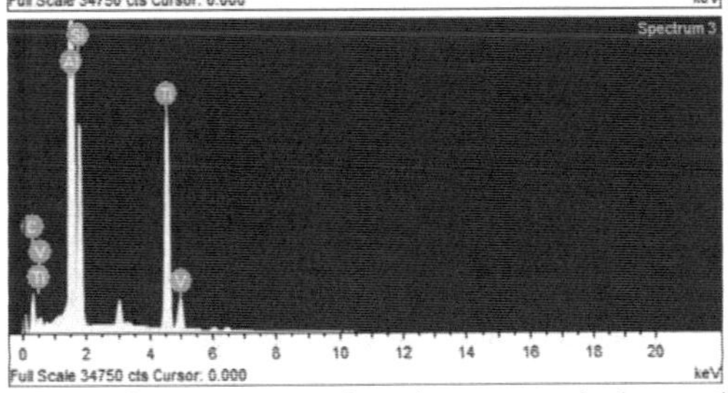

Spectra elemental composition from EDS analysis

Spectra	C	Al	Ti	Si	V
1	40.50			59.49	
2	4.29	5.69	85.77	0.69	3.56
3	23.00	49.67	17.42	9.59	0.32

Figure 7: EDS spectrums of various spots in laser cladded Ti-6Al-4V with 70 wt% SiC + 30 wt% Al

The formation of phases such as $TiSi_2$, $Ti_7Al_5Si_{12}$ and $AlCTi_2$, shows the laser energy and solution condition in the melt pool suffice to decompose the carbide and allow its elements to form separate phases in the cladding. These carbide, aluminide and silicide phases will enhance the surface properties diversely. Figure 7 is the XRD diffractograph of the as-received Ti-6Al-4V showing the various peaks of titanium atoms.

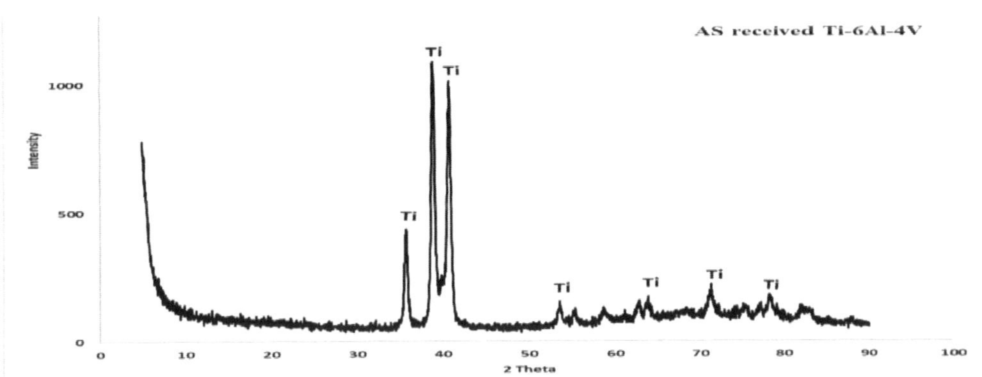

Figure 8. XRD spectral of as received Ti-6Al-4V

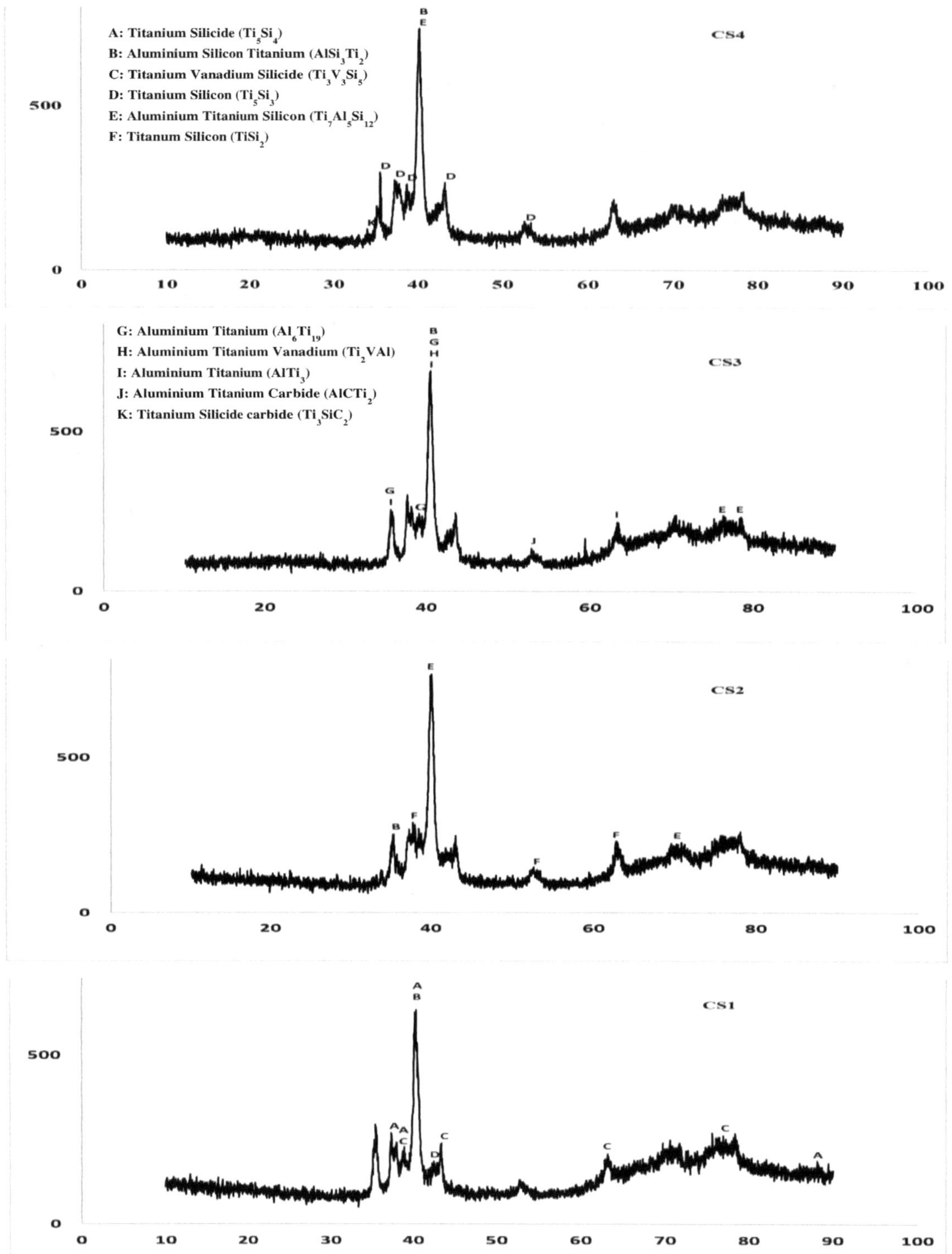

Figure 9. XRD spectral of cladded samples laser cladded Ti-6Al-4V using 100wt % SiC , 90 wt%SiC + 10 wt% Al, 80 wt% SiC + 20wt% Al and 70 wt% SiC + 30 wt% Al.

From the XRD of the surface layer, Ti_5Si_4, $Ti_3V_3Si_5$, Ti_5Si_3, $AlSiTi_2$, $Ti_7Al_5Si_{12}$, $TiSi_2$, $AlSi_3Ti_2$, Al_6Ti_{19}, Ti_2VAl, $AlTi_3$, $AlCTi_2$, Ti_3SiC_2, TiS_2 and Ti_5Si_4 were identified. Figure 8 shows the diffractograph obtained during the XRD study, indicating the angles at which the peaks of these phases were identified. The formation of these phases indicated that SiC and Al properly interacted and got alloyed with the titanium alloy substrate extensively. These intermetallic compounds were generated from chemical reactions in the molten pool involving large amounts of Ti element from the molten part of the matrix and Si, Al, V. From Table 3, it could be observed that the $Ti_7Al_5Si_{12}$ phase was consistent through the binary powder cladded surfaces, the $Ti_7Al_5Si_{12}$ phase has been noted for increasing the wettability of Al- ceramic composites [17]. This is further evidence of an enhanced wettability of the melt pool by the action of the binary powders of SiC and Al.

3.3 *Microhardness and Abrasive Wear*
Table 3 depicts the microhardness, and phases formed at the alloyed zone. The results indicate that CS2. The ternary Ti-Al-Si phase diagram can give insight about possible phases at this range of composition, but there are vanadium and other microelements in the as-received alloy (See Section 2.1). CS2 had the overall best microhardness behaviour, thus indicating the optimum aluminium content of 10% in the matrix. The increase in overall microhardness of the processed alloy could be attributed to the combined effects of the new intermetallic phases and formation of the Ti_5Si_3, Ti_5Si_4 and $Ti_7Al_5Si_{12}$ reinforced phases. The Ti_5Si_3, Ti_5Si_4 and $Ti_7Al_5Si_{12}$ reinforced phases have some outstanding advantages, such as high hardness, excellent wear resistance, high melting point, high thermal stability and excellent high-temperature oxidation resistance and enhanced mechanical properties [17, 18]. More so, laser cladding involves a high rate of heating and cooling which leads to massive grain refinement of the alloyed zone, this, in turn, enhances the surface properties of the surface, which microhardness is one of them.

Table 3. Average microhardness (Hv) under a load of 1000g and phases in the coatings at different powder compositions

Sample	Powder composition	Microhardness (HV)	Phases present (XRD)
As received	-	323.67	Ti (alpha & beta)
CS1	100 wt% SiC	910.25	Ti_5Si_4, $AlSi_3Ti_2$ $Ti_3V_3Si_5$, Ti_5Si_3
CS2	90 Wt% SiC + 10 Wt% Al	1169.95	$AlSi_3Ti_2$, $Ti_7Al_5Si_{12}$ $TiSi_2$
CS3	80 Wt% SiC + 20 Wt% Al	732.30	$AlSi_3Ti_2$, $Ti_7Al_5Si_{12}$ Al_6Ti_{19}, Ti_2VAl, $AlTi_3$, $AlCTi_2$
CS4	70 Wt% SiC + 30 Wt% Al	428.85	$AlSi_3Ti_2$, $Ti_7Al_5Si_{12}$, Ti_3SiC_2, Ti_5Si_4

Figure 10 illustrates the wear rate of the as-received and cladded surfaces under severe three bodies abrasive wear at the load of 1500N. The Ti_5Si_3, Ti_5Si_4 and $Ti_7Al_5Si_{12}$ reinforced phases present in the sample surfaces played a massive role in the wear resistance of the laser cladded surfaces. It was evident that the as-received had higher wear rates than CS1 and 2 for the time interval. However, CS4 had a higher wear rate than the as received. This inferiority of CS4 could be attributed to the quantity of aluminium present at the need like matrix (Figure 7, spectral 3).

The wear tracks of the as-received as shown in Figure 10a and CS4 (Figure 10d) exhibits grooves and ridges in the sliding direction of the stainless-steel wheel, and the presence of agglomerated wear debris was observed in between the grooves and ridges. The agglomeration of the wear debris was caused by the generation of heat involved in the dry sliding, and frictional forces present between two rubbing surfaces. In the three-bodies abrasive tests, the samples were abraded by the action of a stainless-steel wheel and silica sand particles. Plastic deformation occurred on the as-received alloy as a result of the cumulative effects of abrasion caused by the sand and fatigue induced by the rotating stainless-steel wheel. From Figure 10, it can be observed that the worn surface of CS1 and 2, exhibited fewer grooves, pits and accumulated wear debris when compared to the as-received alloy and CS4. It can be concluded that the wear mechanism was characterised by mild abrasive wear behaviour relative to the CS4 and the as received Ti-6Al-4v surface. CS1(100 wt% SiC) had the best abrasive behaviour, due to the abrasive nature of pure SiC used for the cladding experiment.

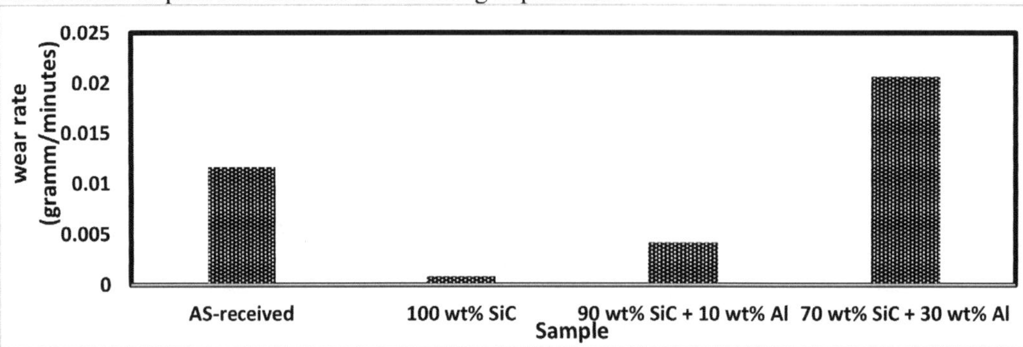

Figure 10. Average wear rates as-received Ti-6Al-4V and laser cladded Ti-6Al-4V using 100 wt% SiC, 90 wt% SiC + 10 wt% Al, and 70 wt% SiC + 30 wt% Al, after severe three bodies abrasive wear at a load of 1500N.

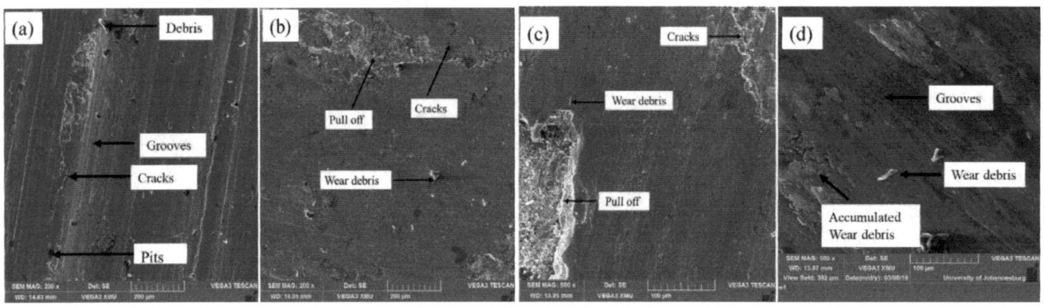

Figure 11. SEM secondary electrons micrograph of the worn surface morphology after severe 3 bodied abrasive wear at a load of 1500N of (a) as-received Ti-6Al-4V (b) laser cladded Ti-6Al-4V using 100 wt% SiC (c) laser cladded Ti-6Al-4V using 90 wt% SiC + 10 wt% Al. (d) laser cladded Ti-6Al-4V using 70 wt% SiC + 30 wt% Al

4. Conclusions

Successful laser cladding was done on Ti-6Al-4V alloy using single and binary powder compositions of SiC and Al; Several microstructural evolutions were observed as the powder compositions were varied. The coating thickness of all the binary cladded samples was higher than the single powder cladding using 100% SiC powder. This was attributed to the enhanced wetting of the binary powders. Wetting of SiC and Al resulted in the formation of various intermetallic as indicated by the XRD micrographs which include $AlSiTi_2$, $Ti_7Al_5Si_{12}$, $TiSi_2$, $AlSi_3Ti_2$, Al_6Ti_{19}, Ti_2VAl, $AlTi_3$, $AlCTi_2$, $T_{13}SiC_2$ and Ti_5Si_4. There was an enhanced microhardness, as all the cladded samples had higher average microhardness (1169.95 – 428.85 Hv) than the as-received (323.67 Hv). The microhardness behaviour of CS2 (90 wt% SiC + 10% Al) was the best, thus indicating that the optimum weight percentage of aluminium in the matrix should be about 10%. CS1(100wt % SiC) recorded better abrasive wear rate than the other cladded surfaces because of the abrasive nature of pure SiC used in the cladding.

Acknowledgements
The authors would like to appreciate the National laser centre (CSIR), National Research Foundation (NRF), Microscopy and Microanalysis Unit University of Witwatersrand and Vaal University of Technology, for the laser time, consumables, equipment and financial support they provided for this project.

References

[1] Zhang Z X, Qu S J, Feng A H and Shen J 2017 Achieving grain refinement and enhanced mechanical properties in Ti–6Al–4V alloy produced by multidirectional isothermal forging *Mater Sci Eng, A* **692** 127-38

[2] Tchein G J, Jacquin D, Aldanondo E, Coupard D, Gutierrez-Orrantia E, Girot Mata F and Lacoste E 2019 Analytical modeling of hot behavior of Ti-6Al-4V alloy at large strain *Mater Des* **161** 114-23

[3] Inagaki I, Shirai Y, Takechi T and Ariyasu N 2014 Application and Features of Titanium for the Aerospace Industry. (http://www.nssmc.com: Nippon Steel & Sumitomo Metal Technical Report)

[4] Molinari A, Straffelini G, Tesi B and Bacci T 1997 Dry sliding wear mechanisms of the Ti6Al4V alloy *Wear* **208** 105-12

[5] Popoola A P I, Fatoba O S, Nkosi H W and Aigbodion V S 2016 Surface hardening of Aluminum by Laser alloying with Molybdenum and Zirconium powder *Int. J. Electrochem. Sci* **11** 13

[6] Nkosi H W, Popoola A P I and Abdulwahab M 2012 Laser surface alloying of Al with Mo for hardness improvement

[7] Vora H D, Rajamure R S, Soundarapandian S, Srinivasan S G and Dahotre N B 2013 Dilution of molybdenum on aluminum during laser surface alloying *J Alloys Compd* **570** 133-43

[8] Hayun S, Paris V, Mitrani R, Kalabukhov S, Dariel M P, Zaretsky E and Frage N 2012 Microstructure and mechanical properties of silicon carbide processed by Spark Plasma Sintering (SPS) *Ceram Int* **38** 6335-40

[9] Touloukain Y S 1967 *Thermophysical properties of high temperature solid materials* vol 5 (New York: MacMillan CO)

[10] Adebiyi D I and Popoola A P I 2015 Mitigation of abrasive wear damage of Ti–6Al–4V by laser surface alloying *Materials & Design* **74** 67-75

[11] Al-Sayed Ali S R, Hussein A H A, Nofal A A M S, Hasseb Elnaby S E I, Elgazzar H A and Sabour H A 2017 Laser Powder Cladding of Ti-6Al-4V α/β Alloy *Materials (Basel, Switzerland)* **10** 1178

[12] Niu H J and Chang I T H 1999 Selective laser sintering of gas and water atomized high speed steel powders *Scripta Mater* **41** 6

[13] AlMangour B, Grzesiak D and Yang J-M 2017 In-situ formation of novel TiC-particle-reinforced 316L stainless steel bulk-form composites by selective laser melting *J Alloys Compd* **706** 409-18

[14] Man H C, Zhang S, Cheng F T and Yue T M 2002 In situ synthesis of TiC reinforced surface MMC on Al6061 by laser surface alloying *Scripta Mater* **46** 229-34

[15] Laurent V, Chatain D and Eustathopoulos N 1987 Wettability of SiC by aluminium and Al-Si alloys *J. Mater. Sci* **22** 244-50

[16] Pleshakov E, Senyavs'kyi Y and Filip R 2002 Laser Surface Modification of Ti-6Al-4V Alloy with Silicon Carbide *Mater Sci* **38** 646-52

[17] Juan L, Kehong W and Deku Z 2016 The Effect of Ti on Microstructural Characteristics and Reaction Mechanism in Bonding of Al-Ceramic Composite *J Mater Eng Perform* **25** 3638-45

[18] Dai J, Zhang F, Wang A, Yu H and Chen C 2017 Microstructure and properties of Ti-Al coating and Ti-Al-Si system coatings on Ti-6Al-4V fabricated by laser surface alloying *Surf Coat Technol* **309** 805-13

Influence of sintering temperature on densification, microstructure and mechanical properties of Ti-6Ni alloy developed via spark plasma sintering

A L Rominiyi [1,*], M B Shongwe [1], N B Maledi [2], S O Jeje [1], B J Babalola [1] and P F Lepele [1]

[1] Department of Chemical, Metallurgical and Materials Engineering, Faculty of Engineering and the Built Environment, Tshwane University of Technology, Pretoria west, Pretoria 0001, South Africa

[2] School of Chemical and Metallurgical Engineering, University of the Witwatersrand, Private Bag 3, Johannesburg, 2050, South Africa

E-mail: rominiyipedi@gmail.com

Abstract. Titanium is used extensively in various engineering applications as a result of its unique physical, chemical and mechanical properties. However, the difficulty encountered in fabricating a fully densified titanium sample informed the introduction of alloying elements into the matrix in order to form a liquid phase and enhance sinterability. This work explored the influence of sintering temperature on the densification, microstructure and mechanical properties of Ti-6Ni alloy developed via spark plasma sintering at varying temperature from 850 °C to 1200 °C in a vacuum, under isothermal holding time, heating rate and applied pressure of 10 min, 100 °C/min and 50 MPa respectively. Characterization of the mixed powder and bulk sintered sample were carried out with the aid of scanning electron microscopy (SEM) coupled with energy-dispersive X-ray spectroscope (EDS) and X-ray diffraction (XRD). Relative density and hardness of the samples were determined following the Archimedes and Vickers method respectively. The SEM results indicated the presence of multiple phases in the microstructure of the sintered samples. These phases were confirmed by the EDS and XRD analysis results, as the intermetallics of Ti-Ni. Densifications above 98%, was achieved and the hardness values were found to vary with the densification of the sintered alloy, across the investigated sintering temperatures.

1. Introduction

Development of new materials that are capable of withstanding the challenges of high temperature applications has been a major challenge to researchers in different fields. However, modification of the properties of the existing materials with known history, has received the attention of researchers as this saves cost, time and accurate prediction of their performance in service can be made. Titanium occupies an important position among the committee of metals capable of performing excellently in high temperature applications [1]. Titanium and its alloys have progressively gaining the attention of the researchers as a result of their attractive intrinsic properties which made them one of the sought after candidate materials in aerospace, automobile, marine, biomedical, chemical and industrial applications, requiring high strength-to-weight ratio, excellent corrosion resistance, biocompatibility and high operating temperature. Despite these attractive properties, further utilization of this material is being threatened by the high cost of its extraction and manufacturing of the final product [2, 3]. In addition, the high level of attraction between titanium and oxygen results in oxide formation on the surface of the pure titanium powder. Consequently, this reduces the chemical reactivity of the surface and creates difficulty for densification during the sintering of titanium [4, 5].

Improvement in the sinterability of pure titanium powder can be achieved through the addition of elements that have the capacity to ease diffusion and improve sinterability through the formation of the liquid phase [6]. The use of powder metallurgy methods such as the spark plasma sintering (SPS) technology [7] which combine joule heating and spark discharge generated between powders due to the applied current, pressure and pulse voltage, to rid the surface of the powders of oxide films and enhance self-diffusion of titanium around the neck area and subsequent growth of this region [8].

It has been confirmed that the addition of a small quantity of nickel to titanium improves the ability of titanium to diffuse and enhances densification [9-11]. Nickel achieves this through reduction of the activation energy needed for self-diffusion of titanium which then leads to enhanced densification [12]. Also, during the spark plasma sintering process, diffusion phenomena increase and transfer of materials take place as a result of the plasma developed between the powders particles. This results in the removal of the oxide impurities from the surface of the materials. Furthermore, the joule heating effect and pressure-assisted plastic deformation of the powders enhanced sinterability and densification [7, 13, 14]. Previously, studies were conducted on dilatometric sintering of titanium with small additions of nickel [12, 15]. The results of these works showed that highest densification was achieved at a temperature just below the solidus line while swelling was observed in samples sintered at temperatures above the solidus line. To this end, this work studied the influence of spark plasma sintering temperature, on the densification behaviour, microstructure and how this affects the mechanical property of Ti-6Ni alloy.

2. Experimental Procedure

2.1. Powder preparation and characterization

Commercially sourced CP-TI Grade1 (TLS Technik GmbH & Co, Germany) and Ni (FlomasterTM metal powder, F.J. Brodmann & Co., ltd., USA) powders of particles size < 25 μm and 0.5-0.3 μm respectively were utilized for this research. Ni and Ti powders were weighed based on the different wt% (Ti-6Ni). The weighed powders were discharged into a 250 ml container, filled up to 20% level and arranged axially into a TF2, Turbular shaker mixer. The powder mixture was blended under a rotational and translational speed of 101 rpm for 8 h.

2.2. Powder consolidation

The admixed powders were weighed and emptied into a non-conducting, graphite die, with an outer diameter of 50 mm and an inner diameter of 30 mm respectively. The graphite die was lined with graphite foil to minimize the heat loss due to radiation and prevent difficulty in removing the samples after sintering. The prepared samples were sintered using spark plasma sintering system HHPD-25, FCT Systeme GmbH, Germany under different sintering temperatures of 850, 1100 and 1200 °C. The applied pressure, isothermal holding time and heating rate were kept at 50 MPa, 10 min and 100 °C/min respectively. The schematic features of the spark plasma sintering equipment are shown in Figure 1.

The temperature was measured with the aid of an external pyrometer with sensitivity at 250 °C, located at a distance of about 3 mm above the sample. At the expiration of the holding time, the samples were allowed to cool down to the ambient temperature in the furnace. Subsequently, the sintered samples were sandblasted to remove the graphite foil layer and the density of the sintered samples was measured by using the Archimedes technique. The relative density, R_ρ was obtained using equation 1.

$$R_\rho = 100\left(\frac{\rho_{exp}}{\rho_{th}}\right) \quad (1)$$

where, ρ_{exp} and ρ_{th} are the experimental and theoretical density respectively.

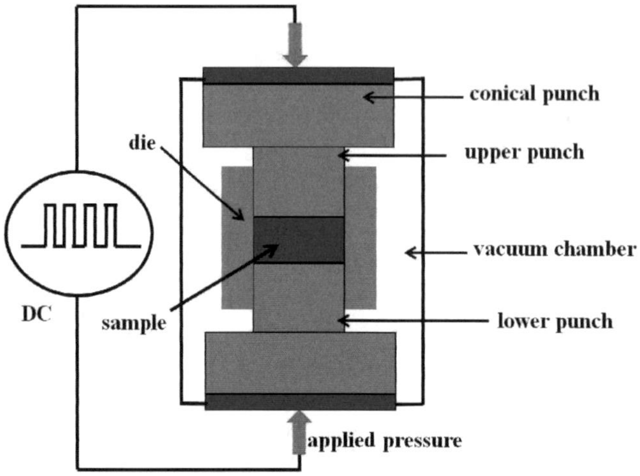

Figure 1. Schematic features of SPS equipment [16].

2.3. Microstructural and phase analysis
Samples for metallographic analysis were sectioned, ground with silicon carbide paper from 320 up to 2400 grit size. Polishing was conducted with diamond pastes of particle sizes 9, 6, 3 and 1 μm. The prepared samples were subsequently etched using Kroll reagent (3% HF, 5% HNO3 and 92% H2O) for 20 s and analyzed by using field electron scanning electron microscope (JOEL JSM-7600F) in conjunction with an energy-dispersive X-ray spectroscope (EDS) detector. Phase Identification was carried out with the aid of PW1710 Philips, X-ray diffractometer, utilizing Cu Kα radiation operated at 40 kV and 40 mA with 2θ ranges from 5 to 90 and a step size of 0.01. The obtained diffraction patterns were compared with the database of the equipment, using X'Pert High Score Plus software.

2.4. Hardness measurement
The Vickers hardness (HV) of the sintered samples was determined under ambient temperature using Future-tech microhardness tester. The test load and dwell time were maintained at 1000 gf and 15 s respectively. The reported microhardness value was obtained as an arithmetic mean of ten consecutive measurements for an individual sample.

3. Results and Discussion
3.1. Microstructure and phase analysis
SEM micrograph of the turbula-mixed powders of elemental titanium and nickel are shown in Figure 2(a). The admixed Ti-6Ni powder consists of regular, spherical and non-porous Ti powders which suggest that the powder was formed by gas atomization while the Ni powders appeared as an agglomeration of small, non-porous particles which may be attributed to water atomization production route. The microstructures of the sintered Ti-6Ni samples sintered at 850, 1100 and 1200 °C are shown in Figure 2b-d respectively. The formation of the lamella and acicular-like structures were observed in the micrographs of the sintered alloys at low and high temperatures respectively. These structures consist of alternating arrangements of dark α-Ti phase and Ti-Ni rich whitish phase, as shown in Figure 2(e) and (f). The intermetallics were formed due to the interactions between nickel and titanium over the sintering temperatures under consideration. Nickel reacts with titanium at temperatures above the transus temperature due to its solubility in β-Ti to form a solid solution of Ti-Ni. However, its limited solubility in α-Ti resulted in the formation of intermetallics on cooling down to the room temperature [12] as observed in XRD patterns presented in Figure 3.
Lamella arrangements of white patches and bright spots within a dark matrix of α-Ti were observed in sample sintered at 850 °C (Figure 2b), this can be ascribed to the interaction between titanium and nickel at low operating temperature, as diffusion also depends on the temperature of the system. Fine and neatly arranged acicular-like structures and bright spots were also observed in Figure 2c as the operating temperature increased to 1100 °C. This can be attributed to the enhanced diffusion of nickel which provided enabling system for self-diffusion of Ti for Ti-Ti interactions and formation of liquid phase

which lowers the energy required for the reaction between Ti and Ni to occur. Hence, the improved nucleation and growth of the acicular-like structure, as shown in Figure 2c. Sample sintered at 1200 °C, shows no noticeable bright spots but distorted acicular-like structure (Figure 2d) was observed. The distorted structure can be traced to the loss of Ti-Ni rich liquid phase [10] which solidified on the graphite punch during sintering, thereby depriving the system the much needed Ti-Ni rich liquid required for the growth of the intermetallic phase(s) within the matrix. The loss of the Ti-Ni rich liquid can be attributed to the composition of the sintered alloy and temperature of the operation. The 6 wt% Ni and the operating sintering temperature of 1200 °C correspond to the boundary between the β single-phase field and the dual-phase field of β+L (solidus line) on the Ti-Ni binary phase diagram. At this location, there is a high possibility of formation of the liquid phase and its subsequent expansion [15] resulted in the observed solidification of the sipped out liquid, on the body of the graphite punch.

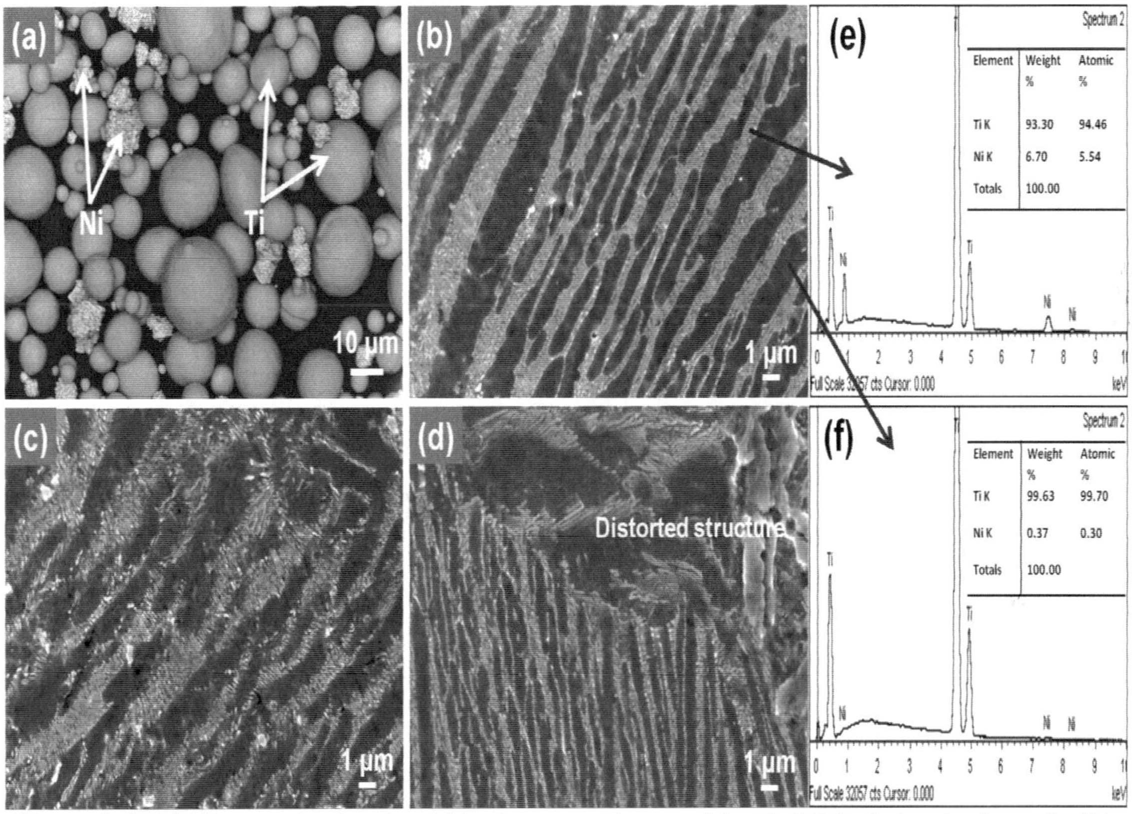

Figure 2. SEM Micrographs showing (a) backscattered image of the Ti-6Ni Turbula mixed powder (b) secondary electron image of Ti-6Ni sintered at 850 °C (c) secondary electron image of Ti-6Ni sintered at 1100 °C (d) secondary electron image of Ti-6Ni sintered at 1200 °C (e) EDS of whitish part of the acicular-like structure (f) EDS of the darkish part of the acicular-like structure.

Identification of phases formed as a result of the reactions during sintering was conducted using X-ray diffraction analysis. The diffraction patterns obtained from analysed Ti-6Ni sintered alloys at different sintering temperature are shown in Figure 3. The patterns show a consistent existence of hcp, α-Ti (dark), $TiNi_2$ (bright) and fcc, Ti_2Ni (whitish) phases within the matrix. The Ti_2Ni and $TiNi_2$ intermetallic phases were formed in the Ti-Ni rich regions as shown in Figure 2(e) and (f) through the diffusion of nickel into the crystalline structure of titanium. The ability of Ni to lower the activation energy of Ti, allows the reaction between Ti and Ni to take place below the eutectoid temperature of 862 °C. Hence, the formation of the solid solution of Ni in β-Ti which upon cooling to ambient temperature forms Ti_2Ni and $TiNi_2$ intermetallics due to the limited solubility of Ni in α-Ti. The phase constitutions of the investigated alloys at different sintering temperature as revealed in Figure 3 are shown in Table 1.

Table 1. Effects of sintering temperature on the phase constitution of the investigated alloys

Sample designation	Sintering temperature (°C)	α-Ti phase content, vol. %	Ti_2Ni phase content, vol. %	$TiNi_2$ phase content, vol. %
Ti-6Ni	850	58	10	32
	1100	50	9	41
	1200	82	18	-

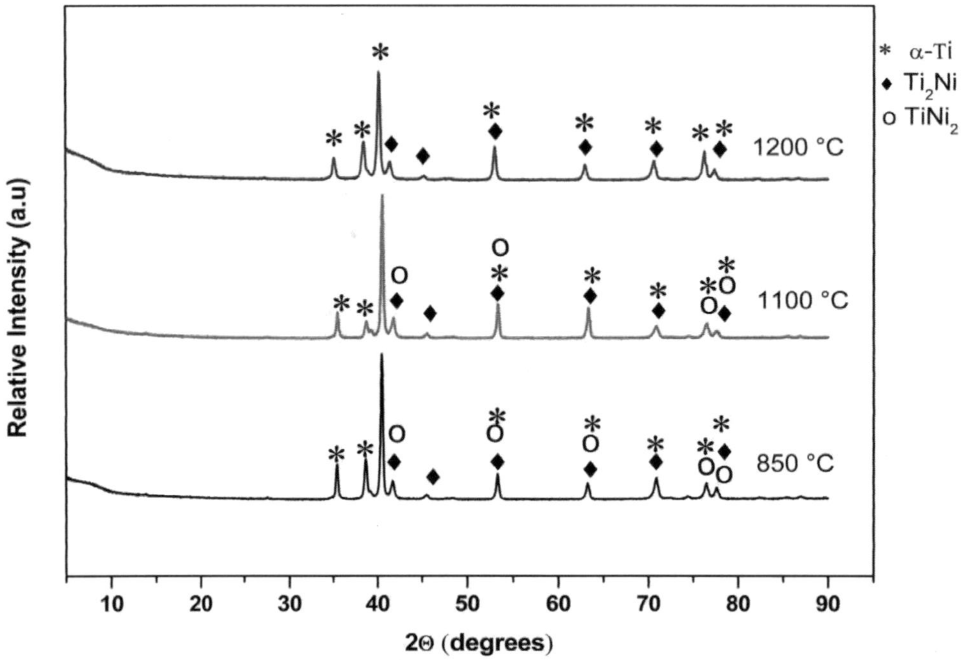

Figure 3. Diffraction patterns of the sintered Ti-6Ni samples under different sintering temperatures.

3.2. Relative density and microhardness

Table 2 shows the summary of the properties of the sintered Ti-6Ni over the investigated sintering temperatures. It can be observed that the sintered samples displayed high level of densification whereas their microhardness values differ. A relative density of 99.79% with microhardness value of 294±3.43 HV was observed in samples sintered at 850 °C. These values increased to 99.87% and 299±5.10 HV for sample sintered at 1100 °C. However, a decline in the relative density (98.6%) and microhardness value (247±12.35 HV) was observed in sample sintered at 1200 °C. The observed trend can be attributed to the influence of nickel addition in aiding self-diffusion of titanium and the formation of hard and brittle Ti_2Ni [9] and $TiNi_2$ intermetallics during the sintering of Ti-6Ni alloys. According to Riva et al. [17], the occurrence of the non-stoichiometric, metastable phases within the matrix of an alloy can be attributed to the rapid cooling of the alloys from the elevated sintering temperatures. The presence of these metastable phases in association with the stable phases of Ti-Ni was confirmed in the work of Fuji and Yamamoto [18], Sadmezhaad and Selahi [19] and Rominiyi et al. [16].

Table 2. Summary of the investigated alloys properties.

Samples designation	Sintering temperature (°C)	Theoretical density, ρ_{th} (g/cm^3)	Experimental density, ρ_m (g/cm^3)	Relative density (%)	Microhardness (HV)
Ti-6Ni	850	4.648	4.638	99.79±0.29	294±3.43
Ti-6Ni	1100	4.648	4.642	99.87±0.21	299±5.10
Ti-6Ni	1200	4.648	4.582	98.58±1.42	247±12.35

The existence of $TiNi_2$ (hcp) and Ti_2Ni (fcc) intermetallic phases within the matrix of the sintered alloys resulted in structure mismatch, distortion and development of lattice strain in the material. The volume of these intermetallics is observed to increase with increasing sintering temperature (Table 1). Consequently, this increased the resistance to the dislocation movement during indentation and this was responsible for the increase in hardness observed in the sintered alloy at 1100 °C when compared with the hardness of sample sintered at 850 °C. However, sample sintered at 1200 °C was expected to possess the highest microhardness value but the loss of Ti-Ni rich liquid, deprived the matrix the needed volume of intermetallics (Table 1 and Figure 3) within the matrix. This resulted in less distortion within the matrix (Figure 2d) and few obstructions are provided in the path of dislocation during indentation. Hence, the lowest microhardness value observed in the sample sintered at 1200 °C.

4. Conclusion

This work investigated the Influence of sintering temperature on densification, microstructure and mechanical properties of Ti-6Ni alloy using spark plasma sintering technique. Based on the extent of work done in this research, it was concluded that increasing the spark plasma sintering temperature from 850 °C to 1100 °C at constant pressure, heating rate and dwelling time of 50 MPa, 100 °C and 10 min respectively improved the microstructure, enhance densification and increase the microhardness value while further increase in temperature up to 1200 °C yielded deteriorated microstructure, decline in densification and the microhardness value of the spark plasma sintered Ti-6Ni alloy. However, it is speculated that lowering the pressure at the same temperature of 1200 °C can prevent loss of the Ti-Ni rich liquid phase during spark plasma sintering thereby achieving higher densification and hardness than what was obtained in the case of sample sintered at 50 MPa and 1200 °C in this work.

Acknowledgement
This study is based on the research supported by the National Research Foundation (NRF) of South Africa under Unique Grant No. 117867.

References

[1] Adrian M. P 2012 "Titanium alloys for aerospace structures and engines A2" in *Introduction to Aerospace Materials*, ed: Woodhead Publishing, pp. 202-223.
[2] Borkar T, Nag S, Ren Y, Tiley J and Banerjee R 2014 "Reactive spark plasma sintering (SPS) of nitride reinforced titanium alloy composites," *J. Alloys Compd.* **617**, pp. 933-945.
[3] Whittaker T M 2015"Titanium Alloys," *Metals,* vol. **5**, pp.1437-1439; doi:10.3390/met5031437.
[4] Panigrahi B B, Godkhindi M M, Das K, Mukunda P G and Ramakrishnan P 2005"Sintering kinetics of micrometric titanium powder," *Mats. Sci. Eng. A,* **396**, pp. 255-262.
[5] Shon J-h, Park J-m, Cho K-s, Hong J-k, Park N-k and Oh M-h 2014"Effects of various sintering methods on microstructure and mechanical properties of CP-Ti powder consolidations," *Trans. Nonferrous Met. Soc. China,* **24**, pp. s59-s67.
[6] Flinn R A and Trojan P K 1990 *Engineering Materials and Their Applications*, 4th ed. One Beacon Street, Boston MA 02108. USA: Houghton Mifflin Company,.
[7] Shahedi Asl M, Sabahi Namini A, Motallebzadeh A and Azadbeh M 2018 "Effects of sintering temperature on microstructure and mechanical properties of spark plasma sintered titanium," *Mater. Chem. Phys.*, **203**, pp. 266-273.
[8] Babalola B J, Shongwe M B, Obadele B A, Olubambi P A, Ayodele O O, Rominiyi A L and Jeje S O 2018 "Comparative study of spark plasma sintering features on the densification of Ni-Cr binary alloys," *MATEC Web Conf.,* **249** (2018) 01004.
[9] Luo S D, Yang Y F, Schaffer G B and Qian M 2013 "The effect of a small addition of boron on the sintering densification, microstructure and mechanical properties of powder metallurgy Ti–7Ni alloy," *J. Alloys Compd.* **555**, pp. 339-346.
[10] Robertson I M and Schaffer G B 2009 "Suitability of nickel as alloying element in titanium sintered in solid state," *Powder Metall.*, **52**, pp. 225-232.
[11] Liu H W, Paul Bishop D and Plucknett K P 2015 "A comparison of Ti–Ni and Ti-Sn binary alloys processed using powder metallurgy," *Mats. Sci. Eng. A,* **644**, pp. 392-404.
[12] Panigrahi B B 2007 "Sintering behaviour of Ti–2Ni and Ti–5Ni elemental powders," *Mater. Lett.*,

61, pp. 152-155.

[13] Feng H, Zhou Y, Jia D and Meng Q 2005 "Rapid synthesis of Ti alloy with B addition by spark plasma sintering," *Mats. Sci. Eng. A,* **390**, pp. 344-349.

[14] Zhang Z-h, Shen X-B, Wen S, Luo J, Lee S -k and Wang F-c 2010 "In situ reaction synthesis of Ti–TiB composites containing high volume fraction of TiB by spark plasma sintering process," *J. Alloys Compd.* **503**, pp. 145-150.

[15] Robertson I M and Schaffer G B 2009 "Swelling during liquid phase sintering of Ti–Ni alloys," *Powder Metall.,* **52**, pp. 213-224.

[16] Rominiyi A L, Shongwe M B, Maledi N, Babalola B J and Olubambi P A 2019 "Synthesis, microstructural and phase evolution in Ti–2Ni and Ti–10Ni binary alloys consolidated by spark plasma sintering technique," *The Int. J. of Adv. Manufac. Technol.,* pp. 1-9. doi.org/10.1007/s00170-019-03950-5.

[17] Riva S, Yusenko K V, Lavery N P, D. Jarvis J and Brown S G R 2016 "The scandium effect in multicomponent alloys," *Inter. Mater. Rev.,* **61**, pp. 203-228.

[18] Fuji A, Horiuchi Y and Yamamoto K 2005 "Friction welding of pure titanium and pure nickel," *Sci. Technol. Weld. Join.,* **10**, pp. 287-294.

[19] Sadrnezhaad S K and Selahi A R 2004 "Effect of Mechanical Alloying and Sintering on Ni–Ti Powders," *Mater. Manuf. Processes.,* **19**, pp. 475-486.

Effect of Temperature on Densification, Microstructural Evolution and Mechanical Properties of Ti-5Al-1Mo Developed via Spark Plasma Sintering

S O Jeje[1,*], M B Shongwe[1], N Maledi[2], A L Rominiyi[1], B J Babalola[1] and P F Lepele[1]

[1] Department of Chemical, Metallurgical and Materials Engineering, Faculty of Engineering and Built Environment, Tshwane University of Technology, Pretoria 0001, South Africa

[2] School of Chemical and Metallurgical Engineering, University of the Witwatersrand, Private Bag 3, Johannesburg, 2050, South Africa

E-mail: jejesamsonn@gmail.com

Abstract. Titanium alloys are ideally suited for advanced aerospace structures and industrial applications because of the combination of high strength, low density, good toughness, stiffness, and good corrosion resistance at cryogenic to elevated temperatures. This paper investigates the effect of temperature on alloy densification, microstructure evolution and mechanical properties of Ti-5Al-1Mo processed by Spark Plasma Sintering (SPS). Ti, Al, Mo elemental powders with 94, 5 and 1 wt.% respectively were mixed thoroughly for 8 hrs to attain homogeneity. The mixed powders were consolidated at sintering temperature of 800°C, 1000 °C, 1100 °C, pressure of 50 MPa, holding time of 10 min, and sintering rate of 100 °C/min. The compacted samples were characterised using scanning electron microscopy (SEM), equipped with energy dispersive X-ray spectroscopy (EDS) and X-ray diffractometry techniques. Compacts hardness was also investigated using Vickers micro-hardness tester. Almost fully densified compacts were obtained for samples sintered at 1000 °C and 1100 °C, the microstructures show that the sintered compacts are chemically homogeneous with the Al fully diffused into the CP-Ti matrix. The microhardness increases with increasing temperature with the maximum value being 290.35 HV.

1. Introduction

Titanium (Ti) matrix composites have attracted significant attention among engineering industries for several applications due to their attractive properties which include high specific strength, low density, stiffness, good toughness, and excellent corrosion resistance at cryogenic and at elevated temperature [1-4]. Ti alloy compositions can be designed to contain varying alpha (α) or beta (β) phases proportions based on application or property requirements [5].

The near-alpha (near-α) alloys are mostly used in aeroengines, particularly at the compressor section because these alloys possess the best creep resistance at temperatures between 400 and 600 °C of all Ti-alloys [6-8]. The near-α contains about 1 to 2 wt.% of β stabilisers (such as Mo, V, Nb and Ta) which stabilises β phase (about 5-10%) into the microstructure of the Ti matrix at ambient temperature [5, 7]. The stabilised β phase allows a two-phase strengthening mechanism by a controlling the morphology, proportion and dispersion of the α/β phases. Aluminium (Al) is the major major α stabiliser known to enhance the creep strength, tensile strength and also the elastic modulus of Ti [7]. The solubility of Al is high both in the α and β-phases of Ti and it also reduces the density of the alloy; However, the amount of Al that can be introduced to the Ti matrix is limited due to the formation of a intermetallic phase(s) of titanium aluminide (Ti-Al) which is brittle when the Al content exceeds 8 wt.% [7]. Considering the ternary system of Ti-Al-X, the Ti-Al-Mo system is of special category because it is composed of Mo atoms which has the best capability for stabilising β phase compared to other transition elements [9]. The addition of Mo in α-Ti grains tends to enhance workability and improves the tribological behaviour of Ti-based alloys while keeping their hardness at very high level [10, 11]; Furthermore, this alloying element also trap vacancies, thereby limiting creep relaxation [12]. Aluminium has a stabilising equivalence exactly inverse to that of Mo, thus the two systems can be combined to provide a qualitative view of Ti alloy stability [13].

A study on conventional sintering technique for titanium powders reveals that sintered sample's full densification in a vacuum environment is feasible only at high temperatures in the range of 1100 to 1400 °C and longer holding time [14]. However, Spark plasma sintering (SPS) has the benefit of rapidly achieving full densification at relatively lower temperatures. SPS entails the application of pressured spark impact, external mechanical pressure and heat (generated by the passage of a pulse current with high energy through the powder particles) simultaneously [15, 16]. Generally, the sintering reaction involves ion diffusion process within an electric field in conjunction with plastic deformation by thermal diffusion.
This study entails the consolidation of Ti-5Al-1Mo powder by SPS method. It investigates the effect of sintering temperature on densification, phase transformation, microstructure evolution and hardness of the metal matrix composites. This route is expected to provide a novel way of rapidly consolidating Ti-5Al-1Mo near alpha Ti-based alloy and promote its usage in industries.

2. Experimental Procedure

2.1 Powder feedstock and preparation

The Ti, Al and Mo powders with properties presented in Table 1 were used in this study. Elemental powders of commercially pure (Cp) Ti matrix consisting of Ti, Al and Mo powders with weight % ratio of 94:5:1 respectively was loaded to cover up to 15% volume of a plastic container. The Ti-Al-Mo powder in the container was blended in a WAB TURBULA SYSTEM SCHATZ turbula mixer at a speed of 48rpm for 8 hrs. Blended powders characterisation was done using Scanning Electron Microscopy (SEM) incorporated with Energy Dispersive X-ray Spectroscopy (EDS) to explore its morphology and confirm the homogeneity of the powders. Figure 1 shows the backscattered electron image (BSE) and the EDS of admixed powder of Ti, Al and Mo.

Table 1. Properties of Starting Powders

Role	Element	Supplier	Purity (%)	Size (μm)	Density (g/cm^3)	Melting Point (°C)
Matrix	Cp Ti	TLS Technik GmbH & Co	99.8	-25	4.5	1670
Reinforcement	Al	TLS Technik GmbH & Co	99.5	-25	2.7	660
Reinforcement	Mo	CERAC	99.9	2-4	10.28	2623

2.2 Sintering

The blended powders were sintered using the SPS machine (model HHPD 5, FCT Germany). Measured weight of the admixed powder was emptied into a graphite mould system containing a die, upper punch, and lower punch. Graphite foil was employed on the inner wall of the die before the

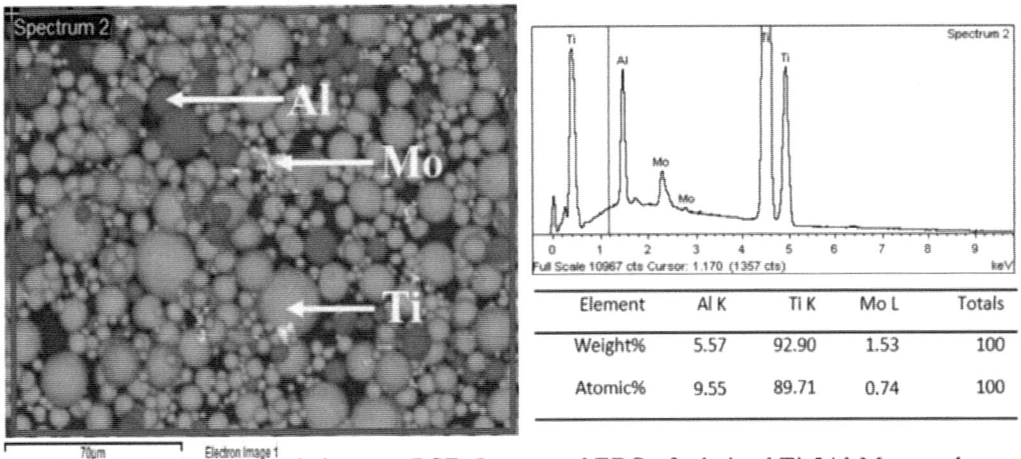

Element	Al K	Ti K	Mo L	Totals
Weight%	5.57	92.90	1.53	100
Atomic%	9.55	89.71	0.74	100

Figure 1. Back scattered electron (BSE) Image and EDS of admixed Ti-5Al-Mo powders

powder was poured to aid easy removal of sintered sample. Ti-5Al-1Mo was sintered at three different temperatures (800, 1000 and 1100 °C), heating rate of 100 °C/min, pressure of 50 MPa and holding time of 10 min under vacuum level of 10^{-2} torr. The temperature during the sintering process was monitored using a pyrometer. The sintered compacts were left to cool to ambient temperature, removed and then sand blasted to remove graphite contaminants.

2.3 Densification
Densities of sintered samples were measured using a densitometer that works using the principle of Archimedes. The density was measured to explore the impact of sintering temperature on the composites. The density value recorded was the average taken from 10 measurements on each sample. The percentage densification was calculated by comparing the measured density to the theoretical density of each compact using equation 1 and 2.

$$\frac{1}{Theoretical\ density\ of\ TiAlMo} = \frac{Mass\ fraction\ of\ Ti}{Density\ of\ Ti} + \frac{Mass\ fraction\ of\ Al}{Density\ of\ Al} + \frac{Mass\ fraction\ of\ Mo}{Density\ of\ Mo}$$

$$\%\ Densification = \frac{Sintered\ sample's\ density}{Bulk\ sample's\ theoretical\ density}\ X\ 100$$

2.4 Microhardness
The Vickers microhardness values of the samples were measured by using microhardness tester (FUTURE-TECH FM 800) using a load of 1 Kgf and holding time of 15 s on well-polished surfaces of the sintered Ti-5Al-Mo. The microhardness value reported was the average taken from 5 measurements on each sample.

2.5 Sintered compact characterisation
The SPS compacts were cut transversely, grounded with Emery papers, polished and etched using Kroll reagent with composition of 92 vol.% H_2O, 2 vol.% HF and 6 vol.% HNO_3 for investigating the microstructure in the alloyed zone. The microstructure evolution and elemental composition of the sintered samples were characterised by SEM incorporated with EDS. X-ray diffractometer ((PW1710 Philips) with Cu K_α radiation of wavelength 0.154nm, set at 10°-90° angular range with a step size of 1°/min at 45 kV and 25 mA) in combination with X'Pert High Score Plus software was employed to analyse the phases present in the sintered compacts.

3. Results and Discussion

3.1. Powders (Elemental and Mixed) Characterisation

The effectiveness of the mixing done by the turbula mixer was observed by the homogeneity (evenly distribution) of the powders as observed in the back scattered electron (BSE) micrograph in figure 1. The analysed weight percentages of Ti, Al and Mo powders are very close to the mixed value of each element. The XRD patterns (Figure 2) for the as-received elemental Ti, Al and Mo powders shows only the peaks of α-Ti, Mo, and Al respectively.

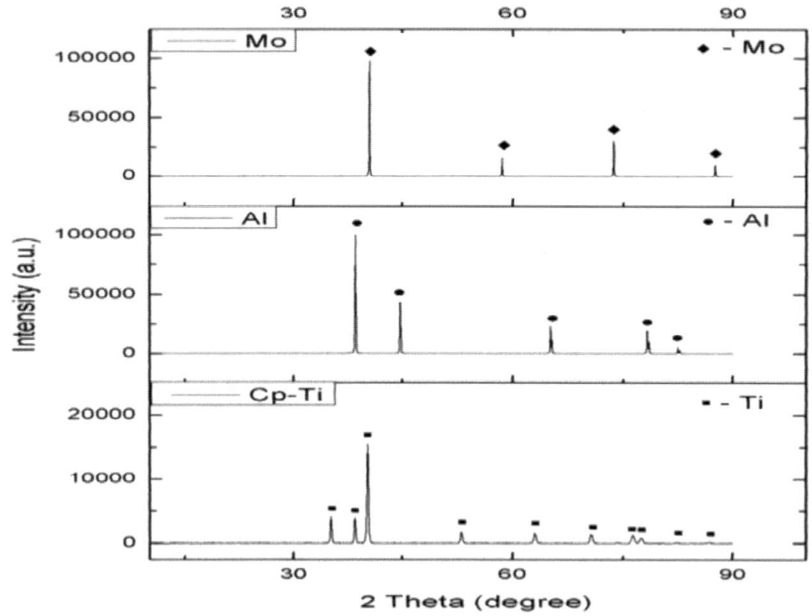

Figure 2. XRD patterns of elemental powders

3.2. Effect of temperature on sintered samples properties

Figure 3 shows the influence of sintering temperature on the spark plasma sintered sample's density and hardness. The sample sintered at 800 °C has a relative density of 81.81%, while fully densified samples of 99.95% and 99.43% relative density were obtained at sintering temperatures 1000 and 1100 °C respectively. Full densification observed in samples sintered at 1000 and 1100 °C are ascribed to the influence of liquid phase sintering that is accompanied with higher uniaxial pressure and heating leading to formation of necks between Ti particles and also the annihilation of porosity as the temperature increases till the end of the sintering holding time. Vickers hardness of the samples was found to increase with decreasing porosity. This is evident as the sample sintered at 800 °C has the least hardness value of 155.15 HV. The measured maximum hardness of 290.35 HV was attained at sintering temperature of 1000 °C. This is in order with the hardness value (245 to 320 HV) reported by Mouritz and Adrian [8] for several α and near α titanium alloys.

From the SEM micrographs shown in Figure 4, it can be observed that the Ti-5Al-1Mo sintered at 800 °C (Figure 4a) has a thin grey Al and white Mo boundary that has locally developed on the surface of nearly all the Ti particles that are in direct contact with the Al particles. This can be attributed to the fact that Al melted to liquid phase and spread throughout the Ti particles which are undergoing necking trying to close up the pores during the sintering process. There was partial interfacial diffusion between the Ti particles and the alloying elements (Al and Mo); this is an indication that sintering (densification) process is not completed. Pores are visible on the micrographs of compacts sintered at 800 °C which agrees with the result of the low densification obtained for the sintered sample. The sintered samples at 1000 °C and 1100 °C sintering temperature are almost fully dense with no evidence of pores nor apparent defects; This is also consistent with the measured values of the density.

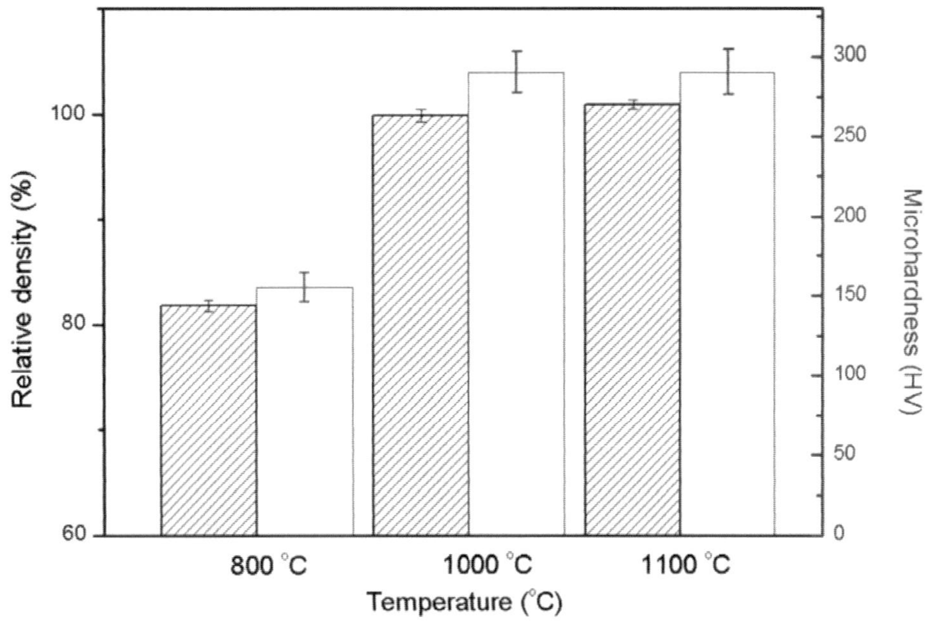

Figure 3. Density and microhardness of bulk alloys sintered at various temperatures

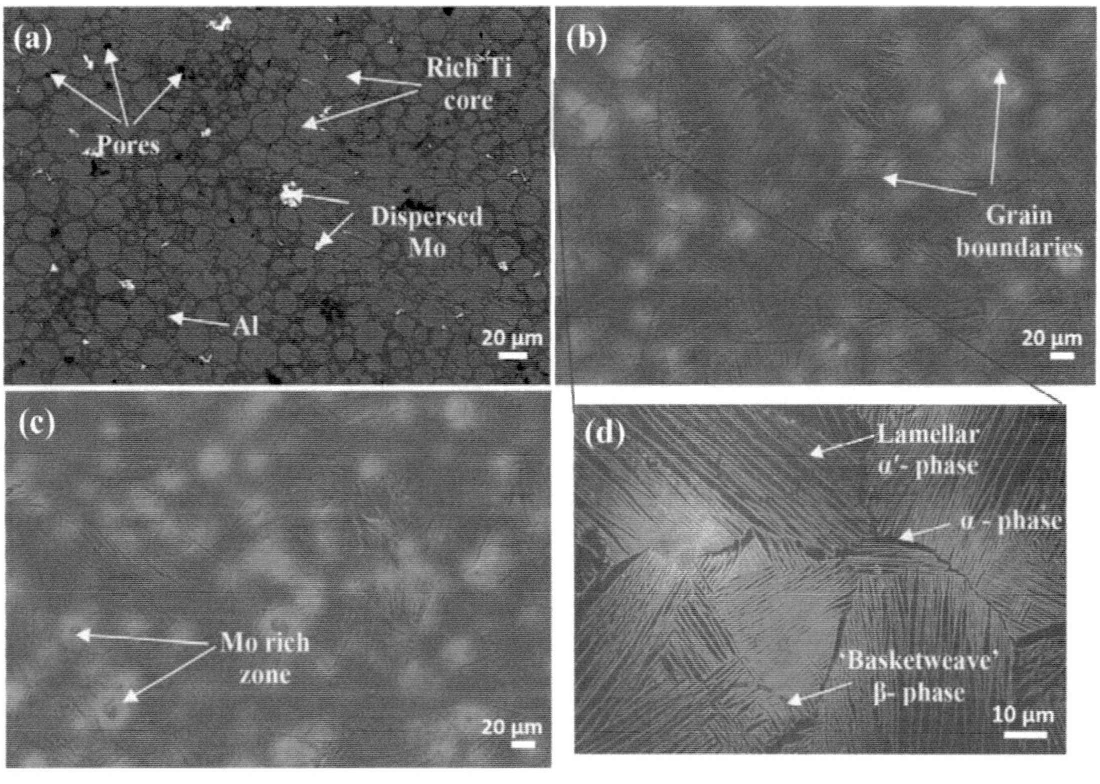

Figure 4. SEM micrograph of spark plasma sintered Ti-5Al-1Mo at various temperatures: (a) 800 °C; (b) 1000 °C; (c) 1100 °C

The alloy compacted at 1000 °C and 1100 °C (Figure 4b and c) exhibits morphology which is comprised of a Widmanstatten lath-like morphology with a lamellar α and small 'basketweave' retained β phase, which is an indication that the produced compacts are near-α titanium alloy, and this is consistent with the phases obtained in the XRD patterns in Figure. 5. The secondary α phase (α'-phase i.e phase exhibiting lighter contrast) represents a typical lamellar lath-like morphology being initiated from the grain boundary of the α layer (phase exhibiting the darker contrast in the micrographs) as shown in Figure 4d. It can also be deduced that Mo particles impeded the kinetics of formation of the α phase.

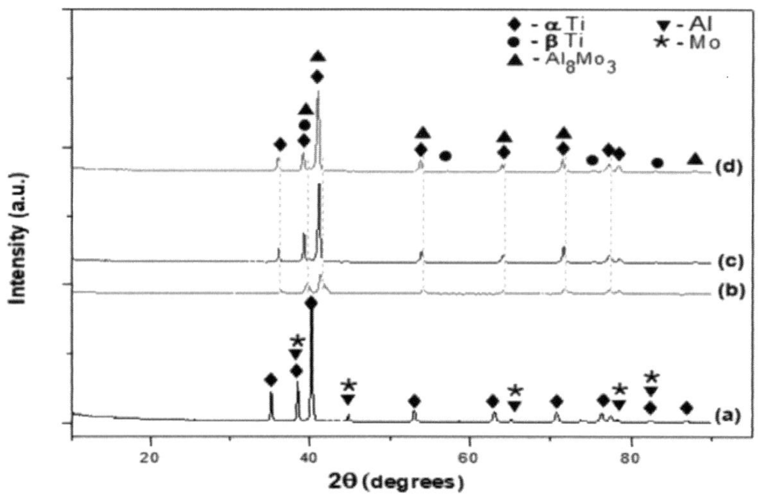

Figure 5. XRD patterns of (a) Admixed Ti-5Al-1Mo powder; Sintered samples at (b) 800 °C; (c) 1000 °C; (d) 1100 °C

This permitted the development of dispersed basketweave microstructures of secondary α-phases at the grain boundaries, which would increase the strength and fatigue resistant of the alloy [17]. It was established by EDS analysis of samples sintered at 1000 °C and 1100 °C that there were Mo rich portion which helped in the pinning down of the β-phase as the sintered samples cooled down to room temperature.

XRD analyses confirmed the presence of α-Ti, Al and Mo phase in the admixed powder (Figure 5a) and there was no formation of any Ti-Al intermetallic phase. The characteristics peaks corresponding to α-Ti, β-Ti and Al_8Mo_3 phases are found in the patterns of the sintered samples with α-Ti as the major phase. There was no noticeable diffraction peak of β-Ti phase in sample sintered at 800 °C; two peaks of β-Ti phase were seen at 1000 °C at 2-theta angle of 38.759° and 83.156° and two other peaks were added for 1100 °C sample at 55.973° and 70.162°. This can be explained in line with the fact the sintering at 800 °C does not reach to the β-transus temperature of the alloy whilst sintering at 1000 °C and 1100 °C was above the β-transus temperature. This implies that Mo as a strong β-stabiliser was able to retain the β-phase as the sample cools to room temperature.

4. Conclusion

The Ti-5Al-1Mo alloy was successfully produced by SPS technique. The sintering temperature effect on density, microstructure evolution, hardness and phase transformation of the alloys were explored. The following deductions can be drawn:

- The density of sintered samples increased with increase in temperature, and fully densified samples were obtained at temperature of 1000 °C and 1100 °C.
- Densification has direct impact on the hardness of the material as it increases as porosity reduces with the highest hardness value of 290.35HV.
- The microstructures of the alloy sintered at temperature of 1000 °C and 1100 °C consisted of a Widmanstatten lath-like morphology with a lamellar α and small 'basketweave' retained β phase.

- The presence of Mo (β-stabiliser) assisted in retaining small portion of β-phase within the microstructure. The development of dispersed basketweave microstructures at the grain boundaries would increase the strength and fatigue resistant of the alloy.
- The obtained results show that the 800 °C SPS temperature regime for near alpha titanium alloy should be avoided.

References

[1] Sankaran K K and Mishra R S 2017 Titanium Alloys *Metallurgy and Design of Alloys with Hierarchical Microstructures* (Elsevier) chapter 5 pp 177-288.

[2] Yang Y, Zhang C, Dai Y and Luo J 2017 Tribological properties of titanium alloys under lubrication of SEE oil and aqueous solutions *Trib. Int.* **109** p 40

[3] Polmear I, StJohn D, Nie J F and Qian M 2017 Titanium Alloys *Light Alloys (Fifth Edition)* (Boston: Butterworth-Heinemann) chapter 7 p 369.

[4] Falodun O E, Obadele B A, Oke S R, Ige O O, Olubambi P A, Lethabane M L, Bhero S 2018 Influence of spark plasma sintering on microstructure and wear behaviour of Ti-6Al-4V reinforced with nanosized TiN *Trans. of Nonferrous Metals Soc. of China* **28** pp 47-54

[5] Zuo Y T, Liu S F, Zhang Z H, Ma H N, Xia W, Liu Y, Zhao Q Q 2013 Microstructure and Texture Characteristics of a Near-Alpha Titanium Alloy Tube *Adv. Mater. Res.* **683** pp 234-7

[6] Sahoo R, Jha B B and Sahoo T K 2013 Experimental study on microstructural variation in near alpha titanium alloy during non-optimal regime of superplastic deformation *Mater. Sci. and Tech.* **29** pp 854-60.

[7] Ahmed Y M, Sahari K S M, Ishak M and Khidhir B A 2014 Titanium and its Alloy *International Journal of Science and Research* **3** pp 1351-61

[8] Mouritz, Adrian P 2012 Titanium alloys for aerospace structures and engines A2 In: *Introduction to Aerospace Materials* (Woodhead Publishing) chapter 9 pp.202-3

[9] Hashimoto K, Kimura M and Suyama R 1994 Alloy Design of Gamma Titanium Aluminide Intermetallic compound *Nippon Steel Technical Report* **62**.

[10] Glatz S A, Koller C M, Bolvardi H, Kolozsvári S, Riedl H and Mayrhofer P H 2017 Influence of Mo on the structure and the tribomechanical properties of arc evaporated Ti-Al-N *Surface and Coatings Technology* **311** pp 330-6.

[11] Witusiewicz V, Bondar A, Hecht U, Stryzhyboroda O, Tsyganenko N, Voblikov V, Velikanova T Y 2018 Thermodynamic re-modelling of the ternary Al–Mo–Ti system based on novel experimental data *J. of Alloys and Compd.* **749** pp 1071-91.

[12] Ready A J, Haynes P D, Grabowski B, Rugg D and Sutton A P 2017 The role of molybdenum in suppressing cold dwell fatigue in titanium alloys *Proceedings of the Royal Society A: Mathematical, Physical and Engineering Sciences* **473** p 189.

[13] Devaraj A, Nag S and Banerjee R 2013 Alpha phase precipitation from phase-separated beta phase in a model Ti–Mo–Al alloy studied by direct coupling of transmission electron microscopy and atom probe tomography *Scripta Materialia* **69** pp 513-6.

[14] Yu C, Cao P and Jones M I 2017 Titanium Powder Sintering in a Graphite Furnace and Mechanical Properties of Sintered Parts (Metals) **7** p 67.

[15] Jeje S O, Shongwe M B, Maledi N, Olubambi P A and Babalola B J 2019 Spark plasma sintering of Ti-48Al intermetallic using elemental powder *The Int. J. of Adv. Manufactur. Technol.* pp 1-8.

[16] Rominiyi A L, Shongwe M B, Maledi N, Babalola B J, and Olubambi P A 2019 Synthesis, microstructural and phase evolution in Ti–2Ni and Ti–10Ni binary alloys consolidated by spark plasma sintering technique *The Int. J. of Adv. Manufactur. Technol.* pp 1-9.

[17] Ackermana A K, Knowles A J, Gardner H M, N´emethb A A, Bantounasa I, Radecka A, Moody M P, Bagot P A, Reed R C, Rugg D and Dye D 2018 The Kinetics of Primary Alpha Plate Growth in Titanium Alloys *cond-mat.mtrl-sci.* preprint arXiv:1805.09885

IOP Conf. Series: Materials Science and Engineering **655** (2019) 012019 doi:10.1088/1757-899X/655/1/012019

Morphological characterization of recycled powder and microstructures of Ti-6Al-4V components synthesized by LENS additive manufacturing

P N Sibisi[1,*], A P I Popoola[1], N K K Arthur[2], S L Pityana[2] and O M Popoola[3]

[1]Department of Chemical, Metallurgical and Materials Engineering, Tshwane University of Technology, Staatsartillerie road, Pretoria 0183, South Africa

[2]Laser Enabled Manufacturing Group, National Laser Centre, Council for Scientific and Industrial Research, Meiring Naudé Road, Brummeria, Pretoria 0185, South Africa

[3]Centre for Energy and Electric Power, Department of Electrical Engineering, Tshwane University of Technology, Staatsartillerie road, Pretoria 0183, South Africa

*Corresponding Author: percivalsibisi@gmail.com

Abstract. Direct Metal Laser Deposition (DMLD) processes such as LENS are gaining ground in commercial manufacturing processes of functional end-products. This is attributed to this process being popularized as a sustainable technology with the potential to substitute the energy intensive, wasteful and environmentally unfriendly traditional techniques. However, during processing a significant fraction of the injected powders do not end up as part of the final component. The properties of these powders change due to heating resulting in coalescence and possible chemical reactions with the surrounding environment. A potential way to overcome this is by adopting an effective powder recycling procedure by which material use efficiency can be improved. Even so, these changes may have detrimental impact on the mechanical properties of the end-product. The objective of this paper is to investigate the effect of powder recycling on mechanical properties as a consequence of a change in powder characteristics. The potential for improving the LENS process by powder recycling process has been evaluated by comparatively studying the feedstock characteristics, i.e. morphology, particle size distribution and microstructure of powders in as received state and after laser exposure during repeated deposition cycles. Microstructures, micro-hardness and Density of the built samples were measured after the first deposition-normalization-reuse cycle. The results revealed a correlation between the powder characteristics and the properties of the corresponding built sample used for that specific deposition.

1. Introduction

The significance of the development of new and existing fabrication techniques in attempt to provide the world with cleaner and cheaper engineering materials comprised of advanced characteristics is forever on the rise. Laser Engineered Net Shaping (LENS) is a cutting edge advanced additive manufacturing technique possessing the potential to change the perception of design and revolutionize manufacturing as a whole [1]. This process involves fabrication of complex 3-D Near Net Shape metallic components directly from CAD models by layer upon layer deposition of metal powders which are transferred into the melt pool created by a laser on the metallic substrate or preceding layers [2], [3]. The operational principle of additive manufacturing processes is in contrast with the conventional subtractive and formative approaches which are wasteful of feedstock and energy intensive as well as limited in freedom of design. Additive fabrication techniques provide various benefits including elimination of forming equipment, high material use efficiency and the flexibility of producing customized parts with high precision which require minimal or sometimes no post process machining; these makes a direct contribution to both the reduction of production cost and delivery time [4]–[6].

Due to the its advantages, LENS is well suitable for critical building and repairing applications such as aerospace, biomedical and architectural industries. Titanium and its alloys are considered expensive materials due to difficulty in refining as a result they are a perfect candidate for near net shape manufacturing techniques [7]. Ti-6Al-4V, the most used titanium alloy has been reported to have applications in many industries because of the excellent combination of attractive properties such as high strength to weight ratio, excellent corrosion resistance and biocompatibility among others. Over the recent years there has been a growing interest in developing additive manufacturing processes of Ti-6Al-4V. The driving force behind this interest is the attempt to reap the benefits provided by this material as it holds a promising future for an extended application scope of Titanium alloys which are commonly limited due to high fabrication cost [8]. However, widespread adoption of LENS fabrication of Ti-6Al-4V is hindered by having most studies concentrating only on process parameters and ignoring the development of powder recycling methods; which can contribute a deal in giving additive manufacturing technologies an upper hand over conventional techniques [9].

According to Renderos et al., [10] the development of recovering, treating and re-using procedures for the powders that remain unfused during deposition eliminates the formation of agglomerations and reduces the risk of clogging as well as maintain the characteristics of the built component while reducing waste. Another study carried out by Asgari et al., [11] involved comparing the characteristics of the virgin, condensate and recycled powder on DMLS process of AlSi10Mg_200C alloy. The authors' observations show that both the virgin and the recycled powder retained relatively identical characteristics, their respective components also revealed similarities in microstructure and mechanical behaviour. From the above literature, it is evident that not much attention has been given to the exploration of recycling methods for the LENS fabrication of Titanium alloys (Ti-6Al-4V in particular) in further assisting in industrialization of the process. The key objective of the present research is to investigate the influence of powder recycling in LENS additive manufacturing.

2. Experimental

2.1. Materials

The present study employs the LENS fabrication for the most widely used titanium alloy, Ti-6Al-4V, which dominates in the aerospace applications. Despite titanium being known as the 4th most abundant metal on the earth's crust, it is still considered expensive due to high refining costs and being difficult to fabricate by subtractive traditional methods which are wasteful of feedstock as machining creates scraps. The above mentioned makes this material and it's based alloys good candidate for exploration of additive manufacturing techniques such as LENS in attempt to reduce fabrication costs and therefore expand their scope application scope in engineering. In this study, the atomized Ti-6Al-4V ELI spherical powders (size range 45-100 microns) were used which were relatively identical in terms of chemistry to the workpiece used as the build platform. The chemical composition of the virgin powder is shown in Table 1.

Table 1. Supplier's chemical composition of the virgin powders.

Element	H_2	N_2	O_2	Fe	C	Al	V	Ti
Composition (%)	0.001	0.006	0.078	0.25	0.006	6.34	3.94	Bal.

2.2. LENS deposition and powder recycling procedure
The Optomec LENS 850-r system mounted with a 1KW IPG fiber laser from CSIR was used for the fabrication of Ti-6Al-4V test samples. The laser deposition processing parameters are presented in Table 2. Ti-6Al-4V plates used as substrates were initially sand blasted and cleaned with acetone in order to remove of dirt/contamination and minimize laser reflection which enhances absorption of the laser energy. The square test samples of 5mm x 5mm x 5mm dimensions were fabricated using virgin powder and then the powder that didn't form part of the test sample was sieved to a particle size range of 45-100 microns using a sieve shaker before it was reused to fabricate the subsequent sample. The samples fabricated from virgin and recycled powder were classified as sample "VP" and sample "RP", respectively. The schematic methodology for the LENS powder recycling procedure followed in this work is shown in Figure 1.

Table 2. Table Showing the Processing parameters, microhardness and the densification of the Ti-6Al-4V deposits using new and aged powder.

Sample	**Laser power (kW)**	**Scan speed (mm/s)**	**Powder feed rate (g/min)**	**Average hardness (HVN)**	**Density (g/cm^3)**
VP	**300**	**16,93**	**1,6**	**367,5**	**4,42**
RP	**300**	**16,93**	**1,6**	**356,22**	**4,32**

The fabricated samples and a powder sample from the first build cycle were taken for characterization in attempt to investigate the changes in both powder and build properties with a successive powder re-use cycle.

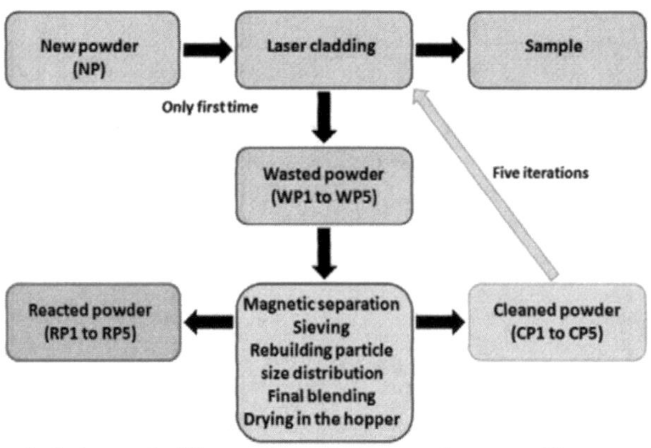

Figure 1. Schematic illustration of the powder recycling procedure.

2.3. Characterization of Powder materials
The characterization of powders carried out after the first re-use cycle has been focused on morphology and particle size distribution of re-used powders in comparison with the fresh virgin powders. The morphological studies and particle size analyses were carried out using SEM/EDS and laser diffraction, respectively.

2.4. Characterization of Built Components

Samples were ground and polished using P320 SiC abrasive/water, Aka-allegran/6micron diamond lubricant and Aka-chemal/Fused Silica Lubricant. Microstructural characterization was carried out by an SEM/EDS. Microhardness and density measurements were done using Vickers micro-hardness tester and the density test rig following Archimedes Principle.

3. Results and Discussions

3.1. Powder Morphology and Particle Size Analyses

The particle size distribution results of the new and re-used powder are presented in the plots in Figure 2 and summarised in Table 3. The particle size of the new powder was found to be different from the particle size range data provided by the supplier (40-100μm) which is confirmed by the presence of particles of up to 119.5 μm size found above the 95th percentile. The analyses of the powder after the exposure to laser radiation shown an increase in particle size of the recycled powder. This serves as proof that despite the potential offered by powder recycling on saving cost through reduction of feedstock required per part built there are issues that may arise due to alterations in powder properties. The particle size found at 95th percentile increases from 119.5 μm to 139.6 which translates to 16.82% increase in particle size.

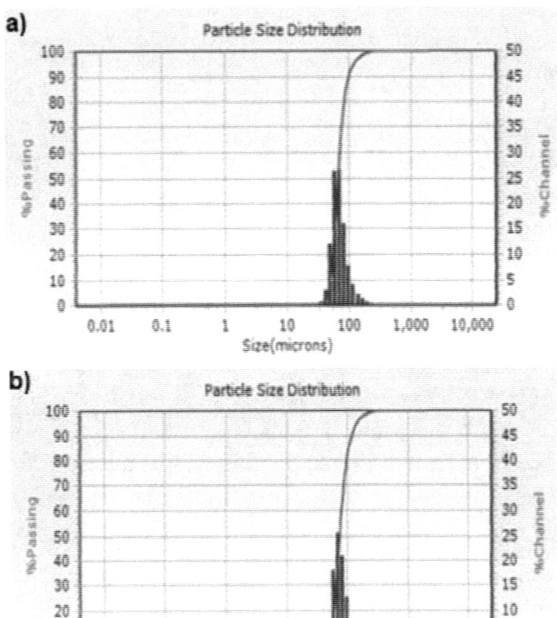

Figure 2. Particle size distribution of a) as-received virgin powder and b) recycled powder.

Table 3. Percentile summary of the particle size distribution

a)

Summary		Percentiles	
Data	Value	%Tile	Size(um)
MV(um):	71.57	10.00	49.51
MN(um):	58.01	20.00	54.32
MA(um):	65.64	30.00	58.06
CS:	9.10E-02	40.00	61.70
SD:	17.66	50.00	65.56
Mz:	68.73	60.00	69.88
si:	19.98	70.00	75.25
Ski:	0.367	80.00	83.32
Kg:	1.335	90.00	99.32
		95.00	119.5

b)

Summary		Percentiles	
Data	Value	%Tile	Size(um)
MV(um):	80.64	10.00	53.31
MN(um):	62.44	20.00	59.04
MA(um):	72.68	30.00	63.79
CS:	8.30E-02	40.00	68.26
SD:	22.48	50.00	72.98
Mz:	77.28	60.00	78.52
si:	24.98	70.00	85.50
Ski:	0.378	80.00	95.73
Kg:	1.304	90.00	115.9
		95.00	139.6

In addition, the average particle size at the 10th percentile which is generaly referred to as D10 and also the lowest in the presented results was seen to increase from 49.51 µm to 53µm after after the first use cycle which suggests that the number of fines is reduced in feedstock therefore flow properties are improved. The d50 and d90 which are represented as the 50th and 90th percentiles, respectively were also found to increase with the former increasing from 65.56 to 72.98µm and the latter increasing from 99.32 to 115µm. This is evident in the micrographs in Figure 3 whereby the smaller particles are seen to have fused onto the larger particle after the first use cycle, which could be the work of coalescence due to the exposure to the laser irradiation.

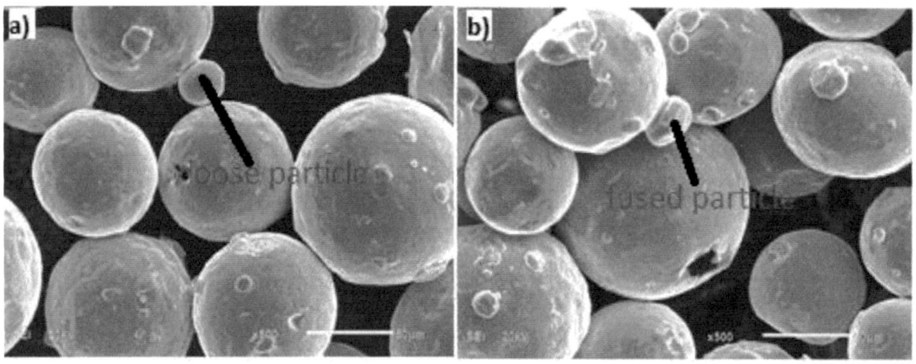

Figure 3. Morphological representation of a) virgin and b) recycled Ti-6Al-4V ELI powder.

This observation serves as evident to the changes in particle size which was seen to increases after the powder was recycled. This can be attributed to the decreasing number of fine sized particles as they are coalesced to the larger particles.

3.2. Microstructures of LENS As-Built Parts

The SEM micrographs of the specimen manufactured using the new and recycled powders in the LENS system are presented in Figure 4. The microstructural features are seen too be relatively similar which is expected since aging of feedstock have very little if not nothing to do with the thermal input and distribution in the system which are determinants of the microstructural evolution during fabrication. This observation is in line with findings reported in literature [12], [13]. However, the re-use of powder induced the formation of defects such as the crack seen in Figure 4b). This observation is attributed to the change in particle size and shape of the feedstock which alter their flowability property. These newly evolved defects may have detrimental effects on the mechanical properties of the built component because they act as the point of weakness where failure propagation is initiated.

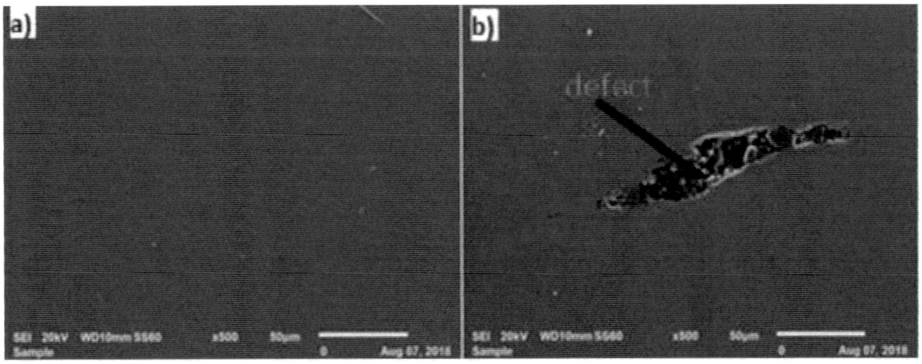

Figure 4. SEM micrographs of the sample built with virgin powder a) and built with recycled powder b).

3.3. Microhardness and Density Measurements

In Table 2, the microhardness and density characteristics of specimen fabricated with feedstock of different characteristics as described above are presented. The results revealed a decrease in both microhardness and density after the first reuse cycle. This decrease in the abovementioned properties can be attributed to an increase in porosity and the existence of defects (e.g. crack in Figure 4b) which are both well known for degrading the properties of the material regardless of the chemical composition.

4. Conclusion

Ti-6Al-4V samples were successfully deposited using recycled powder and results have been reported. The average particle size of powder was found to increase after the first use cycle i.e d10, d50 and d90 increased after exposure to laser radiation during deposition. The morphological micrographs revealed the attachment of fine particles to larger particles which was attributed to coalescence. The decrease in microhardness and densification of deposits after the first re-use cycle was attributed to the evolution of defects as a result of the change in quality of the feedstock after the first re-use cycle. However, a more vigorous study entailing and describing the mechanism of aforementioned changes in powder characteristic is yet to be undertaken in order to correlate the evolution of these changes in powder quality and corresponding deposits.

References

[1] Anderson, I.E., White, E.M. and Dehoff, R., 2018. Feedstock powder processing research needs for additive manufacturing development. *Current Opinion in Solid State and Materials Science*, *22*(1), pp.8-15.

[2] Ahsan, M.N., Pinkerton, A.J., Moat, R.J. and Shackleton, J., 2011. A comparative study of laser direct metal deposition characteristics using gas and plasma-atomized Ti–6Al–4V powders. *Materials Science and Engineering: A*, *528*(25-26), pp.7648-7657.

[3] Qiu, C., Ravi, G.A., Dance, C., Ranson, A., Dilworth, S. and Attallah, M.M., 2015. Fabrication of large Ti–6Al–4V structures by direct laser deposition. *Journal of Alloys and Compounds*, *629*, pp.351-361.

[4] Vrancken, B., Thijs, L., Kruth, J.P. and Van Humbeeck, J., 2014. Microstructure and mechanical properties of a novel β titanium metallic composite by selective laser melting. *Acta Materialia*, *68*, pp.150-158.

[5] Francois, M.M., Sun, A., King, W.E., Henson, N.J., Tourret, D., Bronkhorst, C.A., Carlson, N.N., Newman, C.K., Haut, T.S., Bakosi, J. and Gibbs, J.W., 2017. Modeling of additive manufacturing processes for metals: Challenges and opportunities. *Current Opinion in Solid State and Materials Science*, *21*(LA-UR-16-24513; SAND-2017-6832J).

[6] Liu, L., Shangguan, Y.J., Tang, H.B. and Wang, H.M., 2014. Fretting wear behavior of laser-nitrided Ti–5Al–5Mo–5V–1Cr–1Fe alloy fabricated by laser melting deposition. *Applied Physics A*, *116*(4), pp.1993-2000.

[7] O'Leary, R., Setchi, R. and Prickett, P.W., 2015. An investigation into the recycling of Ti-6Al-4V powder used within SLM to improve sustainability.

[8] Seyda, V., Kaufmann, N. and Emmelmann, C., 2012. Investigation of aging processes of Ti-6Al-4 V powder material in laser melting. *Physics Procedia*, *39*, pp.425-431.

[9] Portolés, L., Jordá, O., Jordá, L., Uriondo, A., Esperon-Miguez, M. and Perinpanayagam, S., 2016. A qualification procedure to manufacture and repair aerospace parts with electron beam melting. *Journal of Manufacturing Systems*, *41*, pp.65-75.

[10] Renderos, M., Girot, F., Lamikiz, A., Torregaray, A. and Saintier, N., 2016. Ni based powder reconditioning and reuse for LMD process. *Physics Procedia*, *83*, pp.769-777.

[11] Asgari, H., Baxter, C., Hosseinkhani, K. and Mohammadi, M., 2017. On microstructure and mechanical properties of additively manufactured AlSi10Mg_200C using recycled powder. *Materials Science and Engineering: A*, *707*, pp.148-158.

[12] Popov Jr, V.V., Katz-Demyanetz, A., Garkun, A. and Bamberger, M., 2018. The effect of powder recycling on the mechanical properties and microstructure of electron beam melted Ti-6Al-4 V specimens. *Additive Manufacturing*, *22*, pp.834-843.

[13] Strondl, A., Lyckfeldt, O., Brodin, H. and Ackelid, U., 2015. Characterization and control of powder properties for additive manufacturing. *Jom*, *67*(3), pp.549-554.

Effect of particle size distribution on green properties and sintering of Ti-6Al-4V

K Ramabulana[1], M Tambani[1], C Machio[1] and M E Makhatha[2]

[1] Light Metals, Materials Science & Manufacturing, Council for Scientific and Industrial Research, Meiring Naudé Road, Brummeria, Pretoria 0185, South Africa

[2] Department of Metallurgy, University of Johannesburg, PO Box 524, Auckland park, 2006, South Africa

E-mail: KRamabulana@csir.co.za

Abstract: In this study, the conventional powder metallurgy route of pressing and sintering was employed to study the effects of the particle size of blended elemental Ti-6Al-4V powders on the properties of powder mixtures, green and sintered densities. The -45 µm and -150 µm 60Al-40V master alloy powder were blended with the titanium powder to produce the -45 µm and -150 µm Ti-6Al-4V blend. Powders were pressed at 580 to 870 MPa and sintered in a tube furnace. Taguchi method was used to optimise the sintering parameters, temperature and time, and ANOVA was employed to determine the significant parameters and their contribution to the sintered density. The optimum parameters maximising the density were found to be sintering temperature of 1300 °C and sintering time of 180 mins, for both particle sizes. The green density was higher for the coarser powder and the sintered density was higher for the finer powder.

1. Introduction

Titanium is the 4th abundant structural metal in the earth's crust. Titanium and its alloys have received interest in a lot of applications such as aerospace, chemical industry, biomedical, and many more because of their high specific strength and excellent corrosion resistance and biocompatibility. They possess the weight half of steels and Ni-based super-alloys. However, because of the extraction and processing costs, titanium and its alloys are expensive and their use is only limited to areas where cost is not an issue. The high cost of titanium-based components can be reduced by changing component production to powder metallurgy [1,2].
The microstructure and properties of sintered powder compacts depend strongly on the quality of the green compacts [3] and in turn, the quality of the green compact is strongly determined by the powder morphology, particle size and size distribution, impurities [4,5] and the densification mechanism [6].
Several authors have studied and established relationships between the compaction pressure, the powder characteristics (such as impurity levels and particle size shape), different compaction route (cold, hot, or dynamic compaction), and the obtained properties of the compacts (density, sinterability and strength) with varying degrees of success [4,5,7,8,9]. Among these, particle size distribution (PSD) is one of the most important, easiest to control and has a direct impact on green compacts and sintering. Properties such as apparent density, tapped density flowability, and compressibility of the powder depend on PSD. Green compacts properties such as porosity, surface roughness and green strength all depend on the PSD [10].
Fine particles generate interparticle friction than coarse powders and this restricts the powder from flowing [10,7]. Moon and Choi [11], have found that the green strength increases with decreasing particle size of the powder due to the larger surface area available for bonding. It is well known that reducing the particle size increases the rate of sintering and densification [5,8].

In this study the effects of the particle size on the green and sintered densities of blended elemental Ti-6Al-4V were investigated. Taguchi method was employed for the design of sintering experiments and analysis of variance (ANOVA) for the optimization process of sintering

2. Experimental Procedure

2.1. Materials

The starting powders used in this work were Ti Hydride-Dehydride (TiHDH) -150 μm and -45 μm and 60Al40V (60 wt.%Al and 40 wt.%V) -150 μm and -45 μm master alloy (MA) both from Baoji Lihua Non-Ferrous Metals Co., Ltd.

2.2. Powder characterisation and mixing

The particle size of the used powders was determined using the Microtrac Bluewave laser particle analyzer (Microtrac Inc, Bluewave, USA) using distilled water as the dispersant. The morphology of all the starting powders was investigated by the JEOL JSM-6510 scanning electron microscope (SEM).
Two Ti4Al6V powder fractions -150 μm and -45 μm were prepared by mixing TiHDH with 10 wt.% 60Al40V master alloy for 30 minutes in a TURBULA® Mixer. The apparent and tapped density of the two powder mixtures was determined according to ASTM D7481-09 and the flowability of powder mixtures was determined according to ASTM B213–03 using a hall flow.

To study the compressibility behaviour of the 2 powder blends, uniaxial die pressing was used as the method of compaction. Powder mixtures were compacted into cylindrical specimens using a 17.5 mm diameter steel die in a single action mode press from 290 MPa – 1160 MPa. No lubricant was applied on the die wall or added in the powder mixture during mixing. The green density of the compacts was calculated from the volume and mass of the pressed compacts. The pressed samples were weighed with a precision mass balance. The volume of the compacted samples was calculated from the diameter and height measured with a vernier calliper. From the mass and calculated volume, the samples' densities were calculated.

Two compaction loads, 580 and 870 MPa were then selected to press compacts for sintering. Taguchi L9 (2^3) orthogonal array was used to design the sintering experiments followed by ANOVA to determine the significant level and contribution of each variables to the sintered density.
Taguchi method is a statistical method for improving the quality of products and process optimization. It provides a simple, efficient and systematic approach to determine optimum conditions with reduced number of experiments. Taguchi identified that there were three specific goals in an experiment, either to i) : minimize the response, The smaller, the better, ii) maximise the response - The larger, the better, or iii) achieving a given target value - target is best. Taguchi uses a performance criteria known as signal-to-noise (S/N) ratio that takes into account both the process mean and the variance. For each of the different goals, the aim is to maximise the S/N ratio [12]. In this study the aim was to maximise the response, larger the better type characteristics was selected and the SN Ratio was determined by:

$$SN = 10log\frac{1}{n}\sum_{n=1}^{n}\frac{1}{y}\dots\dots(1)$$

Where y is the measured response and n is the number of trial runs
Sintering temperature and time were considered as input parameters and the sintered density was considered as the output parameter (the response). The levels at which they were tested is shown in Table 1. Sintering of the pressed compacts was performed in a tube furnace in flowing argon gas and all the runs were conducted at a constant heating rate of 10 °C/min. The sintered density of the specimens was determined by Archimedes principle according to ASTM B311−13. The relative density was calculated as the ratio of the bulk density to the theoretical density and the theoretical density of Ti-6Al-4V was taken as 4.43 g/cm^3.

Table 1: Process parameters and corresponding levels

		levels		
Factors	Unit	1	2	3
Time	min	60	120	180
Temperature	°C	1200	1250	1300

3. Results and discussion

The results of the particle size distribution of the starting powders is shown in Figure 1. The TiHDH -150 µm and MA -150 µm were found to have bimodal powder particle distributions (containing both the fine and coarse particles) and the TiHDH -45 µm and MA -45 µm were unimodal. The measured particle size distribution is in agreement with the observed SEM images as presented in figure 2. The PSD of powders is also presented as D_{10}, D_{50} and D_{90} in Table 2.

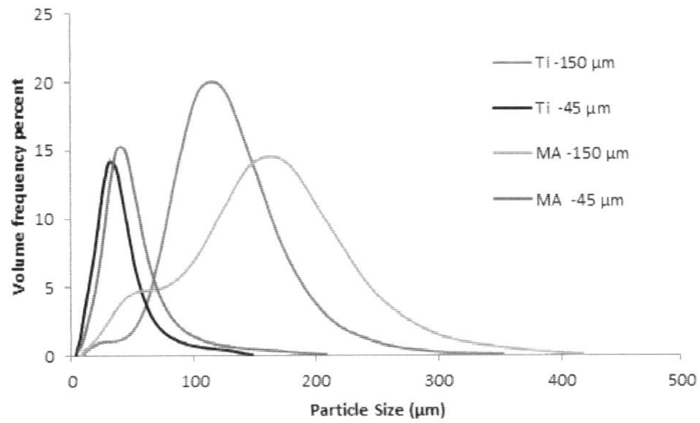

Figure 1. Particle size distribution of starting powders

Table 2: Particle size distribution of starting powders

Powders	d_{10} (µm)	d_{50} (µm)	d_{90} (µm)
Ti -150 µm	54.86	100.2	151.4
Ti -45 µm	11.3	26.68	47.74
MA -150 µm	32.19	114.8	194.9
MA -45 µm	15.16	33.86	59.44

Figure 2 shows the SEM images of the starting materials. The particles of the powders have an angular morphology and for TiHDH, this is consistent with the production process which includes hydrogenation and milling [9].

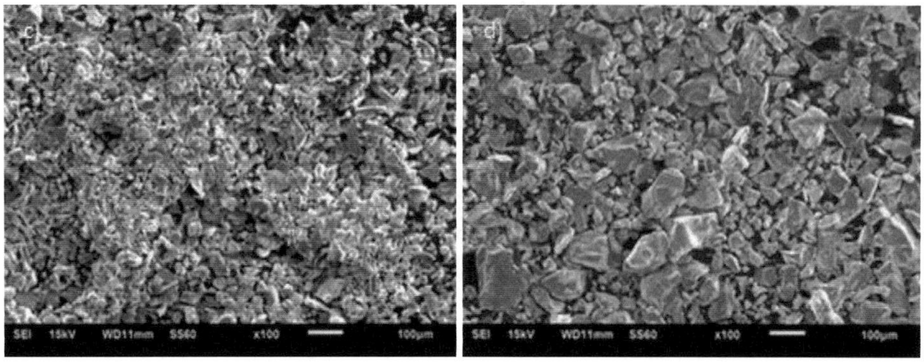

Figure 2. Powder morphology (a) -45 µm Ti HDH (b) -150 µm Ti HDH (c) -45 µm 60Al4V master alloy (d) -150 µm 60Al4V master alloy.

After blending the TiHDH with the master alloy, the properties of the two size fraction were determined and the results are given in Table 3. It can be observed that the decreased size of powders was accompanied with decreased apparent densities and tapped density. The fine powder, Ti6Al4V -45 µm could not flow in the Hall flow meter. This could be attributed to the fact that smaller particles generate greater interpaticle friction and thus particles pack tightly while restricting to flow.

Table 3: Properties of the blended powders

	Apparent Density	Tapped Density	Flowability
Ti6Al4V -150 µm	1.682	2.099	60.12 s
Ti6Al4V -45 µm	1.371	1.924	-

Figure 3 shows the compressibility curves of the investigated powder mixtures. It can be seen in the diagram that the green density of powders increased with the increase in the applied pressure. From 290 MPa to 870 MPa, the green density increases rapidly with the increase in the applied pressure. At this point, the two powders had attained the green densities of 87% for the -45 µm and 92% for the -150 µm. The green density then increased gradually with the increase in compaction pressure up to 1170 MPa, indicating a lower compressibility at higher pressure in the range of this study. The green density increased slightly as the particle size increased. This was expected for the bimodal PSD as the finer particles fill the interstitial voids between coarse particles leading to the reduced porosity and increased green density [13]. Also, there is more inter-particle friction between the -45um powders than the -150um powders. The maximum density obtained at the highest pressure for both powders was 89.9% for the -45 µm and 93.3 % for the -150 µm.

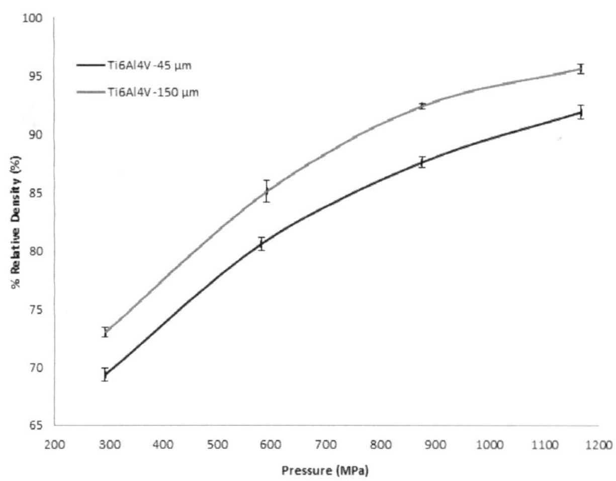

Figure 3. Compressibility curves of the powder mixtures.

In this study 9 sintering trials were done as per Taguchi L9 orthogonal array to study the effects of sintering time and temperature on density at 3 levels on two Ti6Al4V powder mixtures. The effect of sintering time and temperature on density is shown in figure 4. The sintered density was influenced by the sintering temperature and time for both -45 μm and -150 μm particle size at all compaction pressures used, 580 and 870 MPa, but the fine particle size (-45 μm) resulted in higher sintered density despite their low green density as compared to the -150 μm. A comparison of the sintered density for a particular powder pressed at different pressures shows that increasing the compaction pressure increased the sintered densities.

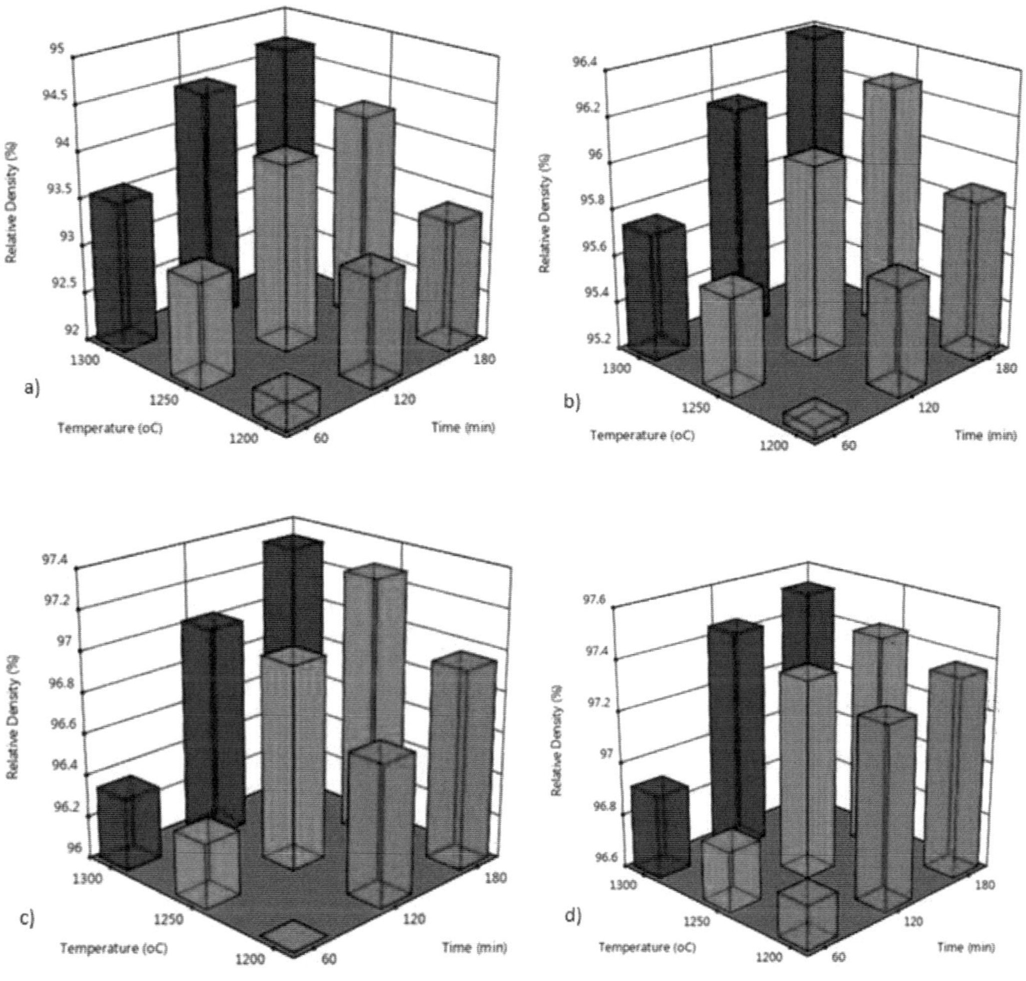

Figure 4. Ti4Al6V interaction plot between Relative Density and Sintering Temperature and Time, a) -150 μm pressed at 580 MPa, b) -150 μm pressed at 870 MPa, c) -45 μm pressed at 580 MPa, d) -45 μm pressed at 870 MPa

The SN equation was used to determine the parameter levels that maximize the density (%) in the sintering of blended elemental Ti6Al4V. SN values were found from equation (1) to determine the parameter levels that maximize the performance response. The results obtained are shown graphically in Figure 5. For both materials, the parameter levels maximising the SN ratio are a sintering temperature of 1300 °C and sintering time of 180 minutes.

IOP Conf. Series: Materials Science and Engineering **655** (2019) 012020 doi:10.1088/1757-899X/655/1/012020

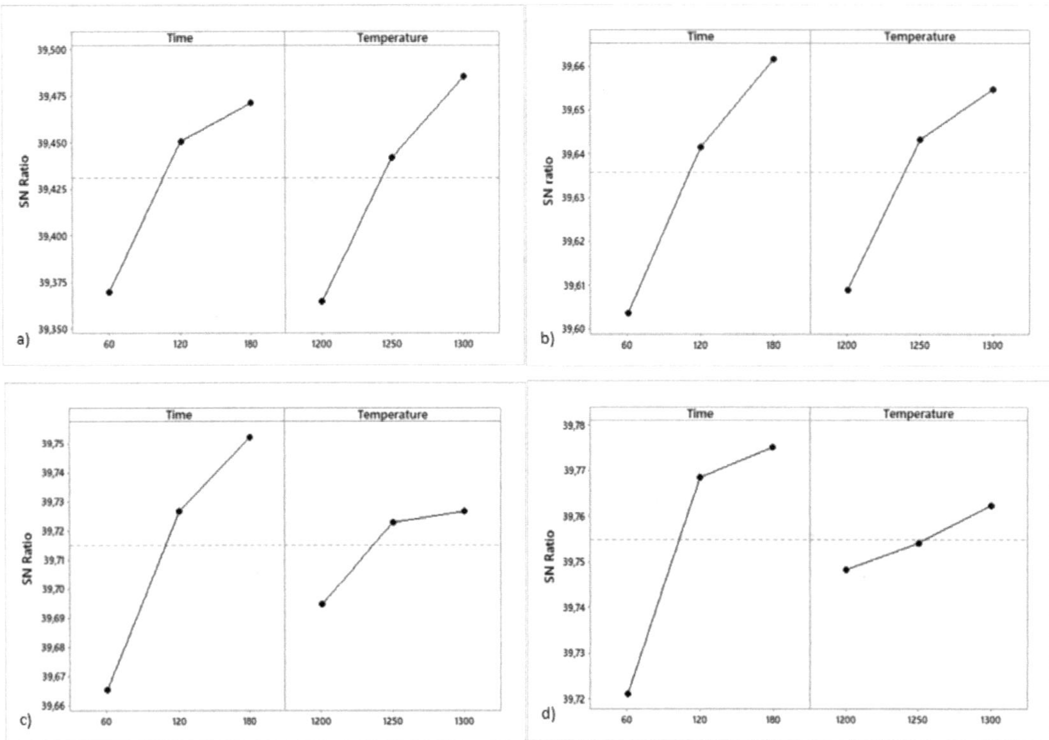

Figure 5. SN Ratio Values for a) -150 μm pressed at 580 MPa, b) -150 μm pressed at 870 MPa, c) -45 μm pressed at 580 MPa, d) -45 μm pressed at 870 MPa

To statistically analyse the significance of process variables the ANOVA was done using Minitab 17 software. Parameters such as P-value, F-value and R-Squared were used to determine the significance of each model parameter. The larger the value of F and the smaller the value of p (< 0.05), the more significant the corresponding coefficient is. Table 4.shows the results of ANOVA analysis for density of the sintered compacts.

From the table of the ANOVA it can be seen by looking at the p-value that for the -150 μm particle size, both temperature and time are significant to maximise the sintered density for all compaction pressures studied, since the p values were very low. However this was not the case for the -45 μm as the p value for temperature increased with the compaction pressure with a p value of 0.681 at 870 MPa and a very low F-Value with a mere contribution of 4.49 to the response. The ANOVA % contribution was supported by the Taguchi Delta value. The Delta Value indicates the relative effect of each factor on the response. It can be seen from Table 4 that the Delta Value for temperature is very small for the -45 μm particle size. The significance of temperature on the sintered density over the ranges investigated could be explained by the fact that sintering densification takes place faster as the particle size decreases because the diffusion distances between particles are shorter and the corresponding curvature stresses are larger [14] and that the samples prepared using fine powder can be sintered at lower temperature than ones prepared by conventional powder no matter by gas-pressure sintering or pressureless sintering techniques [15].

Regression equations were obtained relating input process parameters and the output response Density. The equations are shown in Table 5 with the respective R^2 and Adjusted R^2 values. A model with a good ability to fit the data is required to have the highest R^2 and adjusted R^2 (R^2-adj). This means that the closer the value of R^2 is equal to 1, the model can accurately describe the relationship between the variables and the response variables. Generally, R^2 should be required to be at least 0.8 for the models with well-fitted data [16]. The R^2 and Adjusted R^2 values shown in Table 5 prove that the Taguchi accurately correlates the experimental data for both materials and therefore the regression equations can be used for the prediction of sintered density for confirmation experiments.

Table 4: ANOVA Results

	F-Value	p-value	Contribution (%)	Taguchi Delta Value
-150 µm pressed at 580 MPa				
Time	144.10	0.00020	43.26	1.1
Temperature	187.05	0.00010	56.14	1.3
-150 µm pressed at 870 MPa				
Time	67.36	0.00080	59.50	0.64
Temperature	43.85	0.00190	38.73	0.5
-45 µm pressed at 580 MPa				
Time	33.79	0.0031	82.55	0.97
Temperature	5.14	0.0785	12.56	0.36
-45 µm pressed at 870 MPa				
Time	10.01	0.02780	79.60	0.6
Temperature	0.5648	0.60810	4.49	0.16

Table 5 Regression Equations for the two powders pressed at two different pressures.

Material	Regression Equation	R^2 and Adjusted R^2
150 µm pressed at 580 MPa	$RD\ (\%) = 76{,}26 + 0{,}00914\,t + 0{,}01304\,T$	$R^2 = 0.994$ Adjusted $R^2 = 0.988$
-150 µm pressed at 870 MPa	$RD\ (\%) = 88{,}94 + 0{,}005338\,t + 0{,}005049\,T$	$R^2 = 0.9823$ Adjusted $R^2 = 0.9647$
--45 µm pressed at 580 MPa	$RD\ (\%) = 91{,}37 + 0{,}00806\,t + 0{,}00355\,T$	$R^2 = 0.9511$ Adjusted $R^2 = 0.9023$
-45 µm pressed at 870 MPa	$RD\ (\%) = 94{,}66 + 0{,}00504\,t + 0{,}00156\,T$	$R^2 = 0.8409$ Adjusted $R^2 = 0.6818$

*RD is the Relative density

4. Conclusions

In this study, the effects of particle sizes on the green and sintered were investigated and the experimental results reported here confirm some well-known trends for the effects of particle size on the green density and sintered density.

- The coarse powder packed more densely than the fine powder as it was observed by comparing the apparent density, tapped density of the two powders and the coarse powder consistently exhibited a higher green density for the range of compaction pressures evaluated.
- Fine powder result in higher sintered density than that for the coarse powder for a specific press-and-sinter process scenario.
- Increasing the compaction pressure increased both the green and sintered densities.
- The ANOVA analysis revealed that temperature becomes less significant when sintering fine powder for the temperature range evaluated.

Acknowledgements

The authors would like to acknowledge the Department of Science and Technology and CSIR for funding the project.

Reference

[1] Sachdev A.K., Kulkarni K., Fang Z.Z., Yang R. and Girshov V., (2012) Titanium for Automotive Applications: Challenges and Opportunities in Materials and Processing. JOM, 64, pp: 553-565

[2] Machio C, Machaka R. and Chikwanda H.K., (2016) Consolidation of titanium hydride powders during the production of titanium PM parts: The effect of die wall lubricants. Materials and Design, 90, pp: 757–766

[3] Machaka R, and Chikwanda H (2015). Analysis of the Cold Compaction Behavior of Titanium Powders: A Comprehensive Inter-model Comparison Study of Compaction Equations. Metallurgical and Materials Transactions A, 46, pp: 4286–429.

[4] Ivasishin, O.M, Savvakin, D.G., Froes, F.H., Mokson, V.C, and Bondareva, K.A, (2002). Synthesis of the Ti-6Al-4V alloy having low residual porosity by powder metallurgy method. Powder Metallurgy and Metal Ceramics, 41, pp: 382 – 390.

[5] Qian, M. (2010). Cold compaction and sintering of titanium and its alloys for near-net-shape or preform fabrication. International Journal of Powder Metallurgy 46, pp: 29-44

[6] Panelli R. and Filho F.A. (2001) A study of a new phenomenological compacting equation. Powder Technology, 114, pp: 255–261.

[7] Abdullah E.C., Geldart D. (1999). The use of bulk density measurements as flowability indicators. Powder Technology 102, pp: 151–165

[8] Robertson, I.M. and Schaffer, G.B. (2009). Some Effects of Particle Size on the Sintering of Titanium and a Master Sintering Curve Model. Metallurgical and Materials Transactions A, pp 1968–1979

[9] Robertson, I.M. and Schaffer, G.B (2010). Review of densification of titanium based powder systems in press and sinter processing. Powder Metallurgy, 53, pp: 146-162.

[10] Carson, J.W, and Pittenger, B.H. (1998). Powder Metal Technologies and Applications, Vol 7, ASM Handbook, ASM International, p 287-301)

[11] Moon, I.-H, and Choi, J.S. (!985). Dependence of Green Strength on Contact Area Between Powder Particles for Spherical Copper Powder Compacts

[12] Khuri, A. I. and Mukhopadhyay, S. (2010). Response surface methodology. WIREs Comp Stat, 2, pp: 128-149.

[13] German, R. M. (1992). Prediction of sintered density for bimodal powder mixtures. Metallurgical Transactions A, pp: 1455–1465.

[14] German, R. M., Suri, P., & Park, S. J. (2008). Review: liquid phase sintering. Journal of Materials Science, 44, pp: 1–39.

[15] Eser, O., and Kurama, S. (2012). A comparison of sintering techniques using different particle sized β-SiAlON powders. Journal of the European Ceramic Society, 32, pp: 1343–1347.

[16] Yang, L.Y., Wang, L., Gao, J., Guo, S., Ye, X., Koppala, S., Hu, T., Hou, M.J., & Hu, L. (2018). Optimization of Process Parameters for Preparing Metallic Matrix Diamond Tool Bits by Microwave Pressureless Sintering Using Response Surface Methodology. Materials, 11, p: 2185

IOP Conf. Series: Materials Science and Engineering **655** (2019) 012021 doi:10.1088/1757-899X/655/1/012021

Biocompatibility Study of Ti-based Alloys Fabricated by Spark Plasma Sintering

M R Mahundla[1], W R Matizamhuka [1], R Machaka[2], A Yamamoto[3] and M B Shongwe[4]

[1] Vaal University of technology, Private bag X021, Andries Potgieter Blvd, Department of Metallurgical Engineering,Vanderbijlpark, South Africa

[2] Titanium Centre of Competence, Materials Science & Manufacturing, Council for Scientific and Industrial Research, Meiring Naudé Road, Brummeria, Pretoria 0185, South Africa

[3] National institute for materials Science, 1-1 Namiki, Tsukuba, Ibaraki 305-0044, Japan

[4] Tshwane University of Technology, Staatstirstille Road, Pretoria, South Africa

*mahundla.mr@gmail.com

Abstract. Titanium (Ti) alloys often find their way in orthopaedic applications owing to their superior mechanical properties, biocompatibility, corrosion resistance and low Young's modulus. Nano/sub-micron grain structured commercially pure titanium (CP Ti), Ti-6Al-4V (Ti-64) and Ti-34Nb-25Zr (TNZ) alloys were fabricated via mechanical alloying (MA) of elemental powders followed by spark-plasma sintering (SPS). The powders were milled at 500rpm for 5h and sintered at a temperature of 1200ºC, heating rate of 100ºC/min, dwelling time of 10min and a pressure of 50MPa. The sintered samples were metallographically prepared to a mirror finish surface in 5:2 ratio of colloidal silica and hydrogen peroxide (H_2O_2) mixture. Biocompatibility tests were conducted by water soluble tetrazolium and 1-methoxy phenazinium methylsufate (WST-1 PMS) with sarcoma osteogenetic-2 (SaOS-2) human cell line. High cell proliferation on Ti and Ti-64 was observed after 1 and 4 days of incubation. After 7 days, proliferation on Ti-64 was the highest followed by Ti. However, TNZ alloy have less cell proliferation through the entire period of incubation.

Fibronectin adsorption is less for Ti and TNZ compared to and Ti-64. In decreasing order albumin adsorption was highest on Ti than TNZ alloy followed by less adsorption on Ti-64. Surface wettability of Ti and Ti-64 showed a high contact angle of between 93-98° compared to 86° TNZ alloy. The surface wettability results correlated well to less fibronectin adsorption on Ti-34Nb-25Zr alloy and highest for Ti-6Al-4V. The Ti-34Nb-25Zr showed less cell proliferation which is believed to be linked with fibronectin adsorption results.

1. Introduction

Biometallic materials are increasingly used as implants in medical applications [1]. Titanium alloys are the mostly used metallic biomaterials, specifically for biomedical applications [2]. Owing to their set of suitable properties for those applications, such as low specific weight, high corrosion resistance and biocompatibility [3]. Corrosion resistance is a property of utmost importance in biomedical applications [4]. Biomedical implants are subjected to physiological fluids which are chloride containing and maintained at 37°C. Additionally, corrosion products are responsible mainly for poor biocompatibility and may cause undesirable reactions to the implant-adjacent tissues [5], and corrosion resistance helps in deciding whether the alloy is a best candidate for use as a biomaterial. Some biomaterials that have been used had low biocompatibility and this has encouraged research to improving materials with excellent mechanical behaviour and biocompatibility [5-6].

Ti-6Al-4V is the first titanium alloy used for implant materials, and it is mostly used in implant materials for dental, orthopedic and ostoesynthesis applications [7]. However, studies have indicated that vanadium, a beta-phase stabilizer, produces oxides that are harmful to the human body [7-8]. The toxicity associated with vanadium has encouraged the research for alternative materials to replace Ti-6Al-4V [8]. In order to replace Ti alloys containing vanadium, Ti-6Al-7Nb was developed and it is available commercially [9]. Niobium is a beta-phase stabilizer. However, there has been a concern also associated with Al and Alzheimer's disease [10]. Ti-13Nb-13Zr has been recently developed, a near-beta phase alloy [11-12], the Ti, Nb and Zr elements are not toxic and display excellent biocompatibility [13-15].

The addition of Nb to Ti stabilizes the β phase and it improves the mechanical properties [2].Some of the Ti-based alloys have been produced using arc melting and casting techniques [2, 3; 6–12]. However, when the melting temperatures between the alloy constituents broadly range, the use of conventional casting techniques in the synthesis of Ti-based alloys does not seem to be an optimal route due to the absence of homogeneity in the final alloy [2]. Powder metallurgy (PM) processing is also considered to be an effective in reducing the exorbitant machining costs of Ti alloys. It has the advantage of producing a homogenous alloy compared to other techniques, when the alloying elements possess a higher melting temperature such as Nb [4].

Mechanical alloying (MA) has also been successfully used in the development of nanostructured materials [1]. This technique is known to be simple, multifaceted, and feasible economically [13]. The mechanically alloyed powders are consolidated by spark plasma sintering. Spark plasma sintering is a non-conventional method for consolidating solid powders into a metallic compact at shorter times and low temperatures to prevent melting [1, 2]. It is considered more suitable for the production of nano-materials, since it does not induce undesirable grain growth in comparison to other conventional sintering techniques such as hot isostatic pressing (HIP). It also minimises the formation of undesirable secondary phases that deteriorate the properties [3]. The objective of this study is to evaluate the biocompatibility behaviour of Ti, Ti-6Al-4V and Ti-34Nb-25Zr in SaOs-2 human osteosarcoma cells using WST-1 colorimetric method.

2. Experimental methods

2.1. Biocompatibility evaluation

The cytotoxicity tests were conducted following the procedure for cytotoxicity test methods using a direct contact method for biological safety of medical devices. The sample disks were sterilized using an autoclave for 2h at 120°C.The samples of the 3 alloys were separately placed into the bottom of a 24-well microplate. SaOS-2 osteogenic cells were cultured on each disk at the concentration of 600000 cells/ml of culture medium, Dulbecco's minimum essential medium with 10% fetal bovine serum and 0.2% penicillin streptomycin (D-MEM+10%FBS+0.2%PS). The sample cells were seeded into the well without a sample disk as a control.

The cells were incubated at 37°C in a 5vol% CO_2 atmosphere with relative humidity of 95% for 1, 4 and 7 days. The experiments were performed in triplicates. After incubation, WST-1 PMS in phosphate buffered solution [PBS (-)] colorimetric assay was conducted and the cells were fixed using 25vol% glutaraldehyde solution for 10min and stained using 10vol% of Giemsa's staining solution for 15min. The stained cells were observed by an optical microscope and their images were captured using a closed couple device camera. The stained cells were observed by an optical microscope and their images were captured using a closed couple device camera. The absorbance was then measured using a microplate reader (Multiskan FC Microplate Photometer, Thermo Fisher Scientific, Waltham, USA) at 450nm. Relative cell viability (RCV) was calculated using Equation 1.

$$\boldsymbol{RCV} = \left(\frac{\boldsymbol{Abs.450\ cell\ well - Average\ Abs.450\ medium\ well}}{\boldsymbol{sample\ top\ surface\ area}}\right) \div \left(\frac{\boldsymbol{Abs.450\ cell}}{\boldsymbol{well\ bottom\ area}}\right)$$

Equation 1

2.2. *Protein adsorption analysis*

Protein adsorption analysis was performed for albumin and fibronectin. The surface of the sample was polished until mirror finish and sterilized by an autoclave. After sterilization by an autoclave for 2h at 120°C, the samples were placed in the 24-well microplate and immersed in 1ml of Albumin and 1ml of fibronectin. The samples were placed into an incubator for 3h at 37°C. Afterwards, the samples were rinsed with PBS (-) and fixed using Coomassie plus reagent and mixed in a microplate mixer (AS-ONE, NS-P) for 30s. The samples were left for 10 min at room temperature and adsorption was determined through absorbance reading at 595nm in a microplate reader (Multiskan FC Microplate Photometer, Thermo Fisher Scientific, Waltham, USA).

2.3. *Surface wettability*

Surface wettability tests were performed using a dispensing equipment (Kyowa AD-30, Japan) and ultra-super pure (USP) water (milli-Q, millipore) as a medium for wettability. Contact angle measurements were conducted on the surface of Ti, Ti-6Al-4V and Ti-34Nb-25Zr sample disks. The wettability of each sample was determined by sessile drop measurements of the water contact angle at a contact angle goniometer.

3. Results and discussions

3.1. *Biocompatibility*

Cells are sensitive to the physical and chemical properties of the materials they interact with. Simple modifications of the material, variations in surface texture, topography can significantly affect changes in cellular response [5]. Cell adhesion to implant materials which are artifcial is affected by properties such as surface roughness, surface charge and chemical functionalities [9]. Biocompatibility is the most important property for biomaterials to render them suitable for biomedical applications. The calculated relative cell viability and morphologies of cell adhesion on each substrate are shown in Figure 1 and Figure 2, respectively.

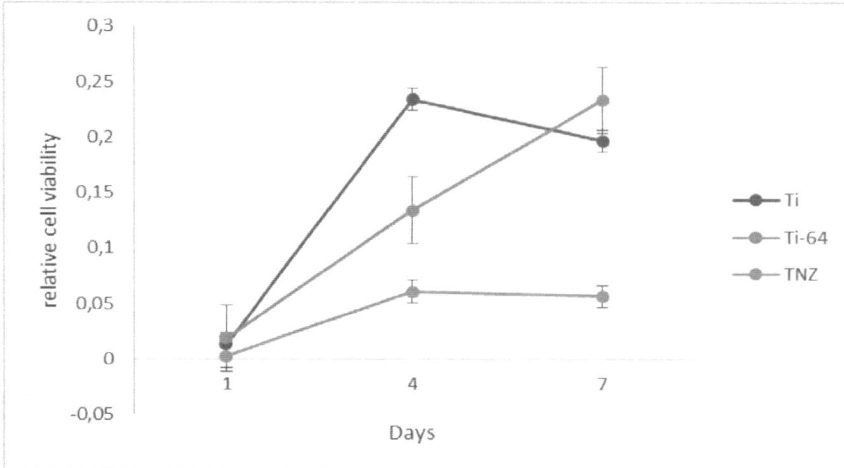

Figure 1. Biocompatibility test for Ti, Ti-6Al-4V and Ti-34Nb-25Zr alloys on SaOS-2 osteosarcoma human cells after 1, 4 and 7 days of incubation.

Figure 1 shows that Ti and Ti-6Al-4V has increasing cell proliferation as observed after 1 and 4 days incubation. After 7 days, cell proliferation on Ti-64 was the highest followed by Ti. However, Ti-34Nb-25Zr alloys have less cell proliferation through the entire period of incubation. This can be attributed by variations in physico-chemical properties of the different substrates in which the cells were cultured as well as the processing technique [5, 11]. Figure 2 shows morphologies of the cells cultured for 1, 4 and 7 days on the three kinds of alloys. There was a notable difference observed in Figure 1 since cell proliferation and adhesion on the samples shown in figure 2 attributed to poor biocompatibility as Ti and Ti-34Nb-25Zr had least cell adhesion. Although Ti-6Al-4V alloy had more cells adhering to the surface and it is promising in terms of biocompatibility which was least expected due to the toxicity of vanadium. However, the non-toxic Ti and Ti-34Nb-25Zr alloys showed a complete opposite in terms of biocompatibility. The calculated relative cell viability compliments the results obtained for cell adhesion as Ti-6Al-4V still showed many cells that grew and spread through the surface.
The main cause for the retardation of the growth pattern of the cells in biomaterial surfaces of the studied alloys are still unknown and will be further investigated.

The microscopic images of cultured cells show that after 1 and 4 days of incubation, the cells visually appeared as scattered stained dots. After 7 days of incubation, the cells grew and spread to cover almost the surface area of Ti and Ti-6Al-4V, while a wide area of empty spaces were observed on the disk of Ti-34Nb-25Zr. Cells on Ti and Ti-34Nb-25Zr showed some sickening rather than the ones which grew and spread smoothly on the surface of Ti-6Al-4V.

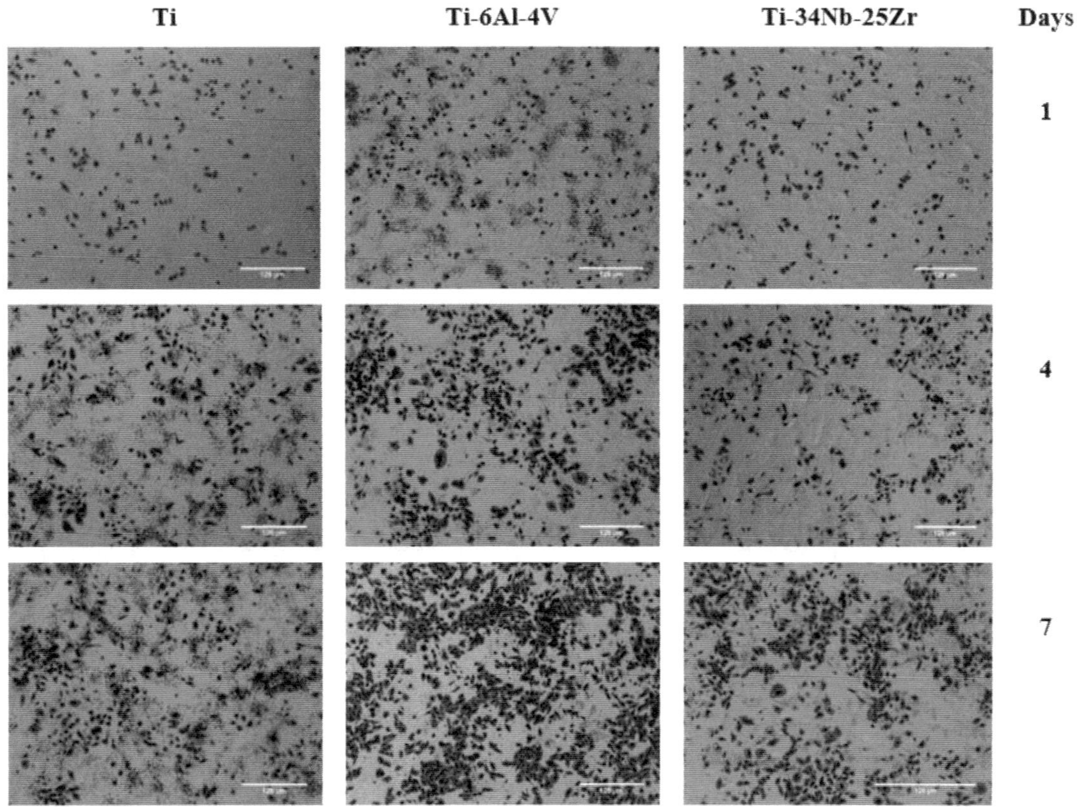

Figure 2. Optical microscopy images of the cultured cells on the disks of Ti, Ti-6Al-4V and Ti-34Nb-25Zr.

3.2. *Protein adsorption*

The adsorption of cell adhesive proteins such as fibronectin, plays an important role in cell adhesion to the biomaterial [7]. During the formation of an adsorbed protein in early stages, albumin is expected to adsorb preferentially onto the surfaces upon exposure. When albumin which acts like a surfactant, adsorbs to the interface there is a large interfacial tension between water and a hydrophobic surface, which eventually reduces the interfacial free energy is reduced [7, 14]. Albumin adsorbs strongly on hydrophobic surfaces (with contact angles above 100°) [15]. When a metallic substrate is exposed to cells suspended in a culture medium supplemented with FBS, protein in the serum are adsorbed rapidly onto the surface before the cells adhere.

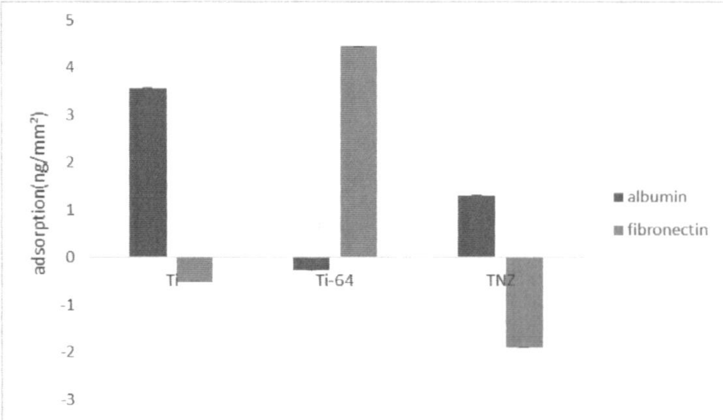

Figure 3. Protein adsorption analysis in albumin and fibronectin

The adsorbed proteins determines the subsequent cell adhesion behaviour [14]. The amount of proteins adsorbed also depend on the water contact angles and also varies with the kinds of surfaces of the substrates. In this study, the amount of proteins adsorbed showed little variations for all the substrates [15]. Ti, TNZ (Ti-34Nb-25Zr) substrates showed good albumin adsorption than Ti-6Al-4V. However, Ti-6Al-4V shows good fibronectin adsorption (Figure.2). Contrarily, there is no obvious correlation between the proteins adsorbed and the number of adherent cells was observed on the studied substrates.

3.3. *Surface wettability*

Generally surface wettability is referred to as hydrophobicity and hydrophilicity. It is one of the most important parameters for materials. Wettability affect protein adsorption, activation, platelet and cell adhesion. Hydrophobic surfaces are generally considered to be more protein adsorbent than hydrophilic surfaces due to the interactions occurring on the surface which has strong hydrophobicity. Ti and Ti-6Al-4V show high contact angle values as compared to Ti-34Nb-25Zr as depicted from Table 1. The alloys showed a hydrophilicity behaviour, since their contact angle values are below 100°.

Table 1. Surface wettability of Ti, Ti-6Al-4V and Ti-34Nb-25Zr using USP water.

Alloy	Contact angle (°)	Type
Ti	93	Hydrophilic
Ti-6Al-4V	98	Hydrophilic
Ti-34Nb-25Zr	86	Hydrophilic

4. Conclusion

Biocompatibility of the spark plasma sintered titanium-based alloys (CP Ti, Ti-6Al-4V and Ti-34Nb-25Zr) was studied by immersion in a simulated body fluid for 1, 4 and 7 days. Surface wettability of water and protein adsorption of albumin and fibronectin was also studied. The results obtained show that all the alloys had fairly poor cell proliferation of SaOS-2 osteosarcoma human cells. The surface wettability results correlated well to less fibronectin adsorption on Ti-34Nb-25Zr solid alloy and highest for Ti-6Al-4V. Ti-34Nb-25Zr showed less cell proliferation which is believed to be linked with fibronectin adsorption and poor cell adhesion. It can be concluded that biocompatibility of SaOs-2 osteosarcoma human cells was not established in the studied alloys, although Ti-6Al-4V proved to be a better candidate in terms of cell adhesion after 7 days incubation. It is recommended that a study in processing technique and surface properties of the alloys be carefully monitored and the biocompatibility tests be conducted using a different kind of cells (i.e. osteoblast (L929) and murine fibroblast).

5. Acknowledgements
The authors would like to thank Vaal University of Technology (VUT), South Africa, Tshwane university of technology (TUT)-INER lab, South Africa, National Research Foundation (NRF) and department of science and technology(DST), South Africa, and National Institute for Materials Science (NIMS), Japan. The support is gratefully acknowledged.

6. References
[1] Hussein M A, Suryanarayana C, Al-Aqeeli N, Fabrication of nano-grained Ti-Nb-Zr biomaterials using spark plasma sintering, *Materials and design.* **87** *(2015)693-700.*

[2] Hussein M A, Suryanarayana C, Arumugam M K, Al-Aqeeli N, Effect of sintering parameters on microstructure, mechanical properties and electrochemical behaviour of Nb-Zr for biomedical applications, *Materials and design.***83** (2015) 344-351.

[3] Mendes M W D, Agreda C G, Bressiani A H A, Bressiani J C, A new titanium-based alloy Ti-27Nb-13Zr produced by powder metallurgy with biomimetic coating for use as a biomaterial, *Mat. Sci and Eng.* C**63** (2016) 671-677.

[4] Niemeyer T C, Grandini C R, Pinto L M C, Angelo A C D, Schneider S G, Corrosion behaviour of Ti-13Nb-13Zr alloy used as a biomaterial, Alloys and compounds, *(2008)172-175.*

[5] Donato T A G, de Almeida L H, Nogueira R A, Niemeyer T C, Grandini C R, Caram R, Schneider S G, Santos Jr A R., Cytotoxicity study of some Ti alloys used as biomaterial, *Materials science and engineering.*C**29** (2009) 1365-1369.

[6] Saji V S, Choe H C, Electrochemical corrosion behaviour of nanotubular Ti-13Nb-13Zr alloy in Ringer's solution. *Corrosion science.* **51**(2009)1658-1663.

[7] Arima Y, Iwata H, Effects of wettability and surface functional groups on protein adsorption and cell adhesion using well-defined mixed self-assembled monolayers. *Biomaterials.***28** (2007)3074-3082.

[8] Hussein M A, Kumar M, Drew R, Al-Aqeeli N, Electrochemical corrosion and in vitro bioactivity of nano-grained biomedical Ti-20Nb-13Zr in a simulated body fluid. *Materials.***11** (2018) 1-15.

[9] Awang-Shri D N, Tsuchiya K, Yamamoto A, Cytocompatibility evaluation and surface characterisation of TiNi deformed by high-pressure torsion. *Material science and Engineering.* **C43** (2014) 411-417.

[10] Suryanarayana C, Al-Aqeeli N, Mechanically alloyed nanocomposites. *Progress in material science.***58** (2013) 383-502.

[11] Uwais Z A, Hussein M A, Samad M A, Al-Aqeeli N, Surface modification of metallic biomaterial for better tribological properties. *A Review. Mechanical Engineering. (2017).*

[12] Bracco G, Holst B, Surface science techniques. **51**(2013).

[13] Bruschi M, Steinmuller-Nethl D, Goriwoda W, Rasse M, Composition and modifications of dental implant surfaces. *Oral implants. (2015).*

[14] Ouberai M M, Xu K, Welland M E, Effect of the interplay between protein and surface on the properties of adsorbed protein layers. *Biomaterials.***35** (2014) 6157-6163.

[15] Xu L C, Siedlecki C A, Effects of surface wettability and contact time on protein adhesion to biomaterial surfaces. *Biomaterials.* **28** (2007) 3273-3283.

[16] Lavos- Velereto IC, Wolynec S, Ramires S, Guastaldi A C, Costa I, *J. Mater. Sci.: Mater. Med* (2004) **55**.

Development of a biocompatible Ti-Nb alloy for orthopaedic applications

L Fikeni[1, 2,*], K A Annan[1], M Seerane[2], K Mutombo [2,]and R Machaka[2, 3]

[1] University of Pretoria, Department of Materials Science and Metallurgical Engineering. Private bag X20, Hatfield 0028, South Africa

[2] Light Metals, Materials Science & Manufacturing, Council for Scientific and Industrial Research, Meiring Naudé Road, Brummeria, Pretoria 0184, South Africa

[3] School of Mining and Metallurgy and Chemical Engineering, University of Johannesburg, Doornfontein Campus, Johannesburg, South Africa

E-mail:,*LFikeni@csir.co.za

Abstract. Metallic biomedical implants such as titanium-based alloys are very useful for orthopaedic applications due to their excellent properties which responds to changes in temperature and other conditions. However, biological toxicity due to alloying elements and relatively high Young's modulus or mechanical incompatibilities of previously used Ti alloys have necessitated the development of biocompatible alloys with compatible mechanical properties such as beta-titanium alloys. This study aims at production of beta-titanium alloy with enhanced properties by varying milling speeds. Ti and Nb powders were mechanically alloyed using the high energy ball-mill Zoz-Simoloyer® to produce Ti-7Nb alloys by varying the milling speed. The milling process produced irregular shaped powders with increasing particles sizes as the milling speed increased due to fragmentation and cold welding during agglomeration. The mechanical alloying process had good yield. The predominant phases of the inhomogeneously milled alloy were alpha and beta phases.

1. Introduction

Titanium-based alloys are used as biomedical implants for orthopaedic applications such as knee or joint replacements due to their high strength, biocompatibility and good corrosion resistance. Ti6Al4V is a commonly used titanium-based alloy and has been reported to release aluminum and vanadium ions which causes health problems such as Alzheimer disease and ostemomalacia [1-2]. The release of these ions and their oxides also causes inflammation leading to aseptic loosening of the implant and possible to a surgery revision [1,3]. The Young's modulus of the bone is about 4-30 GPa, implants with a higher young's modulus may cause stress shielding effect which is when the implant bears most of the mechanical load shielding the bone from bearing the load required for it to maintain strength and healthy structure leading to implant failure[1,4]. The Young's modulus of titanium-based alloys is dependent

on the chemistry and phase of the alloy, with the beta-phase having the lowest and the martensitic phases having relatively low young's modulus and the ω-phase has the highest young's modulus [4-5]. Beta-stabilizing elements such as niobium, zirconium and tantalum have displayed non-toxic behaviour and they show good compatibility with direct contact with cells and relatively good cell growth and mitochondrial activity[3]. Ti-Nb alloys have been prepared by different powder metallurgy routes and casting techniques, producing alloys with different microstructures and mechanical properties [6-8].
Mechanical alloying is a process capable of preparing complex alloys with specific microstructures and good mechanical properties [9-10]. This alloying process produces homogeneous materials that are prepared by mixing pure or pre-alloyed powders and then milled together with a grinding medium such as grinding steel balls [6], [7]. It is a high energy ball milling process in which the alloying of the powder particles occurs due to fracture, plastic deformation and cold welding of particles [11-12].The mechanical alloying process makes possible the production of alloys from elemental powder that are difficult to alloy. It produces novel crystalline phases and a very small range of grain sizes up to nanometres [6]. Important parameters that mostly affect the milled products are the milling speed and time, the grinding ball to powder weight ratio and process environment.
In this study, the mechanical alloying process of Ti-7Nb was studied by varying the milling speed in order to evaluate the alloying of CP-Ti to Nb using the high energy ball milling.

2. Method and materials

The elemental powder of titanium (99.8 % purity, -150 μm), and niobium (99.8 % purity, -25 μm) were used to prepare Ti-7Nb alloys using high-energy ball milling process. The Ti-Nb alloys were milled using the Zoz-Simoloyer® under vacuum in a steel vial and grinding balls. The Zoz-Simoloyer® was purged with argon three times prior to mechanical milling. The Ti-Nb alloys were milled for 6 hours and at milling speed of 800, 900, 100, 1100 and 1200 rpm with a ball-to-powder ratio (BPR) of 10:1.
The as received and milled powders were characterised using the Scanning Electron Microscope (Jeoul™ JSM6510 SEM) equipped with energy dispersive X-ray spectrometry (EDX). The conventional sieving particle analysis were used to determine the morphology and the particle size distribution. Some milled powders were cold mounted, ground to a 4000 grit finish under running water with SiC papers and polished with the colloidal silica to a mirror like surface finish, then characterize using SEM/EDX. The phase analysis of the as received and milled powder was done using X-ray diffraction (XRD) coupled with the PANalytical X'Pert™ Pro powder diffractometer in θ–θ configuration with an X'Celerator™ detector with Fe filtered Co-Kα radiation (λ=1.789Å).

3. Results and discussion

The CP-Ti and Nb elemental powder are shown in Figure 1 a and b respectively. Both powders exhibit irregular morphology. The particle size distribution of the CP-Ti and Nb powder is shown in Figure 1 c and d respectively. CP-Ti powder has larger particles with an average particle size of 92.71 μm and Nb powder has smaller particles with average particle size of 18.61 μm.

(a) (b)

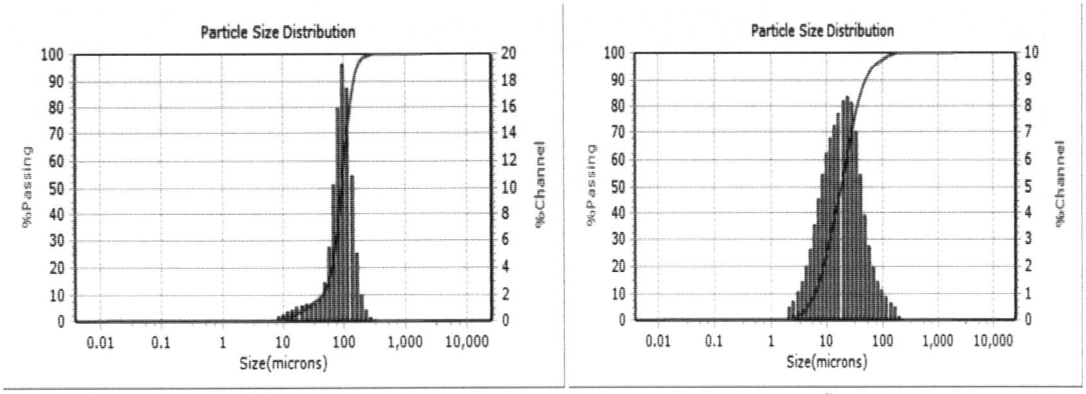

(c) (d)

Figure 1: (a) CP-Ti, (b) Nb elemental powder and, the particle size distribution of (c) CP-Ti and (d) Nb powder.

The yield of the milled powder at different milling speeds varies from 80 to 100 % as shown in Figure 2. The good recovery of the milled powders shows good yield characteristic for the mechanical milling process and non-sticking of the powder to the grinding vial and balls.

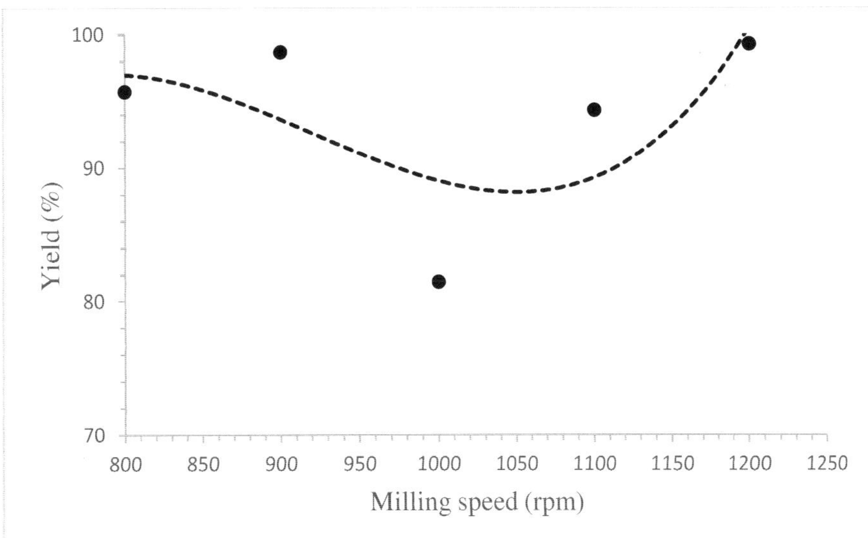

Figure 2: The yield of the milled powders at different milling speeds.

The particle size distribution of the milled powders at different milling speeds is shown in Figure 3. The particle size increases with increasing milling speed due to continuous collisions between the powder and milling medium causing agglomeration and cold welding of the powder particles.

The size of the particles increased with increasing milling speed. There is a wide distribution of size with larger particles which indicates that cold welding and agglomeration of the particles is the prevailing process than fracturing of the powder particles as the milling speed increased. The collision between the grinding balls, the walls of the vial and the powder facilitates the mixing of the powders and gives the kinetic energy for mechanical alloying the powder [9]. The elemental powders were mixed as shown by the elemental distribution in Figure 3 for all the milling speeds

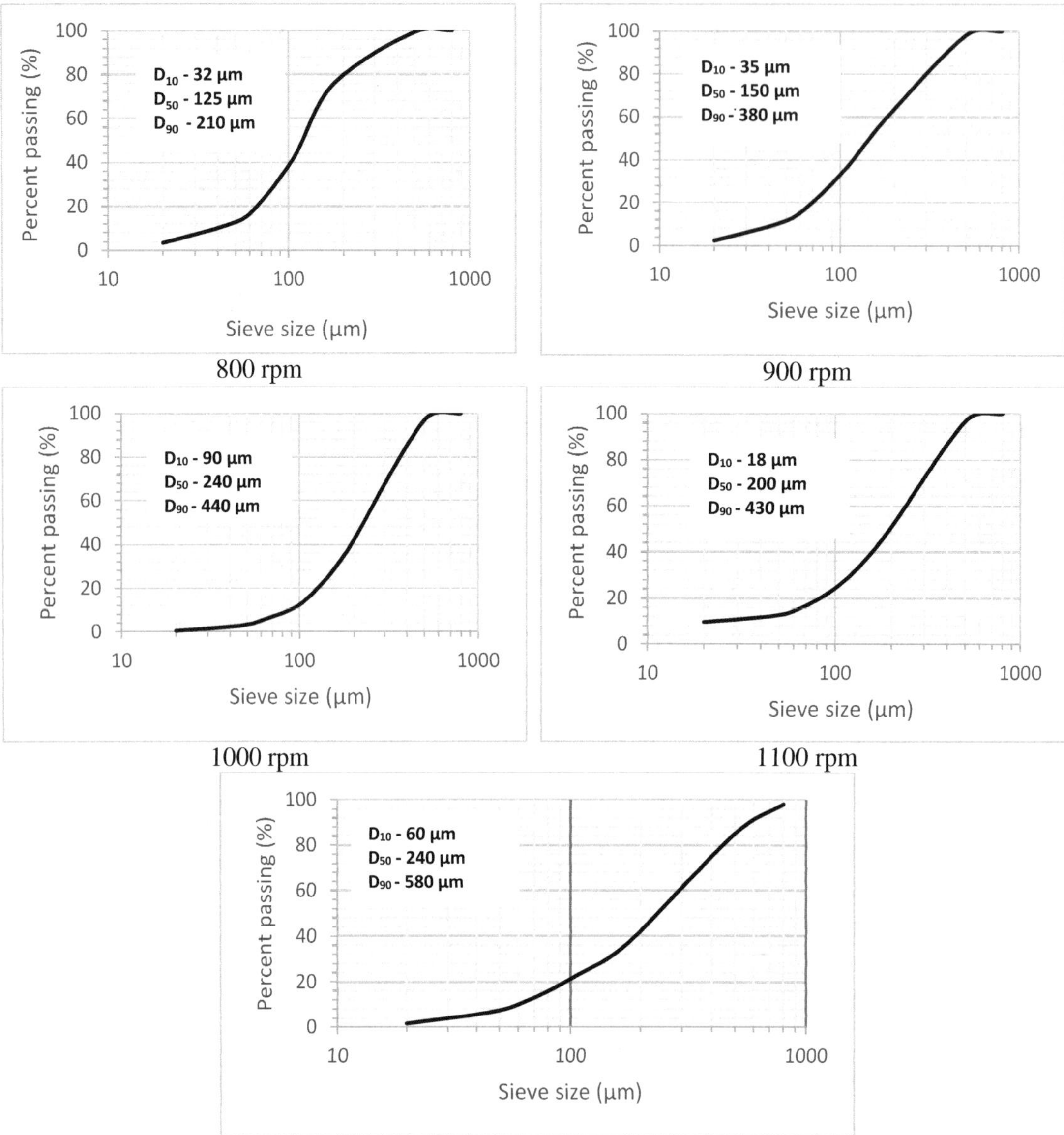

Figure 3: The particle size distribution of the milled powders at different milling speeds.

The morphology and EDS chemical element mapping of the powder milled at 800, 900, 1000, 1100 and 1200 rpm are given in Figure 4. The mechanical milling process produced flakes and irregular shaped granules due to continuous fracture, plastic deformation and cold welding of the powder particles.
The EDS chemical analysis of the powder milled at different speeds is given Table 1. The presence of Ti and Nb is revealed, while the iron (Fe) contamination in the milled powder is peculiarly prevailing at low milling speed.

Figure 4. The morphology and EDX-mapping of the Ti-7Nb powder milled at powder milled at 800, 900, 1000, 1100 and 1200 rpm.

Table 1: The chemical compositions of the Ti-7Nb milled powder.

Milling speed (rpm)	Ti (wt. %)	Nb (wt. %)	Fe (wt. %)
800	91.706	7.656.	0.638
900	93.315	6.385	0.300
1000	92.773	6.960	0.267
1100	94.075	5.709	0.217
1200	94.175	5.570	0.251

The morphology of the mounted Ti-7Nb powder is given in Figure 5. The EDS analysis revealed the Ti-rich phase (light-grey region) and the Nb-rich phase (white region). The Ti and Nb were blended uniformly from 900 rpm milling speed, at 800 rpm the Ti and Nb rich phases are not uniformly distributed as shown in Figure 5 (800 rpm). The Nb-rich phase became predominant in the Ti-rich phase matrix as milling speed increased.

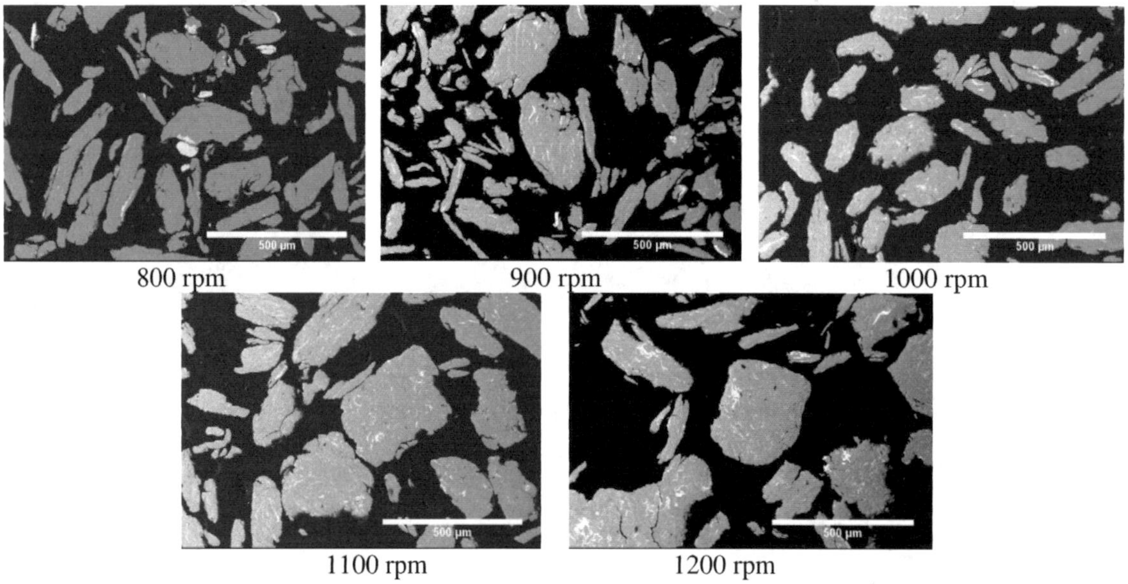

Figure 5: The morphology of the Ti-7Nb milled powder at 800, 900, 1000, 1100 and 1200 rpm.

The XRD patterns of the CP-Ti and Nb mixed elemental powder and Ti-7Nb milled powder at 800, 900, 1000, 1100 and 1200 rpm are shown in Figure 6. The hexagonal close packed crystal structure (hcp) or α-phase (Ti-rich phase, Figure 5), and face centred cubic structure (bcc) or β-phase (Nb-rich phase, Figure 5) which is stabilised by Nb element are revealed. The elemental Nb peaks (as mixed elemental powder) disappeared with increasing milling speed which suggests the occurrence of phase formation and consequently the alloying process during the milling of elemental powder of Nb and CP Ti. The peaks broaden as the milling speed increases which is due to the refinement of the crystallite size and the strains due to plastic deformation in powder particles [10].

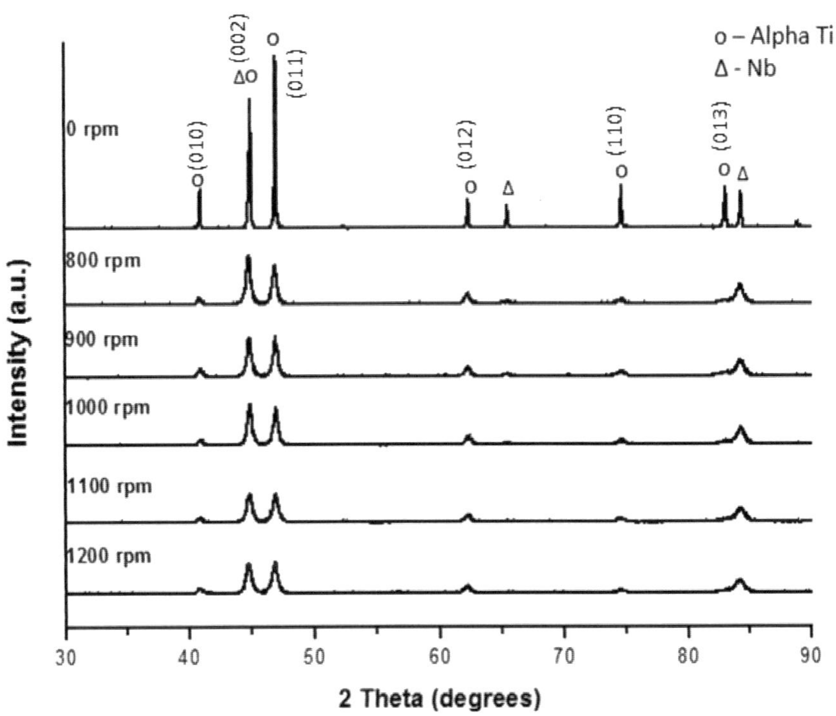

Figure 6: The XRD patterns of the CP Ti and Nb mixed elemental powder and Ti-7Nb milled powder at 800, 900, 1000, 1100 and 1200 rpm.

4. Conclusions

A mixture of CP-Ti and 7 % wt. Nb was mechanically milled using the high energy ball-mill Zoz-Simoloyer®, the following conclusions were drawn:

- The mechanical alloying process had good yield for the milled powder with little to no sticking of the powder to the grinding medium which means there is no need for a process control agent.
- The particle size increased with increasing milling speed due to fragmentation and cold welding during agglomeration of particles.
- The elemental powder of CP-Ti and Nb inhomogeneously alloyed in order to form single phase of alpha and beta phase in the milled powder.

Acknowledgements
Authors acknowledge the financial support of the department of Science and Technology (DST) South Africa. The technical support of the Council of Scientific and Industrial Research (CSIR) and the contribution of the University of Pretoria is acknowledged as well.

References

[1] M. Geetha, A. K. Singh, R. Asokamani, and A. K. Gogia, "Ti based biomaterials, the ultimate choice for orthopaedic implants - A review," *Prog. Mater. Sci.*, vol. 54, no. 3, pp. 397–425, 2009.

[2] M. Abdel-Hady Gepreel and M. Niinomi, "Biocompatibility of Ti-alloys for long-term implantation," *J. Mech. Behav. Biomed. Mater.*, vol. 20, pp. 407–415, 2013.

[3] E. Eisenbarth, D. Velten, M. Müller, R. Thull, and J. Breme, "Biocompatibility of β-stabilizing elements of titanium alloys," *Biomaterials*, vol. 25, no. 26, pp. 5705–5713, 2004.

[4] C. M. Lee, C. P. Ju, and J. H. C. Lin, "Structure ± property relationship of cast Ti ± Nb alloys," pp. 314–322, 2002.

[5] B. Sharma, S. K. Vajpai, and K. Ameyama, "Microstructure and properties of beta Ti-Nb alloy prepared by powder metallurgy route using titanium hydride powder," *J. Alloys Compd.*, vol. 656, pp. 978–986, 2015.

[6] C. Suryanarayana, "Mechanical Alloying and Milling Mechanical Engineering," *Prog. Mater. Sci.*, vol. 46, pp. 1–184, 2001.

[7] P Daswa; Z Gxowa; MJI Monareng; K Mutombo, "Effect of milling speed on the formation of Ti-6Al-4V via mechanical alloying Effect of milling speed on the formation of Ti-6Al-4V via mechanical alloying," in *Series, I O P Conference Science, Materials*, 2018.

[8] D. L. Zhang, "Processing of advanced materials using high-energy mechanical milling," in *Progress in Materials Science*, 2004, vol. 49, no. 3–4, pp. 537–560.

[9] M. S. Khoshkhoo *et al.*, "Mechanical Alloying of β-Type Ti-Nb for Biomedical Applications," *Adv. Eng. Mater.*, vol. 15, no. 4, pp. 262–268, 2012.

[10] Z. Gao, H. Luo, Q. Li, and Y. Wan, "Preparation and Characterization of Ti-10Mo Alloy by Mechanical Alloying," *Metallogr. Microstruct. Anal.*, vol. 1, no. 6, pp. 282–289, 2012.

[11] D. L. Zhang, "Processing of advanced materials using high-energy mechanical milling," *Prog. Mater. Sci.*, vol. 49, no. 3–4, pp. 537–560, 2004.

[12] J. S. Benjamin and T. E. Volin, "Mechanism of Mechanical Alloying.," *Met. Trans*, vol. 5, no. 8, pp. 1929–1934, 1974.

Tensile properties and microstructural characterization of additive manufactured, investment cast and wrought Ti6Al4V alloy

K Beyl[1, 2], K Mutombo[1] and C P Kloppers[2]

[1] Light Metals, Materials Science & Manufacturing, Council for Scientific and Industrial Research, Meiring Naudé Road, Brummeria, Pretoria 0184, South Africa

[2] School for Mechanical Engineering, North-West University, Potchefstroom Campus, 11 Hoffman Street, Potchefstroom 2531, South Africa

E-mail: KBeyl@csir.co.za, KMutombo@csir.co.za

Abstract. This paper presents an evaluation of the tensile properties and microstructural characterization of Ti6Al4V alloy manufactured with three different processing routes; traditional wrought processing, investment cast and Laser Engineered Net Shaping (LENS). Tensile specimens were machined from each process and tensile tested at room temperature. Fractured specimens were characterized using light optical microscopy, stereo microscopy, microhardness and Scanning Electron Microscopy (SEM) to investigate the microstructural morphology and the structural hardness variation. The investment cast Ti6Al4V alloy microstructure revealed large equiaxed grains containing various orientated lamellar colonies. The additive manufactured microstructure revealed long columnar grains with Widmanstätten α′ martensite laths and retained β grain boundaries. While the wrought Ti6Al4V microstructure was observed as smaller equiaxed grains with large colonies of fine lamellar and transformed β. Additive manufactured specimens had higher yield strength, ultimate tensile strength and hardness compared to the investment cast and wrought manufactured specimens.

1. Introduction

Titanium alloys have found many uses in the industries of aerospace, medical implants, medical equipment and chemical plants, this is because of its high strength to weight ratio and corrosion-resistant properties [1]. The Ti6Al4V alloy is commonly used in these industries for complex components, currently often manufactured with investment casting (IC) or thermomechanical processing routes [2]. The IC process is the most developed near-net-shape manufacturing process [3] and the main route for complex titanium products. Thermomechanical processing such as forging, rolling or compression carried out at elevated temperatures can improve the ingot's structure from brittle and porous with a coarse microstructure into a wrought structure with a finer microstructure giving better mechanical properties [4]. Laser Engineered Net Shaping (LENS), an additive manufacturing (AM) method has become a strong competitor in the production of titanium products due to its ability in creating near-net-shape products in less time. The process is cost-

effective in producing near-net-shape products, it has excellent material utilisation and gives an appealing fine microstructure [5], [6]. Each of these manufacturing methods has its own advantages and shortcomings. The challenge is to determine which manufacturing process is best suited for a specific component based on the mechanical properties required. The mechanical properties are dependent on the microstructures, while the microstructures are influenced by the manufacturing process. The Ti6Al4V alloy's microstructure is particularly sensitive to the manufacturing process and thermal history [5]. Thus, it is vital to study the effect of manufacturing processes on the microstructure of the alloy to understand and predict the mechanical properties of the alloy.
Although the microstructures and mechanical properties of IC, AM and wrought manufacturing have been studied individually, little attention has been given to an analytical comparison study between these manufacturing processes with respect to a single material, Ti6Al4V. Therefore, this paper is aimed at comparing the microstructures and tensile properties of wrought, IC and AM Ti6Al4V products.

2. Experimental

Ti6Al4V wrought bars were bought from Boati in China (Figure 1 a), the chemical composition for the as-received bar is given in Table 1. Wrought rods were wire cut from the as-received Ti6Al4V wrought bar (Figure 1 a).

Table 1. Chemical composition of the as-received wrought Ti6Al4V bar (wt. %)

Al	V	Fe	Y	C	H	N	O	Other	Ti
6.75	4.5	0.30	≤0.005	≤0.08	≤0.0125	≤0.05	≤0.2	≤0.4	Bal

The investment cast (IC) started with a wax tree pattern for eight rods of 212 mm length and 15 mm diameter (Figure 1 b) which was dipped into two alternating ceramic slurries, colloidal ZrO_2 and stuccoing with ZrO_2 stucco. The mould (Figure 1 c) was dried for 24 hours at room temperature. The wax burned-out was performed at 200°C and 8 bars pressure using a Leeds and Bradford Boiler (LBBC) Steam Boilerclave® for 15 minutes and then fired for two hours at 800°C. The as-received Ti6Al4V wrought bars (Figure 1 a) were melted using an induction melting furnace, the molten alloy was poured into the ceramic shell and left to solidify. The shell was broken off revealing the runners/sprue and Ti6Al4V investment cast rods (Figure 1 d). The IC rods were cut-off from the runners and sprue (Figure 1 e). The chemical composition of the Ti6Al4V IC rods is given in Table 2. Between

Table 2. Chemical composition of Ti6Al4V investment casting (wt. %)

Al	V	Fe	Y	C	Cr	Mo	Ni	S	Sn	Cu	Zr	Ti
6.76	4.51	0.12	ppm≤7	0.018	0.12	0.009	0.007	≤0.005	ppm≤50	0.021	ppm≤100	Bal

Figure 1. a) As received Ti6Al4V wrought bar, b) the wax tree pattern, c) ceramic mould, d) runners and Ti6Al4V investment cast rods, and e) Ti6Al4V investment cast rods.

Two as-received wrought Ti6Al4V alloy bars were supplied for atomisation to TLS (Technik GmbH & Co. Spezialpulver) an international supplier of titanium powder in Germany. Electrode Inert Gas Atomisation (EIGA) was used to produce spherical powder for additive manufacturing. EIGA is done by induction melting of a rotating electrode and atomised by inert gas [7]. The spherical atomized Ti6Al4V powder (Figure 2) and Energy Dispersive X-ray Spectrometer (EDS) analysis of some single particles of the atomized Ti6Al4V powder is given in Table 3.
The powder produced is spherical with a large range of particle sizes.

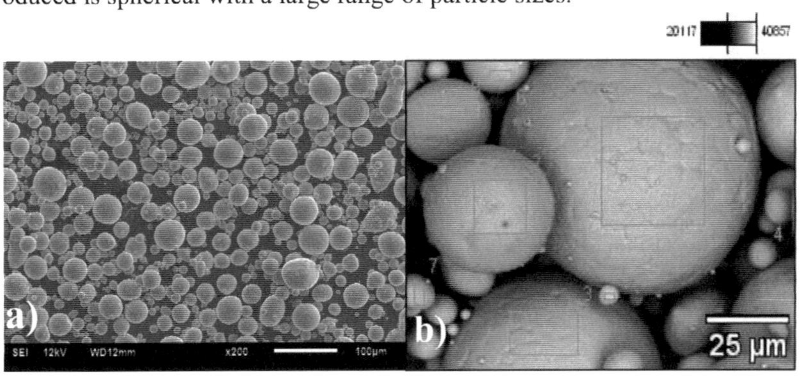

Figure 2. a) Spherical Ti6Al4V atomized powder, and b) Electron microscopy EDS analysis on some single particles

Table 3. EDS analysis of some single particles of atomized Ti4Al4V powder

Element	Al-K	Ti-K	V-K
Atomized Ti6Al4V _pt1	5.5	90.1	4.4
Atomized Ti6Al4V _pt2	5.5	90.1	4.3
Atomized Ti6Al4V _pt3	4.7	89.6	5.6
Atomized Ti6Al4V _pt4	4.3	91.5	4.2
Atomized Ti6Al4V _pt5	1.3	93.5	5.1
Atomized Ti6Al4V _pt6	3.5	92.6	4.0
Atomized Ti6Al4V _pt7	4.4	91.0	4.6
Atomized Ti6Al4V _pt8	6.1	89.5	4.4
Atomized Ti6Al4V _pt9	5.8	90.2	3.8

The atomised powder was used with the Optomec LENS™ 850-R system to laser print Ti6Al4V rods. The rods were built in the Y direction as seen in Figure 3 b. The Optomec LENS™ used a 1 kW IPG high energy fibre laser and three co-axial powder nozzles which are set at 8 ± 2 mm from the substrate (stand-off distance). The substrate was a 5 mm thick Grade 5 titanium plate which was sandblasted and cleaned with acetone to remove any surface contamination. The chamber was filled with argon gas to reduce oxidation, this argon gas also played the role of quenching medium as it was fed at room temperature. An energy density range of 180 J/mm^3 to 315 J/mm^3 was used to 3D print the Ti6Al4V rods.
The as-received Ti6Al4V wrought rods were heat-treated at 1050°C for 30 minutes then furnace cooled to generate colony lamellar structure, however, the IC Ti6Al4V rods were cut and hot isostatically pressed (HIPed) to reduce the porosity. While the AM rods were cut from the substrate in as 3D printed conditions. The tensile specimens were machined from IC and HIPed, heat-treated wrought and as 3D printed AM rods according to ASTM E 8M-04 standards (Figure 3 b). Tensile tests were carried out using a servo-hydraulic fluid-controlled machine (Instron 1342). Specimens were pulled at 0.25mm/min at room temperature (19.80°C). The initial length, area, continuous load and extension values were captured, and the stress-strain curves drawn.

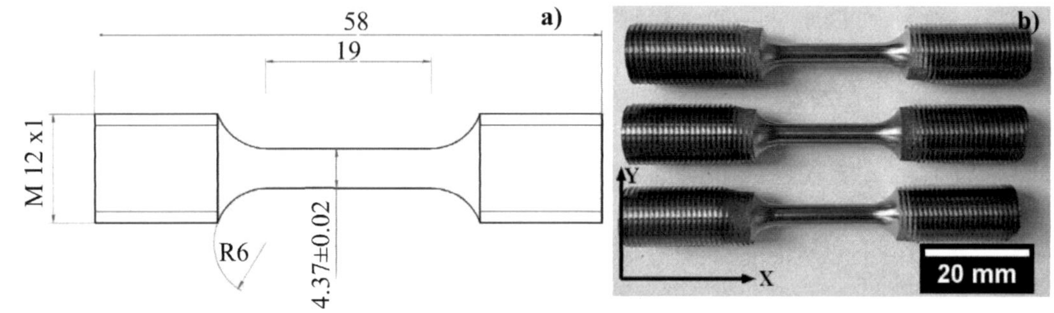

Figure 3. a) Schematic of the tensile specimen, and b) machined tensile specimens

Wrought, IC and AM samples were cut, mount and ground then polished and etched using Kroll's reagent to reveal the microstructure. The microstructure analysis was performed using the Leica DMI5000M Optical Microscope (OM) equipped with an image analysis software and Scanning Electron Microscope (SEM) equipped with energy-dispersive X-ray spectroscopy (EDX).
Microstructures were quantified using an image processing program (ImageJ). The grain sizes were measured using the line intercept method. The volume fraction of alpha-phase (α-phase) and beta-phase (β-phase) were measured using the segmentation method.
An automatic Vickers micro-hardness tester FM-700 was used to measure the change in hardness to investigate the structural homogeneity of the different manufacturing processes. Indentations were made on wrought, IC and AM specimens using a load of 500 g for a dwell time of 15 seconds.
Tensile fractures were studied using the Leica MZ16 Stereo-microscope and Scanning Electron Microscopy (SEM) to reveal the fracture mode.

3. Results

The microstructure of wrought, investment cast (IC) and additively manufactured (AM) Ti6Al4V specimens is shown in Figure 4 a, b, and c, respectively. The wrought Ti6Al4V microstructure shows small equiaxed grains, with blocky plate-like α/β-phase inside the grains (Figure 4 a). The IC grains have α-phase grain boundaries (prior β), the grains are equiaxed with diameters up to 1 mm (Figure 4 b). The grain boundaries, colonies and layer-bands of the AM microstructure are revealed in Figure 4 c.

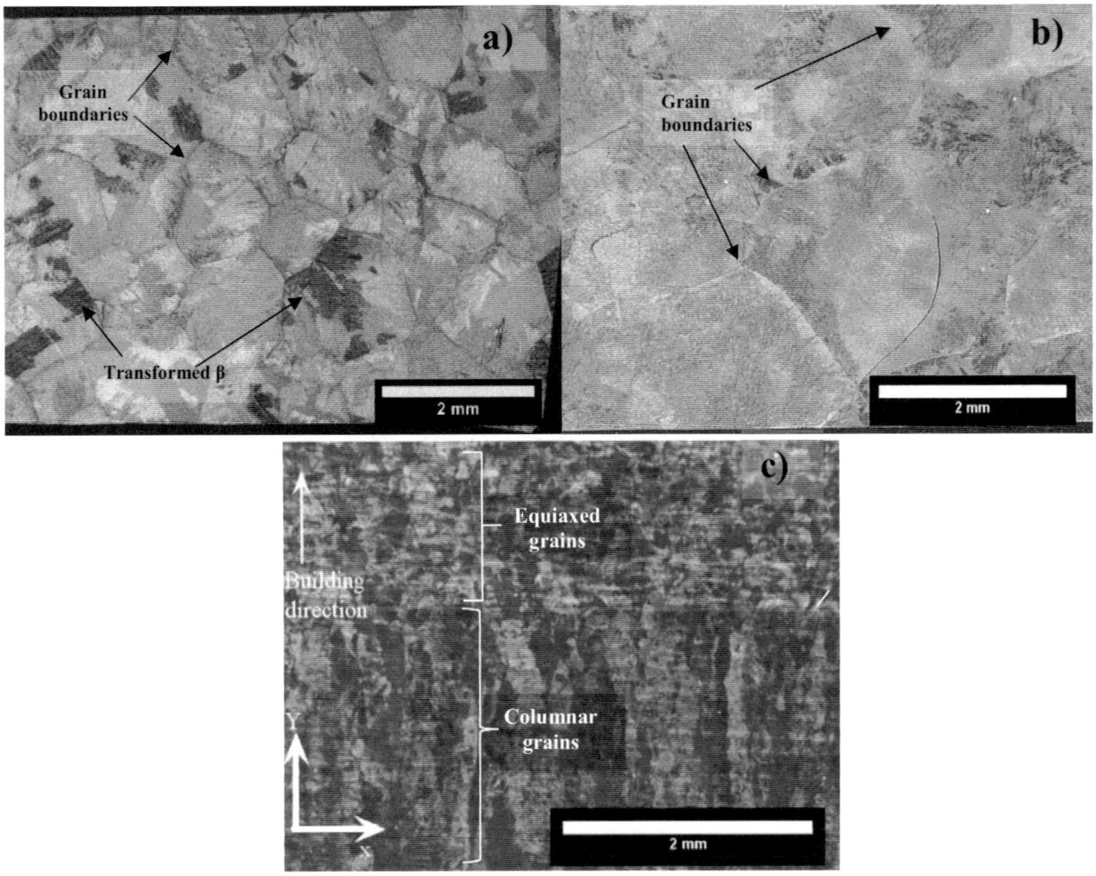

Figure 4. Grain boundaries, equiaxed and columnar grains respectively observed in a) wrought b) investment cast, and c) additively manufactured Ti6Al4V specimens.

The average grain size and the volume fraction of α and β-phase in wrought, investment cast and additively manufactured Ti6A4V specimen are given in Table 4.

Table 4. Average grain size and volume fraction of α and β-phase

Manufacturing process	Average Grain size (μm/grain)	Average % α	Average % β
Wrought	551.1	65.0585	34.9415
Investment Cast	1109.2	64.273	35.727
Additively manufactured	146.7	65.214	34.786

The microstructural features of the wrought, investment cast and additively manufactured Ti6A4V are revealed in Figure 5 a, b and c, respectively. Fine α/β lamellae, blocky plate-like and α/β colony structure are observed in Figure 5 a. Coarse α/β-lamellae colonies in various directions with blocky plate-like α (transformed β) are observed in IC Ti6Al4V Figure 5 (b). Columnar grains, Widmanstätten α′ martensite laths with retained β- grain boundaries are revealed in AM Ti6Al4V.

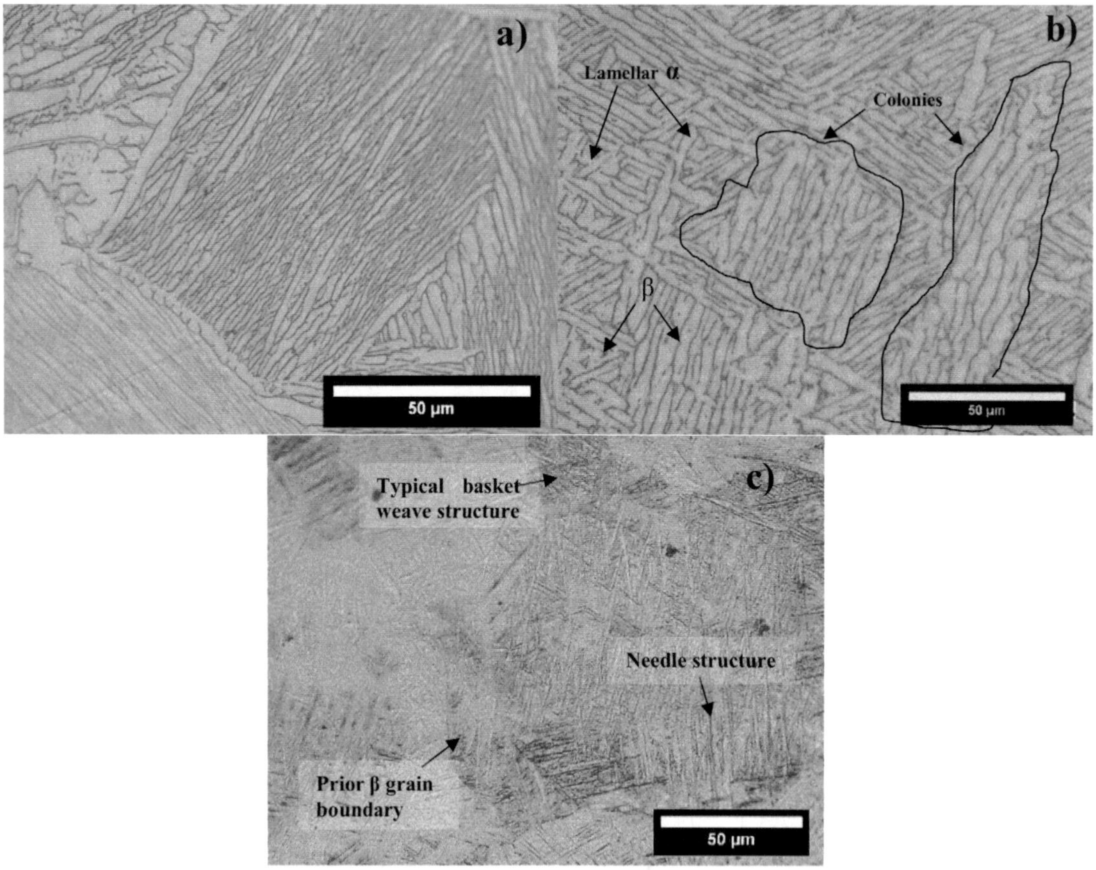

Figure 5. Optical microscope images of a) Wrought b) Investment cast and c) Additive manufactured

The Vickers micro-hardness profile of the wrought, IC and AM Ti6Al4V are given in Figure 6. The results clearly illustrate that the hardness varies for different manufacturing processes. The AM with a partial martensitic microstructure gave the highest average hardness value (408.8 HV0.5). The wrought and IC Ti6Al4V alloy specimens with colony lamellar structure gave similar average hardness of 341.8 HV0.5 and 320.3 HV0.5, respectively.

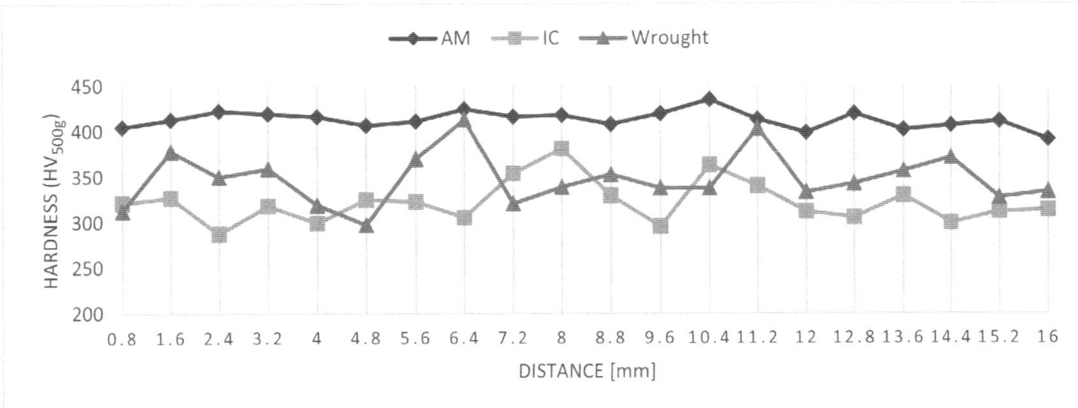

Figure 6. Hardness profile of AM, IC, and wrought Ti6Al4V structures

The stress-strain curves for tensile tests of wrought, IC and AM Ti6Al4V are given in Figure 7. The AM Ti6Al4V specimen had superior tensile properties with leading values in ultimate tensile strength (UTS) and yield strength (YS). However, the tensile elongation for AM is low (less than 10 %) as compared to the wrought and IC Ti6Al4V tensile specimens (higher than 10 %). The slopes are very similar, with the AM slope observed to be somewhat lesser steep compared to wrought and IC. The gradual slope of the AM specimen indicates a low modulus of elasticity. The wrought and IC specimens had the steepest slopes and relatively high % El (Percentage elongation), which is in agreement with Figure 8 a and b showing a larger necking for wrought and IC specimens and not any for AM specimens (Figure 8 c).

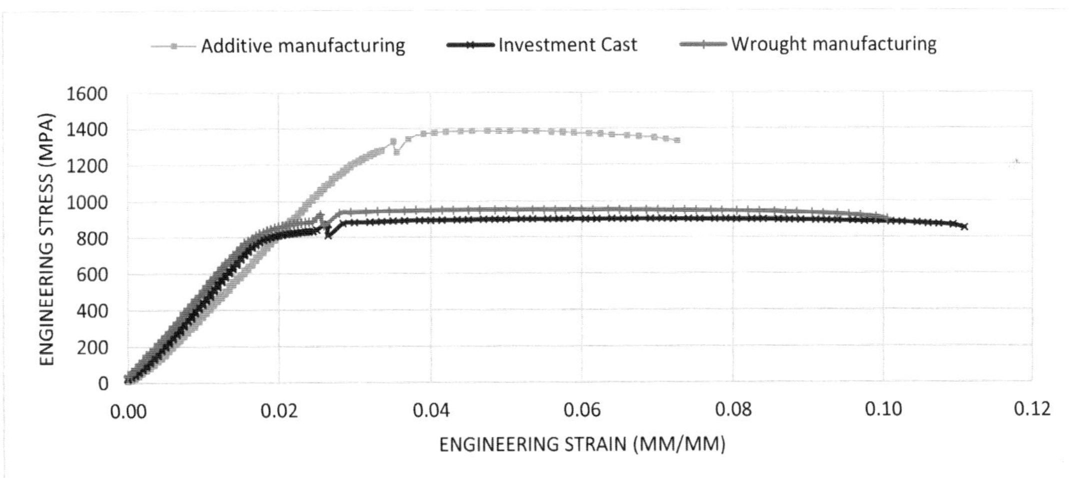

Figure 7. Stress-Strain plots of the IC, AM and wrought Ti6Al4V specimens

The tensile fractures of wrought, IC and AM Ti6Al4V tested specimens are shown in Figure 8. The wrought and IC fractures have somewhat necking (Figure 8 a, b), while the AM fracture has no necking (Figure 6 c).

Figure 8. Stereo-microscope images of cross-sectioned tensile test fractures fabricated by a) wrought b) IC and c) AM processes

Fracture cross-sections of the wrought, IC and AM Ti6Al4V are given in Figure 9. Figure 9 a and b show kinking of α/β lamellae, the lamellar lines tend to deflect towards a fractured edge, in the ductile IC and wrought structure especially in the necking region of the tensile specimens. Figure 9 c illustrated a coalescence of microcracks along α′-laths in the partial martensitic AM structure.

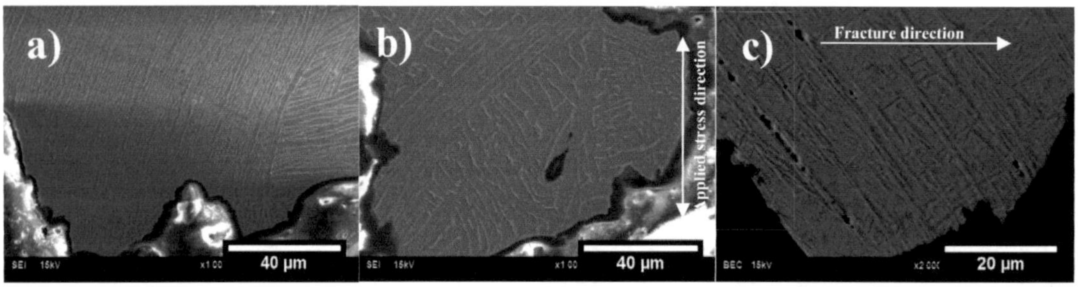

Figure 9. SEM micrographs of fracture cross-sections a) wrought, b) IC, and c) AM tensile test specimens

The fracture surfaces of the wrought, IC and AM are shown in **Figure 10**. The wrought and AM fracture surfaces (**Figure 10** a) and c), respectively) showed relatively rougher surfaces, while the IC fracture surface appeared to be almost flat (**Figure 10** b).

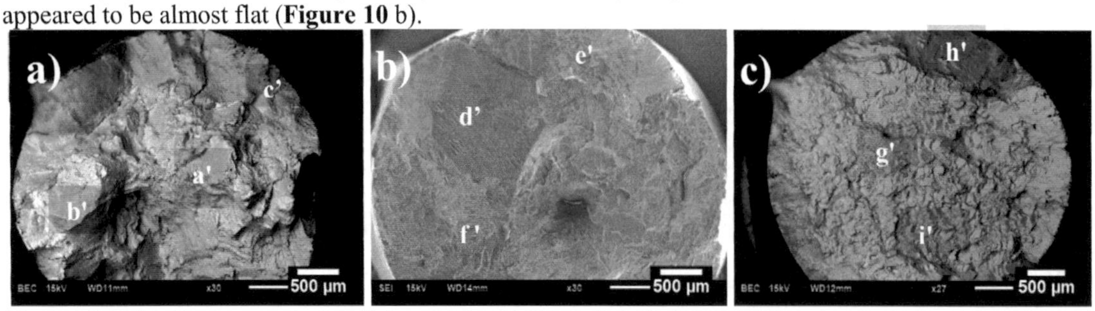

Figure 10. The fracture surfaces of a) wrought, b) IC, and c) AM Ti6Al4V

Figure 11 gives a higher magnification of the numbered areas in **Figure 10**.
The fracture surfaces of a specific area as shown in **Figure 10** of the wrought, IC and AM Ti6Al4V are given in Figure 11 a)-c), Figure 11 e)-f) and Figure 11 g)-i), respectively. The wrought and IC fracture surfaces revealed transgranular quasi-cleavage surface areas (larger in IC fracture) with ridges, microvoids

and laths. However, transgranular quasi-cleavage surface areas were relatively small (size of grains) in the AM fracture surface.

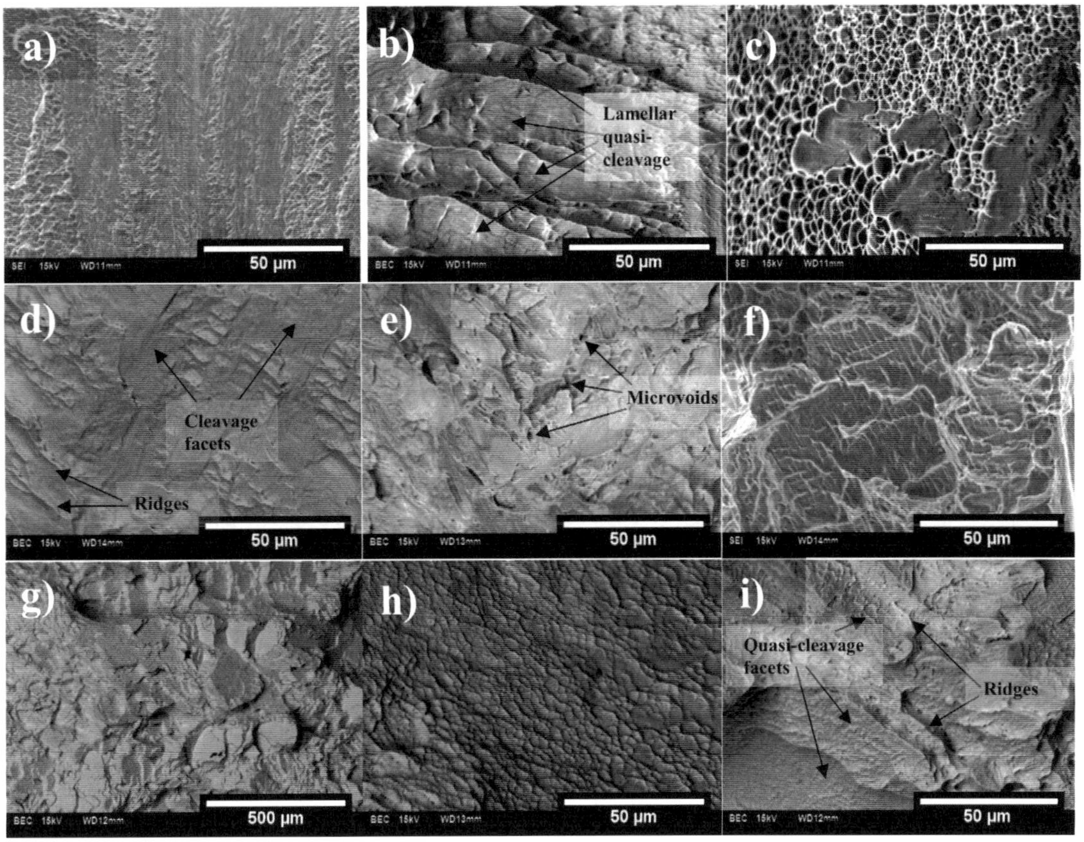

Figure 11. Fracture surfaces of a)-c) Wrought, d)-f) IC and g)-i) AM Ti6Al4V tensile specimens

4. Discussion

The columnar grains of the additive manufactured (AM) microstructure were oriented in the building direction (Y direction, see Figure 4 c) and crossing multiple-layer bands. The grain growth direction is counter to the heat flux. Since heat rapidly dissipates into the substrate (horizontally) when the process starts, forcing the grains to grow vertical [5]. The layer bands are the consecutive horizontal lines in Figure 4 c, the lines are visible due to the re-melting of previously solidified layers when a new layer is added. The re-melting also allows for epitaxial grain growth [8]. A change in the geometry of the grains in the top layers is visible in Figure 4 c. The grains changes from columnar to coarse equiaxed grains, this is due to a slower cooling rate caused by the additional heat of the previous layers [9]. As the number of layers increases the heat inside the chamber rises, resulting in slower solidification of the top layers.

The smaller grains in the wrought microstructure led to a slight increase in hardness over the IC Ti6Al4V hardness [5],[10],[11]. However, the AM Ti6Al4V with partial martensitic microstructures proved to be superior in hardness. The high hardness of AM specimens is attributed to the presence of partial martensitic structure due to the high cooling rate during the 3D printing process [12].

The microstructure of Ti6Al4V influences the tensile properties, hardness and fracture modes of Ti6Al4V components. Depending on the manufacturing processes, Ti6Al4V can be of equiaxed, lamellar or martensitic microstructures. The martensitic microstructure showed high YS, UTS and hardness. While lamellar structures have a higher ductility and resistance to crack propagation.

When grain sizes decrease the number of grains in the same area increase, thus causing an increase in the number of grain boundaries. Grain boundaries are obstacles for a fracture movement causing entanglement and resulting in increased material strength [13]. This is in agreement following the comparison of microstructures and YS for IC and wrought Ti6Al4V tensile specimens. The low AM elongation is an indication of a brittle fracture mode, this was confirmed by Figure 8 c indicating a brittle fracture with no necking or permanent deformation. The Hall-Petch equation (1) indicates that smaller grain sizes promote higher YS and UTS [10].

$$\sigma_y = \sigma_0 + \frac{K_y}{\sqrt{d}} \quad (1)$$

Where parameter σ_y is the yield strength, σ_0 material constant, K_y strengthening coefficient and d the diameter of the average grain.

The fracture started with coalescence of microvoids, opening small pores between grain boundaries, or at any defect inside the specimen. The fracture propagates with increased applied stress. A 45° fracture deflection is observed towards the edge of the specimen (Figure 8) at the maximum shear stress (shear lips), this fracture segment is transgranular. Transgranular fractures are usually caused by catastrophic failure. Catastrophic failure occurs when necking or fracture already started, and the remaining intact specimen's area is reduced, thus decreasing the applied stress required for failure. The failure is sudden, fast and complete, cutting directly through grains.

The wrought manufactured specimen's overall fractured surface (Figure 8 c) resembles a partial inter-crystalline fracture, where some fractures occur between grains [14]. The IC fracture showed cleavage facets, microvoids and ridges on cleavage facets, resulting in an overall quasi-cleavage fracture. The quasi-cleavage fracture is common in high strength materials [14]. Most of the fracture surface of AM was observed as quasi-cleavage facets of a brittle nature. Secondary cracks as seen in Figure 9 c can only follow grains or lamella, and never be catastrophic [15], often corresponding to microvoid coalescence (MVC) [16].

5. Conclusions

The microstructure, hardness, tensile properties were investigated for additively manufactured (AM), investment casting (IC) and wrought Ti6Al4V alloy rods. The following conclusions were drawn:

- The investment cast Ti6Al4V had a colony lamellar structure in large equiaxed grains. The additively manufactured Ti6Al4V had long columnar grains with a Widmanstätten α' martensitic laths and retained β-grain boundaries. While the wrought Ti6Al4V microstructure was observed as smaller equiaxed grains with large colonies of fine lamellar and transformed β.
- The additive manufactured Ti6Al4V alloy specimen proved to be the hardest. The investment cast and wrought Ti6Al4V alloy specimens had similar hardness.
- The additively manufactured's martensitic microstructure led to a higher ultimate tensile strength and yielding strength, compared to investment cast and wrought Ti6Al4V. However, additively manufactured Ti6Al4V gave the lowest percentage elongation. The tensile properties of investment cast and wrought Ti6Al4V were similar.
- The investment cast fractures exhibited cleavage facets and microvoids, leading to a quasi-cleavage fracture. The additively manufactured fractures showed areas of intergranular fracturing and a large area of quasi-cleavage fracture. Wrought manufactured tensile tested specimens resulted in fracture

areas of quasi-cleavage and lamellar cleavage patterns. Thus, wrought, investment cast and additively manufactured Ti6Al4V revealed a typically brittle failure.

6. References

[1] G Lütjering and J C Williams, *Titanium*. Berlin, Heidelberg: Springer Berlin Heidelberg, 2007.
[2] S Liu and Y C Shin, "Additive manufacturing of Ti6Al4V alloy: A review," *Mater. Des.*, vol. 164, p. 107552, Feb. 2019.
[3] K Mutombo, P Rossouw, and G Govender, "Mechanical Properties of Mill-Annealed Ti6Al4V Investment Cast," *Mater. Sci. Forum*, vol. 690, pp. 69–72, 2011.
[4] S Kalpakjian, S R Schmid, and V. Sekar, *Manufacturing Engineering and Technology*. 2013.
[5] A Bagheri, N Shamsaei, and S. Thompson, *Microstructure and Mechanical Properties of Ti-6Al-4V Parts Fabricated by Laser Engineered Net Shaping*. 2015.
[6] L E Murr, S M Gaytan, D A Ramirez, E Martinez, J Hernandez, K N Amato, P W Shindo, F R Medina, R W Wicker, "Metal Fabrication by Additive Manufacturing Using Laser and Electron Beam Melting Technologies," *J. Mater. Sci. Technol.*, vol. 28, no. 1, pp. 1–14, Jan. 2012.
[7] N Arthur, M Khodja, H Moller, and & P Rossouw, "Performance evaluation and characterisation of eiga produced titanium alloy powder for additive manufacturing processes", *RAPDASA Conference*, 2017
[8] G Kühn, "W Kurz, D J Fisher, Fundamentals of Solidification. Trans Tech Publications, Switzerland-Germany-UK-USA, 1986 (Erstauflage 1984), 242 Seiten, zahlreiche Abbildungen und Tabellen, Sachwortindex, SFr 54.00, ISBN 0-87849-523-3," *Cryst. Res. Technol.*, vol. 21, no. 9, p. 1176, Sep. 1986.
[9] P A Kobryn and S . Semiatin, "The laser additive manufacture of Ti-6Al-4V," *JOM*, vol. 53, no. 9, pp. 40–42, Sep. 2001.
[10] M Shunmugavel, A Polishetty, and G Littlefair, "Microstructure and mechanical properties of wrought and Additive manufactured Ti-6Al-4V cylindrical bars," *Procedia Technol.*, vol. 20, pp. 231–236, 2015.
[11] N K K Arthur and S Pityana, "Microstructure and material properties of LENS fabricated Ti-6Al-4V components ," *R&D Journal* , vol. 34. scieloza , pp. 33–36, 2018.
[12] J S Zuback and T. DebRoy, "The Hardness of Additively Manufactured Alloys," *Mater. (Basel, Switzerland)*, vol. 11, no. 11, p. 2070, Oct. 2018.
[13] J Dong, F Li, and C Wang, "Micromechanical behavior study of α phase with different morphologies of Ti–6Al–4V alloy by microindentation," *Mater. Sci. Eng. A*, vol. 580, pp. 105–113, Sep. 2013.
[14] V A Joshi, *Titanium alloys : an atlas of structures and fracture features*. Boca Raton: CRC Taylor & Francis, 2006.
[15] W Zhou and K G Chew, *The rate dependent response of a titanium alloy subjected to quasi-static loading in ambient environment*, vol. 37. 2002.
[16] W F Hosford, *Mechanical Behavior of Materials*. Cambridge University Press, 2010.

IOP Conf. Series: Materials Science and Engineering **655** (2019) 012024 doi:10.1088/1757-899X/655/1/012024

The effect of ternary alloying with vanadium on Ti-Pt shape memory alloys

P Daswa[1,2], S Chikosha[1,a], M L Mahlatji[1], C W Siyasiya[2]

[1]Light Metals Impact Area, Manufacturing Cluster, Council for Scientific and Industrial Research (CSIR), Meiring Naudé, Pretoria, South Africa

[2]Department of Materials Science and Metallurgical Engineering, University of Pretoria, Pretoria, South Africa

[a]SChikosha@csir.co.za

Abstract. Ti-Pt alloys have attracted much research due to high transformation temperatures (1050°C) which makes them suitable for high temperature applications in automotive and aerospace industries. The binary alloy exhibits negligible shape memory effect and the amount of platinum (Pt) required to produce this alloy makes it very expensive which limits its practical application and commercialization. Ternary alloying of Ti-Pt could lead to solid solution strengthening and improved shape memory properties. Furthermore, ternary alloying by replacing Pt could reduce the cost of the alloy. However, it could lead to changes in the microstructure, crystal structure, transformation temperatures and transformation temperature hysteresis that would significantly affect the shape memory behaviour of the alloy. This paper investigates the effect of ternary alloying of Ti-Pt with varying vanadium contents (0-18.25 at.%) on the as-cast condition. The effect of ternary alloying on the microstructure, transformation temperature and transformation temperature hysteresis were studied. The Ti_{50}-Pt_{50-x}-V_x ($x = 6.25$ to 18.25 at.%) alloys were produced by the button arc melting method using elemental powders of titanium, platinum and vanadium. Results showed the formation of martensitic α-$Ti_{50}(Pt,V)_{50}$ with Ti_3Pt, $(Ti,V)_3Pt$, Ti_4Pt_3-like and oxide phases in Ti-Pt-(V) alloys. The phase transformation temperature of Ti-Pt alloys decreased with increasing vanadium content up to 10.2 at.% V.

1. Introduction

Shape memory alloys (SMA) are a type of smart material that have attracted much research in the automotive, biomedical, civil and aerospace industries [1,2,3]. They display a peculiar capability of recovering large deformations upon heating through a reversible martensitic transformation that enables them to generate an actuating force [1,2]. This characteristic makes SMAs very attractive for actuating applications in aerospace, automotive and biomedical industries.

Ti-Ni, the most successfully commercialized SMA, has been used for varying applications due to its excellent properties such as superelasticity, shape memory property, biocompatibility and good corrosion resistance [4, 5]. However, it is limited to lower temperature applications of less than 100°C due to its related low transformation temperatures [4-5]. Although research had proposed that alloying of the binary alloy system could be used as an alternative way of increasing the transformation temperature, the maximum martensitic transformation temperatures for the Ti-Ni based alloys achieved to date still remain below 750°C [1,6–8]. This still limits their application in the aerospace industry. The Ti-Pt alloy has been suggested as a high temperature shape memory alloy due to its high martensitic transformation temperature of about 1050°C [9]. It undergoes a cubic austenitic B2 to an orthorhombic martensitic B19 phase transformation. Ti-Pt alloys exhibit negligible shape memory effect and the high cost of Pt could limit the practical application of these alloys. Ternary alloying of Ti-Pt alloys by substituting Pt with a cheaper element has attracted research as this could be an alternative of improving the shape memory behavior of Ti-Pt alloys to improve work output (actuating ability) and reduce cost of the alloys [10-12]. Chikosha *et al.* [12] investigated the effect of ternary alloying Ti-Pt alloys with vanadium replacing Pt by 6.25 at.% V. They reported that the transformation temperature decreased by the addition of 6.25 at.% V in the alloy. None of the published work has considered the effect of adding varying contents of vanadium and its effect on transformation temperature. Therefore, this study investigated the effect of ternary alloying with varying vanadium contents (0-18.25 at.% V) on the microstructure and transformation temperature of as-cast Ti-Pt shape memory alloys.

2. Experimental Procedure

Elemental powders of high purity Ti grade 1 (99.9 %), Pt (99.9 %) and V (99. 5%) were prepared in desired quantities to achieve homogeneous powder blends with compositions: Ti_{50}-Pt_{50}, Ti_{50}- $Pt_{43.75}$-$V_{6.25}$, Ti_{50}-$Pt_{37.5}$-$V_{12.5}$ and Ti_{50}-$Pt_{31.25}$-$V_{18.75}$ (at.%). Ti elemental powder was sourced from TLS, Technik, GmbH & Co, Germany. LGC Industrial Analytical (Pty) Ltd, South Africa, supplied the Pt and V elemental powders. The powder blends were cold compacted at room temperature using an ENERPAC VLP 100 tonne press at 2842 MPa (100 bars) with a die size of 10 mm diameter to produce green compacts. The green compacts were then arc-melted in an argon atmosphere to produce 10 mm diameter alloy buttons. In order to achieve homogeneity, the binary Ti-Pt compact was remelted three times and the ternary Ti_{50}-Pt_{50-x}-V_x compacts were remelted six times. All the characterisation and analysis was conducted on as-cast samples. Metallographic analysis was done on the sectioned, mounted and polished as-cast samples. Scanning electron microscopy (SEM) with an energy-dispersive X-ray spectrometer (EDS), using a JEOL JSM-6510 instrument was used to determine microstructure, composition and phases of the alloys. Differential scanning calorimetry (DSC), using a Netzsch STA449F3 was done to investigate the thermal behaviour such as transformation temperatures, hysteresis and latent heats of transformation. The heating and cooling rate was 20°C/min and the temperature range was from room temperature to 1400°C Samples were subjected to two thermal heating and cooling cycles.

3. Results and Discussion

Figures 1-4 show SEM backscattered electron images (BEI) and EDX results of Ti_{50}-Pt_{50}, Ti_{50}-$Pt_{43.75}$-$V_{6.25}$, Ti_{50}-$Pt_{37.5}$-$V_{12.5}$ and Ti_{50}-$Pt_{31.75}$-$V_{18.25}$ alloys in the as-cast condition. The chemical compositions shown for all phases are the average values taken from three different sections of the sample. The overall compositions (Figures1-4) were slightly different from the initial measured compositions, but the difference is within the error margin of EDX measurements. However, they were still within the TiPt single phase region [13].

The resultant microstructures were inhomogeneous and consisted of at least three contrast areas. All of the alloys followed a similar solidification sequence during cooling. On solidification, above the eutectic temperature, the alloy consisted of the liquid and β-TiPt dendrites. At the eutectic temperature, the residual liquid transformed to the eutectic Ti_3Pt and β-TiPt phases. Below the peritectoid temperature, the β-TiPt dendrites disintegrated into Ti_4Pt_3 at the β-TiPt/interdendritic interface which grew towards the core of the dendrite [14]. At low martensitic transformation temperature, the β-TiPt in the dendrite and interdendritic regions transformed to the low temperature α-TiPt phase. The formation of an oxide phase in Ti_{50}-Pt_{50} and Ti_{50}-$Pt_{31.75}$-$V_{18.25}$ was observed. The latter is stabilised by oxygen picked up during sample processing [15,18].

Binary Ti-Pt was comprised of α-TiPt, which transformed martensitically from primary β-TiPt dendrites, and an interdendritic region with Ti_4Pt_3, oxide phase suspected to be Ti_3Pt (stabilised by oxygen) and α-TiPt (Figure 1).

The Ti_{50}-$Pt_{43.75}$-$V_{6.25}$ alloy consisted of α-Ti(Pt,V) dendrites and the interdendritic regions were comprised of $(Ti,V)_4Pt_3$ and Ti_3Pt phases (Figure 2). The EDX results from the SEM revealed that the transforming Ti(Pt,V) phase contained 5.8 at.% V.

The Ti_{50}-$Pt_{37.5}$-$V_{12.5}$ alloy consisted of α-Ti(Pt,V) dendrites while the interdendritic regions were composed of Ti_4Pt_3 and Ti_3Pt phases (Figure 3). It was suspected that the Ti_4Pt_3 phase has vanadium substituting for both Ti and Pt, its full characterisation is part of ongoing work. The EDX results from the SEM revealed that the transforming Ti(Pt,V) phase contained 10.2 at.% V.

The Ti_{50}-$Pt_{31.75}$-$V_{18.25}$ alloy had the most complex microstructure, consisting of Ti_4Pt_3 and oxide phase dendrites interspersed with a low volume fraction of Ti(Pt,V) phase in the interdendritic regions (Figure 4). During solidification, the alloy consist of liquid and β-TiPt dendrites above eutectic temperature and the liquid further transforms to the Ti_3Pt + β-TiPt eutectic at the eutectic temperature. Below the peritectoid temperature, the β-TiPt dendrites transform partitionlessly to Ti_4Pt_3 upon cooling [14]. The β-TiPt phase in the eutectic transforms to α-TiPt at the martensitic transformation temperature. The EDX results showed that the Ti(Pt,V) phase formed contained only 8.2 at.% V which is far off from the intended composition of 18.25 at.% V, as such it will not be reported on further.

The volume fraction of the interdendritic regions increased with increasing vanadium addition, thereby reducing volume fraction of the transforming TiPt phase. During transformation, the β-TiPt dendrite decomposed into $Ti_4(Pt,V)_3$ and α-TiPt. However, with the higher contents of vanadium, the transformation of $Ti_4(Pt,V)_3$ was favoured and resulted in the low volume fraction of the TiPt phase in Ti_{50}-$Pt_{31.25}$-$V_{18.25}$.

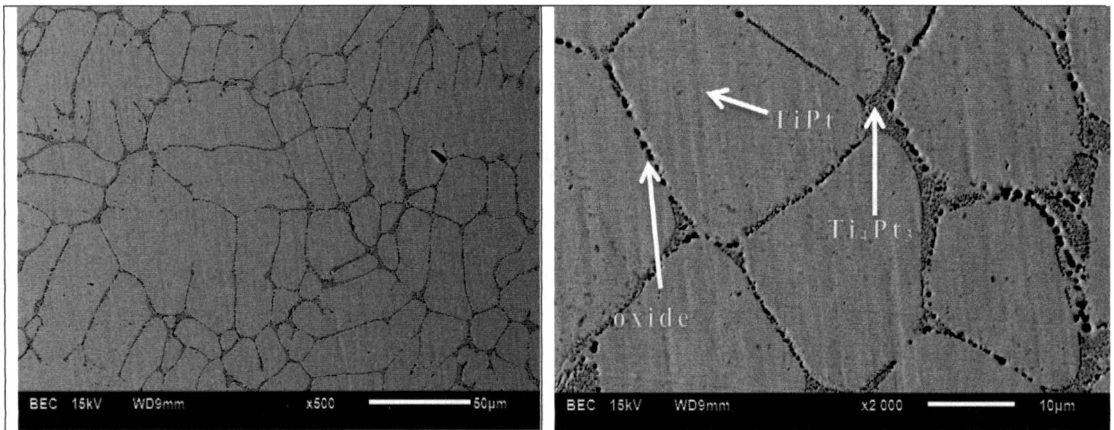

(a) (b)

	O-K		Ti-K		Pt-K	
	wt%	at.%	wt%	at.%	wt%	at.%
Overall	-	-	20.5±0.1	51.3±0.3	79.5±0.3	48.7±0.2
TiPt	-	-	20.0±0.3	50.5±0.7	80.0±0.5	49.5±0.3
Ti_4Pt_3	-	-	24.0±0.5	56.3±1	76.0±0.8	43.7±0.5
oxide	15.7±0.9	44.0± 2	51.7±0.4	48.5±0.3	32.6±0.3	7.5±0.1

Figure 1. SEM backscattered electron images (BEI), (a) low and (b) high magnification with EDX results of Ti_{50}-Pt_{50} in the as-cast condition.

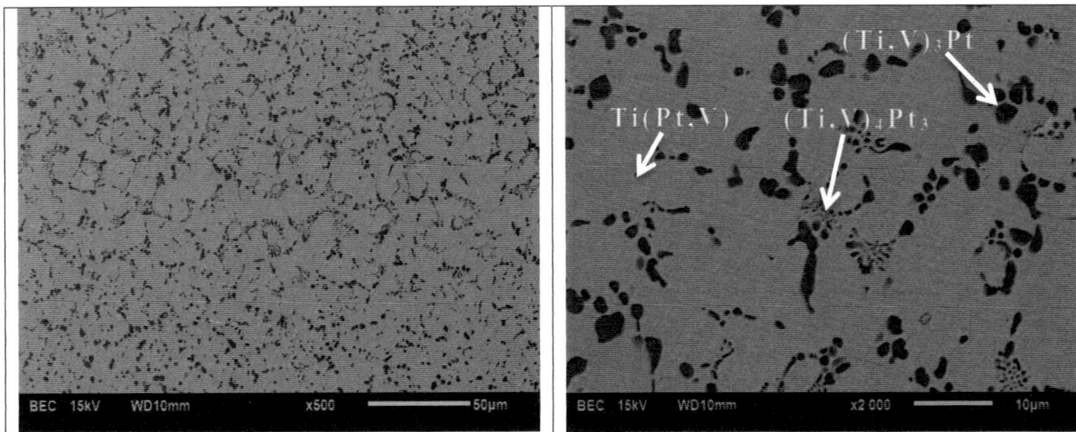

(a) (b)

	Ti-K		V-K		Pt-K	
	wt%	at.%	wt%	at.%	wt%	at.%
Overall	21.8±0.2	50.7±0.4	2.9±0.1	6.4±0.2	75.3±04	42.9±02
Ti(Pt,V)	20.6±0.4	49.1±-1	2.6±0.22	5.8±0.5	76.8±0.7	45.1±0.4
$(Ti,V)_4Pt_3$	21.6±0.2	50.28±0.6	2.9±0.2	6.4±0.4	75.5±0.4	43.3±0.3
$(Ti,V)_3Pt$	38.6±0.6	67.5±0.9	5.1±0.5	8.4±0.8	56.2±0.7	24.1±0.3

Figure 2. SEM backscattered electron images (BEI), (a) low and (b) high magnification with EDX results of Ti-$Pt_{43.75}$-$V_{6.25}$ in the as-cast condition.

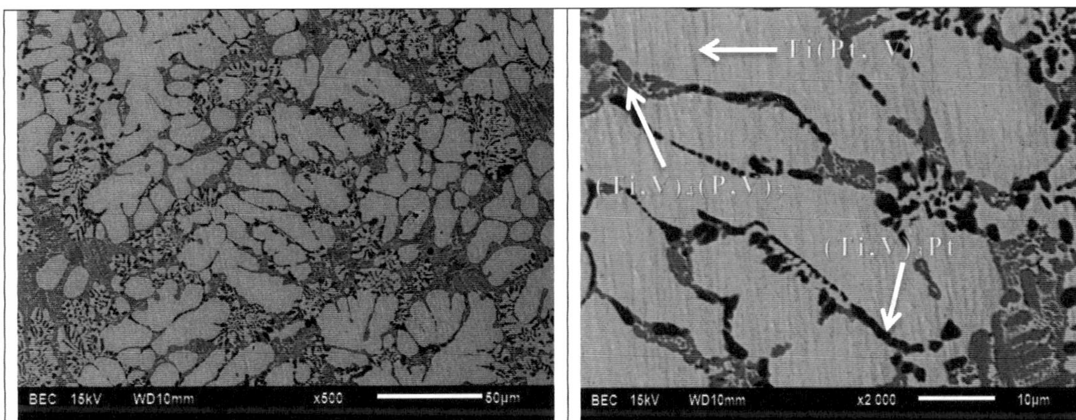

(a) (b)

	Ti-K		V-K		Pt-K	
	wt%	at.%	wt%	at.%	wt%	at.%
Overall	24.4± 0.1	51.4±0.2	6.5±0.1	12.9±0.2	69.1±0.2	35.8±-0.1
Ti(Pt,V)	22.1±0.3	49.5±0.7	4.8±0.2	10.2±0.4	73.1±0.5	40.3±0.3
$(Ti,V)_4(Pt,V)_3$	22.6±0.3	42.2±0.6	17.1±0.3	30.1±0.6	60.3±0.5	27.7±0.2
$(Ti,V)_3Pt$	38.6±0.4	67.9±0.6	4.5±0.2	7.5±0.4	56.9±0.5	24.6±0.2

Figure 3. SEM backscattered electron images (BEI), (a) low and (b) high magnification with EDX of Ti-$Pt_{37.5}$-$V_{12.5}$ in the as-cast condition.

IOP Conf. Series: Materials Science and Engineering **655** (2019) 012024 doi:10.1088/1757-899X/655/1/012024

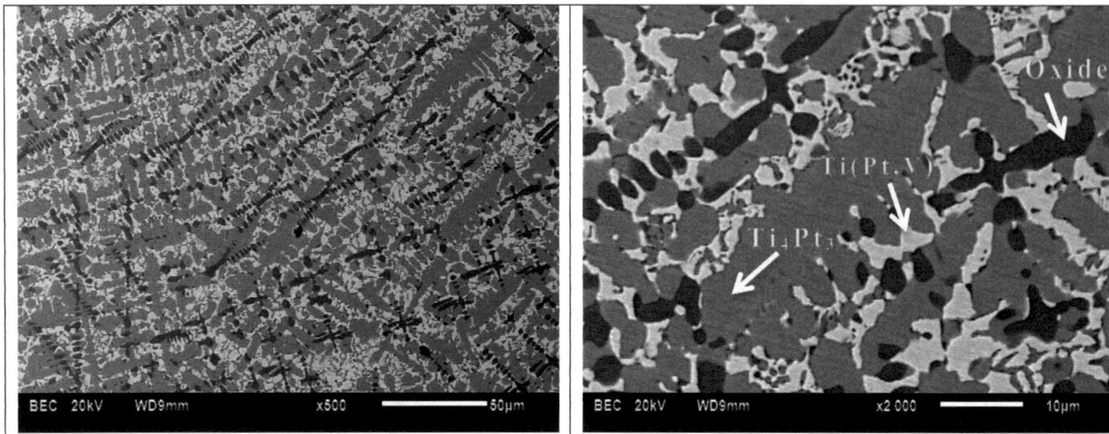

(a) (b)

	O-K		Ti-K		V-K		Pt-K	
	wt%	at.%	wt%	at.%	wt%	at.%	wt%	at.%
Overall			25.6±0.1	50.2 ±0.3	10.2 ±0.1	18.9± 0.2	64.2± -1	30.9±0.5
Ti(Pt, V)	-	-	23.7± 0.5	51.4±1.1	3.9±0.4	8.2±0.8	72.4±0.4	40.4±1.9
$(Ti,V)_4(Pt,V)_3$	-	-	22.2± 0.4	41.4±0.7	17.7±0.4	31.1±0.7	60.1±4.3	27.5±2
oxide	8.3±0.5	32.4 ± 2	34.4± 0.5	44.8±0.8	5± 0.3	6.1±0.4	52.3±4	16.7±1.3

Figure 4. SEM backscattered electron images (BEI), (a) low and (b) high magnification with EDX of $Ti\text{-}Pt_{31.75}\text{-}V_{18.25}$ in the as-cast condition

The SEM-EDX results revealed that the exact compositions of the transforming TiPt phases deviated from the intended compositions as follows: The $Ti_{50}\text{-}Pt_{50}$ sample formed $Ti_{50.5}Pt_{49.5}$, the $Ti_{50}\text{-}Pt_{43.75}\text{-}V_{6.25}$ sample formed $Ti_{49.1}(Pt_{45.1},V_{5.8})$, the $Ti_{50}\text{-}Pt_{37.5}\text{-}V_{12.5}$ sample formed $Ti_{49.5}(Pt_{40.3},V_{10.2})$, and the $Ti_{50}\text{-}Pt_{31.75}\text{-}V_{18.25}$ sample formed $Ti_{51.4}(Pt_{40.4},V_{8.2})$. Due to the large deviation of the transforming phase from the intended composition in the $Ti_{50}\text{-}Pt_{31.75}\text{-}V_{18.25}$ sample, its DSC result was not presented. Furthermore, its transformation temperatures did not vary significantly from those of $Ti_{49.5}\text{-}Pt_{40.33}\text{-}V_{10.2}$.

Figure 5 shows two heating and cooling DSC cycles of the as-cast Ti-Pt alloy. The first heating segment displayed a single step reverse transformation labelled M→A, with austenite start and finish temperatures of A_s = 1031.8°C and A_f = 1046.5°C. During cooling, the forward transformation was believed to proceed via a multi-stage transformation sequence, similar to what was observed by Biggs *et al.* and proposed that the behaviour was analogous to the Ni-Ti system [13,14]. The first cooling peak, labelled $A \rightarrow M_1$, with martensite start and peak temperatures of M_s = 992.3°C and M_p = 983.2°C was the partial transformation of austenite to B19 martensite phase. On further cooling, the remaining austenite transformed into an unidentified intermediate phase (I) in a reaction characterised by a peak temperature I_p of 971.0°C. With further cooling, the intermediate phase transformed into a B19 martensite phase according to the reaction $I \rightarrow M_2$ with peak and finish temperatures of M_p = 951.5°C and M_f = 935.7°C. On the contrary, Karem [14] reported the occurrence of a decomposition reaction of β-TiPt phase below the peritectoid isotherm into Ti_4Pt_3 and Pt-enriched β-TiPt. It is not yet clear whether such decomposition reaction takes place in the present alloy, but if it does, it would account for the intermediate exothermic peak attributed to the A→I reaction. In this case, the Pt-rich decomposition β-TiPt product would transform to martensite at a slightly lower temperature. Further work is required to fully understand the mechanism behind this observed transformation sequence.

The second heating segment proceeded through two steps. Based on the intermediate theory analogous to the Ti-Ni system, the first peak was reverse transformation of the B19 martensite (M_2) into austenite, $M_2 \rightarrow A$, followed by reverse transformation of B19 martensite (M_1) in the reaction $M_1 \rightarrow A$. The characteristic transformation temperatures and thermal hysteresis of the second cycle heating $M_1 \rightarrow A$ transformation were close to those of the first cycle heating M→A transformation (Figure 5 and Table

1). This was an indication that the martensite types M and M_1 comprised martensite variants of similar characteristics, most likely forming freely within dendrites in a similar manner to Ti-Ni [16]. The reaction $M_2 \rightarrow A$ was associated with the forward reaction $A \rightarrow I \rightarrow M_2$. Similarly, the B2→B19′ transformation in Ni-Ti alloys can also be supressed by conditions such as compositional inhomogeneity, local stress fields around precipitates and high density dislocation networks produced by thermomechanical treatments [13-14]. This changes the transformation sequence to B2→R→B19′, where the R-phase is a trigonal-like martensite phase that forms in competition with the monoclinic B19′ phase.

The thermal hysteresis was defined as the difference between the forward (A→M) and reverse (M→A) transformation temperatures using equation [18]:

$$\Delta T = Af - Ms \quad (1)$$

A thermal transformation hysteresis of ΔT = 65.7°C on the first cycle and 60.7°C on the second cycle was achieved. This was higher than previously reported for Ti-Pt [19]. The transformation enthalpies of the forward and reverse transformations were not easily discernible due to the overlapping of intermediate phase peaks. The exception was the reverse transformation on the first cycle where the transformation enthalpy was determined -33.2J/g, which was higher than reported for Ti-Pt [19].

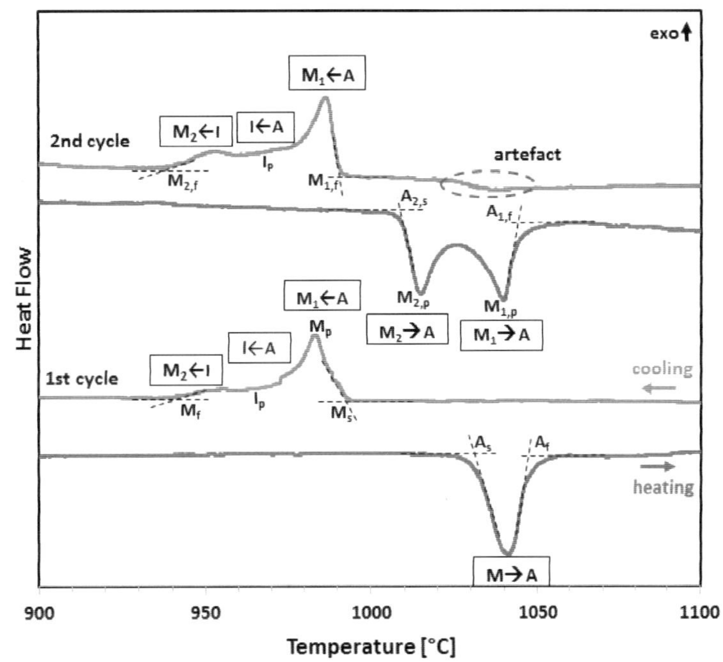

Figure 5. Two DSC cycles of as-cast Ti-Pt alloy sample showing the transformation temperatures of the $Ti_{50.5}Pt_{49.5}$ (at.%) phase. A = B2 Austenite. M = B19 Martensite. I = Intermediate

Table 1. Transformation temperatures of $Ti_{50.5}Pt_{49.50}$ phase during heating and cooling for two DSC thermal cycles.

Transformation Temperatures (°C) Heating					**Transformation Temperatures (°C) Cooling**				
				Cycle 1					
		M→A start	M→A peak	M→A finish	A→M_1 start	A→M_1 peak	A→I peak	I→M_2 peak	I→M_2 finish
		1031.8	1040.8	1046.5	992.3	983.2	971.0	951.5	935.7
				Cycle 2					
M_2→A start	M_2→A peak		M_1→A peak	M_1→A finish	A→ M_1 start	A→ M_1 peak	A→I peak	I→M_2 peak	I→M_2 finish
1009.5	1014.7		1039.6	1043.5	993.1	986.5	971.9	952.6	937.7

Figure 6 shows DSC second cycle curves of Ti-Pt_{50-x}-V_x phase ($x = 0$, 5.8 and 10.2 at.% V). The binary Ti-Pt second cycle curve was replotted for ease of comparison. First and second cycle heating curves for $TiPt_{50-x}V_x$ phase ($x = 5.8$, 10.2) did not display significant variation, so only second heating cycle results were plotted. Ternary alloying of Ti-Pt with V by replacing Pt changed the transformation from multiple to single step and reduced the martensitic transformation temperatures. The M_s temperatures decreased from 992°C (binary Ti-Pt) to 925°C for $Ti_{49.1}$-$Pt_{45.1}$-$V_{5.8}$ and 820°C for $Ti_{49.5}$-$Pt_{40.33}$-$V_{10.2}$. The reduction in transformation temperatures was predicted by first principles calculations, which showed that ternary alloying of V to Ti-Pt alloy promoted the stability of the B2 phase by increasing the C′ shear modulus, thus decreasing the transformation temperature [20]. This was in agreement with transformation temperature reduction trends observed in ternary alloying when Ni was partially substituted with V, Cr, Mn, Fe or Co in Ni-Ti alloys [3,4].

The $Ti_{49.1}(Pt_{45.1},V_{5.8})$ phase showed no significant difference on the start and finish temperatures of the reverse transformation. However, it significantly depressed the forward transformation, widening the transformation hysteresis to $\Delta T = 142.2$°C. The $Ti_{49.5}(Pt_{40.3},V_{10.2})$ phase resulted in further widening of the thermal hysteresis to $\Delta T = 191.4$°C. This would be a disadvantage in actuation applications where narrow thermal hysteresis is required for fast response times [3].

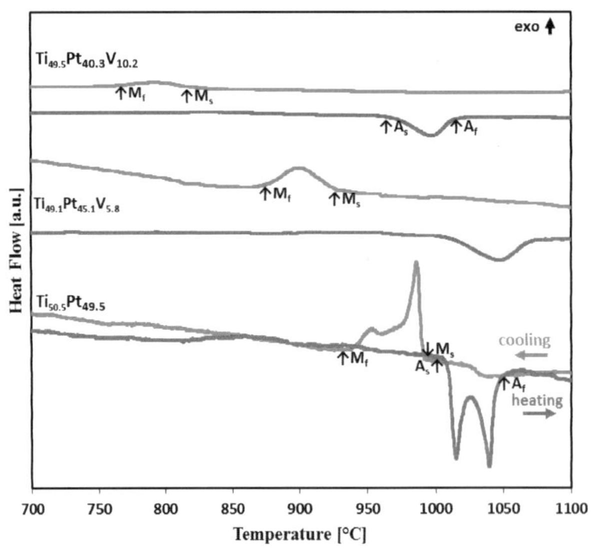

Figure 6. Second cycling DSC curves of as-cast $Ti_{50}Pt_{50}$, Ti_{50}-$Pt_{43.75}$-$V_{6.25}$ and Ti_{50}-$Pt_{37.5}$-$V_{12.5}$ (at.%) alloy samples showing the transformation temperatures of the $Ti_{50.5}Pt_{49.5}$, $Ti_{49.1}(Pt_{45.1},V_{5.8})$, $Ti_{49.5}(Pt_{40.3},V_{10.2})$ (at.%) phases.

Table 2. Transformation temperatures and hysteresis of second thermal cycles on $Ti_{49.1}$-$Pt_{45.1}$-$V_{5.8}$ and $Ti_{49.5}$-$Pt_{40.3}$-$V_{10.2}$ phases of as-cast Ti_{50}-$Pt_{43.75}$-$V_{6.25}$ and Ti_{50}-$Pt_{37.5}$-$V_{12.5}$ (at.%) alloy samples.

Phase	A_s (°C)	A_f (°C)	M_s (°C)	M_f (°C)	ΔT (°C)
$Ti_{49.1}$($Pt_{45.1}$,$V_{5.8}$)	1012.2	1067.4	925.0	878.0	142.4
$Ti_{49.5}$($Pt_{40.3}$,$V_{10.2}$)	974.6	1011.4	820.0	766.9	191.4

Ternary alloying of Ti-Pt by substituting Pt with varying contents of vanadium resulted in the formation of an inhomogeneous microstructure with different phases in the dendrites and inter-dendritic regions. The binary Ti-Pt alloy showed multiple transformation peaks due to intermediate phases being confined to interdendritic regions. The addition of V changed the transformation sequence from multiple steps to single step transformation, lowered the transformation temperature and widened the transformation hysteresis. Although the martensitic transformation temperatures of $Ti_{49.1}Pt_{45.1}V_{5.8}$ and $Ti_{49.5}Pt_{40.3}V_{10.2}$ phases were decreased, they were still above 750 °C. Therefore, these alloys could still be potentially used for high temperature shape memory applications (>750°C), while lowering cost by reducing the Pt content.

4. Conclusions

- Multiple phases were observed in as-cast Ti-Pt-V alloys.
-The amount of transforming Ti(Pt,V) phase formed reduced with increasing vanadium content.
-Binary Ti-Pt showed multiple thermal transformation peaks, this changed to single step transformation in ternary Ti_{50}-Pt_{50-x}-V_x.
- The addition of vanadium reduced the transformation temperatures of the Ti-Pt alloy and increased transformation hysteresis.
- Transformation temperatures higher than 750°C were achieved in alloy with 10.2 at.% V, retaining potential for high temperature applications.

5. References

[1] Saghaian S M, Karaca H E, Souri M, Turabi A S and Noebe R D 2016 Tensile shape memory behavior of $Ni_{50.3}Ti_{29.7}Hf_{20}$ high temperature shape memory alloys *Mater. Des.* **101** 340–5
[2] Mamiko K, Madoka T, Satoshi T and Yoko Y-M 2012 Effect of Zr on phase transformation and high temperature shape memory effect in TiPd alloys *Mater. Lett.* **89** 336–8
[3] Kamila S 2013 Introduction, classification and applications of smart materials: An overview *Am. J. Appl. Sci.* **10** 876–80
[4] Hee Y K, Tatsuhito F, Pio John B, Tae-Hyun N and Shuichi M 2011 Martensitic transformation and shape properties of Ti-Ta-Sn high temperature shape memory alloys *Mater. Sci. Eng.* A **528** 7238–46
[5] Ma L W, Cheng H S, Cao C W and Chung C Y 2012 Study of thermal scanning rates on transformations of Ti-19Nb-9Zr (at.%) by means of Differential Scanning Calorimetry analysis *J. Mater. Eng. Perform.* 2675–9
[6] Rios O, Noebe R, Biles T, Garg A, Palczer A, Scheiman D, Seifert H J and Kaufman M 2005 Characterization of ternary NiTiPt high-temperature shape memory alloys *Smart Structures and Materials 2005: Active Materials: Behavior and Mechanics* vol 5761 p 376
[7] Karakoc O, Hayrettin C, Bass M, Wang S J, Canadic D, Mabe J H, Lagoudas D C and Karaman I 2017 Effects of upper cycle temperature on the actuation fatigue response of NiTiHf high temperature shape memory alloys *Acta Mater.* **138** 185–97
[8] Canadinc D, Trehern W, Karaman I, Sun F and Chaudhry Z 2019 Ultra-high temperature multi-component shape memory alloys *Scr. Mater.* **158** 83–7
[9] Biggs T, Cortie M B, Witcomb M J and Cornish L A 2001 Martensitic Transformations , Microstructure , and Mechanical Workability of TiPt *Metall. Mater. Trans.* A **32** 1881–6
[10] Yamabe-Mitarai Y, Arockiakumar R, Wadood A, Suresh K S, Kitashima T, Hara T, Shimojo M, Tasaki W, Takahashi M, Takahashi S and Hosoda H 2015 Ti(Pt, Pd, Au) based high temperature shape memory alloys *Mater. Today Proc.* S **2** 517–22

[11] Bozzolo G, Mosca H O and Noebe R D 2007 Phase structure and site preference behavior of ternary alloying additions to PdTi and PtTi shape-memory alloys *Intermetallics* **15** 901–11

[12] Chikosha S, Mahlatji M, Modiba R and Chikwanda H 2018 The effect of vanadium on structure and martensitic transformation temperature of TiPt alloy *Conference of the South African Advanced Materials Initiative (COSAAMI-2018)*

[13] Biggs T, Cornish L A, Witcomb M J and Cortie M B 2004 Revised phase diagram for the Pt-Ti system from 30 to 60 at.% platinum *J. Alloys Compd.* **375** 120–7

[14] Karem E T A 2012 *Phase transformations and equilibria of titanium platinum alloys the composition range 30-50 atomic percent platinum*

[15] Lin H-C, Lin K M, Lin C-S and Chang S K 1999 A study of TiNiV ternary shape memory alloys *J. Alloys Compd.* **284** 213–7

[16] Otsuka K and Ren X 2005 Physical metallurgy of Ti-Ni based shape memory alloys *Prog. Mater. Sci.* **50** 511–678

[17] Liu Y and McCormick P 1996 Criteria of transformation sequence in NiTi shape memory alloys *Mater. Trans.* **37** 691–6

[18] Sehitoglu H, Hamilton R, Maier H J and Chumlyakov Y 2004 Hysteresis in NiTi alloys *J. Phys.* **IV** 3–10

[19] Rotaru G M, Tirry W, Sittner P, Van Humbeeck J and Schryvers D 2007 Microstructural study of equiatomic PtTi martensite and the discovery of a new long-period structure *Acta Mater.* **55** 4447–54

[20] Modiba R, Baloyi E, Chikosha S, Chauke H . and Ngoepe P E 2018 Computational modelling of Ti50Pt50Vx potential shape memory alloy *Conference of the South African Advanced Materials Initiative (COSAAMI-2018)*

[21] Wadood A, Takahashi M, Takahashi S, Hosoda H and Yamambe-Mitarai Y 2013 High temperature mechanical and shape memory properties of TiPt-Zr and TiPt-Ru alloys *Mater. Sci. Eng. A* **564** 34–41

[22] Tello K E, Noebe R D, Garg A, Allaz J and Kaufman M J 2017 Materials Characterization Revisiting the Ti-Pt system relevant to high temperature shape memory alloys **130** 97–104

The microstructural and mechanical characterization of the β-type Ti-11.1Mo-10.8Nb alloy for biomedical applications

M L Raganya[1, 2, 3*], N M Moshokoa[1, 2, 3], B Obadele[3], P A Olubambi[3] and R Machaka[1, 2, 3]

[1] Light Metals, Manufacturing Cluster, Council for Scientific and Industrial Research, Meiring Naudé Road, Brummeria, Pretoria 0184, South Africa

[2] Titanium Centre of Competence, Manufacturing Cluster, Council for Scientific and Industrial Research, Meiring Naudé Road, Brummeria, Pretoria 0185, South Africa

[3] Centre of Nanoengineering and Tribocorrosion , School of Mining and Metallurgy and Chemical Engineering, University of Johannesburg, Doornfontein Campus, Johannesburg, South Africa

E-mail: leeragm@gmail.com

Abstract. Ti-Mo alloys are promising biocompatible materials with lower elastic modulus than the extensively used Ti-6Al-4V alloy. However, research work done on these alloys indicate that their elastic modulus is still higher than that of the bone, even after the execution of numerous heat treatment procedures. Therefore, this study was aimed at adding Nb (non-toxic β stabilizer) to Ti-Mo system followed by the characterization of its microstructural and mechanical properties in the as-cast condition. This study will provide systematic preliminary information towards the design and development of novel biomedical components. The microstructure and phase analysis were carried out using optical microscope, SEM, and XRD. Mechanical tests were conducted using the uniaxial tensile test machine and Vickers microhardness tester. The as-cast Ti-11.1Mo-10.8Nb alloy consisted primarily of β phase and a possible small volume fraction of ω phase. The Vickers micro-hardness, elastic modulus, bending strength were measured as 311.62 $HV_{0.5}$, 56.9 GPa 1671.4 MPa, respectively. The Ti-11.1Mo-10.8Nb alloy also exhibited ductile fracture behaviour during bend testing. The Ti-11.1Mo-10.8Nb design is a promising alloy for biomedical applications.

1. Introduction

The most widely used biomaterial today is probably Ti6Al4V, owing to its low-weight, high corrosion resistance, high specific strength and availability [1–4]. However, in addition to low wear resistance [5], two particularly serious concerns are often raised in more recent years. Firstly, the Ti6Al4V alloy has an elastic modulus of 110 GPa, which is significantly higher than that of the human bone (10 – 40 GPa) [1–4,6]. This difference in the elastic moduli between an implant and a bone can cause bone degradation ('stress shielding effect') and resorption. This drawback puts serious limitations on their performance as implant materials [6]. Secondly, The release of Al and V ions from the alloy into the human body might cause genotoxic and cytotoxic responses from the human body such as the potential for the development of Alzheimer's disease, peripheral neuropathy, and osteomalacia especially in long-term implantation cases [5,7,8].

Therefore, strategies to design next-generation Ti-based biomaterials with lower elastic modulus from biocompatible elements are developed [1–4,9]. Global research into next-generation Ti alloys has focussed on the lower elastic modulus β-type than α-type Ti alloys - their cubic crystal symmetry incidentally makes them processable too [9]. The β stabilizing elements are classified as isomorphous (form continuous solid solution with limited solubility in α phase like Mo, Ta, Nb, and V) and eutectoid (form continuous solid solubility in α and β phases like Zr and Hf) [6,10]. Amongst these β-stabilizers, Mo is the strongest β stabilizer, is the cheapest, can improve strength and corrosion resistance at lower elastic modulus and it can regulate the pH balance in the body [11].

The design of these β-type alloys can be carried out using several approaches amongst them being the Mo equivalence (Mo_{eq}) and the average valence electron concentration (e/a) ratio approaches. The Mo_{eq} is a parameter used in Ti alloy compositions to characterize the contribution of alloying elements on the β-phase stability [10], [12]–[15]. The proposal of the Mo_{eq} formula as a measure to compare the elemental contributions to that of Mo, the most β-stabilizing element, was presented by Bania as follows:

$$Mo_{eq} = 1.0\,Mo + 0.67\,V + 0.44\,W + 0.28\,Nb + 0.22\,Ta + 2.9\,Fe + 1.6\,Cr + 0.77\,Cu + 1.11\,Ni + 1.43\,Co + 1.54\,Mn + 0.0\,Sn + 0\,.0\,Zr - 1.0\,Al\,(wt\%) \quad \textbf{Eq. 1}$$

For each alloying element, the coefficient value is the least amount of an alloying element that can suppress the formation of martensitic phases (α' and α") and retain a single β-phase microstructure at room temperature upon water quenching – it's also known as a critical value (β_c) for that alloying element. The value of β_c experimentally determined by Bania was is approximately 10 wt% Mo [14–16]; that is a single β phase with equiaxed grains can be stabilized at Mo content of 10 wt% or more [7,17,18]. Hence, Wang *et al* (2015) proposed a new Mo_{eq} model for the characterization of the critical stability limit of multicomponent β-type Ti-based alloy systems, where the critical lower limit for β stabilization is 11.78 wt% [14].

$$Mo_{eq} = 1.0\,Mo + 1.25\,V + 0.59\,W + 0.28\,Nb + 0.22\,Ta + 1.93\,Fe + 1.84\,Cr + 1.5\,Cu + 2.46\,Ni + 2.67\,Co + 2.26\,Mn + 0.30\,Sn + 0.47\,Zr + 3.01\,Si - 1.47\,Al\,(wt\%) \quad \textbf{Eq. 2}$$

The average valence electron concentration (e/a) approach is also used to design alloys. The e/a ratio is defined as the average number of valence electrons in each atom of an alloy and the precipitation of the metastable omega (ω) phase occurs between the range 4.13 and 4.30 [14,19]. This phase is formed under the three conditions: upon quenching above the β transus via a diffusionless (shuffle) mechanism, during ageing in the temperature range 100 °C to 500 °C in the regions lean in the alloying elements through a diffusion-controlled process and deformation at room temperature [20]. The formation of the β-phase is dominant at e/a ratios above 4.30 [14,19].

Mo and Nb are known to segregate easily in β-Ti alloys during solidification and extensive research is being carried out to hinder this problem [21,22]. In this study, a novel β-type Ti-Mo-Nb alloy with high strength and low elastic modulus was designed using the Mo equivalence and the average valence electron concentration ratio approaches. The microstructural characteristics and mechanical properties of the alloys in as-cast condition were investigated. This study will provide systematic preliminary information towards the design and development of novel biomedical components.

2. Materials and Methods

2.1. Alloy design

The β-type Ti-11.1Mo-10.8Nb alloy used for this study is shown in Table 1 below. The alloy formulation was designed using a combination of the Mo equivalence and the average valence electron concentration ratio approaches. The investigation will also be carried out on CP Ti and Ti-11.78Mo binary alloy. Their calculated Mo equivalence and e/a ratio values are listed in table 1. The designed alloy is classified in the β metastable category [7] and according to references [17,23], the precipitation of ω phase is expected in Ti-11.78Mo and Ti-11.1Mo-10.8Nb alloys, but not in CP Ti.

2.2. Material Preparation

The three Ti-based alloys of 100 g each were prepared from elemental powders of CP Ti (99.9%), molybdenum (99.5%) and niobium (99.8%). Green compacts of each alloy composition were first cold-pressed and then melted in a water-cooled copper crucible with a tungsten electrode using a commercial arc melting vacuum-pressure casting system. The evacuation and purging of the melting chamber with argon was carried out before melting. The ingots were each turned-over and remelted three times to promote chemical homogeneity [24].

Table 1. Compositions of CP Ti, Ti-11.78Mo and Ti-11.1Mo-10.8Nb alloys.

Alloy Name	(Mo_{Eq})Bania [wt%]	(Mo_{Eq})Qing [wt%]	e/a Ratio
CP Ti	0	0	4
Ti-11.78Mo	11.78	11.78	3.8
Ti-11.1Mo-10.8Nb	14.19	13.65	4.3

2.3. Microstructural Characterization

An optical microscope (Leica CTR4000) and a Scanning electron microscope (SEM) (JEOL: JSM-6510) were used to observe the microstructure and elemental composition in the alloys. Samples were precision cut from the vacuum arc-melt as-cast ingots, mounted, ground using silicon carbide papers up to 2400 mesh, polished using 3 µm diamond suspension and colloidal silica (final polishing) and etched with Kroll's reagent (85 ml of distilled water, 15 ml of Nitric acid and 5 ml of hydrofluoric acid). The polished samples were then etched with Kroll's reagent (85 ml of distilled water, 15 ml of nitric acid and 5 ml of hydrofluoric acid), consecutively. Phase identification was carried out via X-ray diffractometry (XRD) XPERT-PRO diffractrometer with Cu K_α radiation and graphite monochrometer operated at an accelerating voltage of 45 kV and a current of 40 mA. The various Ti phases were identified by matching the observed XRD profile peaks with the Joint Committee on Powder Diffraction Standards files or in prior published work.

2.4. Mechanical Testing

The density measurements produced specimens were carried out using Archimedes method. The Vickers micro-hardness measurements were conducted using a Vickers micro-hardness tester (FM-700) with a load of 500 g for 15 s. At least ten different indentation measurements were taken and the average was calculated. Tensile specimens with the dimensions illustrated in Figure 1 were prepared from the vacuum arc-melt as-cast ingots via electrical discharge machining (EDM). The stress-strain data was recorded at room temperature using the Instron^{TM} 1342 model apparatus.

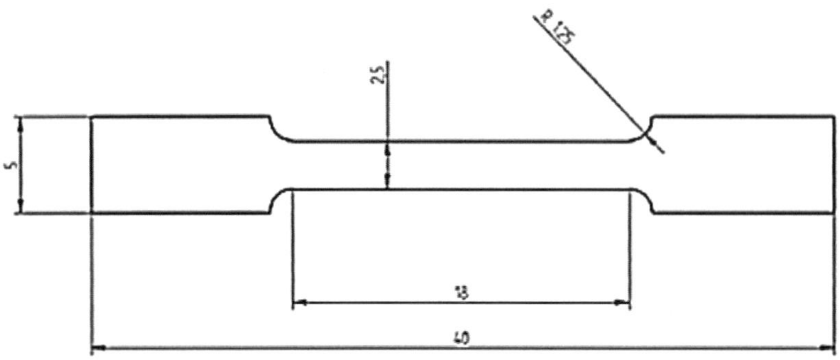

Figure 1. Schematic dimensions of the tensile specimen.

3. Results and Discussion

3.1. Microstructural Characterization

The XRD patterns of the as-cast CP Ti, Ti-11.78Mo and Ti-11.1Mo-10.8Nb with identified phases and indexed diffraction planes are exhibited in Figure 2. The as-cast CP Ti XRD patterns are characterised by martensitic orthorhombic α' phase peaks and these results are in agreement with Ho *et al*'s observations [25]. The optical micrographs of the as-cast CP Ti alloy presented in Figure 3(a) show coarse primary plates with partitioned regions filled with progressing small secondary plates and matches the one obtained by Davis *et al* [17]. The microstructure is characteristic of a fast-cooled metastable feather-like microstructure of hexagonal martensitic α' phase was observed in the CP Ti as in Ref [3].

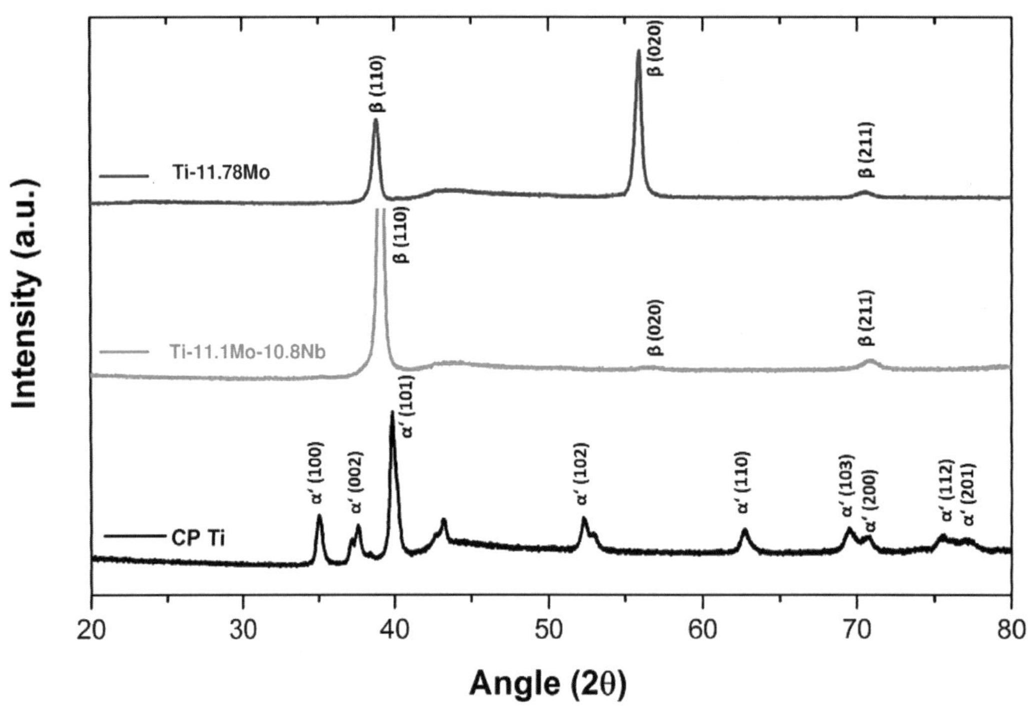

Figure 2: X-ray Diffraction (XRD) patterns of CP Ti, Ti-11.78Mo and Ti-11.1Mo-10.8Nb

Both the XRD patterns of alloys Ti-11.78Mo and Ti-11.1Mo-10.8Nb exhibit the existence of a single cubic β phase. The optical micrographs of alloys Ti-11.78Mo and Alloy are presented in Figures 3(b and c), respectively corroborate the experimental XRD results. With the addition of 11.78 wt% Mo to CP Ti in Ti-11.78Mo, the microstructure was composed of a significant amount of retained β-phase with equiaxed grains of various sizes. The grains of Ti-11.1Mo-10.8Nb were finer than those of Ti-11.78Mo. Therefore while the alloying of Ti-Mo alloy with Nb also contributed to the stabilization of the β phase it tentatively results in grain refinement the microstructure possibly due to the interaction of grain boundary that minimizes grain growth [21].

No peaks of martensitic or ω phases were observed on the XRD patterns. This is probably due to two tentative reasons; (i) the amount of the formed ω phase might be below the detection limits of the diffractometer and/or (ii) that the provision of sufficient beta-stabilizers suppressed the formation of the ω phase as well as other martensitic phases [10]. Our observations and interpretation are consistent with findings also available in the literature, specifically by Ho *et al.* [25], Gabriel *et al.* [26] and Xu *et al.* [27]. That means that Mo and Nb alloying elements contributed to the stabilization of the β phase by suppressing the formation of the martensitic phases. However, the single peak exists in the XRD patterns of all the alloys at 2θ ≈ 44.3° could not be conclusively identified.

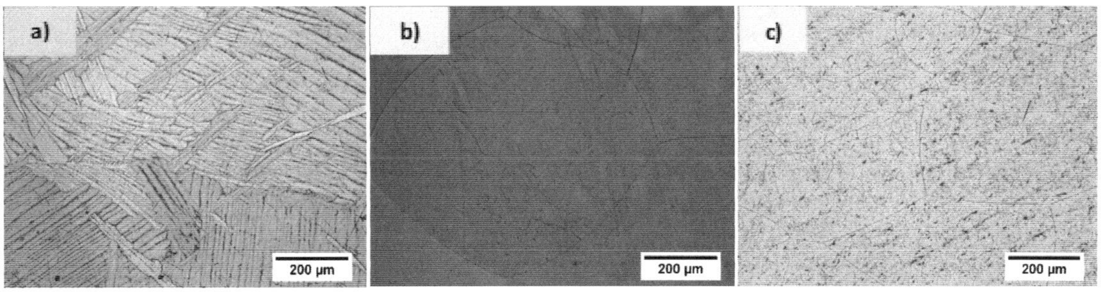

Figure 3: Optical micrographs of as-cast a) CP Ti, b) Ti-11.78Mo and c) Ti-11.1Mo-10.8Nb.

Mo and Nb are known to segregate easily in β Ti alloys during solidification [21]. Figure 4 shows the distributions of the key elements in the as-cast alloy Ti-11.1Mo-10.8Nb analysed using EDS mapping. Both the β stabilizing elements, Mo and Nb, are present and homogeneously distributed, with no significant enrichment or depletion of both beta stabilizers in the as-cast alloy.

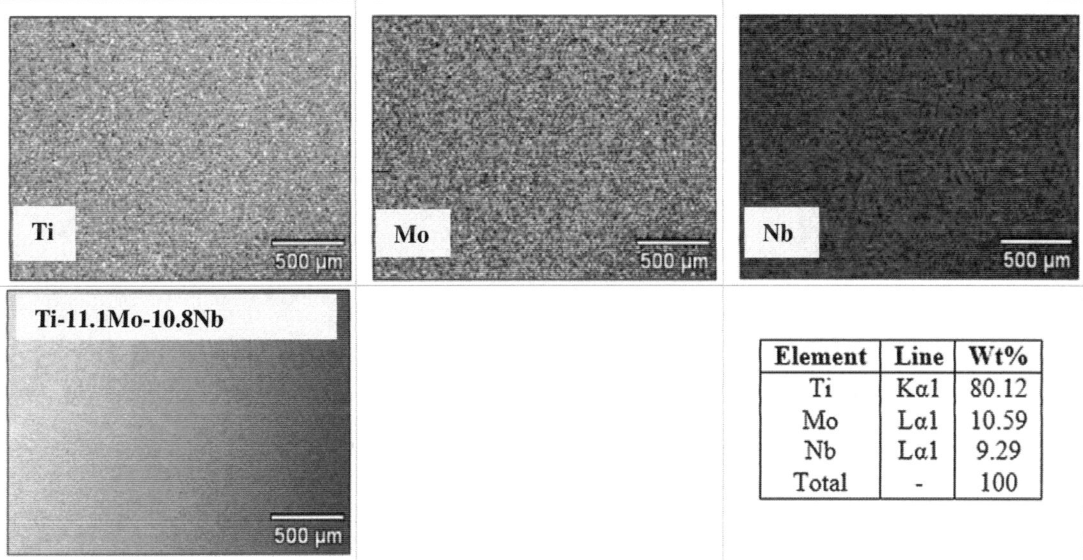

Element	Line	Wt%
Ti	Kα1	80.12
Mo	Lα1	10.59
Nb	Lα1	9.29
Total	-	100

Figure 4: Energy dispersive spectroscopy (EDS) elemental mapping of Ti, Mo and Nb distribution and spectra of Ti-11.1Mo-10.8Nb

3.2. Mechanical Properties

Table 2 summarises the physical and mechanical properties of the studied alloys. The addition of 11.78 wt% Mo to CP Ti contributed to a significant increase in the micro-hardness while with Nb addition, the micro-hardness of Ti-11.1Mo-10.8Nb was significantly lower than of Ti-11.78Mo, but higher than that of CP Ti. The general increase can be attributed to the possible solution strengthening effect of the alloying elements as well as the tentative precipitation of the ω phase during cooling [10,17,23,25]. The addition of Nb possibly contributed towards the stabilization of β phase and the suppression of the formation of ω and/or the α' phases (in the Ti-11.1Mo-10.8Nb alloy), which has been reported to have lower hardness [25,28].

Table 2: Comparison of the mechanical properties of the designed alloys (Ti-11.78Mo and Ti-11.1Mo10.8Nb) with other biomaterials.

Alloy Name	Rel. Density [g/cm³]	Hardness [Hv]	Elastic Modulus [GPa]
Ti6Al4V ELI wrought [29]	-	-	110
Ti-12Mo-13Nb (hot swaged + aged) [30]	-	-	110 ± 2.62
Ti-10Mo-20Nb (cold swaged +/or aged) [31]	-	267.36 - 333.28	100.12 - 101.32
Ti-12Mo-8Nb (hot swaged + annealed) [32]	-	275 ± 9	91 ± 0.53
CP Ti (this study)	4.5 ± 0.01	234.22 ± 14.77	133.7 ± 0
Ti-11.78Mo (this study)	4.8 ± 0.01	389.62 ± 32.22	133.15 ± 10.52
Ti-11.1Mo-10.8Nb (this study)	5.1 ± 0.01	311.62 ± 5.44	56.9 ± 3.08

The elastic modulus generally decreases with the increasing Mo equivalence [14]. It is evident that the increased Mo_{eq} of Ti-11.78Mo did not have a significant influence on the elastic modulus of CP Ti. The elastic modulus of Ti-11.1Mo-10.8Nb alloy was decreased significantly by the Nb that contributed to increase in Mo_{eq} and the stabilization of β phase, thereby suppressing the formation of ω phase. The mechanical properties show that this Ti-11.1Mo-10.8Nb is a promising β-type alloy for biomedical applications.

4. Conclusions

The design of Ti-11.1Mo-10.8Nb β-type alloy for biomedical applications was carried out using the Mo equivalence and e/a ratio approaches. Based on the microstructures, phase constitution and mechanical properties of the alloys studied, the following conclusions can be drawn:

a. The microstructure of as-cast Ti-11.1Mo-10.8Nb alloy was composed primarily of β and possibly some ω phase, which were characterized by optical microscopy and XRD.
b. The distribution of the β stabilizing elements (Mo and/or Nb) in the alloys was homogeneous and there was no significant enrichment or depletion of both beta stabilizers.
c. The tentative precipitation of ω phase in the alloy contributed to the significant increase in micro-hardness compared to CP Ti.
d. The addition of 11.78 wt% Mo decreased the elastic modulus due to the precipitation of ω phase during fast cooling, while the micro-alloying of Ti-Mo alloy with Nb contributed to the decrease in elastic modulus from 133.15 ± 10.52 GPa to 56.9 ± 3.08 GPa.
e. This β-type Ti-11.1Mo-10.8Nb composition design is a promising alloy for biomedical applications.

5. Acknowledgments

This research is financially supported by the National Research Foundations (NRF) and the Council for Scientific and Industrial Research (CSIR), South Africa, through Thuthuka Grant No. 115859.

6. References

[1] Hagihara K, Nakano T, Maki H, Umakoshi Y and Niinomi M 2016 Isotropic plasticity of β-type Ti-29Nb-13Ta-4.6Zr alloy single crystals for the development of single crystalline β-Ti implants *Sci. Rep.* **6** 1–11

[2] Guo S, Meng Q, Cheng X and Zhao X 2015 Microstructural evolution and mechanical behavior of metastable β-type Ti-30Nb-1Mo-4Sn alloy with low modulus and high strength *Prog. Nat. Sci. Mater. Int.* **25** 414–8

[3] Hagihara K, Nakano T and Todai M 2017 Unusual dynamic precipitation softening induced by dislocation glide in biomedical beta-titanium alloys *Sci. Rep.* **7** 1–9

[4] Li Y, Yang C, Zhao H, Qu S, Li X and Li Y 2014 New developments of ti-based alloys for biomedical applications *Materials (Basel).* **7** 1709–800

[5] Lin C-W, Ju C-P and Chern Lin J-H 2005 A comparison of the fatigue behavior of cast Ti–7.5Mo with c.p. titanium, Ti–6Al–4V and Ti–13Nb–13Zr alloys *Biomaterials* **26** 2899–907

[6] Mohammed M T, Khan Z A and Siddiquee A N 2014 Beta Titanium Alloys : The Lowest Elastic Modulus for Biomedical Applications : A Review

[7] Ho W F, Ju C P and Chern Lin J H 1999 Structure and properties of cast binary Ti-Mo alloys *Biomaterials* **20** 2115–22

[8] Bălţatu M S, Vizureanu P, Tierean M H, Minciună M G and Achiţei D C 2015 Ti-Mo Alloys Used in Medical Applications *Adv. Mater. Res.* **1128** 105–11

[9] Kolli R and Devaraj A 2018 A Review of Metastable Beta Titanium Alloys *Metals (Basel).* **8** 506

[10] Lin J, Ozan S, Li Y, Ping D, Tong X, Li G and Wen C 2016 Novel Ti-Ta-Hf-Zr alloys with promising mechanical properties for prospective stent applications *Sci. Rep.* **6** 1–11

[11] Li Y H, Wang F and Li J J 2017 Current developments of biomedical porous Ti-Mo alloys *Int. J. Mater. Res.* **108** 619–24

[12] Gordin D M, Gloriant T, Texier G, Thibon I, Ansel D, Duval J L and Nagel M D 2004 Development of a β-type Ti-12Mo-5Ta alloy for biomedical applications: Cytocompatibility and metallurgical aspects *J. Mater. Sci. Mater. Med.* **15** 885–91

[13] Zhang L B, Wang K Z, Xu L J, Xiao S L and Chen Y Y 2015 Effect of Nb addition on microstructure, mechanical properties and castability of β-type Ti-Mo alloys *Trans. Nonferrous Met. Soc. China (English Ed.* **25** 2214–20

[14] Wang Q, Dong C and Liaw P K 2015 Structural Stabilities of β-Ti Alloys Studied Using a New Mo Equivalent Derived from [β/(α + β)] Phase-Boundary Slopes *Metall. Mater. Trans. A Phys. Metall. Mater. Sci.* **46** 3440–7

[15] Wang Q, Zhang X D, Li X N, Ji C J and Dong C 2013 Designing Multi-Component β-Ti Alloys with Low Young's Modulus *Mater. Sci. Forum* **747–748** 885–9

[16] Bania P J 1994 Beta titanium alloys and their role in the titanium industry *Jom* **46** 16–9

[17] Davis R, Flower H M and West D R F 1979 Martensitic transformations in Ti-Mo alloys *J. Mater. Sci.* **14** 712–22

[18] Oliveira N T C, Aleixo G, Caram R and Guastaldi A C 2007 Development of Ti-Mo alloys for biomedical applications: Microstructure and electrochemical characterization *Mater. Sci. Eng. A* **452–453** 727–31

[19] Nnamchi P S, Obayi C S, Todd I and Rainforth M W 2016 Mechanical and electrochemical characterisation of new Ti – Mo – Nb – Zr alloys for biomedical applications *J. Mech. Behav. Biomed. Mater.* **60** 68–77

[20] Hickman B S 1969 The formation of omega phase in titanium and zirconium alloys: A review *J. Mater. Sci.* **4** 554–63

[21] Gao J, Nutter J, Liu X, Guan D, Huang Y, Dye D and Rainforth W M 2018 Segregation mediated heterogeneous structure in a metastable β titanium alloy with a superior combination of strength and ductility *Sci. Rep.* **8**

[22] Min X H, Tsuzaki K, Emura S and Tsuchiya K 2012 Heterogeneous twin formation and its effect on tensile properties in Ti-Mo based β titanium alloys *Mater. Sci. Eng. A* **554** 53–60

[23] Ugender S, Kumar A and Reddy A S 2014 Experimental Investigation of Tool Geometry on Mechanical Properties of Friction Stir Welding of AA 2014 Aluminium Alloy *Procedia Mater. Sci.* **5** 824–31

[24] Li S J, Cui T C, Hao Y L and Yang R 2008 Fatigue properties of a metastable β-type titanium alloy with reversible phase transformation *Acta Biomater.* **4** 305–17

[25] Ho W ., Ju C . and Chern Lin J . 1999 Structure and properties of cast binary Ti–Mo alloys *Biomaterials* **20** 2115–22

[26] Gabriel S B, Panaino J V P, Santos I D, Araujo L S, Mei P R, de Almeida L H and Nunes C A 2012

Characterization of a new beta titanium alloy, Ti–12Mo–3Nb, for biomedical applications *J. Alloys Compd.* **536** S208–10

[27] Xu L J, Chen Y Y, Liu Z G and Kong F T 2008 The microstructure and properties of Ti–Mo–Nb alloys for biomedical application *J. Alloys Compd.* **453** 320–4

[28] Zhang E, Zheng L, Liu J, Bai B and Liu C 2015 Influence of Cu content on the cell biocompatibility of Ti-Cu sintered alloys *Mater. Sci. Eng. C* **46** 148–57

[29] Geetha M, Singh A K, Asokamani R and Gogia A K 2009 Ti based biomaterials, the ultimate choice for orthopaedic implants - A review *Prog. Mater. Sci.* **54** 397–425

[30] Gabriel S B, Dille J, Rezende M C, Mei P, Almeida L H de, Baldan R and Nunes C A 2015 Mechanical Characterization of Ti–12mo–13nb Alloy for Biomedical Application Hot Swaged and Aged *Mater. Res.* **18** 8–12

[31] Gabriel S B, Rezende M C, Almeida L H de, Nunes C A, Dille J and Soares G de A 2015 Control of the Microhardness to Young Modulus Ratio by Mechanical Processing of a Ti-10Mo-20Nb Alloy *Mater. Res.* **18** 39–42

[32] Nunes A R V, Gabriel S B, Nunes C A, Araújo L S, Baldan R, Mei P, Malet L, Dille J and Almeida L H de 2017 Microstructure and Mechanical Properties of Ti-12Mo-8Nb Alloy Hot Swaged and Treated for Orthopedic Applications *Mater. Res.* **20** 526–31

IOP Conf. Series: Materials Science and Engineering **655** (2019) 012026 doi:10.1088/1757-899X/655/1/012026

Effect of powder bed preheating on distortion and mechanical properties in high speed selective laser melting

L P Motibane[1, 2], L C Tshabalala[1], N R Mathe[1], S Hoosain[1], R D Knutsen[2]

[1] National Laser Centre, Council for Scientific and Industrial Research, Pretoria, South Africa

[2] Centre for Materials Engineering, Department of Mechanical Engineering, University of Cape Town, South Africa

Email: Lmotibane@csir.co.za

Abstract Selective Laser Melting (SLM) is known to cause residual stresses due to the inherent large thermal gradients from high heating and cooling rates during manufacturing. The residual stresses tend to induce distortion, delamination of parts from the base plate as well as cracking because they reduce the threshold flaw size required for crack initiation. These challenges form a barrier to the use of this additive manufacturing method for structural applications for the aerospace industry where high part integrity is a critical requirement. The aim of the study was to evaluate the degree of distortion and crack growth resistance during increasing monotonic loading and cyclic loading of As-built Ti6Al4V parts produced on the Aeroswift high speed SLM process preheated at 200°C. The monotonic loading results are comparable to those of commercial SLM systems but are lower when compared conventional manufacturing methods. The crack growth resistance of the As-built specimen is lower than that of heat treated specimen. Distortion at this preheating temperature is evident at from 12cm away from the base of the cantilever and spreads to height of 3.2mm.

1. Introduction

Additive Manufacturing according to ASTM is a process of joining materials to make objects from 3D model data, usually layer upon layer, as opposed to subtractive manufacturing methodologies [1]. Classification of the processes is based on feed material, energy source and on how the feed is introduced into the machine. Selective Laser Melting (SLM) is a powder bed fusion type of process where a thin layer of metallic powder is deposited on a bed and a laser selectively melts the powder according to a 3D model. Selective Laser Melting has become a promising processing method for the manufacture of parts in the aerospace, automotive, biomedical and other industries. Its advantage is in its ability to produce complex and customised parts [2-5]. The SLM process involves fast heating and cooling of material by a laser creating high local thermal gradients. This results in highly nonequilibrium microstructures (e.g. Martensite), along with substantial residual stresses [6]. The residual stresses are associated with pronounced deformations especially for thin-walled features [7]. Distortion or even cracking of the part can sometimes happen during the process when the stresses reach critical levels [3]. Moreover, residual stresses strongly influence crack initiation and growth [8,9]. All this will affect the part's functionality and integrity. Post processing methods have been employed to reduce residual stresses and increase elongation to failure. This though, implies additional time or step and importantly, the post build heat treatments do not address part distortions during the build, and any distortions will still need to be dealt with [10]. A solution that can be implemented during manufacture is needed to address this.

Work has been done on preheating with promising results. Residual stresses have been effectively reduced and mechanical properties improved by preheating in different materials by a number of authors, Vracken et al.[11] and Ali et al.[12] on Titanium alloys, Buchbinder et al.[13] on Aluminium alloys and Mertens et al. on steel[14] .

This study will evaluate the degree of distortion and crack growth resistance during increasing monotonic loading and cyclic loading of Ti6Al4V parts produced on the Aeroswift high speed SLM process preheated at 200 degrees Celsius

2.Methods

2.1 *Materials and Process*

The material composition of powder determined by Electron Dispersive Spectroscopy (EDS) was 87.18wt% titanium, 8.45 wt% aluminium and 4.36 wt% vanadium. The specimen were manufactured on the Aeroswift high speed Selective Laser Melting machine at the CSIR National Laser Center. The laser power used was 2kW under an argon filled atmosphere and a preheating temperature of 200°C. The build samples are shown in fig. 1. The Archimedes method was used to measure the density of the specimen and it yielded >99% for all the specimen.

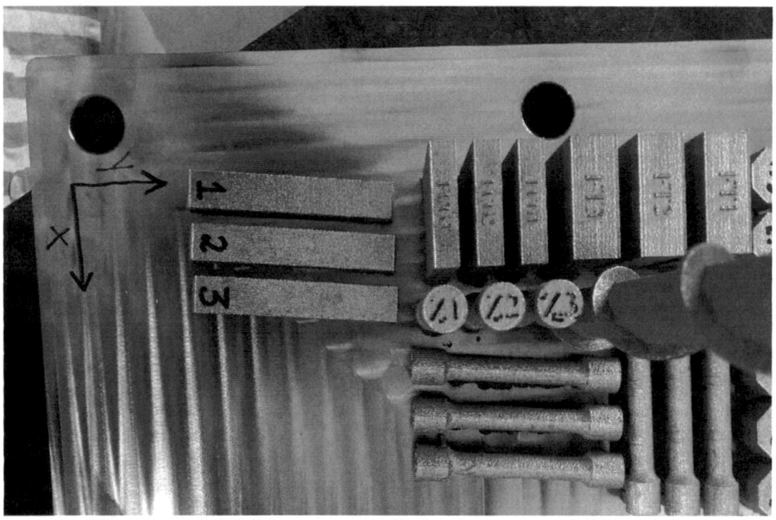

Figure 1. Ti6Al4Vspecimen built for the study on a base plate.

2.2 *Mechanical Testing*

Standard samples (ASTME8) were machined and tensile tests were conducted using an Instron universal testing machine with a crosshead velocity of 1 mm/min. A set of three tensile specimens were tested. Blocks manufactured for fracture toughness and fatigue crack growth rate specimen and were machined to size according to ASTM standards E399-17 and E647-15 respectively. Their direction was based on previous work done by Cain et al. [8]. This study is based on the worst performing direction which was the ZX from the Cain et al. study. This direction is designated by the ASTM E399 standard: axis direction perpendicular to the notch plane followed by the axis direction in which the crack is expected to propagate. The ESH 45kN servo-hydraulic machine is used for Fatigue Crack Growth and Fracture Toughness testing

IOP Conf. Series: Materials Science and Engineering **655** (2019) 012026 doi:10.1088/1757-899X/655/1/012026

2.3 *Distortion Measurements*
Three cantilevers were manufactured to measure distortion, hteir geometry and dimensions is shown in figure 2 below.. While they were still attached to the base plate, their height was measured relative to the base plate. They were then wirecut up to the base of the cantilever and the height was measured again. . The MetraScan 750 Optical Elite System 3D scanner was used to do the measurement. The spread of the cantilever Δl is calculated by equation 1.

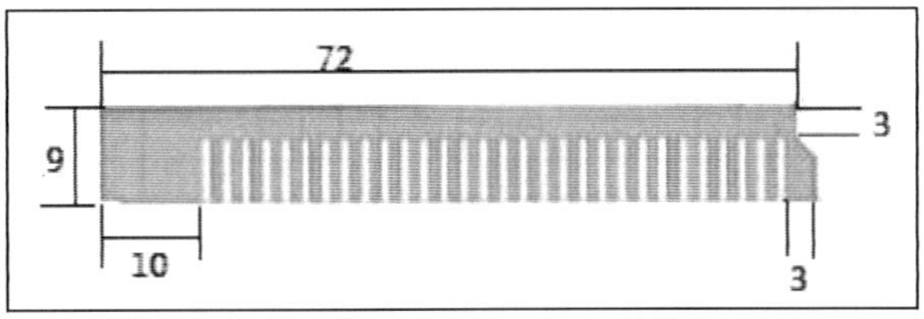

Figure 2. Cantilever geometry and dimensions

$$\Delta l = l_h - l_0 \qquad \text{(eq. 1)}$$

Where l_h is the height of the cantilever after cutting off the base plate and l_0 is the height before cutting

3.Results

3.1 *Tensile Testing*
The tensile mechanical properties of Aeroswift As-built specimen, commercial SLM systems As-built specimen and ASTM cast and wrought standard results are shown in table 1. The Aeroswift results are comparable with those of the commercially available SLM systems and are characteristic of the SLM process that has an inherent brittle martensitic microstructure resulting in a high yield stress, a high ultimate tensile strength but a relatively low ductility (<10%) [15-17]. The UTS and YS of the Aeroswift as-built specimens are higher than those of cast and wrought but their elongation is far less, this however should be improved by post processing [3,5,11]. Conventional manufacturing methods produce <1000MPa UTS and YS and >8% elongation.

Table 1: Tensile mechanical properties of SLM as-built Ti6AlV specimen.

	UTS (MPa)	YS(MPa)	Elongation (%)
Aeroswift	1211.66 ± 3.11	1067.91 ± 26.31	4.34±2.33
SLM [5]	1216+-8	1125+-22	6+-0.4
EOS M270[5]	1219+-20	1143+-30	4.89+-0.6
ASTM F1472 As-cast[18]	>758	>860	>8
ASTM F1108 Wrought[18]	>930	>860	>10

The SEM images of the fracture surfaces are shown in figure 3 below. Low magnification shows a cup and cone fracture. Pores are visible and contribute to the cracking of the material. At high magnification figure 2b) dimples are visible, these are associated with pore coalescence and ductile type of fracture [19]

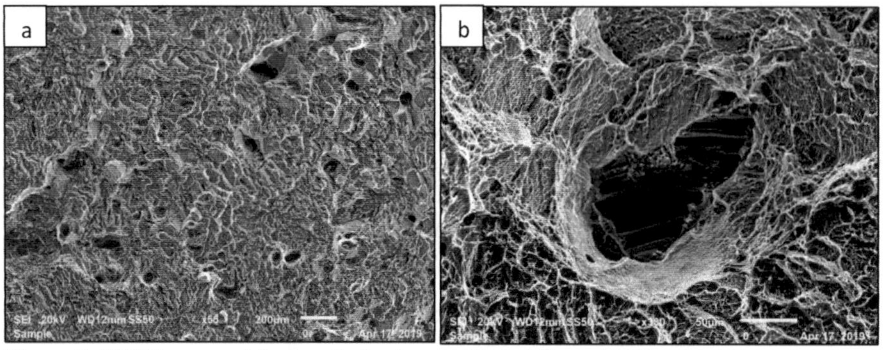

Figure 3: SEM images of the tensile fracture surface a) Pores b) Dimples

3.2 Crack growth properties

The crack growth of the Aeroswift as built specimen as well as the stress relieved specimen is given in figure 4. The as built specimen fractured after 41000 cycles and the crack had grown up to almost 8mm. The stress relieved specimen fractured only after 220000 cycles and had a crack growth length of 19mm. This indicates the influence of temperature on crack growth properties and the improvement of the properties temperature provides.

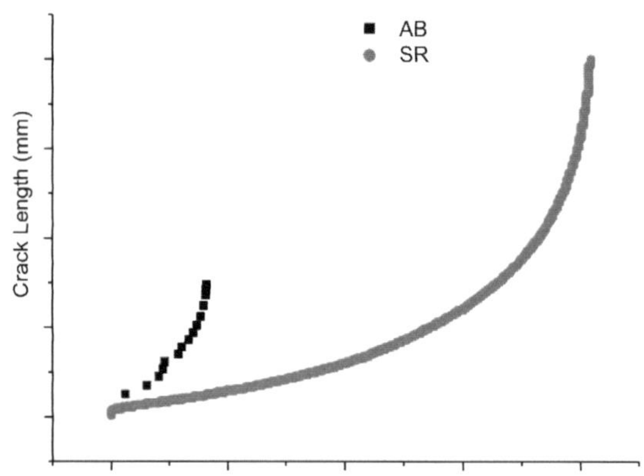

Figure 4. Graph showing the relationship of crack growth in distance with the number of cycles.

Testing for this material processed by additive manufacturing had not been done before on this machine, thus the testing load had to be determined using the specimen. After three rounds of testing that included increasing and decreasing the load depending on the rate at which the crack was growing, the load needed for this material for a recordable rate was established. Two of the test specimen were used for the load determination and thus only one set of results is reportable at the time of submission of this manuscript.

3.3 *Distortion Results*

The distortion results of the Aeroswift cantilevers are compared to a study by Vrancken at al. [11] and are depicted in figure 5. The Vrancken at al results represent two sets of data, one is of 25mm cantilevers built at no preheating (NP) and the other built at 200°C preheating temperature, the same temperature as this study. The results show that spreading (distortion) of the cantilever for the Aeroswift specimen (72 mm) starts from 12cm from the cantilever base and the spread, Δl is 3.526 mm. From the Vrancken et al. study, the amount of spreading between no preheating and 200°C preheating is evident but is not substantial. The distance ratios are shown in table 2, they were used to normalize the distance at which distortion begins in each case. This helps in comparing the data as parallels since the cantilever lengths were different. The bigger the ratio, the larger the distance between the cantilever base and the point at which distortion starts.

At the same preheating temperature, distortion starts after a larger distance for Aeroswift specimen than Vrancken et al. results. The test qualified the preheating temperature as no delamination of the parts occurred, even thin parts like those of the cantilever teeth which are very prone to delamination [6].

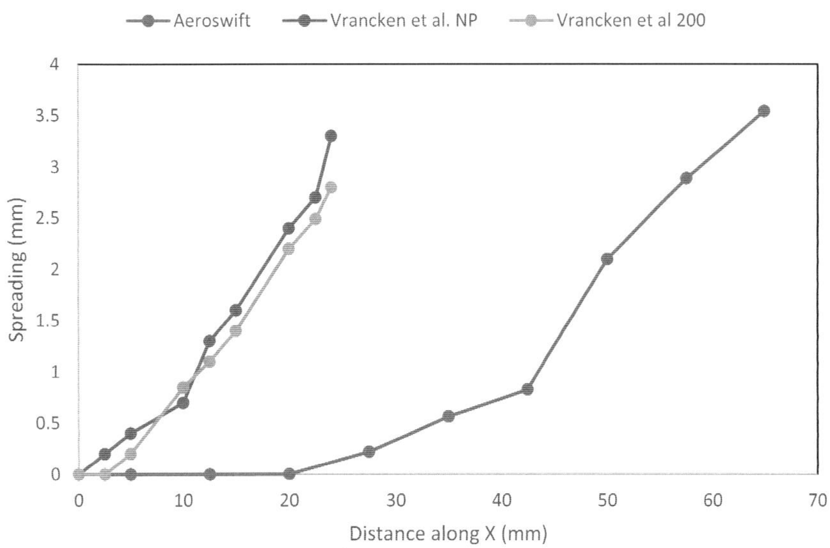

Figure 5. Cantilever spread results at 200˚C preheating

Table 2. Start of distortion ratios

	Tot. Length (mm)	Start of distortion from base (mm)	Ratio
Aeroswift	67	12.5	0.186567
Vrancken et al. NP	25	1	0.04
Vrancken et al. 200˚C	25	2.5	0.1

4. Conclusion
Preheating at 200˚C on the Aeroswift yields tensile properties similar to those of other SLM commercial systems. The UTS and YS are better than those produced by conventional methods but the elongation is less than that of cast and wrought parts. The cantilever spread indicates the presence of residual stresses at this temperature but no delamination was experienced. Further testing is to be done to confirm crack growth behavior at this preheating temperature. The study was useful in informing the operational ability of the Aeroswift machine and to the roadmap to producing capable and high integrity parts. Future work will look into implementing higher preheating temperatures to effectively reduce distortion causing residual stresses and improve static and dynamic mechanical properties without post process heat treatments.

5. References
[1] ASTM F2924-14 Standard Specification for Additive Manufacturing Titanium-6 Aluminum-4 Vanadium with Powder Bed Fusion1

[2] Liu Y,Yang Y and Wang D, , International Journal of Advanced Manufacturing Technology, **87**, 647–656

[3] Le Roux S, Salem A,and Hor A, Additive manufacturing, **22**, 320-329

[4] Mercelis P, Kruth J, Rapid Prototyping Journal, **12 (5)**, 254–265, 2006

[5] Shipley H, McDonnell D, Culleton M, Coull R, Lupoi R, O'Donell G and Trimble D, 2018International Journal of Machine Tools and Manufacture, **128**, pp 1-20

[6] Lewandowski J and Seifi M., 2016 Annual Review of Materials Research, **46**, pp 151-186

[7] Mukherjee T, Zhang W and DebRoy T, *Computational Materials Science*, **126**, pp 360–372

[8] Cain V, Thijs L, Van Humbeeck J, Van Hooreweder B and Knutsen R, , *Additive Manufacturing*, **5**, pp68-76

[9] Leuders S, Thöne M,, Riemer A,Niendorf T, Tröster T, Richard H and Maier H, *International Journal of fatigue*, **48**, pp 300- 307

[10] Roehlinga J, Smith W, Roehlinga T, Vrancken B, Guss G, McKeowna J, Hill M, and Matthews M,*Additive Manufacturing*, **28**,pp 228-235

[11] Vrancken B, Buls S, Kruth J-P and Van Humbeeck, Proceedings of the 13th World Conference on Titanium

[12] Ali H,Mab L, Ghadbeigia H,and Mumtaza K, *Materials Science & Engineering A*, **695**, pp 211-220

[13] Buchbinder D, Meiners W, Pirch N and Wissenbach J, *Journal of Laser Applications*, **26(1)**, pp 012004-1 – 012004-10.

[14] Mertens R, Vracken B, Holmstock N,Kinds Y, Kruth J-P and Van Humbeeck J, *Physics Procedia*, **83**, pp 882-890.

[15] Te Haar, T., Becker, T., 2018, *Materials*, **11(46)**, pp 1-15

[16] Tao P,Li H-X, Huang B-Y, Hu Q-D, Gong S-L, and Xu Q-Y. 2018, *China Foundry*,**15**, pp243-252

[17] Vrancken B,Thijs L, Kruth J-P and Van Humbeeck J,*Journal of Alloys and Compounds*,**541**, pp 177-185

[18] Kok Y, Tana X, Wang P, Nai M, Lohb N, Liua,E and Tor S, *Materials and Design*, **139**, pp 565-58

[19] Gorsse S, Hutchinson C, Goune M and Banerjee R 2017, *Sci and Tech of Adv. Mat,* **18, pp** 584-610

Influence of stress relieving thermal cycles on AISI10Mg specimens produced by selective laser melting

B J Mfusi[1,2a], N R Mathe[1b], P A I Popoola[2c] and L C Tshabalala[1d]

[1] National Laser Centre, Council for Scientific and Industrial Research, Meiring Naudé Road, Brummeria, Pretoria 0185, South Africa

[2] Chemical, Metallurgical and Materials Engineering Department, Tshwane University of Technology, Staatsartillerie Rd, Pretoria West, Pretoria, 0183

[a]mfusibusisiwe08@gmail.com*, [b]nmathe@csir.co.za*, [c]PopoolaAPI@tut.ac.za, ltshabalala1@csir.co.za

Abstract: The AlSi10Mg alloy is known for its castability, corrosion resistance, high heat conductivity, low weight and good weldability. Applications of the aluminium alloy components includes heat exchangers that are subjected to high temperatures and corrosive environments. The production process of AlSi10Mg alloy using selective laser melting has been extensively explored because of its hypoeutectic composition of Al and Si. However, it has been observed that stress relieve has a negative consequence on tensile properties of SLM built samples, while retaining and improving ductility. The thermal behaviour of AlSi10Mg alloy fabricated by selective laser melting with the specimen heat treated at 300°C for 2 hours investigated in this work. The specimens exhibited mass gain during the thermalgravimetric analysis as a result of the precence of oxygen as one of the elements present as detected by the elemental analysis. The fatigue results post thermal treatment demonstrated improved fatigue resistance qualities comparable to die-casted AlSi10Mg.

1. Introduction

Conventional methods such as casting for component production are slowly being substituted by additive manufacturing (AM) methods which are currently under investigation especially in the fabrication of multifaceted structures [1]. As a category of the AM technology for the manufacturing of the metal components from powder, selective laser melting (SLM) makes an interesting study for aluminium alloys. These alloys are becoming the most desirable metals to different industries due to its properties such as high thermal and conductivity of electricity, low specific weight, low power radiation, high plasticity and ductility, high weldability as well as toxicity [2]. AlSi10Mg as one of the composite matrix, is acquiring a reputation in the automobile, aerospace and rail applications to substitute the monolithic counterparts [3]. Components such as heat exchangers and propellers, which are used in numerous mechanical engineering systems have depended on conventional manufacturing methods for production, which due to manufacturing confinements have affected the designs complexities [4] [5, 6]. These components can be produced by AM as an alternative manufacturing process. However, one of the main drawbacks of AM is the need for post built thermal treatment to relieve the inherent residual stresses experienced during the rapid heating and cooling cycles of the process.

In this case, the modifying of the mechanical properties of SLM materials by thermal treatment customarily is based on the techniques which have been established for materials that are manufactured using conventional methods [7]. Oxidation of aluminium alloys during AM processing has also been studied [8], where it was determined that the oxygen in the chamber is capable of forming thin oxide films on the molten and solid material which may lead to porosity in the material. Brandl et al. suggested that when aiming to increase fatigue resistance, combining different heating profiles such as stress relieving and peak-hardening is a good idea [9-11]. Francolin et al. [12], used thermal gravimetric analysis (TGA) to prove the influence of oxygen on physical and mechanical properties of scaffold sheets. However little information is available on the TGA analysis of SLM produced AlSi10Mg.

Since some components like heat exchangers are subjected to high temperatures and cycling loads, the kind of temperatures and fatigue conditions that these sample can withstand, as well as the improvements that needs to be done, will be studied. Therefore, this study aims to investigate the thermal and fatigue behaviour of stress relieved AlSi10Mg manufactured by selective laser melting.

2. Experimental Procedure

2.1. Samples production

The AlSi10Mg samples were built using the SLM Solutions M280, with fixed parameters of; 150W power, 1000 mm/s scan speed, 50 µm hatch spacing and 50 µm powder layer thickness (see [13] for base plate images). The samples were built in three different orientations namely, 0° horizontal (A), 45° oblique (B) and 90° vertical (C). The samples were stress relieved using a vacuum furnace at 300°C for 2 hours, and furnace cooled to room temperature in an inert atmosphere, which is a casted AlSi10Mg parts heat treatment profile.

2.2. Characterization

JEOL JSM 6010 Plus/LA, scanning electron microscope (SEM), fitted with Electron Diffraction Spectroscopy (EDS) detector, was used for EDX analysis at an accelerating voltage of 20 kV. The samples were cut, mounted, polished and etched using Keller's reagent prior to SEM and EDX characterisation. For thermal analysis, samples of 20 mg filings were prepared for the Perkin Elmer Pyris 1 TGA thermogravimetric analyser, under conditions of 50 to 900 °C at a rate of 10 °C/min for thermogravimetric analysis in a Nitrogen environment. The 20kN Zwick/ Roell Tensile tester was used applying ASTM E466 standard for axial fatigue cycle tests, with a load varied from 80 – 120 MPa, and 0.1stress ratio. The number of fatigue cycles ranged from 230 000 to 2 000 000 at the frequency of 10 Hz. The samples were stopped after 2 million cycles.

3. Results and discussion

3.1. Energy Dispersive Spectroscopy (EDS)

The EDS results presented in table 1 demonstrates the atomic percentages of the elements present in the samples. Some of the areas of the samples also comprised of oxygen as one of the elements present in them. Louvis [8], attributed this presence of as a result of a system chamber that is not completely oxygen free. This oxygen level of 0.1-0.2% is high enough to affect aluminium alloys, [2], and capable of forming oxide layers on two different scanned tracks.

Table 1: EDS results of the SLM produced AlSi10Mg samples showing atomic% of element content in each microstructure (AB-as-built, SR-Stress relieve)

	ATOMIC %					
ELEMENTS	A-AB	B-AB	C-AB	A-SR	B-SR	C-SR
Mg	1.07	1.22	1.21	1.00	1.39	1.02
Al	89.9	92.14	92.14	85.10	89.53	89.47
Si	7.72	5.66	5.49	12.3	9.09	9.51
O	1.33	0.99	1.16	1.60	0.00	0.00
TOTAL	100.00	100.00	100.00	100.00	100.00	100.00

Figure 1 and 2 presents the SEM images and EDX spectra respectively, of the samples before and after stress relieving, which were focused on certain areas of the microstructure at 60X and 500X magnification. It was observed that silicon is the second major element in all the samples. It was also observed that post thermal treatment, there is an increase in silicon elemental composition which precipitate to the surface while aluminium composition decreased. This is because during the cooling and solidification, silicon contracted and during post thermal treatment, it precipitated as silicon and Mg_2Si [14, 15]. Aboulkhair [1], stated that during rapid cooling, α-aluminium is permitted to solidify first into a cellular structure while residual silicon is left to form at the boundaries of the grain. The silicon is only allowed to disperse during the coarsening of the microstructure subsequent to post thermal treatment.

(a) *(b)*

Figure 1: SEM-EDS area images of the SLM produced AlSi10Mg samples showing (a) 60X and (b) 500X magnifications

Due to the higher electron backscattering coefficient, pores and oxide particles are easily distinguished from one another through the brightness threshold which makes oxide particles appear brighter than pores [16]. In Figure 1b, silicon is observed to have precipitated to the surface after stress relieve (indicated by the light spots on the surface).

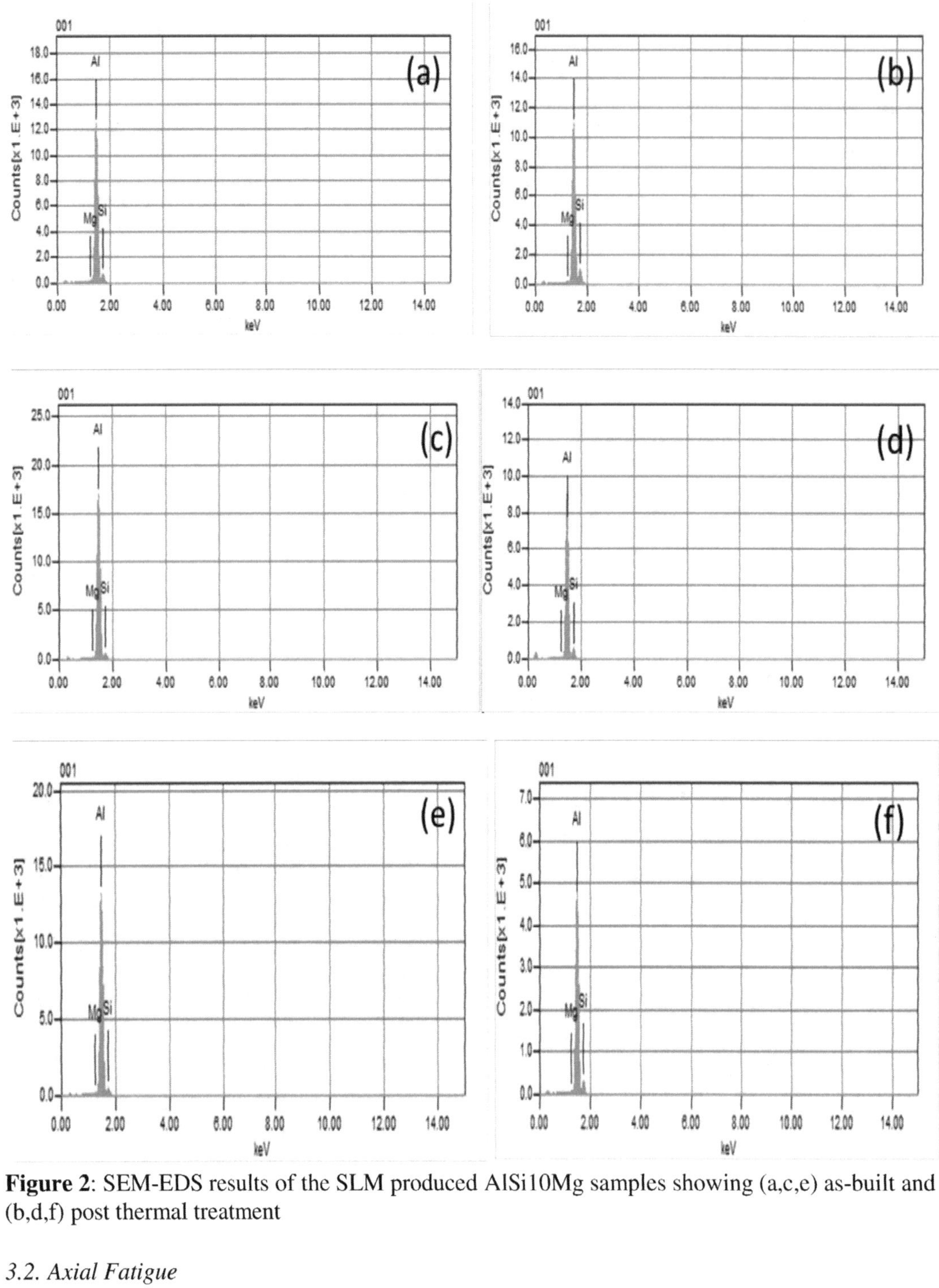

Figure 2: SEM-EDS results of the SLM produced AlSi10Mg samples showing (a,c,e) as-built and (b,d,f) post thermal treatment

3.2. Axial Fatigue

The stress relieved samples for batch A (0°) and C (90°) were further analysed for axial fatigue in order to determine the influence that the thermal treatment process has on the properties. Figure 3 a&b presents the dog-bone samples used in this test. It should be noted that not all samples fractured during the test, some samples were run out after 2 000 000 cycles. The axial fatigue results are presented in Figure 4, with the data in this work indicated in yellow, showing the number of cycles as a function of the maximum stress. The results were compared to the literature data of AlSi10Mg and T6 heat treated AlSi10Mg, [17]. Comparing the obtained results with the literature values for SLM processed AlSi10Mg (red) and die-casted Al6061 that was T6 heat treated (blue), it was observed that the samples can withstand higher fatigues and are comparable to literature values of the same load range of 80 – 100 MPa [18].

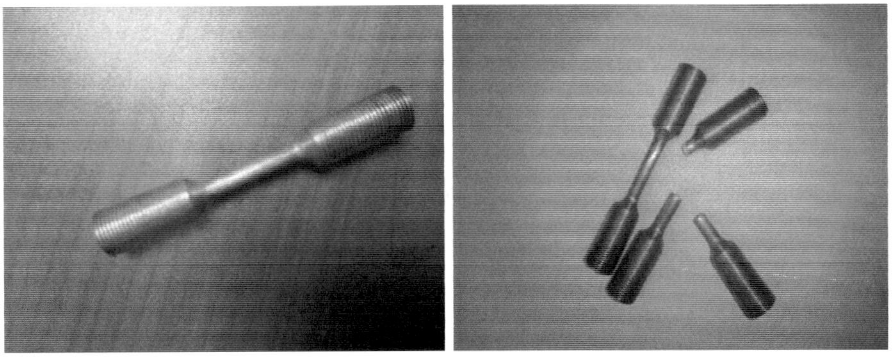

(a) *(b)*

Figure 3: (a) Dog-bone sample used for axial fatigue tests (b) Fractured fatigue samples after testing.

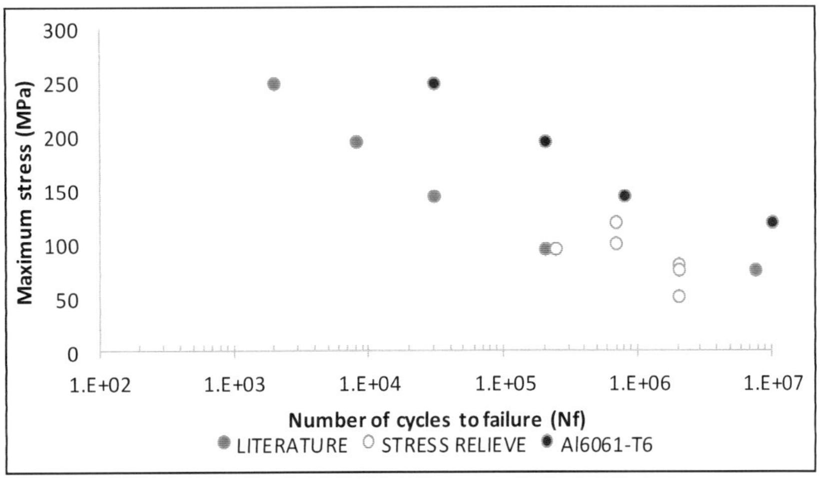

Figure 4: Axial fatigue results of the SLM produced AlSi10Mg samples

These results show that the even though the stress relieving profile used in this case resulted in a decline in strength, it increased the fatigue life of the samples. Most studies on fatigue life are based on the behaviour of cast aluminium alloys such as A360 which was believed to be affected by microstructural defects such as porosity, spacing of cellular and dendrite arms as well as grain size and inclusions [9, 16, 19]. Major and serious fatigue cracks found at the extension of the defect size dissemination are found to be the cause of failure in die-casting [20], whereas with AM components defects that lead to probable fractures are generally caused by oxide films. Alumina is easily formed as oxide film on the powder particles since aluminium has a great affinity to oxygen. Therefore, the residual porosity exposed on the surface of the component is the substantial to dictate of fatigue life for SLM produced AlSi10Mg [16].

3.3. Thermogravimetric Analysis (TGA)

Figure 5 presents the TGA results of the as-built and post heat treatment samples. In the Figure 5a, the anisotropy in build direction is observed in the variation of the graphs. The as-built thermal behaviour of orientation B (45°) stands out from A (0°) and C (90°), in the range of temperatures between 25 and 400 °C, showing weight loss that is associated with the material defects decomposition. At 595 °C all the specimens loose mass due to melting, which is the melting point of aluminium. On the other hand, the post stress relieving plots presented in Figure 5b showed no variation as a function of build orientation, which made all the structures homogeneous. The weight loss and subsequent gain was also observed at the same temperature, as seen for the as-built samples. The specimen showed a weight gain due to aluminium reacting with oxygen and forming aluminium oxide as Tang noted the high affinity they have to one another [16]. In this case, the AlSi10Mg can withstand operational conditions of temperatures above its melting point of 595 °C due to the reaction of aluminium with oxygen to form aluminium oxide which melts at 2072 °C [16].

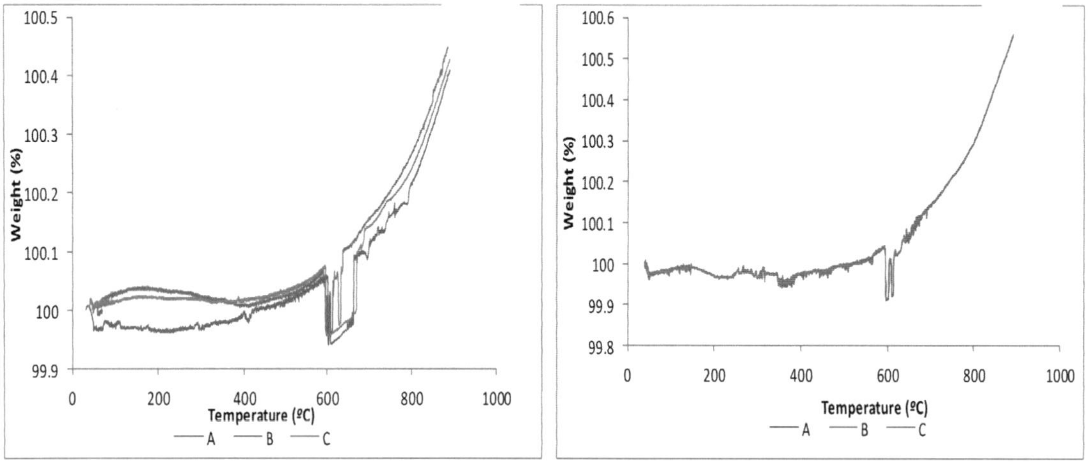

(a) *(b)*

Figure 5: TGA results of the SLM produced AlSi10Mg samples showing (a) As-built and (b) Post thermal treatment

4. Conclusion

The EDX elemental analysis showed a change in the Si and Al content before and after stress relieving, where Si increased and Al decreased. This was also seen from surface morphology of Si. Axial fatigue measurements on the stress relieved samples were comparable to T6 heat treated and die-casted results obtained in literature. The thermal behaviour of the SLM produced AlSi10Mg changed with the post build stress relieving, but there was anisotropy observed only for the as-built samples. In this case, the presence of oxygen content in the microstructure as observed through EDX, contributed a positive influence on the TGA results of SLM produced AlSi10Mg components. It can thus be concluded that although the stress relieving of components results in drastic strength loss, it improved the fatigue life, where the residual stresses were modified.

5. Acknowledgement

Tshwane University of Technology and Council for Scientific and Industrial Research (South Africa) are appreciatively recognized. The authors would also like to extend their gratitude to the National Laser Centre (NLC) Metallurgical Laboratory and Material Science and Manufacturing Mechanical Testing Laboratory for sample preparation and characterization. The National Research Foundation Grant No. 114675 is acknowledged for conference funding. Acknowledgement also goes to the following individuals for their support, Mr. Duncan Ramulifho, Mrs. Londiwe Motibane and Mr. Thabo Lengopeng.

6. References

[1] N.T. Aboulkhair, C. Tuck, I. Ashcroft, I. Maskery, N.M. Everitt, On the precipitation hardening of selective laser melted AlSi10Mg, Metallurgical and Materials Transactions A 46(8) (2015) 3337-3341.
[2] E. FELLA, Selective laser beam melting of AlSi10Mg alloy, (2016).
[3] S. Vasamsetti, S.V. Lingaraju Dumpala, Synthesis, Characterization and Hardness Studies of Nano Rice Husk Ash Reinforced Al6061 Nanocomposites, Journal of Engineering Science and Technology 13(9) (2018) 2916-2929.

[4] G. Staiano, A. Gloria, G. Ausanio, A. Lanzotti, C. Pensa, M. Martorelli, Experimental study on hydrodynamic performances of naval propellers to adopt new additive manufacturing processes, International Journal on Interactive Design and Manufacturing (IJIDeM) 12(1) (2018) 1-14.
[5] K. Nishimura, H. Matsuura, S. Tanaka, A Review of Recent Advances in Additively Manufactured Heat Exchangers, (2018).
[6] K. Wong, A. Hernandez, A Review of Additive Manufacturing. ISRN Mechanical Engineering, 2012, DOI http://dx. doi. org/10.5402/2012/208760 (2012).
[7] N.T. Aboulkhair, I. Maskery, C. Tuck, I. Ashcroft, N.M. Everitt, The microstructure and mechanical properties of selectively laser melted AlSi10Mg: The effect of a conventional T6-like heat treatment, Materials Science and Engineering: A 667 (2016) 139-146.
[8] E. Louvis, P. Fox, C.J. Sutcliffe, Selective laser melting of aluminium components, Journal of Materials Processing Technology 211(2) (2011) 275-284.
[9] F. Trevisan, F. Calignano, M. Lorusso, J. Pakkanen, E.P. Ambrosio, M. Lombardi, M. Pavese, D. Manfredi, P. Fino, Effects of Heat Treatments on A357 Alloy Produced by Selective Laser Melting, European Congress and Exhibition on Powder Metallurgy. European PM Conference Proceedings, The European Powder Metallurgy Association, 2016, pp. 1-6.
[10] A.T. J. Fiocchi, P. Bassani, C.A. Biffi, Low Temperature Annealing Dedicated to AlSi10Mg Selective Laser
Melting Products, Journal of Alloys and Compounds 695 (2017) 3402-3409.
[11] E. Brandl, U. Heckenberger, V. Holzinger, D. Buchbinder, Additive manufactured AlSi10Mg samples using Selective Laser Melting (SLM): Microstructure, high cycle fatigue, and fracture behavior, Materials & Design 34 (2012) 159-169.
[12] I. Francolini, E. Perugini, I. Silvestro, M. Lopreiato, A. Scotto d'Abusco, F. Valentini, E. Placidi, F. Arciprete, A. Martinelli, A. Piozzi, Graphene oxide oxygen content affects physical and biological properties of scaffolds based on chitosan/graphene oxide conjugates, Materials 12(7) (2019) 1142.
[13] B. Mfusi, L.C. Tshabalala, N.R. Mathe, The effect of selective laser melting build orientation on the mechanical properties of AlSi10Mg parts, IOP Conference Series: Materials Science and Engineering, IOP Publishing, 2018, p. 012028.
[14] M. Tang, Inclusions, porosity, and fatigue of AlSi10Mg parts produced by selective laser melting, Carnegie Mellon University, 2017.
[15] M. Vončina, P. Mrvar, J. MEDVED, Thermodynamic Analysis of AlSi10Mg Alloy Termodinamična analiza zlitine AlSi10Mg, RMZ-Materials and Geoenvironment 52(3) (2006) 621-633.
[16] M. Tang, P.C. Pistorius, Oxides, porosity and fatigue performance of AlSi10Mg parts produced by selective laser melting, International Journal of Fatigue 94 (2017) 192-201.
[17] J.N.D. Ngnekou, Y. Nadot, G. Henaff, J. Nicolai, L. Ridosz, Influence of defect size on the fatigue resistance of AlSi10Mg alloy elaborated by selective laser melting (SLM), Procedia Structural Integrity 7 (2017) 75-83.
[18] A.B. Anwar, Q.-C. Pham, Selective laser melting of AlSi10Mg: Effects of scan direction, part placement and inert gas flow velocity on tensile strength, Journal of Materials Processing Technology 240 (2017) 388-396.
[19] A. Yadollahi, M. Mahtabi, A. Khalili, H. Doude, J. Newman Jr, Fatigue life prediction of additively manufactured material: Effects of surface roughness, defect size, and shape, Fatigue & Fracture of Engineering Materials & Structures 41(7) (2018) 1602-1614.
[20] N.T. Aboulkhair, I. Maskery, C. Tuck, I. Ashcroft, N.M. Everitt, Improving the fatigue behaviour of a selectively laser melted aluminium alloy: Influence of heat treatment and surface quality, Materials & Design 104 (2016) 174-182.

Preparation and microstructural characterization of the 60Al-40V master alloy

T O Mapoli[1, a], K A Annan[1], C W Siyasiya[1] and Kalenda Mutombo[2]

[1] University of Pretoria, Department of Material Science and Metallurgical Engineering, Lynnwood Road, Hatfield, Pretoria 0002, South Africa.

[2] Light Metals, Materials Science & Manufacturing, Council for Science and Industrial Research, Meiring Naude Road, Brummeria, Pretoria 0185, South Africa.

E-mail: [a]tresormapoli@gmail.com

Abstract. The present study focused on preparation and characterization of 60Al-40V master alloy which was produced via aluminothermic process. The study assessed the production in order to provide the optimum Al requirement for maximum recovery of the V. V_2O_5 (Vanadium pentoxide) and Al metal were mixed in the proportion of producing 60Al-40V and Al_2O_3 (slag) through an exothermic reaction. Microstructural analysis and phase identification were done using Optical Microscopy (OM), Scanning Electron Microscopy (SEM) equipped with Energy Dispersive Spectroscopy (EDX) and X-Ray Diffraction (XRD) analyses. The XRD revealed diffraction peaks mainly of the Al_3V phase and some peaks of slightly homogenized Al_8V_5 intermetallic compounds. The chemical composition was found to be 65 ± 0.03 and 35 ± 0.03 wt. % Al and V respectively. Good agreement between the experimental results and the predicted phases using Thermo-Calc. Software was observed.

1. Introduction

Ti and its alloys are sought after for their mechanical and thermal properties [1]. For example, the Ti-6Al-4V alloy is one of the most widely used in applications requiring excellent oxidation resistance and high-temperature strength retention such as automobile and aerospace field especially [2]. Al-V alloys are important master alloys, they are used as Al and V sources to produce Ti-6Al-4V alloy. Two classical powder metallurgy (PM) methods for producing useful Ti alloys, such as Ti-6Al-4V are the blended elemental and pre-alloyed approach [3]. In the pre-alloyed method, the 60Al-40V master alloy can be added to pure Ti in order to produce the Ti-6Al-4V alloy, which is considered as the semi-finished product that is used as raw materials in the Ti industry. These pre-alloyed elements are produced via aluminothermic (thermite) smelting process within a refractory-lined or copper vessel. However, the aluminothermy processes are usually used in the manufacture of master alloys in the steel and superalloy industry. In this process refractory metal oxides or ores are reduced thereby offering a production method free of carbon. The greater difference in O affinity of Al and metal oxide, the higher is the ease of reaction [4]. For the 60Al-40V master alloy been in considerable demand to develop Ti-6Al-4V alloy; in this study, the aluminothermy technique has been proposed as a method to produce the master alloy. The produced pre-alloy must content about 60 at % Al. and 40 at % V. In the production of the master alloy, V_2O_5, and Al metal need to be mixed in such proportion to produce an exothermic reaction capable of generating the master alloy [5]. The exothermicity of the reaction is dependent on the level of V_2O_5 and the concentration levels of the starting materials [6]. This makes it difficult to produced pure V metal because V reacts and absorbs readily O, N, and C considered as impurities. The level of impurity pronounced the strengthening effect, therefore affecting the mechanical properties of V. Furthermore, dissolved C, O, and N levels of the order of 2000 to 4000 raises the ductile-brittle transition temperature

(DBTT) in unalloyed V at room temperature [7]. Hence, the presence of these impurities renders V brittle. The aluminothermy is a process for reducing metal oxides using finely divided Al powder. The mixture of Al and oxides is ignited, causing the Al to be oxidized and the metal oxide to be reduced to the metal [8]. The reduction of V_2O_5 in the electrical furnace leads to high carbon content, therefore, the aluminothermy technique may be an effective way of producing 60Al-40V master alloy if a suitable reaction vessel is used. In the current study, the 60Al-40V master alloy was produced through the aluminothermic reaction process. Analytical techniques were then employed to assess the aluminothermic process in terms of chemical element contamination and thermodynamic feasibility in order to provide the optimum Al requirement for maximum recovery of V.

2. Experimental procedure

The 60Al-40V master alloy supplied by the CSIR was produced through aluminothermic reaction. V_2O_5, pure Al and flux are produced in water cooler copper vessel. The different constituents were mixed in such proportion that the reaction is exothermic and reduction of V_2O_5 to V by Al is thermodynamically favourable, equation 1.

$$3V_2O_5+16Al \rightarrow 6AlV+5Al_2O_3+heat \quad \text{(equation 1)}$$

The Thermo-Calc. software with Al database was used for thermodynamic calculation and predictions of Al-V phases. The specimens of 60Al-40V produced were prepared for metallography investigation according to the ASTM standard E3-11 [9] (i.e. mounting, grinding, polishing and etching). The mounted samples were wet ground on SiC abrasive papers (P400-P1200) and polished up to 0.25 μm finished with colloidal silica (OP-S) suspension. The polished samples were thereafter etched in diluted Kroll's reagent followed by immediate rinsing under running water to avoid corrosive action of the etchant. The etched specimens were then examined using an optical microscope equipped with a high-resolution digital camera and Olympus Stream Essentials™ software. The phase identification was performed using XRD analysis with Cu Kα radiation and the results interpreted using X'Pert High Score Plus software. Zeiss Ultra PLUS FEG SEM, coupled with an Aztec X-Max energy dispersive spectroscopy (EDS) technique was used for imaging, and determination of chemical elements present in the different phases identified.

3. Results

3.1. Thermodynamic predictions

The equilibrium phase diagram of the given system is key in understanding the phase equilibria between constituents. The phase diagram of the Al-V binary and expected intermetallic phases have been investigated in this work using Thermo-Calc. software. Table 1. gives the list of phases of the Al-V predicted using Thermo-Calc. computer software.

Table 1. Phases in the Al-V system using Thermo-calc. computer software in this work.

Phase names	Other names	Weight range	strukturbericht	Space group	Pearson symbol	Model
liquid						(Al, V)
(Al)	fcc	0-0.05V	A1	Fm-3m	cF4	(Al, V)
$Al_{21}V_2$	$Al_{10}V$	8.7-4.9		Fd-3m	cF184	$(Al)_{10}(V)_1$
$Al_{45}V_7$	Al_7V	16.1V		C2/m	Mc104	$(Al)_7(V)_1$
$Al_{23}V_4$		21-24V		P63/mm	hP54	$(Al)_{23}(V)_4$
Al_3V		24-38V	D022	I4/mmm	tI8	$(Al)_3(V)_1$
Al_8V_5		38V	D82	I4-3m	cI2	$(Al)_8(V)_5$
(V)	bcc	51.5-100V	A2	Im-3m	cI2	(Al, V)

In the Al-V binary system, 8 known stable phases are found: the Al (fcc) liquid phase, in which a small amount of V is dissolved, the V (bcc, A2) dissolving a large amount of Al and 5 intermetallic compounds [10] [11]. In the specific composition of the pre-alloyed 60Al-40V study in this work, two intermetallic phases are expected (Al_3V and Al_8V_5). Figure 1. Shows the phase diagram corresponding to the Al-V binary system constructed and used in this work using Thermo-Calc.

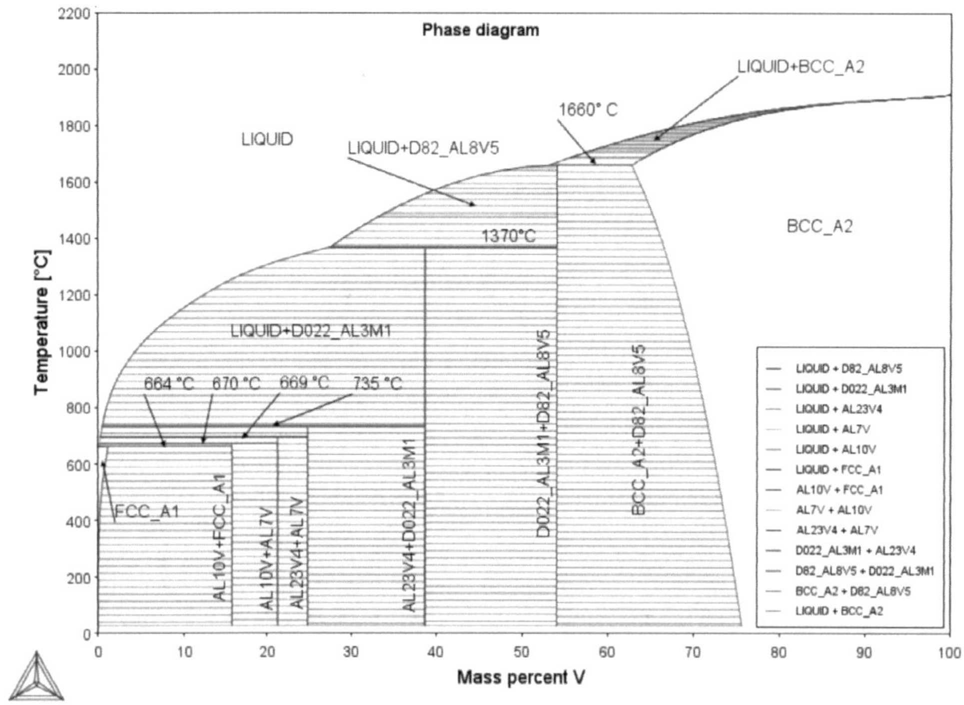

Figure 1. Al-V equilibrium binary phase diagram showing the various phases.

The Thermo-calc software was used to construct the Al-V phase diagram as well as the property diagram (Figure 1 and Figure 2). The intermetallic phases in this master alloy were predicted using the TC Binary Solutions V1.1 and SSOLOL4: Alloys Solutions Database V4.9 g. The predicted phases were calculated based on 60 wt. % Al. and 40 wt. % V. composition and were as follows: As expected, intermetallic phases namely $Al_{23}V_4$, Al_3V, and Al_8V_5 were obtained. The $Al_{23}V_4$ dissolve completely at a temperature of 800°C while the Al_3V started to dissolve at a temperature of 800°C and ended at 1380°C. The high-temperature Al_8V_5 phase only existed between 1380°C to 1450°C. Therefore, for high applications two intermetallic phases, namely Al_3V, and Al_8V_5 can be expected for temperatures above 800°C.

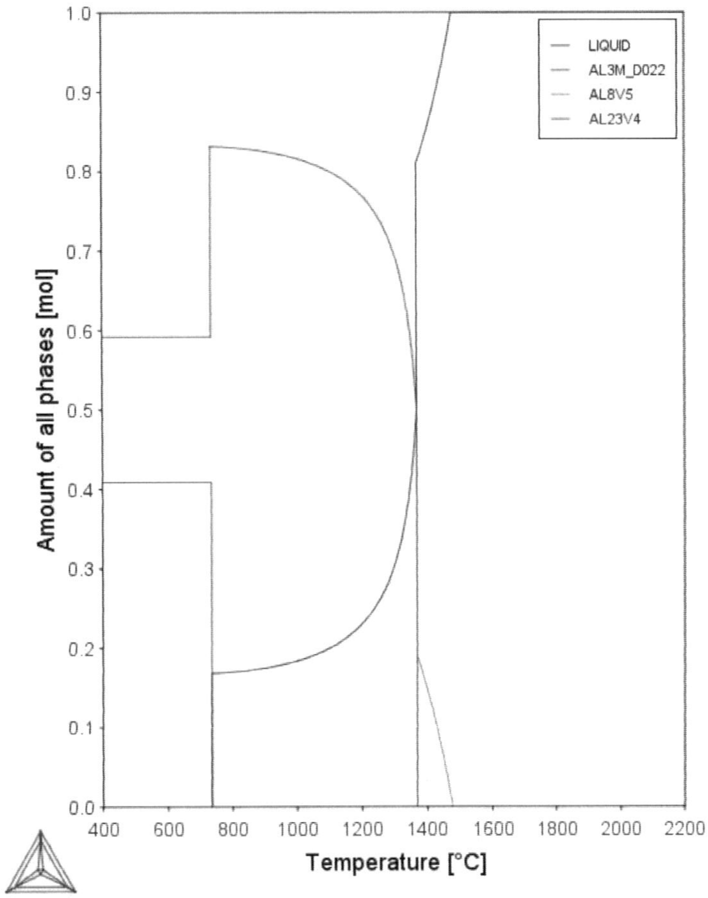

Figure 2. The amount of phases as a function of temperature in the 60Al-40V binary alloy system.

3.2. Phase identification

Figure 3 illustrates the XRD patterns of the 60Al-40V master alloy showing the phases identified in the alloy. The XRD patterns presented peaks of diffraction which corresponded to the crystalline intermetallic phases. The identified dominant peaks were the Al_3V (illustrated by high-intensity diffraction peaks) and some peaks of the Al_8V_5.

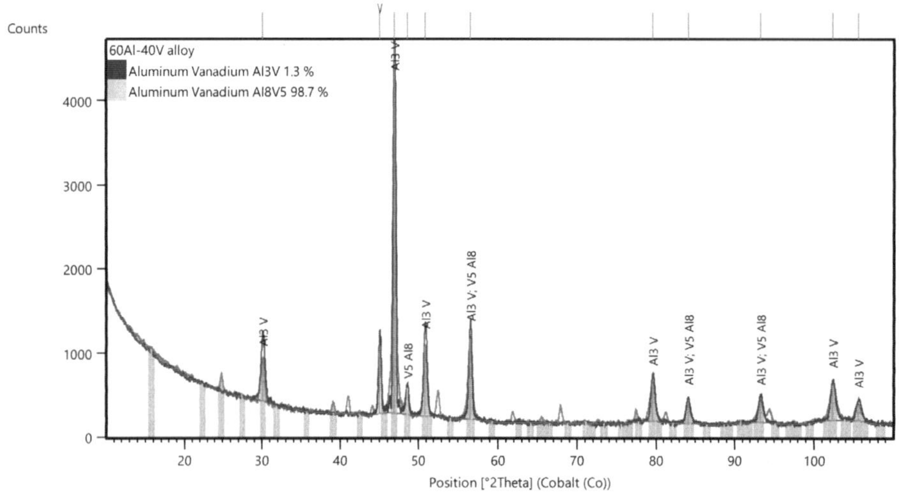

Figure 3. XRD patterns of the 60Al-40V master alloy showing the dominant phases in the alloy

3.3. Microstructural analysis

Figure 4.a shows SEM micrograph of the 60Al-40V master alloy, showing the microstructure which is typically dendritic structure. As may be seen, the microstructure of the alloy is composed of three phases i.e. the light grey, the dark grey, and the dark phases. The quantitative determination of the composition of each phase was done by EDS analysis (Figure 4.b.) illustrating the macrographs-EDS of the alloy by point scanning. The chemical compositions and the atomic percentage of the corresponding elements found in the phases are summarized in Table 2.

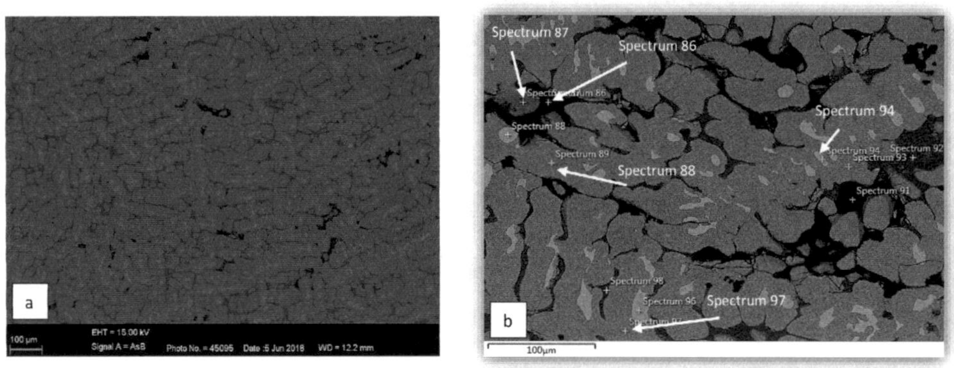

Figure 4.a) Electron Backscattered image; b) macrograph-EDS (point scanning) of the 60Al-40V master alloy. Magnification at 100 µm.

Table 2. Phase color and chemical composition

Phase color	Element content	
	Al [At. %]	V [At. %]
Dark	35.4	64.6
Light grey	63.4 / 65.3	36.6 / 34.7
Grey	75.0 / 74.9	25.0 / 25.1

Figure 5 (b) Shows the EDS elemental map showing that the alloy consisted of Al and V elements. The average proportions of these elements in the bulk material were 65 ± 0.03 and 35 ± 0.03 wt. %.; 77.81 and 22.19 at. %. respectively (Table 3).

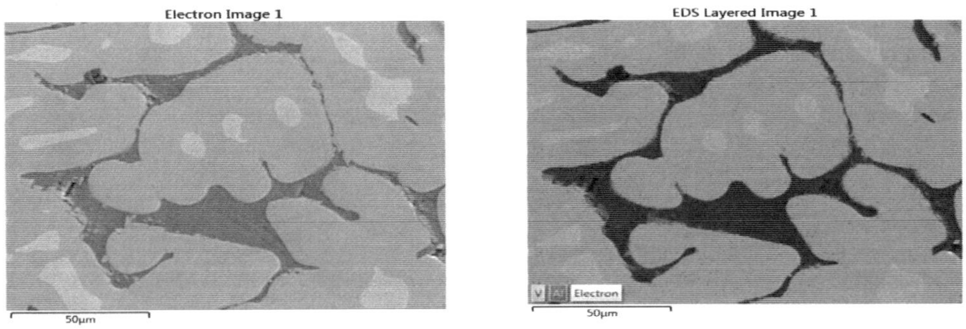

(a) **(b)**

Figure 5. (a) SEM micrograph, (b) SEM-EDS layered image.

Table 3. EDS analysis-overall chemical composition

Elements	mass %	mass % sigma	At. %
Al	65.00	0.03	77.81
V	35.00	0.03	22.19
Total	100		100

4. Discussion

The Al-V binary phase diagram and property diagram calculated in this work using Thermo-calc, the proposed diagrams in Figure 1 and Figure 2, gave information on the phases formed (Table 1), and invariant temperatures as compared to the work of others [2][10] [12]. The solubility of V in liquid Al calculated from the temperature of 1100°C to 2200 °C, the liquidus was nicely reproduced as well as invariant temperatures. As can be observed, the corresponding phase diagrams and property diagrams of the system are similar in all features. Two intermetallic phases, namely the Al_3V and Al_8V_5 with the highest melting composition in Al-V to be the Al_8V_5 were obtained. It was therefore, observed that the melting reactions of peritectic type gave as transformation temperatures for Al_3V and Al_8V_5 of 1370°C and 1660°C respectively, while in their studies Richter gave peritectic temperatures of 1270°C and 1408°C for Al_3V and Al_8V_5 [13], and Massalki of 1360°C and 1670°C for Al_3V and Al_8V_5 respectively [14]. As can be seen, the results agree with those found in literature, the main difference of the phase diagram data is in the solubility of Al in bcc at the invariant reaction liquid +bcc $\rightarrow$ Al_8V_5 which agrees with results observed by Massalki but deviate from the findings of Richter and Ipser. The XRD patterns revealed the presence of high-intensity diffraction Al_3V and the Al_8V_5 intermetallic phases (Figure 3) with Al_3V (cubic I4/mmm tL8) and the Al_8V_5 (cubic I-43m cI2) as illustrated in the ICSD data, which indicates that the system is not amorphous but crystalline. The qualitative determination of the alloy through optical microscopy, the SEM micrograph in Figure 4.a showed a predominance of grey phases typically dendritic structure which was quantitatively identified by EDS technique through point scanning (Figure 4.b) and map scanning (Figure 5). The validation of alloying, by EDS analysis, showed that the aluminothermy-produced alloy consisted of both Al and V and present in the form of Al_3V and Al_8V_5 intermetallic phases. The mass ratio distribution of Al and V of the grey phases in the following spectrums 89 and 97 were 75.0 at. % and 25.0 at.%; and 74.9 at.% and 25.1 at.% respectively. Furthermore, the EDS map scanning revealed the average proportion of Al and V as approximatively 77.81 at % and 22.19 at. % respectively (Table 3), the phase equilibria being the Al_3V as also determined in [15]. The chemical composition found to be 65 ± 0.03 and 35± 0.03 wt.% Al and V respectively. Although the starting materials were in the right proportions, the final alloy of 65A-35V is close to the theoretical elemental composition of the 60Al-40V master. This indicates that the composition of atoms is close to 3:1 (Al: V), and Al_3V is the most expected compound. Hence, the phase constitutions which were determined by XRD conform to EDS results.

5. Conclusion

The aluminothermy process was adopted in this experiment to produce the intermetallic 60Al-40V master alloy. The microstructural characterization of the 60Al-40V master alloy produced was carried out. According to the XRD analysis, coupled with SEM-EDX technique, the following conclusions were drawn:

- The alloy consisted of both Al and V elements with the elemental chemical composition was 65 wt. % and 35 wt. % respectively.
- Two intermetallic phases namely Al_3V and the Al_8V_5 were formed in the final product as predicted by Thermo-Calc.
- The aluminothermic process can be used to produce the 60Al-40V master alloy to produce the famous Ti-6Al-4V alloy.

6. References

[1] Y. M. Ahmed, K. Salleh, M. Sahari, M. Ishak, and B. A. Khidhir, "Titanium and its Alloy," no. October 2014.
[2] A. Kostov and D. Zivkovi, "Thermodynamic analysis of alloys Ti-Al, Ti – V, Al – V and Ti-Al – V ˇ," vol. 460, pp. 164–171, 2008.
[3] M. A. Masikane, "Investigation into the microstructure and tensile properties of unalloyed titanium and Ti-6Al-4V alloy produced by powder metallurgy, casting and layered manufacturing," 2015.
[4] E. Gock and B. Friedrich, "ALUMINOTHERMIC PRODUCTION OF TITANIUM ALLOYS (PART 2): IMPACT OF ACTIVATED RUTILE ON PROCESS SUSTAINABILITY," 2015.
[5] A. Oxide, "A thermite reaction (sometimes called a "," vol. 3, no. Iii,
[6] M. K. G. Vermaak, "Vanadium recovery in the electro-aluminothermic production of ferrovanadium," University of Pretoria, 2000.
[7] T. Division, "1. Introduction," vol. 143, pp. 1117–1124, 1986.
[8] D. D. Akerkar, "Physico Chemical Aspects Of Alumino-Thermic Reduction In The Production Of Low Carbon Ferro-Alloys by," in *4th Refresher Course on Ferro Alloys*, 1994, p. 7.
[9] ASTM International, "ASTM E3-11(2017) Preparation of Metallographic Specimens 1," *Area*, vol. 03, no. July, pp. 1–17, 2001.
[10] B. Lindahl, X. L. Liu, Z. Liu, and M. Selleby, "CALPHAD : Computer Coupling of Phase Diagrams and Thermochemistry A thermodynamic re-assessment of Al – V toward an assessment of the ternary Al – Ti – V system," vol. 51, pp. 75–88, 2015.
[11] A. Kroupa, M. Mazalová, and K. W. Richter, "CALPHAD : Computer Coupling of Phase Diagrams and The reassessment of the Al-V system and new assessment of the Al-Si-V system," vol. 59, no. July, pp. 47–60, 2017.
[12] H. Okamoto, "Al-V (Aluminum-Vanadium)," *J. Phase Equilibria Diffus.*, vol. 33, no. 6, p. 491, 2012.
[13] K. W. Richter and H. Ipser, "Al-V phase diagram between 0 and 50 atomic percent vanadium,"

Zeitschrift fuer Met. Res. Adv. Tech., 2000.
[14] V. Raghavan, "Al-Ti-V (aluminium-titanium-vanadium)," *J. Phase Equilibria Diffus.*, vol. 26, no. 3, pp. 276–279, 2005.
[15] Q. Luo, K. Li, and Q. Li, "Thermodynamic investigation of phase equilibria in Al-Si–V system," *J. Mater. Sci. Technol.*, 2018.

Investigation of graphene-based nanocomposite for hydrogen storage

S J Baloyi [1,*], A Ngqalakwezi [1, 2], D Nkazi[2] and T A Ntho [1]

[1] Advanced Materials Division, Mintek, Private Bag X3015, Randburg 2125, Southern Africa

[2] School of Chemical and Metallurgical Engineering, Wits University Braamfontein, South Africa

E-mail: jeffreyb@mintek.co.za

Abstract. In this study, a simple, fast, and green one-pot method was developed for the preparation of Ni/Graphene nanocomposite for the storage of hydrogen. The precursor graphene oxide was prepared from expanded graphite flakes using a modified Hummer's method. Graphene oxide was reduced effectively using natural reducing agents such as lemon juice as eco-friendly potential reducing agents instead of harmful hydrazine (N_2H_4). The characterization of as-synthesized graphene oxide and Ni/graphene nanocomposites were performed by powder X-ray diffraction (p-XRD), high-resolution electron microscopy (HRSEM) with energy dispersive spectroscopy (EDS) analysis and transmission electron microscopy (TEM) before investigating the performance for hydrogen storage. It was observed from the results obtained that the synthesized graphene contains thin stacked flakes of shapes having a well-defined multi-layered structure at the edge. And the round shaped Ni nanoparticles of 13 nanometres in size were dispersed around the graphene structure. The hydrogenation capacity of the Ni/graphene nanocomposites were tested using thermogravimetric analysis (TGA). The prepared Ni/graphene nanocomposites exhibited good hydrogen storage capacities.

1. Introduction

The heavy use of carbon based material and their negative effects on the atmosphere have pushed the world into a paradigm shift where greener systems and methods have to be implemented. Environmental issues like global warming and acid rain are direct consequences of using fossil fuels, as a result, intensified research to find more environmentally friendly sources of energy and in the long run, eradicate the current environmental issues have been prioritized. The use of hydrogen as a substitute source of energy has gained favorable consideration because of its minimal emissions to the environment [1-3]. Hydrogen can form solid through physical or chemical combination with materials like complex hydrides, carbon materials, and metal hydrides. The United States Department of Energy requires these materials to have high volumetric and gravimetric densities at low temperatures and pressures [4]. Metal hydrides garnered a lot of attention due to their high gravimetric hydrogen capacities however their slow kinetics have hindered the practical application of these materials for hydrogen storage. Carbon materials have restricted hydrogen uptake due to weak Van Der Waals forces. However, researchers have suggested that decorating these materials with metal nanocrystals could improve hydrogen uptake in these materials [5].

Graphene amongst other material has been considered as a potential material for hydrogen storage because of its excellent onboard reversibility, high specific surface area, lightweight characteristic, and chemical stability. Graphene is a 2D carbon material with a 2sp honeycomb lattice structure [6]. Graphene is a very versatile material as such it can be manipulated to form hybrid materials by integrating it with nanomaterial. Hybrid materials are achieved through the decoration of the surface with metal nanocrystals resulting in improved catalytic activity and good electrochemical properties. Chen et al. [7] have reported success in attaching platinum, iron, and palladium onto the surface of Graphene. Zhou et al. [5] suggested that Ni makes the sorbent structure stable by re-arranging the geometry of graphene to catalytically improve the hydrogenation and dehydrogenation reaction in nanocomposites. In addition, reduced graphene oxide decorated with Ni shows improved cycling performance.

In this work, we synthesized Ni/Graphene nanocomposites, with precursor graphene prepared using Modified Hummers method and reduced with eco-friendly reducing agents. It was noted that the uptake of hydrogen is relatively proportional to the pressure of hydrogen.

2. Experimental procedures

2.1. Materials and chemicals

Nickel (II) acetate tetrahydrate (Ni $[(OAc)_2]_3$ and expandable graphite flakes (99.99%, 200 mesh) used to prepare graphene oxide was purchased from Sigma Aldrich. Lemon juice that was used as a natural reducing agent to reduce graphene oxide to graphene was purchased from Woolworth retailer store. Analytical grade potassium permanganate ($KMnO_4$), concentrated sulphuric acid (98%), hydrogen peroxide aqueous solution (H_2O_2) (30%), hydrochloric acid (HCl) and sodium hydroxide (NaOH) were obtained from Merck Chemicals (Pty) Ltd. All chemicals were utilized as such without further purification. High purity water (resistivity 18.2 MΩcm) decontaminated using a Milli-Q water system Millipore in the Mintek laboratory was used to prepare all aqueous solutions throughout.

2.2. Preparation of graphene oxide and graphene

Graphene was prepared from graphene oxide (GO) by reduction of graphene oxide using a natural reducing agent such as lemon. Therefore, GO was prepared first from natural expandable graphite flakes using a modified Hummers method. In a typical preparation, approximately 3 g of expandable graphite flakes were added into concentrated H_2SO_4 (70 mL) and stirred while the temperature was kept low at 10 °C in a beaker with ice water. $KMnO_4$ (9 g) was slowly added to the mixture while stirring at a constant speed. The temperature was raised to 35 °C, after the complete addition of $KMnO_4$ and stirred for a further 30 min. The mixture was diluted with high purity water (150 mL) and the temperature was raised to 95 °C. Additional high purity water (500 mL) and 30% H_2O_2 (20 mL) were added to the mixture (producing a dark brown to the yellow mixture). The H_2O_2 solution was consequently added to the mixture to stop the reaction. The obtained graphite oxide was exfoliated to GO through sonication in water for 10 min at room temperature. The subsequent mixture was washed with 15 mL of 1M HCl aqueous solution to remove metal ions and centrifuged at 4000 rpm for 30 minutes. The mixture was added to 5000 mL of high purity water and left stirring overnight. Then followed by filtering and washed again to further remove acid then oven-dried overnight at 30 °C. The solid obtained was then reduced with lemon juice as an environmentally friendly natural reducing agent.

2.3. Preparation of graphene-based nanocomposite

Ni/Graphene nanocomposite was prepared from precursor graphene oxide prepared by a modified Hummers method. Nickel (II) acetate tetrahydrate and graphene oxide gel were utilized as starting materials at a ratio of 5%:95% Ni to C atomic ratio. Typical, about 2.51 g of (Ni $[(OAc)_2]_3$ were disintegrated in 50 mL of high purity water. The (Ni [(OAc) 2]3 mixture was added into 20 mg/mL of graphene oxide gel. The mixture turned a dark shade of brown and was stirred for a few minutes before being ultra-sonicated for 10 min at 25 °C to attain a homogenous solution. Exactly 10 mL of lemon juice was slowly added to the mixture. The pH of the mixture was maintained at 10-12 through the addition of 1M NaOH solution while stirring. The solution was stirred again for 10 min before the mixture was placed in a tightly closed container and heated at 70 °C for 60 min, then after the heating was raised to 95 °C for another 60 min. Subsequently, the uniform solution was centrifuged and the remaining contaminants were washed repeatedly with high purity water and centrifuged again. Then the obtained precipitate was vacuum dried to yield a powder of $Ni(OH)_2$/graphene precursor. Lastly, the $Ni(OH)_2$/graphene was reduced to Ni/graphene nanocomposite by hydrogen thermal treatment with hydrogen at 350 °C for 3 h.

2.4. Characterization

X-ray diffraction (XRD) patterns were obtained by a high-resolution Bruker AXS D8 X-ray advanced powder diffractometer with CuKα-radiation (40 kV, 40 mA, 1.78897 Å) using a LinxEye detector. The XRD patterns were recorded with a step size of 0.02° in the 2θ range of 5 to 80° diffraction angle and using a standard speed with an equivalent counting time of 1 second per step. The specific surface area (SBET) was calculated by the Brunauer-Emmett-Teller (BET) method, using the Micromeritics ASAP TriStar 3000 unit, which operates with physical adsorption of a monolayer of nitrogen gas (area = 0.162 nm2) at –196° and at a pressure of about 2 μmHg for porosity measurement. HRTEM images were obtained using a JEM2100F electron microscope from JEOL, Ltd. fitted with an EDX, wavelength dispersive spectroscopy (WDS) and electron beam backscattered diffraction (EBSD) operating on Oxford Instruments software. The instrument was operated at an accelerating electron beam of 120 kV. The images were photographed using a Proscan CCD camera and image analysis was done using the Soft Image System (SIS) software. Samples were prepared by suspending the powder in methanol and applying the suspension to a carbon coated copper grid. Elemental analysis of Ni/graphene was performed using an EDX equipped with TEM. The samples were then allowed to dry at 25°C and the grid viewed under the microscope. The Raman characterization of the samples was conducted using a PerkinElmer Raman spectroscopy with a wavelength of 785 nm laser. The lateral resolution was measured at ~ 1 μm. An accumulation time of 3 seconds with a grip spacing of 0.25 mm was used in the mapping of each spot in the sample. The PerkinElmer (Bucks, UK) spectrum software was utilized to process the spectra.

2.5. Hydrogen storage

The hydrogen storage uptake of Ni/graphene was analyzed using high-pressure autoclave made of stainless steel and simultaneous thermogravimetric analysis (STA 6000; Perkin–Elmer model). The sample inside high-pressure autoclave reactor was charged with hydrogen at different pressures (0 bar, 5 bar, and 10 bar) and standard temperature. The reaction temperature inside the reactor was measured using a thermocouple and controlled by a PID regulator. The samples were each charged with hydrogen for 30 min and then placed in a firmly sealed container at room temperature for TGA analysis. Approximately 5 - 10 mg of the sample weight was used, the sample was kept for 10 min at 20 mL/min and then heated at a regular increase of temperature from 25°C to 900°C with a heating rate 10°C/min while the nitrogen flow rate was 20 mL/min. Hydrogen release was measured with an increase in temperature. A hydrogen storage capacity in the sample was calculated from the weight percentage of the sample which was taken as a function of temperature. The sample not charged with hydrogen (0 bar) was used as a reference to quantify for other oxidation products, possible water adsorption and other impurities and other oxidation products that might have been attached to the sample during the transfer step.

3. Results and discussion

3.1. Oxidation reaction color change

There was an observable change in the color of expandable graphite blend during the oxidation reaction. The dark green color of the graphite blend was observed before oxidation reaction and the graphite blend became dark brown to bright-yellow in color after oxidation reaction. The dark green color before oxidation reaction might be attributed to contact between H_2SO_4 and Mn_2O_7 from $KMnO_4$ used as starting material. At room temperature, Mn_2O_7 appeared as dark red oil before contacting H_2SO_4. The Mn_2O_7 result from a reaction between reaction $KMnO_4$ and H_2SO_4 as illustrated in Eq. (1).

$$2KMnO_4 + 2H_2SO_4 \rightarrow Mn_2O_7 + H_2O + 2KHSO_4 \quad (1)$$

Then the subsequent dark brown to bright-yellow suspension was terminated by addition of more high purity water followed by hydrogen peroxide solution.

3.2. Physiochemical and morphological characterization

Transmission electron microscopy (TEM) is used to produce sample images by scanning with a focused electron beam. Sample images are taken by a beam of electrons transmitted through a sample which interacts with the sample and forms an image. Figure 1a-d shows the TEM images of the $Ni(OH)_2$/graphene before reduction as a reference and the surface morphologies of the precursor material after reduction. The particles size of the Ni(OH)2 on the graphene were observed to have an average size between 23 - 70 nm. After the thermal reduction of $Ni(OH)_2$/graphene to Ni/graphene using hydrogen, the Ni particles in Ni/graphene nanocomposite were homogenously dispersed over the graphene surface with a mean particle size of about 13 nm. The result of chemical analyses of the Ni/graphene shown in Table 1 indicates a successful preparation of Ni/graphene nanocomposite. Table 1 reveals the presence of C (86%), O (10.29%) and Ni (4.58%) on the Ni/graphene surface, which correspond to the component ratio used for the preparation. The BET surface area of graphene was found to be 711.2 m^2/g with adsorption pore width of 5 nm.

Table 1. Chemical analysis of Ni/graphene sample.

Sample	**C(%)**	**O (%)**	**Cu (%)**	**Cl (%)**	**Ni (%)**	**Total (%)**
Ni/Graphene	86	10.29	0.08	0.27	4.58	100

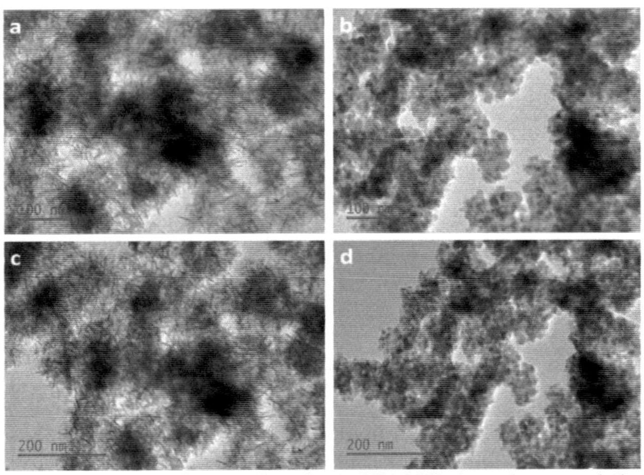

Figure 1. TEM images at 100 nm and 200 nm of (a & c), $Ni(OH)_2$/graphene and (b & d), Ni/graphene.

The XRD characterization gives an indication of the phase composition, crystal structure of materials and interlayer distance between graphene oxide and graphene. This analysis technique is based on Bragg's law [5]. Figure 2(a) indicates the XRD patterns of graphene, graphene oxide, and graphite. While Figure 2(b) shows a Ni/graphene before and after reduction and metallic Ni. Figure 2(a) demonstrates that graphite exhibited a sharp and high-intensity diffraction peak at $2\theta = 31^\circ$. The interlayer distance (*d*-spacing) of 0.33 nm along the (002) orientation was obtained. The graphite peak completely disappears when oxidation has taken place as can be observed from the red and blue lines representing graphene and graphene oxide respectively. As shown in Figure 2(b) after hydrogen thermal reduction, the complete disappearance of the $Ni(OH)_2$ peak ($2\theta = 30.2^\circ$) in the Ni/ graphene (black line) indicates that it was reduced to Ni in the graphene structure. Three diffraction peaks can be observed in the diffraction patterns Ni (blue line) at $2\theta = 52.8^\circ$, 61° and 93°, which corresponds with the decrease peaks of Ni/graphene (black line) diffraction pattern. Again this results confirm the successful reduction of $Ni(OH)_2$ to Ni in the graphene surface. These results suggest that lemon juice successful reduced graphene oxide to graphene. Other researchers proposed three possible routes for reduction of O group from graphene oxide; epoxide from graphene oxide are reported to be attacked by lemon juice from back side of the epoxide ring, the sp^2 C can be attacked by lemon juice from the front side of the epoxide ring and the sp^2 C at the meta position of epoxide can be attacked by lemon juice [7].

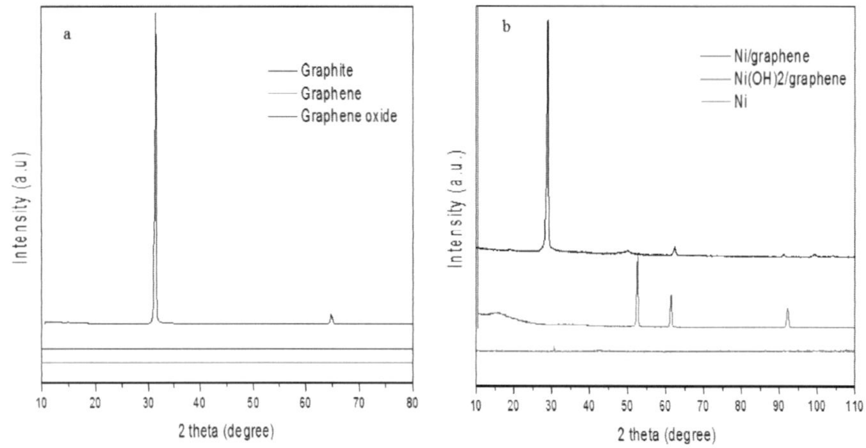

Figure 2. XRD spectra of graphite (black line), graphene oxide (blue line), and graphene (red line) and (b) Ni (blue line), Ni/graphene nanocomposite (black line), and precursor $Ni(OH)_2$/graphene (red line).

3.3. Electronic structure determination

Raman spectroscopy identifies the vibrations of interatomic bonding and structure that come up due to structural differences [8]. This technique utilizes a monochromatic laser to make interactions with the phonons and molecular vibrational modes which leads to the laser energy being moved up and down through inelastic scattering [9
]. For the graphene structure, the D and G bands are the principal characteristics in the spectrum. The G represents and includes sp^2 hybridized carbon in plan vibrational mode while the D band represents carbon hybridized in the sp^3 band as a defected band [8]. As shown in Figure 3, the D and G are located at 1578 cm^{-1} and 1658 cm^{-1} respectively. The Ni/graphene spectra depict no shift in the peak position however the Ni/graphene charged with hydrogen depicts a shift. From the Ni/graphene spectra, it can be observed that sp^2 hybridization was strengthened while the sp^3 was weakened due to the decreased D/G intensity ratio. High-frequency peaks are proportional to the number of modes which increase as the number of graphene sheets increase thus low intensity peaks are caused by the packing of graphene platelets [10-1]. The restacking of the graphene platelets in the Ni/graphene charged with H_2 results in the shift of the D peak which can be observed in the spectra.

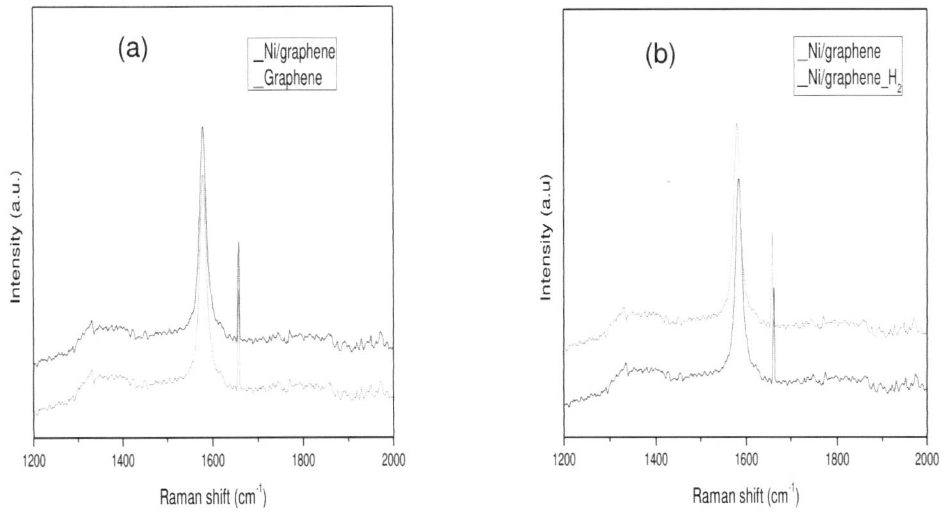

Figure 3. Raman spectra of graphene, Ni/graphene, and Ni/graphene_H_2.

3.4. Hydrogen storage capacity of Ni/Graphene composite

Thermogravimetric analysis was used to verify the hydrogen uptake and release in as-prepared Ni/Graphene. The sample was heated at a rate of 10°C/min under nitrogen conditions. A non-hydrogen charged sample was first run as a reference for verifying hydrogen uptake in Ni/Graphene sample charged at 5 and 10 bars respectively. The non-charged Ni/Graphene sample had a lower mass loss compared to the charged sample due to the eradication of gas impurities on the sample and adsorbed moisture. The hydrogen charged samples showed a faster decrease in mass with increasing temperature as depicted from Figure 4. A continuously decrease in a weight loss for hydrogen charged samples with an increase in hydrogen charging pressures were observed. Zhou et al. [4] reported on the characteristics that could affect the mass changes of the samples charged with hydrogen when using TGA: elimination of gas toxins and adsorbed water during the transfer process, hydrogen release and the reduction of hydrogen in NiOx, removal of water that resulted from the potential of the hydrogen oxidizing due to the existence of Ni particles during the transfer step and an irreversible interaction between Ni/graphene nanocomposite and hydrogen upon thermal treatment [11 12]. Therefore, these assumptions were considered when calculating the hydrogen storage capacity using uncharged nanocomposite as a reference. Typical, the amount of released hydrogen was calculated by subtracting the weight loss of the uncharged sample (0 bar) from weight loss of each sample charged with hydrogen [13]. The Ni/graphene nanocomposite takes up hydrogen with a capacity of 0.25 wt% at standard temperature and at standard pressure of hydrogen. The hydrogen storage capacity increased with an increase in hydrogen charging pressure to 10 bar, the hydrogen storage capacity reached 2.78 wt% in the Ni/graphene nanocomposite.

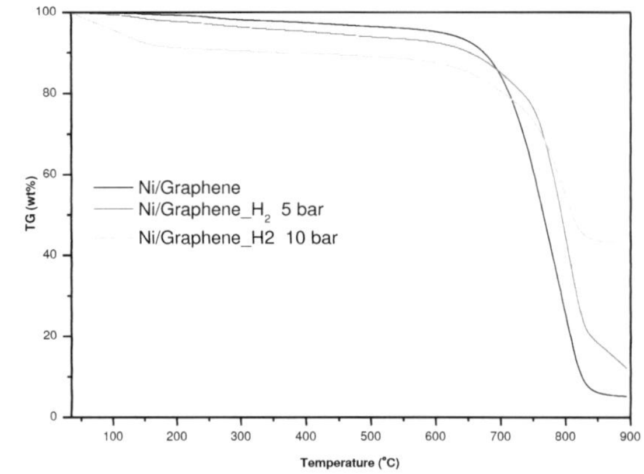

Figure 4. TG analysis of hydrogen uptake in the Ni/graphene nanocomposites with a loading amount of 5 at. % Ni at different hydrogen charging pressures (5 and 10 bar).

4. Conclusions

The Ni/graphene nanocomposites were successfully prepared for hydrogen storage using a natural environmental friendly reducing agent. It was demonstrated in this study that lemon juice can efficiently reduce graphene oxide to graphene. This method can be considered as eco-friendly and environmental friendly since the method used one-pot preparation process and natural reducing agent. The Ni/graphene nanocomposite achieved a hydrogen storage capacity of 0.25 wt. % at standard temperature and standard hydrogen charging pressure. The amount of hydrogen stored in the Ni/graphene nanocomposite was found to be directly proportional to the hydrogen charging pressure. The hydrogen gravimetric density of 2.78 wt. % was achieved with Ni/graphene nanocomposite charged with hydrogen at 10 bar. The morphologies of nanocomposites will be studied further by using a scanning electron microscope (SEM). Auto Chem connected to mass spec will be used to support claims made using TGA as mass spec can give exact type of gas being released in the nanocomposite. Furthermore, a high-pressure volumetric analyzer (HPVA) will be used to determine hydrogen storage capacity.

Acknowledgment

The authors wish to acknowledge financial support from the Mineral Science Council of South Africa (Mintek), South Africa through research Grant ADR 31901.

References

[1] Schlapbach L and Züttel A 2001 *Nature*. **414** 353
[2] Wagemans RW, van Lenthe JH, de Jongh PE, Van Dillen AJ, K. de Jong 2005 *J Am Chem Soc*. **127** 16675
[3] Xu E, Li H, You X, Bu C, Zhang L, Wang Q, Zhao Z 2017 *International Journal of Hydrogen Energy*. **1** 6
[4] Zhou C, Szpunar J.A., Cui X 2016 *Applied materials and interfaces*. **8** 15232
[5] Chong S.W., Lai C.E., Hamid S.B.A 2015 *Ceramics International*. 41 9505
[6] Chen M-L, Park C-Y, Choi J-G, Oh W-C 2011 *Journal of the Korean Ceramic Society*. **48** 147
[7] Kim Y, Lee J, Yeom M.S, Shin J.W, Kim H, Cui Y, Kysar J.W, Hone J, Jung Y, Jeon S 2013 *Nature Communications*. **4**
[8] Wall M 2012 *Adv. Mater. Process*. **170**
[9] Gardiner D.J 1989 *Springer*. **1**
[10] Childres L, Jauregui L.A., Park W, Cao W, Chen Y-P 2013 *Development in Photon and Materials Research*. **978**
[11] Thakur S, Karak N 2012 *Carbon*. **50** 5331
[12] Thomsen C, Reich S 2000 *Phys. Rev. Lett*. **85** 5214

[13] Zhou C and Szpunar J.A 2016 *ACS Applied Materials & Interfaces.* **8** 25933

IOP Conf. Series: Materials Science and Engineering **655** (2019) 012030 doi:10.1088/1757-899X/655/1/012030

Influence of Heat Treatment on Microstructure and Mechanical Properties of Ni-Fe-Co Ternary Alloy Prepared via Spark Plasma Sintering

I M C Bopape[1,*], M B Shongwe[1], E N Ogunmuyiwa[2], A L Rominiyi[1], S O Jeje[1]

[1] Department of Chemical, Metallurgical and Materials Engineering, Faculty of Engineering and Built Environment, Tshwane University of Technology, Pretoria 0001, South Africa

[2] Department of Chemical, Materials and Metallurgical Engineering, Botswana International University of Science and Technology, Palapye, Botswana

E-mail: imcbopape@gmail.com

Abstract. The use of nickel-based superalloys has extended to different fields such as turbines, rocket motors, chemical equipment, space vehicles and power plants due to their excellent mechanical properties. This study investigates the effect of heat treatment on the microstructure and microhardness of spark plasma sintered Ni-Fe-Co ternary alloy. 50wt.% Ni and varying percentages of Fe and Co powders were milled and fabricated by Spark Plasma Sintering (SPS) technique at a temperature of 900 °C, pressure and holding time were kept at 50 MPa and 10 min respectively. The sintered compacts were heat-treated at 1000 °C, soaked for 1 hr and quenched in distilled water for 5 min. Subsequently, the sintered and heat-treated samples were characterized by scanning electron microscopy (SEM) equipped with energy-dispersive spectroscopy (EDS) and X-ray Diffraction (XRD) to determine the microstructural evolution and phase transformation accompanying the process. XRD results revealed the evolution of strengthening phases in the heat treated of the samples. The as-sintered and heat-treated compacts Vickers hardness was also investigated before and after the heat treatment. The results show a general improvement in the microstructure after the heat treatment which translated to the observed increase in the hardness of the heat-treated samples.

1. Introduction

Superalloys are materials of special interest for applications in corrosive and elevated temperature due to their high strength, excellent resistance to wear, fatigue and creep. Generally, most of the superalloys used in power generation and aerospace industries are Ni, Fe and Ni-Fe based [1, 2]. Ni-Fe-Co ternary alloys are specifically developed for usage in the turbine hottest and high pressure region e.g. the combustor [3]. Ni-Fe-Co ternary alloys are of use in other industrial applications which include power plants, chemical plants, petrochemical plants, and reciprocating engines because of their excellent combination of mechanical properties such as high hardness, strength, good wear resistance, corrosion resistance and hot resistance [4]. Even though Ni-Fe-Co possesses excellent mechanical properties, thermal treatment can be employed to alter the microstructure to produce enhanced strength, hardness, ductility and creep resistance [5-8]. In hardening, the alloy is heated to a temperature high enough to

promote the formation of supersaturated solid solution, hold at that temperature and then quenched in distilled water to a suitable rate to allow martensitic transformation to take place.
Spark plasma sintering (SPS) technique has the advantage of low sintering temperature, short sintering time and lower pressure relative to conventional powder metallurgy sintering processes like hot pressing and hot isostatic pressing [9-11]. The short sintering time for consolidation of powders during SPS process inhibits microstructure coarsening thus retaining nearly the grain size of the starting powder and achieving enhanced mechanical properties. Hence, SPS technique overcomes the challenges of grain coarsening and poor mechanical properties encountered during the production of nickel-based superalloy using the conventional production route.
Several works have been conducted on the heat treatment of Ni-based superalloys [6, 12, 13]. However, few studies are available in the literature on the properties exhibited by Ni-Fe-Co ternary alloy due to solution treatment of this alloy. Hence, in this study, Ni-Fe-Co powder with varying composition of Fe and Co was consolidated through SPS technique and the effect of heat treatment (hardening) on the microstructure, phase evolution and mechanical properties of the sintered alloy were evaluated.

2. Experimental Procedure

2.1 Powder preparation

Ni, Fe and Co powders with characteristics shown in Table 1 were used as starting materials. Each powder was milled in a high-energy planetary ball mill (Retsch, PM 400, Germany) with steel balls and vial as the grinding media for particle size reduction. Milling was done for 15hrs at a vial rotation speed of 350rpm and ball to powder ratio of 10:1. To prevent powder agglomeration, wet milling was employed using ethanol as the suspension media. The milled powder mixture having composition Ni-20Fe-30Co (wt.%) and Ni-20Fe-30Co (wt.%) were separately mixed by subjecting the powders to translational and rotational motion in a turbula mixer at 49 rpm for 8 h. The ad-mixed powders were analysed with the use of scanning electron microscope (SEM) incorporated with energy dispersive x-ray spectroscopy to check its morphology and elemental composition. The SEM micrograph and the EDS of the milled powders are shown in Figure 1.

Table 1: Particle size and purity of starting powders.

MATERIALS	PARTICLE SIZE (μm)	PURITY (%)	SUPPLIER
Nickel (Ni)	≥0.5 to 30	99.5	WEAR TECH PTY (LTD)
Iron (Fe)	<44	99.9	WEAR TECH PTY (LTD)
Cobalt (Co)	<44	99.9	WEAR TECH PTY (LTD)

2.2 Sintering of powder mixtures

The mixed powders were compacted by spark plasma sintering (SPS) technique using machine model HHPD 5, FCT Germany. Predetermined weight of the mixed powder was poured into a graphite die with a diameter of 20 mm and sintered at 900 °C in a vacuum. The Optical Pyrometer was used to measure the sintering temperature which was implanted in the SPS apparatus at 3 mm from the top of the sample surface. The pressure was maintained at 50 MPa and heating rate of 150 °C/min throughout the whole process. The sintering temperature was 900 °C with dwell time of 10 min. After the completion of the sintering process, the compacted sample was removed at room temperature. The sintered samples were sand blasted to remove graphite impurities.

IOP Conf. Series: Materials Science and Engineering **655** (2019) 012030 doi:10.1088/1757-899X/655/1/012030

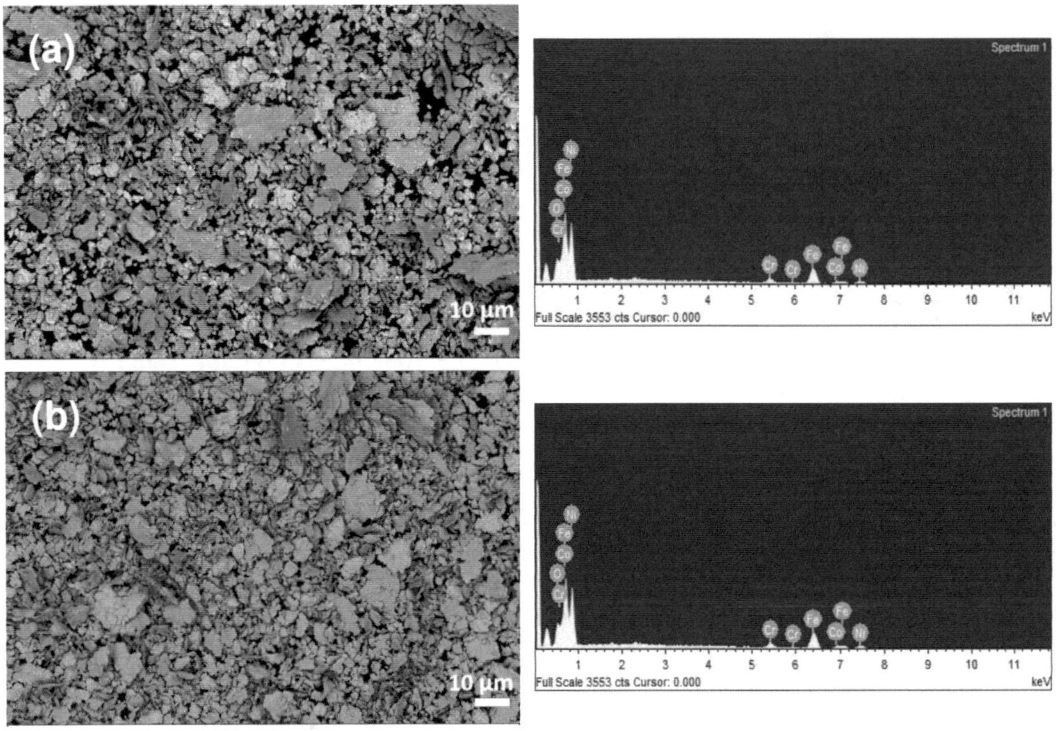

Figure 1. SEM micrograph and EDX of (a) Ni-20Fe-30Co (b) Ni-15Fe-35Co

2.3 Heat treatment
To study the effect of heat treatment on microstructural and mechanical properties, hardening of the sintered Ni-20Fe-30Co and Ni-15Fe-35Co samples was carried out at a temperature of 1000 °C in a muffle furnace. The compacts were held for 1 hour and quenched in distilled water for 5 minutes.

2.4 Relative density
To study the effect of heat treatment on the as-sintered alloy, a densitometer which employs Archimedes' principle was used to determine the density. The densitometer uses deionised water as the immersion liquid and automatically calculates the density of the sintered compact by comparing its weight in water and air using Archimedes' principle. The average of five measurements done on each sample was recorded as the density value. The relative density was determined by comparing the measured and theoretical density of the sintered sample.

2.5 Microstructural Characterization
The microstructure of the as-sintered and heat-treated polished surface was examined using SEM (JSM-7600F, Joel, Japan) incorporated with EDS facilities (Oxford X-max). The phases prior and after heat treatment were analyzed by XRD (PANalytical Empyrean model) with Cu kα radiation at 40kV and 20mA and analysed Xpert High Score software.

2.6 Hardness
Hardness testing was done using a Vickers hardness tester (FUTURE-TECH FM 800). The as-sintered and heat-treated samples were subjected to a load of 300 gf for a dwelling time of 10 seconds. The recorded hardness in HV for each compact is the average of 10 indentations.

3. Results and Discussion

3.1 Powder characterization

The SEM micrographs of the mechanically milled powders are shown in Figure 1. The powder particles were observed to be flattened out into sheets due to ball-ball and ball-wall of the vial interaction during the milling operation for 15 hrs. This resulted in severe plastic deformation of the powder particles. According to Karimbeigi et al. [14] and Rominiyi et al. [15], particle disintegration takes place at longer milling time as a result of fracturing and welding cycle induced by strain hardening of the materials.

3.2 Microstructure

The SEM (EBSD) micrographs of the sintered and heat-treated samples are shown in Figure 2. Similar microstructures comprising light and dark grey phases were observed in Figure 2a and b. These phases were revealed by the EDS analysis of the microstructures to be Ni and Co-Fe rich phases respectively. as, were observed in as-sintered Ni-20Fe-30Co (Figure 2a) and Ni-15Fe-35Co (Figure 2b). In addition to this, few pores were observed. This may be as a result of the sintering temperature (900 °C), which caused poor metallurgical bonding between the powders particle [16]. After solution treatment for 2 hrs and quenching with distilled water to room temperature, the microstructure of the heated-treated samples presented Figure 2c and d showed a more refined and homogeneous distribution of these phases.

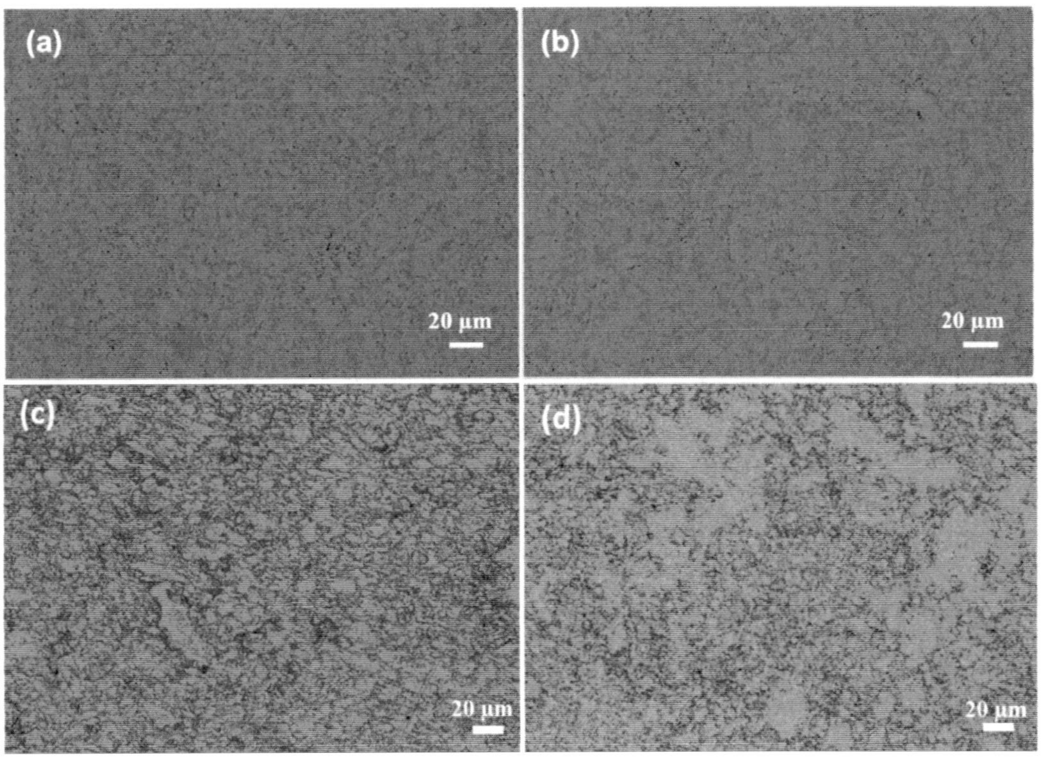

Figure 2. SEM micrograph of Ni-20Fe-30Co sample as-sintered (a) Ni-20Fe-30Co (b) Ni-15Fe-35Co and heat treated (c) Ni-20Fe-30Co (d) Ni-15Fe-35Co.

3.3 Phase Analysis

The X-ray diffraction analysis results of the admixed powder, as-sintered and heat-treated alloys are presented in Figure 3. The diffraction pattern of the admixed powder consisted of the peaks of Ni, Fe, Co and C. The peak of C observed, though weak, may be as a result of the release of carbon into the milled powder during stainless steel ball-ball and ball-wall of the vial interaction that characterized the milling operation. The predominance of peaks corresponding to that of ordered B2 structure of CoFe (BCC) and (γ′) $FeNi_3$ (BCC) with few peaks of γNiFe (FCC) [17] was observed in the diffraction patterns of as-sintered samples (Figure 3a and b). This confirmed the compositions of the light and dark grey regions in the microstructure of the as-sintered samples (Figure 2a and b) to be a combination of $FeNi_3$ (BCC) and γNiFe (FCC) and CoFe (BCC) phases respectively. γNiFe phase exists as a high temperature phase, its presence in the sintered samples can be ascribed to the fast cooling rate of the SPSed samples which prevented its transformation to a more stable phase of $FeNi_3$. The weak peak of Fe_3C observed was as a result of the interaction of ingressed C during the milling operation and the elemental Fe in the admixed powder.

The diffraction patterns of the heat-treated samples showed in Figure 3a and b revealed the presence of trapped high-temperature FCC phases such as γ(Ni,Fe), γ(Co,Fe) and orthorhombic, Fe_3C after solution treatment as the dominant phases. This can be attributed to the dissolution of the low temperature intermetallics such as (γ′) $FeNi_3$ and CoFe and subsequent quenching during the heat treatment at 1000 °C. This is in agreement with the results of the study conducted by Turchanin et al. [18]. Furthermore, narrower and higher intensities of the peaks were observed in heat-treated samples relative to the as-sintered samples. This can be attributed to the occurrence of grain refinement within the matrix of the heat-treated samples.

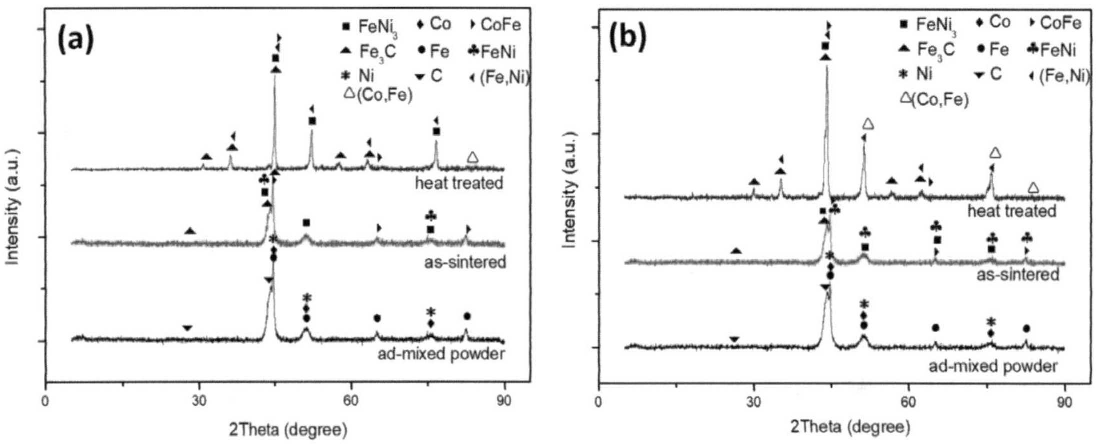

Figure 3. XRD patterns of (a) Ni-20Fe-30Co (b) Ni-15Fe-35Co

3.4 Relative density and hardness

Figure 4a showed the variation of the relative density of the as-sintered and heat-treated samples. The as-sintered Ni-20Fe-30Co and Ni-15Fe-35Co has a relative density of 84.2% and 87% respectively. These relative density values were found to increase to 86 % and 89 % respectively as the samples were heat-treated. Improvement in the densities of the heat-treated samples can be attributed to their close-structure dominated matrix compared to the pronounced open-structure that controlled the matrix of the as-sintered samples.

The microhardness test carried out before and after solid solution treatment of the investigated alloy indicated that the hardness of Ni-20Fe-30Co sample increased from 174.59 to 210.09 HV while an improvement in hardness from 198.45 to 298.09 HV was obtained for Ni-15Fe-35Co sample (Figure 4). It can also be observed that as the wt.% Co increased, the hardness value increased. This can be attributed to its higher relative density [9] and the fact that Co might have reduced the solubility of Fe in the matrix, causing significant increase in the volume fraction of the γ′ phase [19]. Also, grain refinement due to solution treatment of the as-sintered samples increased the volume of the grain boundaries within the matrix. This created impediments to the dislocation movement, hence enhancing the mechanical properties of the heat-treated samples [20].

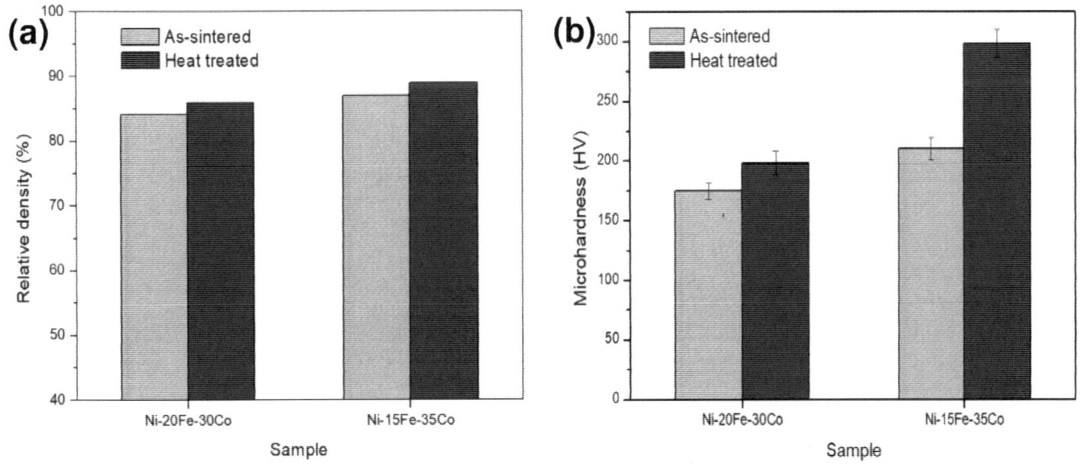

Figure 4. (a) Relative density (b) Hardness of as-sintered and heat-treated samples.

4. Conclusion

Heat treatment of Ni-Fe-Co ternary alloys was conducted. The impact of this operation on the microstructure, phase evolution and mechanical properties were investigated. The following were concluded from this work:

1. NiFe, $FeNi_3$ and FeCo were the predominant precipitates found in the as-sintered samples while γ(Ni,Fe) , γ(Co,Fe) and orthorhombic, Fe_3C constituted the quenched solid solution in the heat-treated sample.

2. The relative density and microhardness values of the investigated alloy increases due to an increase in the volume of $FeNi_3$ and FeCo intermetallics precipitated as the wt.% Co increases in the as-sintered samples.

3. The predominance of FCC structure in the quenched solid solution was responsible for the increase in the relative density and hardness observed in the heat-treated samples. These also increased with increasing wt.% Co.

4. The outcome of this work shows that applications of post sintering heat treatment enhanced the precipitation of hardening phases which eventually improved the mechanical properties of the sintered alloys. This makes the developed material a potential candidate in the applications that require high temperature strength and hardness such as in the production of turbine discs in aero-engines.

5. Acknowledgements

This work is based on the research supported by the National Research Foundation of South Africa for the grant, Unique Grant No 99348.

6. References

[1] Pollock T. M and Tin S 2006 "Nickel-based superalloys for advanced turbine engines: chemistry, microstructure and properties," *Journal of propulsion and power,* **22** pp. 361-374.
[2] Lu X, Miki T and Nagasaka T 2018 "High-temperature oxidation paths, according to the oxide-alloy phases equilibria in the ternary Ni-Co-Fe system," *Corrosion Science,* **139**, pp. 114-123.
[3] Singh J and Singh 2017 "A Wear Characterization of High Temperature Oxidized Ni Based Superalloys" *Indian Journal of Science and Technology,* **10(25)** , pp. 105-113.

[4] Nowotnik A 2016 "Nickel-Based Superalloys," in *Reference Module in Materials Science and Materials Engineering*, ed: Elsevier.

[5] Whitmore L, Ahmadi M R, Guetaz L, Leitner H, Povoden-Karadeniz E, Stockinger M, *et al.* 2014, "The microstructure of heat-treated nickel-based superalloy 718Plus," *Materials Science and Engineering: A,* **610**, pp. 39-45.

[6] Wu H, Zhuang X, Nie Y, Li Y, and Jiang L 2019 "Effect of heat treatment on mechanical property and microstructure of a powder metallurgy nickel-based superalloy," *Materials Science and Engineering: A,* **754**, pp. 29-37.

[7] Osada T, Gu Y, Nagashima N, Yuan Y, Yokokawa T and Harada H 2013 "Optimum microstructure combination for maximizing tensile strength in a polycrystalline superalloy with a two-phase structure," *Acta Materialia,* **61**, pp. 1820-1829.

[8] Li H, Sun J, Hardy M, Evans H, Williams S, Doel T, *et al.* 2015 "Effects of microstructure on high temperature dwell fatigue crack growth in a coarse grain PM nickel based superalloy," *Acta Materialia,* **90**, pp. 355-369.

[9] Babalola B J, Shongwe M B, Obadele B A, Olubambi P A, Ayodele O O, Rominiyi A L, *et al.* 2018. "Comparative study of spark plasma sintering features on the densification of Ni-Cr binary alloys," *EDP Sciences MATEC Web of Conferences* ***249***.

[10] Jeje S O, Shongwe M B, Maledi N, Olubambi P A, and Babalola B J 2019 "Spark plasma sintering of Ti-48Al intermetallic using elemental powder," *The International Journal of Advanced Manufacturing Technology,* pp. 1-11.

[11] Rominiyi A L, Shongwe M B, Maledi N, Babalola B J, and Olubambi P A 2019 "Synthesis, microstructural and phase evolution in Ti–2Ni and Ti–10Ni binary alloys consolidated by spark plasma sintering technique," *The International Journal of Advanced Manufacturing Technology* pp. 1-9.

[12] Li J, Zhao Z, Bai P, Qu H, Liu B, Li L, *et al.* 2019 "Microstructural evolution and mechanical properties of IN718 alloy fabricated by selective laser melting following different heat treatments," *Journal of Alloys and Compounds,* **772**, pp. 861-870.

[13] Wang Z, Guan K, Gao M, Li X, Chen X and Zeng X 2012 "The microstructure and mechanical properties of deposited-IN718 by selective laser melting," *Journal of Alloys and Compounds,* **513**, pp. 518-523.

[14] Karimbeigi A, Zakeri A and Sadighzadeh A 2013 "Effect of Composition and Milling Time on the Synthesis of Nanostructured Ni-Cu Alloys by Mechanical Alloying Method" *Iranian ournal of Materials scirnce and Engineering,* **10**, pp. 27-31.

[15] Rominiyi A L, Shongwe M B and Babalola B J 2018 "Development and characterization of nanocrystalline cobalt powder prepared via high energy ball milling process," *IOP Conference Series: Materials Science and Engineering,* **430**, p. 012029.

[16] Makena I M, Shongwe M B, Ramakokovhu M M and Lethabane M L 2017 "Influence of Temperature on Microstructure and Mechanical Properties of Ni-40Fe-10Co Alloy Consolidated by Spark Plasma Sintering," *Procedia Manufacturing,* **7**, pp. 708-713.

[17] Dreval L, Turchanin M, Agraval P and Du Y 2018 "Cu–Fe–Co System: Verification of the High-Temperature Phase Equilibria and Thermodynamic Modeling of the Low-Temperature Phase Relations Involving Ordered Phase" *Powder Metallurgy and Metal ceramics,* **56**, pp. 546-555.

[18] Turchanin M A, Dreval L A, Abdulov A R and Agraval P G 2011 "Mixing enthalpies of liquid alloys and thermodynamic assessment of the Cu–Fe–Co system," *Powder Metallurgy and Metal Ceramics,* **50**, pp. 98-116.

[19] Retima M S, Bouyegh and Chadli H, 2011 "Effect of the heat treatment on the microstructural evolution of the nickel based superalloy," *Metalurgija,* **17**, pp. 71-77.

[20] Ishikawa T 2012 "1 - Understanding and controlling microstructural evolution in metal forming: an overview," in *Microstructure Evolution in Metal Forming Processes*, J. Lin, D. Balint, and M. Pietrzyk, Eds., ed: Woodhead Publishing, pp. 3-16.

IOP Conf. Series: Materials Science and Engineering **655** (2019) 012031 doi:10.1088/1757-899X/655/1/012031

Effect of nickel powder particle size on the microstructure and thermophysical properties of spark plasma sintered NiCrCoAlTiW-Ta superalloy

O F Ogunbiyi[1*], T Jamiru[1], E R Sadiku[2], O T Adesina[1], S A Salifu[1], L W Beneke[1]

[1]Department of Mechanical Engineering, Mechatronics and Industrial Design, Tshwane University of Technology, Pretoria South Africa

[2]Institute for NanoEngineering Research (INER) and Department of Chemical, Metallurgical and Materials Engineering, Tshwane University of Technology, Pretoria South Africa

Corresponding author*: olugbengaogunbiyi@gmail.com

Abstract. The spark plasma sintering technique (SPS) was used to consolidate admixed elemental powder in order to fabricate NiCrCoAlTiW-Ta superalloy. Nickel which is the key element that formed the matrix powder particle size was varied in the range of 3-44, 45-106 and 106-150 μm. The effect of varying the starting powder particle size, on the formation of intermetallic phases was investigated. Also, important thermophysical properties in the development of superalloys for high temperature applications were investigated by using laser flash machine, in the range of 100-800 °C. The constituents of the microstructure for the sintered alloy includes, the gamma (γ) matrix phase (Ni), precipitated intermetallic gamma prime (γ') phase $Ni_3(Al, Ti)$ and the precipitated solid solution strengthening elements (Cr and W). There was an increase in the density with decreasing powder particle size. Also, the averave grain size increased with increasing powder particle size. The thermophysical properties obtained, which include thermal conductivity and diffusivity increase with increasing temperature. Non-linear behaviour and inflection points were observed as the temperature increases due to short range order and disorder phenomenon which is associated with superalloys. The powder particle size has little or no significant effect of the thermophysical behaviour as the pattern observed were almost the same.

1. Introduction

NiCrCoAlTiW-Ta alloy belongs to the class of nickel based-superalloy that has been extensively used in gas/steam turbines. This is largely due to the appreciable mechanical properties and structural stability strength that the alloy posed when used for high-temperature applications. The alloy is unique due to its high specific Young's modulus, coupled with excellent oxidation, corrosion, creep and fatigue resistance properties. Oil and gas, chemical, power generation and aerospace industries, etc are great beneficiaries of the outstanding qualities of this alloy. Valuable components and equipment like heat exchanger, valve, turbine blade, driller and gas-cooled reactors have been developed by using NiCrCoAlTiW-Ta alloys. The microstructure formation of superalloy is quite complex, and this is due to the combination of different constituent elements with different material properties. Often, some major microstructural phases are developed in superalloys due to the reaction between powder elements during sintering [1, 2]. For NiCrCoAlTiW-Ta alloy, the constituent phases that are normally present include the matrix gamma (γ) phase, the ordered intermetallic gamma prime (γ') phase and the precipitate disordered solid solution

strengthening phase. These phases have a face-centered cubic crystal structure which aid structural stability at ambient and high-temperature environments. A numbers of literature provide substantial information on the evolution of the three phases and their influence on the resulting material properties [3-5]. The fluctuation in temperature applied to the alloy while in service, especially turbine blades often cause thermal shock to the material. The shock is often due to thermally induced stresses, which leads to extensive thermal and stress gradients in the alloy. Consequently, a potential thermal gradient is developed which pose a threat to the efficiency and life span of a turbine system.
Hence, a careful selection of both processing routes and alloy composition has helped in reducing the occurrence of material and microstructural defects. By using spark plasma sintering (SPS) technique for the fabrication of superalloys, the limitations such as fluctuating mechanical properties across the bulk volume of the resulting alloy, which are largely due to chemical and structural heterogeneities are addressed. It has been reported that the wide variation in the particle size and shape of the starting elemental powder contribute to material densification [6, 7]. This in turn has significant effect on the thermal and material responses of the alloy when in service. A number of methods such as hot press, selective laser melting, spark plasma sintering (SPS) and injection moulding are available in powder metallurgy (PM), and among these, SPS technology is used in this study. SPS is a PM technique which consolidates powder by the simultaneous application of direct pulsed current and uniaxial force [8, 9]. Using this technique, an electric current density is supplied to induce high temperature within the green powder via the Joule heat effect thereby resulting in the sintering of the powder. SPS technique is a fast sintering process that densifies material in one single operation. The detailed SPS mechanism, advantages and disadvantages have been reported in the literature [10, 11]. This study is aimed at developing NiCrCoAlTiW-Ta superalloy via SPS technique and investigating the effect of the varied starting powder particle size (nickel matrix) on the microstructure and thermal properties of the sintered alloy.

2. Experiment

The sintering of NiCrCoAlTiW-Ta superalloy was carried out by combining seven elemental powders in accordance with the stoichiometric composition standard of the superalloy [12]. The elements that make up the composition include; nickel which constitutes 64.58 wt% of the alloy, with 99.8% purity (varied over three powder particle sizes: 3-44 μm, 45-106 μm and 106-150 μm). Others are chromium (16 wt%, 99.9% purity), cobalt (8.3 wt%, 99.8 purity), titanium (3.4 wt%, 99.8 purity), aluminium (3.4 wt%, 99.9% purity), tungsten (2.6 wt%, 98.5 purity) and tantalum (1.72 wt%, 98.5% purity). All the elements were supplied by TLS-TechnikGmbh (Ni, Ti and Co) Sigma Aldrich (Cr, Al and Ta) and Alfa Aesar (W) companies. They were weighed by using an electric weighing machine to accuracies of $\pm$0.001, poured into a plastic container and pre-alloyed in a tubular mixer. The pre-alloyed process was done over a period of 10 hours at a speed of 49 rpm. The green powder was poured into the graphite die with 30 mm diameter. The thickness of the samples was fixed at 5 mm for each of the alloys. Prior to sintering, 0.2 mm thick graphite paper was used to line the interior of the graphite die and the faces of the punches that made contact with the green powder. The purpose of the graphite paper was to prevent reaction between the graphite die and the powder at high temperature and to enable easy removal of the alloy after sintering. The parameters used for the sintering process are, 1100 °C sintering temperature, 32 MPa pressure, 100 °C/min heating rate and 5 min holding time. It was ensured that the process was done in a vacuum atmosphere. At the completion of the sintering process, the furnace was switched off and the temperature was allowed to cool down to room temperature before the alloy was removed. A sand blasting process was carried out to remove carbon deposited on the surface of the sintered alloy from the graphite paper.
The material was later sectioned into smaller pieces for metallographic investigations. Series of emery paper was used (320, 600, 800 and 200 grit size) to remove patches and scratches on the surface of the test samples. The samples were polished by using diamond paste and etched afterwards. The etchant used was kalling reagent (2 g of $CuCl_2$, 50 ml of HCl, 25 ml of HNO_3, and 200 ml of H_2O). The microstructure was investigated by using a scanning electron microscope (SEM) (JEOL JSM-7600F SEM) equipped with an EDS detector with INCA X-Stream2 pulse analyzer software. Polarized optical microscope was used to calculate the average grain size of the sintered alloys and this follows ASTM E112 standard [13].

Chemical phases present in the alloy were analyzed by X-Ray diffraction (XRD), PANalytical Empyrean model with CuKα radiation. The thermal properties of the sintered alloys were also evaluated. The density and thermal diffusivity values used in this study were measured by the Archimedes method and NETZSCH model 427 laser flash diffusivity respectively. The specific heat capacity used was taken from the literature [14, 15]. The thermal diffusivity study was conducted from 100 - 800 °C and the thermal conductivity properties was calculated by using equation 1. The dimensions of the sample used on the laser flash equipment is 10 mm x 10 mm x 4 mm.

$$\lambda = \rho \, . \, C_p \, . \, a \tag{1}$$

Where: λ - thermal conductivity (w/(m.k))

ρ - Bulk density (g/cm^3)

C_p- Specific heat capacity (j/(g.k))

a - Thermal diffusivity (mm^2/s)

The laser flash analysis (LFA 427) procedure used in this study was in accordance with ASTM E-1461 standard and the thermal diffusivity measurement was conducted under a nitrogen environment. Graphite solution (Al_2O_3-graphite) was used to coat the surface of the test sample to facilitate fast absorption of the flashlight. This, in turn, leads to heat emission on the backside of the test piece, which is recorded by an attached sensor and the values were recorded.

3. Results and discussion

3.1 Relative density and average grain size

The relative density and the average grain size of the sintered NiCrCoAlTiW-Ta superalloy is presented in figure 1. These were for the three sintered alloys (106-150 and 45-106 and 3-44 µm initial powder particle sizes) and their porosities were observed to be 11.27 %, 8.43 % and 3.9 % respectively. The highest relative density of 96.1% was observed for an alloy with 3-44 µm as the initial powder particle size. The theoretical density was calculated to be 7.894 g/cm^3. Decrease in the average grain size was obsevered with decreasing powder particle sizes. The biggest average grain size was obtained for alloy with 106-150 µm (21 µm) particle size. The porosity observed in the alloy with fine particle size is reduced compare to the coarse particle size. This implies that finer particles size are closely packed with reduced initial porosity between powder compact. This enhances densification of the sintered alloy. The increase in the average grain size suggest that the sintered alloy keep memory of the starting powder particle size [16]. This means that localized deformation of particles (plastic flow) during densification could not reduce the coarser particle size to form grains of the same size with alloy of finer particle. Further work is in progress to establish particle behaviour and densification kinetics during the sintering process of NiCrCoAlTiW-Ta superalloy. However, relatively dense NiCrCoAlTiW-Ta superalloy with refine grains can be produced via SPS technique.

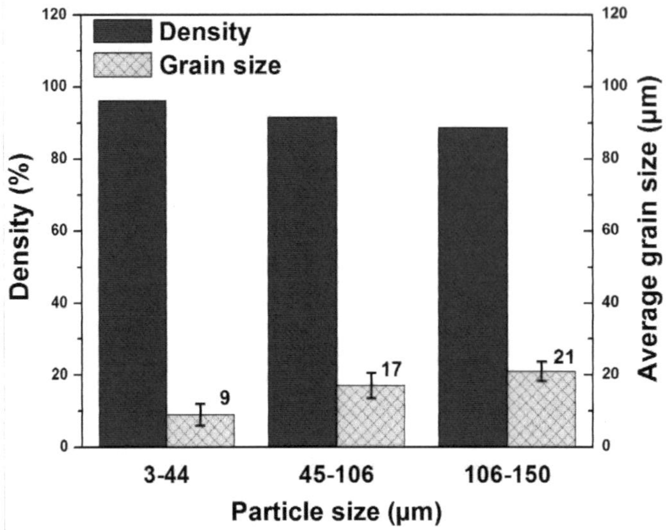

Figure 1. Relationship between relative density and grain size of sitered sintered NiCrCoAlTiW-Ta superalloy.

3.2 Microstructure analysis

SEM images of NiCrCoAlTiW-Ta superalloy samples, sintered at 1100 °C with the three different starting powder particles sizes, are given in figure 2. The EDX graphs adjacent to the SEM image show the elemental composition of the sintered alloys and their respective weight percent. The sintering has resulted in the formation of different phases and grain shapes. The three different colours on the SEM images, light grey, dark grey and the white dotted patches that are heterogenosly distributed within the microstructures. Further analysis shows that the light grey area represents the $\gamma + \gamma'$ phase, the dark grey represents precipitated chromium-rich zones and the white represent the precipitated tungsten solid solution. Careful observation shows that chromium precipitates were largely formed on the grain boundary of the γ matrix and γ' intermetallic phases. However, the overlap observed in the formation of the γ matrix phase and the γ' intermetallic phase is expected since these phases are both know to be face center cubic crystal structures. Hence, the γ' intermetallic phase forms precipitates within the γ matrix phase due to the similar cubic lattice and the closeness in their lattice parameters. The XRD results also confirmed the presence of the overlapped phases that was observed on the microstructure. An increase in average grain size was observed with increasing powder particle sizes. The effect of this observations was captured in section 3.1. Apart from chromium, tungsten was also observed to form precipitated solid solution strengthened elements within the microstructure, and this is sparsely distributed within the structure of the sintered alloys. Tungsten is known as a hard metal with high melting temperature and by forming solid solution strengthened precipitate in the alloy, the strength of the sintered alloy is improved. Idowu et al. [17], in their study on cast IN738LC superalloy observed the formation of precipitated γ and γ' phase alongside the formation of solid solution strengthened elements. These results are similar to that obtained in this study.

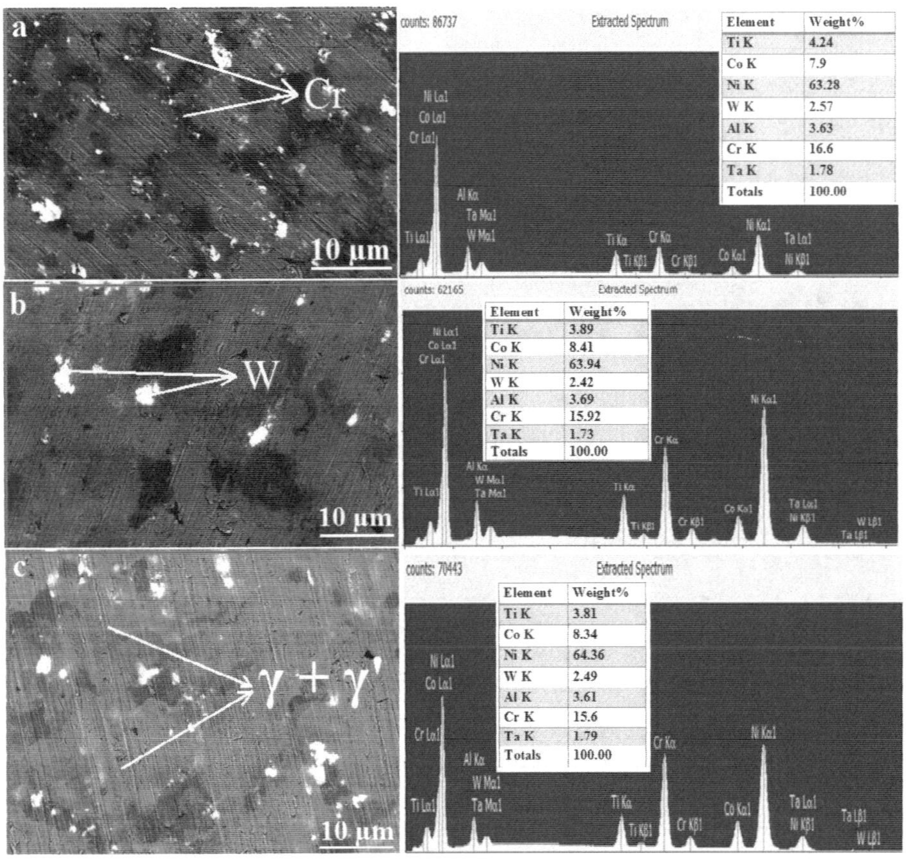

Figure 2. SEM images of spark plasma sintered NiCrCoAlTiW-Ta superalloy on the left, with EDX analysis results on the right. (a) Ni-matrix powder particle size (3-44 µm) (b) Ni-matrix powder particle size (45-106 µm) (c) Ni-matrix powder particle size (106-150 µm).

3.3 XRD analysis

Figure 3 shows the XRD patterns for the starting powder and the three sintered (different particle size range of matrix) samples. The XRD analysis on the SPS solid samples shows the formation of gamma (γ) matrix phases with prominent peaks observed at 42, 52 and 76^0 2θ diffraction angles. In general, the gamma (γ) matrix phase is the dominant phase that is present in all the sintered samples. The pre-alloyed powder has sharp minor peaks reminiscent of the constituent elements. Intermetallic phases were not present in the pre-alloyed powder, since there was no reaction yet between powder particles. The analysis of the three sintered alloys shows the presence of intermetallic phases. This was formed from the application of thermal energy which triggered chemical reaction within the structure of the sintered alloy. In addition, the type of secondary intermetallic phase formed was gamma prime γ' phase (Ni_3Al and Ni_3Ti). These two γ' phases overlap in the peak with the highest intensity at 42^0 angle 2θ, while at 52 and 78^0 diffraction angles, they are separated respectively for Ni_3Al and Ni_3Ti. The intensity of the peak formed at angle 42^0 angle 2θ was observed to decrease with decreaseing starting powder particle size. Marginal decrease was observed with the peak formed at 42^0 angle 2θ, for alloy with c and d. This suggests that there was an increased formation of intermetallic precipitate phase, with decreasing starting powder particle sizes [18]. This is due to reduction in the volume of matrix elements (nickel) that dissociated to form secondary phases [19]. For this reason, it was considered that more intermetallic phases were present in the alloy sintered with smaller initial powder particle size.

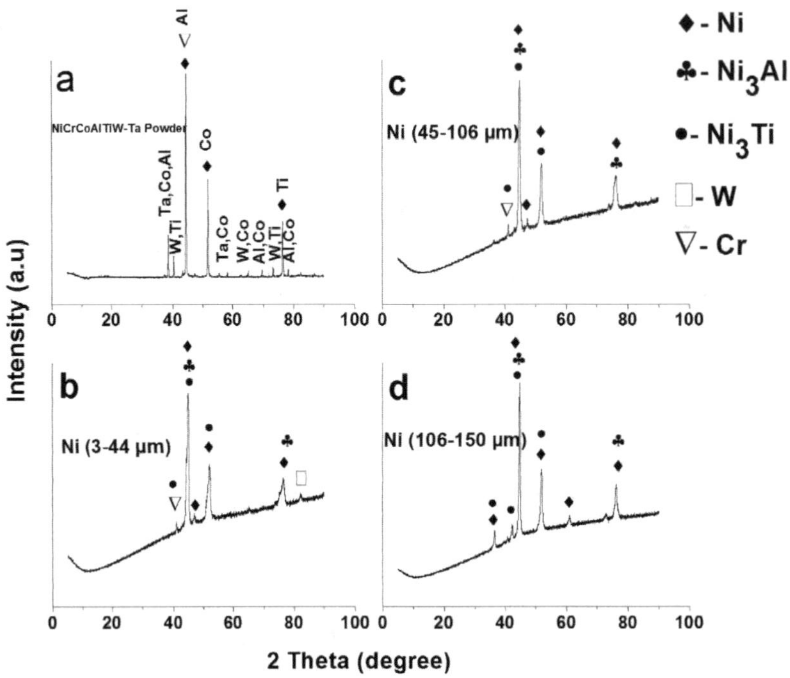

Figure 3. XRD patterns of NiCrCoAlTiW-Ta samples (a) Pre-alloyed powder and b-c is the spark plasma sintered superalloys; (b) Ni-matrix powder particle size (3-44 µm) (c) Ni-matrix powder particle size (45-106 µm) (d) Ni-matrix powder particle size (106-150 µm).

3.4 Thermal diffusivity and thermal conductivity measurement

The response of the thermal diffusivity and thermal conductivity of the sintered NiCrCoAlTiW-Ta superalloys as a function of temperature is shown in figure 4. Table 1 shows the data as obtained from the laser flash analyzer, while the specific heat capacity data used were from the literature [14]. In general, figure 4 shows that the thermal diffusivity, conductivity and specific heat capacity increased with increasing temperature for the three sintered alloys. For sample with 3-44 µm initial powder particle sizes (figure 4a), there is a slight decrease in the thermal properties between 200 and 300 °C. Thereafter, the thermal properties start to increase with increasing temperature. A plateaus pattern was observed in the thermal behaviour of alloy with 106-150 µm initial powder particle size (figure 4c) with increasing temperature. The points where the patterns deviate from linear progression with increase in temperature is refered to as an inflection point, and this is more pronounced in sample C (106-150 µm initial powder particle size). The point of inflection was observed at 600 °C for sample B. Several reasons can be responsible for the deviation in the patterns as the temperature increases, especially the occurrence of inflection points. For this type of alloy, the deviation is attributed to phase transformation and structural adjustment. The structural reconfiguration is a result of atom redistribution which causes disorder in the matrix phase [20]. Also, at high temperature state, dissolution of precipitated γ' intermetallic and γ matrix phases might occur which may equally lead to inflections. However, it was reported that chromium precipitate at high temperature, usually above 477 °C, while it depletes at a temperature above 800 °C [20]. The absorption of heat will create thermal stress within the microstructure which is also a factor that could result in inflection due to the precipitation of chromium along grain boundaries (figure 1). Similar reports in the literature [20-22], stated that the occurrence of inflection point was as a result of dissolution of γ' intermetallic precipitates, which usually occur at high temperatures.

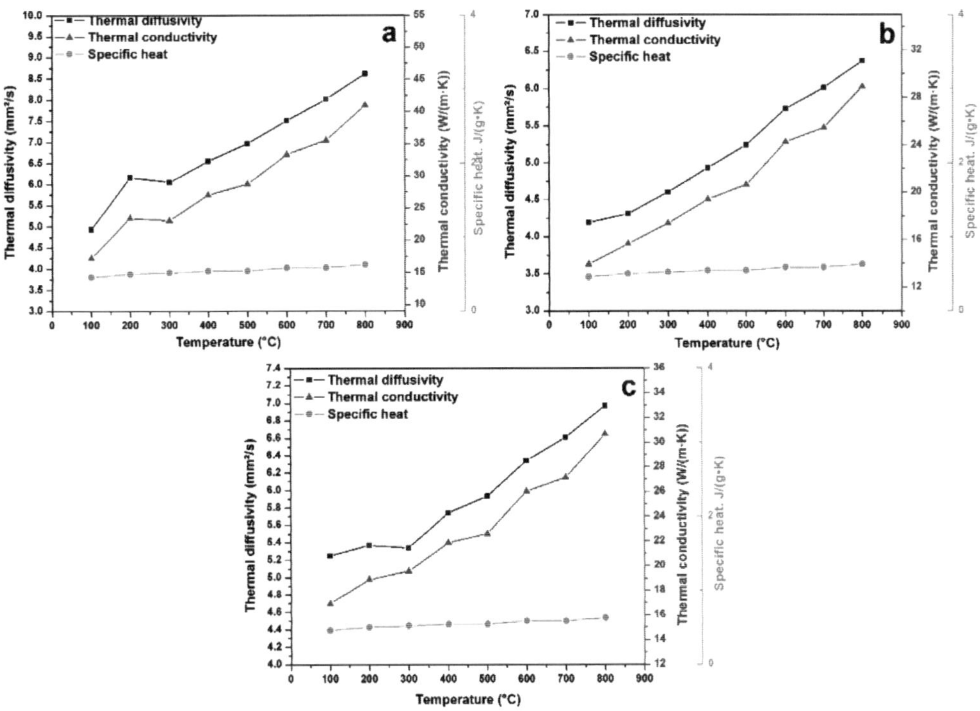

Figure 4. Thermal diffusivity, thermal conductivity and specific heat of NiCrCoAlTiW-Ta superalloy: (a) Ni-matrix powder particle size (3-44 µm), (b) Ni-matrix powder particle size (45-106 µm), (c) Ni-matrix powder particle size (106-150 µm).

Table 1: Thermal diffusivity, thermal conductivity and specific heat of spark plasma sintered NiCrCoAlTiW-Ta superalloy as a function of temperature

(a) Ni (3-44 µm) particle size. Density: 7.586 g/cm³			
Temperature, °C	Thermal diffusivity, mm²/s	Specific heat, J/(g·K)[14, 15]	Thermal conductivity, W/(m·K)
100	4.93	0.461	17.24
200	6.16	0.502	23.47
300	6.05	0.523	23.10
400	6.55	0.544	27.04
500	6.97	0.544	28.76
600	7.51	0.586	33.38
700	8.01	0.586	35.60
800	8.61	0.628	40.99
(b) Ni (45-106 µm) particle size. Density: 7.229 g/cm³			
100	4.19	0.461	13.96
200	4.31	0.502	15.65
300	4.60	0.523	17.37
400	4.93	0.544	19.40
500	5.24	0.544	20.62
600	5.73	0.586	24.27
700	6.01	0.586	25.47
800	6.37	0.628	28.90

(c) Ni (106-150 μm) particle size. Density: 7.004 g/cm³			
100	5.25	0.461	16.94
200	5.37	0.502	18.90
300	5.34	0.523	19.58
400	5.74	0.544	21.88
500	5.93	0.544	22.60
600	6.34	0.586	26.03
700	6.61	0.586	27.14
800	6.97	0.628	30.68

4. Conclusions
NiCrCoAlTiW-Ta superalloy with strengthening intermetallic and precipitated solid solution phase were established by using Spark Plasma Sintering. Also, the thermal conductivity was calculated from the knowledge of thermal diffusivity, heat capacity and alloy density over temperatures of 100-800 °C. Highly dense alloys were achieved for the sintered alloys with relative densities in the range 88 – 96%. The alloy with starting powder particle range between 3-44 µm has the highest relative density of 96.1% and the least density of 88.73% was obtained for alloy with 106-150 µm starting powder paticles (coarser). The starting powder particle sizes was observed to have significant effect on the density of the sintered product, even at the same processing parameters. The non-linear pattern observed from the thermophysical properties is due to atomic redistribution and short-range order/disorder phenomena which are known to occur in the Ni–Cr alloy system. The variations in the thermophysical properties are not so significant, however, the thermal conductivity of the superalloys depends on temperature and the phases formed. Thermal conductivity and diffusivity were discovered to increase with increasing temperature.

5. Acknowledgments
This work is funded by the Research and Innovation Directorate of Tshwane University of Technology and supported, in part, by the Department of Mechanical Engineering, Mechatronics and Industrial Design, Institute for Nano Engineering Research (INER), Department of Chemical, Metallurgical and Materials Engineering and the Faculty of Engineering and Built Environment of the Tshwane University of Technology, Pretoria, South Africa.

6. References
[1] Mills K. C, Youssef Y. M, Li Z and Su Y 2006 Calculation of thermophysical properties of Ni-based superalloys *ISIJ. int.* **46** pp. 623-632.

[2] Ogunbiyi O, Jamiru T, Sadiku E, Beneke L, Adesina O and Adegbola T 2019 Microstructural characteristics and thermophysical properties of spark plasma sintered Inconel 738LC *The Int. J. Adv. Manuf. Tech.* pp. 1-12.

[3] Jalilvand V, Omidvar H, Shakeri H and Rahimipour M 2013 Microstructural evolution during transient liquid phase bonding of Inconel 738LC using AMS 4777 filler alloy *Mats. Charact.* **75** pp. 20-28.

[4] Maleki V, Omidvar H and Rahimipour M. R "Influences of gap size and cyclic-thermal-shock treatment on mechanical properties of TLP bonded IN-738LC superalloy 2018 *Trans. Nonf. Metals Soc. China* **28** pp. 920-930.

[5] Balikci E and Raman A 2000 Characteristics of the γ′ precipitates at high temperatures in Ni-base polycrystalline superalloy IN738LC *J. mats. scie* **35** pp. 3593-3597.

[6] Nickel H, Clemens D, Quadakkers W and Singheiser L 1999 Development of NiCrAlY alloys for corrosion-resistant coatings and thermal barrier coatings of gas turbine components *J. pressure vessel tech.* **121** pp. 384-387.

[7] García J. M. J, Li Z, Vahlas C, de La Torre S.D and López F. J 2010 Spark plasma sintering and characterization of NiCoCrAlY-Ta superalloy powder *J. Mats. Scie. Eng.* **4** pp. 57-63.

[8] Munir Z, Anselmi-Tamburini U and Ohyanagi M 2006 The effect of electric field and pressure on the synthesis and consolidation of materials: A review of the spark plasma sintering method *J. Mats. Scie.* **41** pp. 763-777.

[9] Borkar T and Banerjee R 2014 Influence of spark plasma sintering (SPS) processing parameters on microstructure and mechanical properties of nickel *Mats. Scie. Eng. A* **618** pp. 176-181.

[10] Guillon O, Gonzalez-Julian J, Dargatz B, Kessel T, Schierning G, Räthel J *et al.* 2014 Field-assisted sintering technology/spark plasma sintering: mechanisms, materials, and technology developments *Adv. Eng. Mats.* **16** pp. 830-849.

[11] Dudina D. V and Mukherjee A. K 2013 Reactive spark plasma sintering: successes and challenges of nanomaterial synthesis *J. Nano-mats.* ***2013*** pp. 5.

[12] Rani S, Agrawal A.K and Rastogi V 2017 Failure analysis of a first stage IN738 gas turbine blade tip cracking in a thermal power plant *Case stud. Eng. failure anly.* **8** pp. 1-10.

[13] A. Standard 2013 E112-13 *Standard Test Method for Determining Average Grain Size* (West Conshohocken PA) pp. 1-28.

[14] IN-738 Alloy - Preliminary Data (497) - Nickel Institute (One New York plaza: New York) N.Y 10004.

[15] Technical Data - NiPERA (One New York plaza: New York) N.Y 10004.

[16] Yu J, Huang L and Luo H 2019 Effects of Cu particle size on CuSnFeNi/diamond composite processed using hybrid microwave sintering *Powder Met.* **62** pp. 124-132.

[17] Idowu O, Ojo O and Chaturvedi M 2006 Microstructural study of transient liquid phase bonded cast INCONEL 738LC superalloy *Met. Mats. Trans* **37** pp. 2787-2796.

[18] Ozgün O, Gülsoy H. O, Yılmaz R and Fındık F 2013 Microstructural and mechanical characterization of injection molded 718 superalloy powders *J. Alloys Compds* **576** pp. 140-153.

[19] Rai S. K, Kumar A, Shankar V, Jayakumar T, Rao K. B. S and Raj B 2004 Characterization of microstructures in Inconel 625 using X-ray diffraction peak broadening and lattice parameter measurements *Scripta materialia* **51** pp. 59-63.

[20] Rai A. K, Trpathy H, Hajra R, Raju S and Saroja S 2017 Thermophysical properties of Ni based super alloy 617 *J. Alloys Compds* **698** pp. 442-450.

[21] Yang L, Chang K. M, Mannan S and DeBarbadillo J 1997 A new DTA approach for verifying precipitate solvus in Inconel alloy 718 *Superalloys 718, 625, 706 and various derivatives,* pp. 353-365.

[22] Zieliñska M, Yavorska M, Porêba M and Sieniawski J 2010 Thermal properties of cast nickel based superalloys *Archs. Mats. Scie. Eng.* **44** pp. 35-38.

IOP Conf. Series: Materials Science and Engineering **655** (2019) 012032 doi:10.1088/1757-899X/655/1/012032

Influence of spark plasma sintering temperature on the densification and micro-hardness behaviour of Ni-Cr-Al alloy

B J Babalola[1]*, M B Shongwe[1], S O Jeje[1], A L Rominiyi[1], O O Ayodele[2], and P A Olubambi[2].

[1]Institute of Nanoengineering Research (INER), Department of Chemical, Metallurgical, and Materials Engineering, Faculty of Engineering and Built Environment, Tshwane University of Technology, Pretoria, South Africa.

[2]Center for Nanoengineering and Tribocorrosion, School of Mining, Metallurgy and Chemical Engineering, University of Johannesburg, Johannesburg, 2094, South Africa.

Corresponding author: beejegs@gmail.com*

Abstract. Sintering temperature is essential towards attaining desired densification and formation of phases which in turn influences the microstructure and properties of a material. In this study, the densification and microhardness behaviour of Ni-Cr-Al alloy prepared by spark plasma sintering (SPS) at different sintering temperatures were investigated. After sintering operation, the density, hardness, phase analysis, and microstructural evolution were investigated using the Archimede's principle, hardness tester, X-ray diffraction (XRD), and scanning electron microscopy (SEM) respectively. The nickel based alloy was sintered at temperatures of 600, 750, 950 and 1100 °C. The results indicated that the densification, microstructure, and hardness values were influenced by changes in the sintering temperature. The relative density increased from 73.89 % at 600 °C to 99.89 % at 1100 °C, while the hardness value was enhanced from 131.9 ± 2.8 HV to 404 ± 1.2 HV respectively.

1. Introduction

Spark plasma sintering (SPS), otherwise referenced as the field assisted sintering technique (FAST) has been recognized for efficient consolidation of materials ranging from metallic alloys, composites, ceramics, functionally graded materials [1] and polymers [2]. The consolidation to near theoretical density is enhanced by spark discharge which is formed between the voids of elemental particles. The spark discharge effect is notable for the purification of the particles' surface from potential oxides and impurities present and its activation which promotes the generation of joule heating. The joule heating effect on the surface of the particles which initiates liquid state particle surfaces results into necking of one particle to another and subsequently aid densification with applied pressure [1, 3]. Shorter sintering times [4], easy control of sintering parameters, inhibiting little or no grain growth [5] and the possibility to consolidate constituent powders at a temperature relatively below melting point [6] forms part of the significant benefits of the spark plasma sintering process when compared to other processing routes. In addition, conventional processing routes have been reported to induce defects such as microsegregation of alloying elements [7, 8], inclusion of low melt inclusion which can

compromise the structural integrity of engineering materials in service and subsequently lead to catastrophic failure. One of the materials which have gained attention over the years are nickel and its alloys owing to their superior ability to retain their strength at elevated temperatures, resistance to oxidation and corrosion among others. They find application in aggressive environments such as aerospace engines, thermal plants, marine [9] and other elevated temperature demanding environments. For instance, Ni-Cr-Al system accounts for the most critical ternary system for nickel based alloys, utilized for elevated temperature application and has attracted research attention [10].
However, defects which arise during conventional fabrication of these alloys have plagued their excellent performance in service and thus necessitate the production of these alloys via novel approach in order to meet increasing performance and efficiency demands. This investigation forms part of research work conducted on the development of nickel-based superalloy via spark plasma sintering technique in which some have been reported [4, 6, 11].

2. Material and methodology

The features of the elemental powders utilized for this research are presented in Table1. The nickel powder used in this study was further subjected to mechanical milling process through high energy ball milling process (PM400) and a nanocrystalline size of 7.867 nm after 10 hr milling duration at 350 rpm, 10:1 ball to powder weight ratio (BPR), was attained. The mechanical milling of the nickel powder was carried out in a stainless steel milling vial and balls with dimensions of 250 ml and 2 mm diameter respectively. Ethanol was used the process control agent (PCA). The detail report on the milling experiment has been reported elsewhere [11].

Table 1: Alloy elemental constituent's details

Element	Particle size (µm)	Purity (%)	Supplier
Al	25	99.8	TLS Technik GmbH & Co.
Cr	10	99.2	FloMaster Metal Powder
Ni	4-7	99.8	Goodfellow metals

The milled nickel powder after drying was blended with chromium (Cr) and aluminium (Al) in their as received state according to Ni-17Cr-10Al composition by wt.% and mixed thoroughly in a Turbula mixer for 8 h in order to attain homogenized alloy powder prior to sintering process. Mixed starting powder (containing constituent elements) was emptied into a graphite die which have 20 mm internal diameter, lined with graphite foil to ensure lubrication and easy removal after sintering.

The schematic diagram which illustrates the graphite die and sample assembly within SPS process is shown in Figure 1. The mixed powder was sintered at different sintering temperatures (600, 750, 950 and 1100), at 50 MPa sintering pressure, 10 min holding time and 100 ºC/min heating rate in vacuum using HHPD-25 SPS model (FCT Germany).
After sintering, 20 mm diameter and 5 mm thickness samples were obtained, followed by sand blasting to remove the remains of graphite foil on the surface of the alloy. This was followed by density evaluation by Archimedes principle; an average of 5 readings for each sintered alloy was recorded. Thereafter, the samples were then prepared for microstructural examination by polishing following standard procedure and using a solution which contains 50 ml of distilled water, 10 g of $CuSO_4$, few drops of H_2SO_4 and 50 ml of HCl as etchant.

The characterization of phases formed in the nickel-based alloys was done by utilizing a PANalytical Empyrean model X-ray diffraction (XRD), with Cu Kα radiation. The diffractographs of the sintered alloys obtained over a range of 2θ between 5 and 90°, were studied through the use of Highscore plus software. The microstructural evolutions due to increasing sintering temperature of the alloy were characterized by JOEL JSM-7600F field emission scanning electron microscope furnished with an energy-dispersive X-ray spectroscopy (EDS) detector.

The hardness value of the sintered alloy was evaluated at room temperature by utilizing Future-tech micro-hardness tester at a test load of 1000 gf for a dwell time of 10 s. The average of 10 indentations was recorded for hardness value.

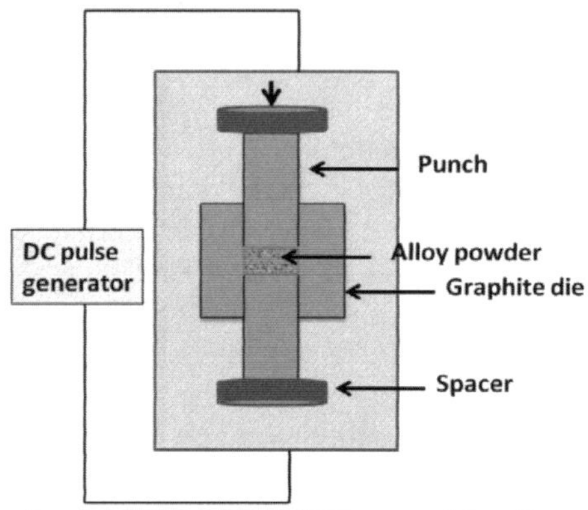

Figure 1. Schematic diagram illustrating SPS assembly of sample and graphite die.

3. Results and discussion

3.1. Densification and hardness behaviour

Figure 2 shows the densification and hardness behaviour of sintered Ni-Cr-Al alloy at increasing sintering temperatures. The relative displacements which occur between the upper punch and the lower punch, within the graphite wall due to the consolidation effect of different sintering temperatures on the powder is presented in Figure 2(a). It can be seen that the displacement increases with corresponding increase in the sintering temperature, while at sintering temperature of 950 °C and 1100 °C attains highest displacement at 3.74 mm and 3.76 mm respectively. This is a typical representation of the densification mechanism of the alloy powder at different sintering temperatures. The relative displacement experienced by the alloy at different sintering temperatures clearly indicates the reduction in the height of powder sample during sintering with respect to time. Similar account on densification analysis has been given in the literature to better understand the consolidation of powders with respect to sintering parameters [12, 13].

Similar behaviour was observed for the relative density and hardness, which agrees with corresponding increase in the sintering temperature, as shown in Figure 2(b). The relative density trend with respect to increasing sintering temperature in this study is in agreement with a study conducted by Zhou *et al.* [14].

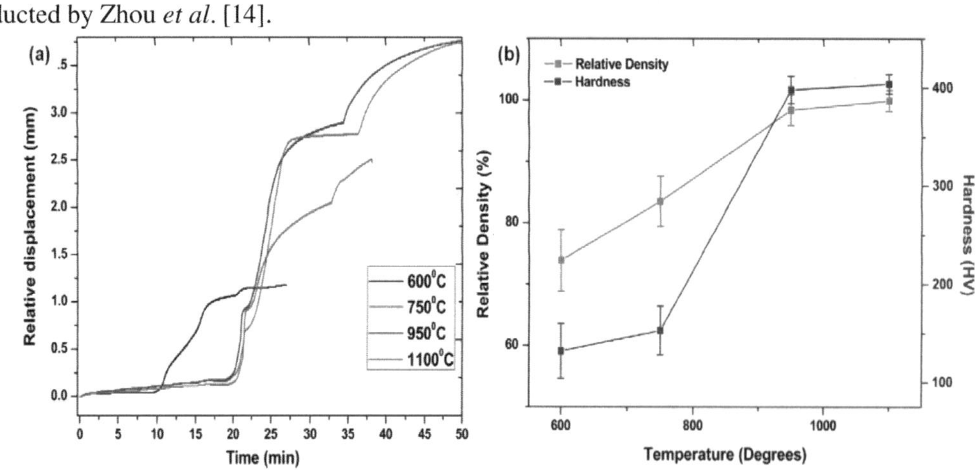

Figure 2. Densification features of sintered Ni-Cr-Al alloy (a) relative displacement against time and (b) comparison of relative density with hardness at different sintering temperatures.

It can also be seen that the hardness value increases as the densification increases (Figure 2b), this can be attributed to the effect of reduced porosity, increased bonding of constituent particles as sintering temperature increases. The lowest hardness value, 131.9 ± 2.8 HV attained at sintering temperature of 600 °C can be attributed to its poor densification while the highest hardness value of 404 ± 1.2 HV was obtained at 1100 °C.

3.2. Phase analysis and microstructural behaviour of sintered alloys.

The XRD patterns of the sintered Ni-Cr-Al alloy according to sintering temperatures (600, 750, 950 and 1100 °C) are shown in Figure 3. There were no foreign inclusions or elements detected, which indicates that spark plasma sintering technique aids the fabrication of materials towards attaining intrinsic properties. As shown in the XRD patterns, the occurrence of phases, as the sintering temperature increases. Ni_3Al phase occurred only at 1100 °C, Ni_2Al_3 phase evolved at 950 °C, NiAl and Ni_2Al_3 evolved at 750 °C and NiAl evolved at 600 °C. According to Mitra [15] and Dey [16], these are the phases responsible for the strengthening of nickel based superalloys. The increase in hardness value can be ascribed to the occurence of these phases in the alloy, in accordance to the densification of the alloy at different sintering temperature.

Figure 4 shows the microstructural evolution of Ni-Cr-Al alloy obtained at different sintering temperatures. As shown in Figure 4(a), the initiation of necking formation of the constituent particles of the alloy is evident which clearly reflects that the alloy is yet to attain full densification at 600 °C. At increased sintering temperature, 750 °C (Figure 4b), more bonding between the particles exist and consequently improves the densification. This is in agreement with the relative density results, as discussed in Section 3.1.

However, at higher sintering temperature (950 °C, 1100 °C), finer microstructures were attained (see Figure 4c-d) as a result of effective bonding of particles, reduction of porosity, which in turn enhances the hardness value as discussed above. Elemental spot analysis was conducted on the alloy consolidated at 950 °C and 1100 °C, the dominant peaks in the analysis were mainly Ni and Al. This is in confirmation to the XRD analysis which reveals the presence of strengthening phase which is nickel aluminides. Similar report relating to the impact of sintering temperature on density and its associated properties have been reported by [4, 17, 18].

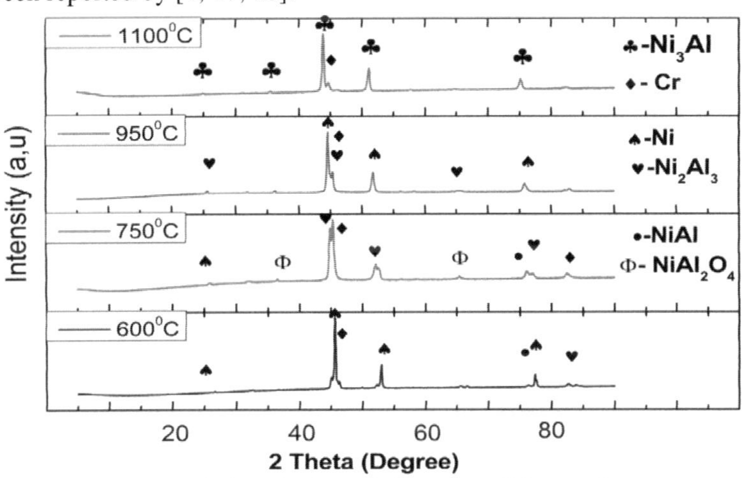

Figure 3. XRD pattern of Ni-Cr-Al alloy fabricated at different sintering temperatures.

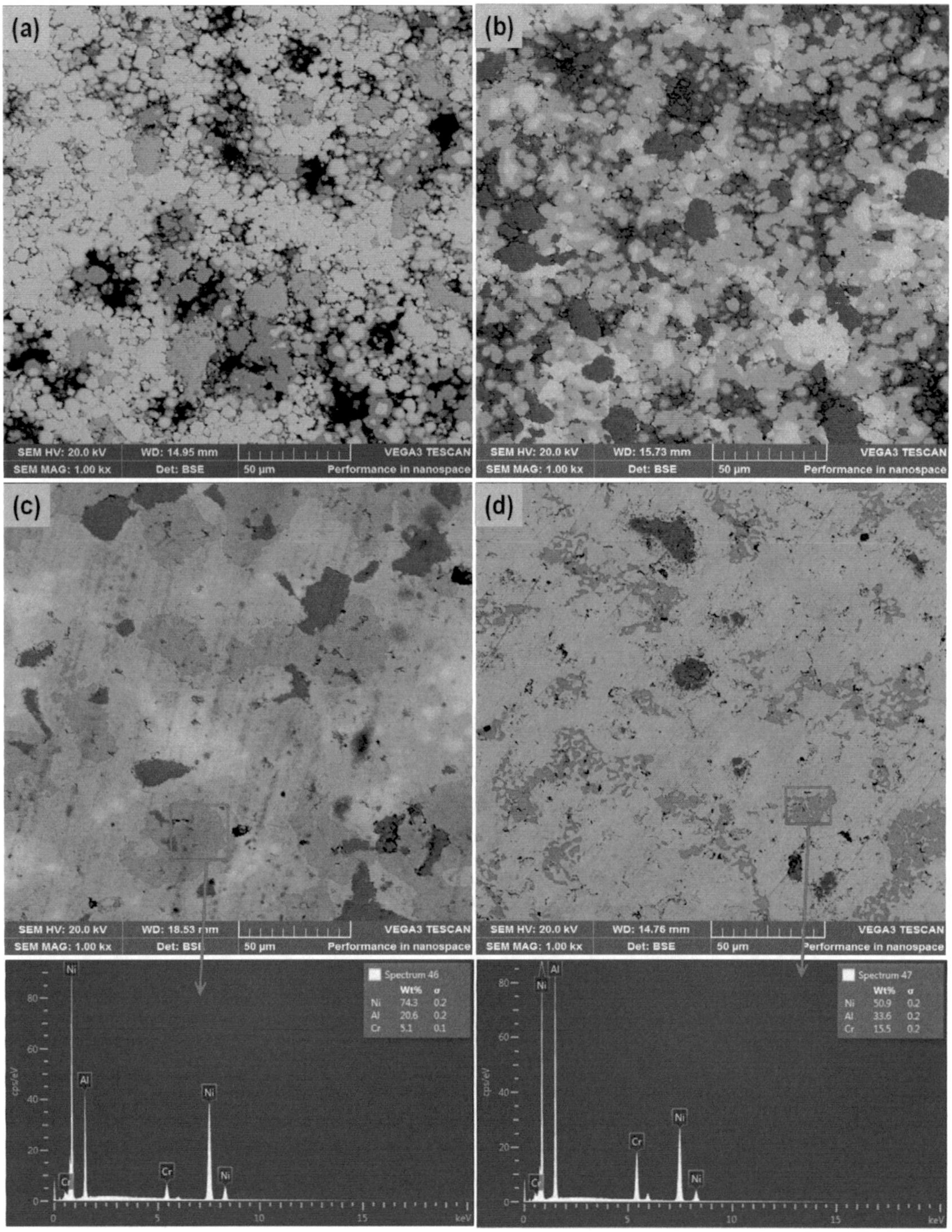

Figure 4. SEM images of sintered Ni-Cr-Al alloy at (a) 600 °C, (b) 750 °C, (c) 950 °C with EDS analysis and (d) 1100 °C with EDS analysis.

4. Conclusion

In this research, an investigation on the role of sintering temperature on the densification and micro-hardness on Ni-Cr-Al alloy was conducted. The following key conclusions were deduced;

1. The relative density and hardness of the Ni-Cr-Al alloy increases due to increasing sintering temperature.

2. The sintering temperature also influences the microstructure of the alloys, more bonding of the particles and finer microstructure were obtained at higher sintering temperature which in turn enhances the hardness value.

3. The formation nickel aluminide phases, which are the strengthening phases in nickel based alloys, were detected. Hence, this production technique is efficient towards the fabrication of advanced material void of defects which can impair its structural integrity while in service.

5. Acknowledgement
Financial support was made available by the National Research Foundation (NRF) of South Africa, Grant No. 99348.

6. References
[1] Yamanoglu R., Bradbury W., Karakulak E., Olevsky E., and German R., 2014 "Characterisation of nickel alloy powders processed by spark plasma sintering," *Powder Metallurgy,* vol. 57, pp. 380-386.
[2] Adesina O., Sadiku E., Jamiru T., Ogunbiyi O., Beneke L., and Adegbola A., 2019 "Optimization of SPS processing parameters on the density and hardness properties of graphene reinforced polylactic acid nanocomposite," *The International Journal of Advanced Manufacturing Technology,* pp. 1-12.
[3] Babalola B. J., Shongwe M. B., Obadele B. A., Olubambi P. A., Ayodele O. O., Rominiyi A. L.,and Jeje S. O., 2018 "Comparative study of spark plasma sintering features on the densification of Ni-Cr binary alloys," *MATEC Web Conf.,* vol. 249, (2018) p. 01004.
[4] Babalola B. J., Shongwe M. B., Obadele B. A., and Olubambi P. A., 2018 "Densification, microstructure and mechanical properties of spark plasma sintered Ni-17%Cr binary alloys," *The International Journal of Advanced Manufacturing Technology,* vol. 101, p. 8, 2018/11/22 2018.
[5] Rominiyi A. L., Shongwe M. B., Maledi N., Babalola B. J., and Olubambi P. A., 2019 "Synthesis, microstructural and phase evolution in Ti–2Ni and Ti–10Ni binary alloys consolidated by spark plasma sintering technique," *The International Journal of Advanced Manufacturing Technology,* pp. 1-9, 2019.
[6] Babalola B. J., Maledi N., Shongwe M. B., Bodunrin M. O., Obadele B. A., and Olubambi P. A., 2019 "Influence of nanocrystalline nickel powder on oxidation resistance of spark plasma sintered Ni-17Cr6.5Co1.2Mo6Al4W7.6Ta alloy," *Journal of King Saud University - Engineering Sciences,* 2019/01/21/ 2019. https://doi.org/10.1016/j.jksues.2019.01.002
[7] Dash K., Chaira D., and Ray B. C., 2013 "Synthesis and characterization of aluminium–alumina micro- and nano-composites by spark plasma sintering," *Materials Research Bulletin,* vol. 48, pp. 2535-2542, 2013/07/01/ 2013.
[8] Zhang Y., Song A., Ma D., Zhang X., Ma M., and Liu R., 2014 "Sintering characteristics and grain growth behavior of MgO nanopowders by spark plasma sintering," *Journal of Alloys and Compounds,* vol. 608, pp. 304-310, 2014/09/25/ 2014.
[9] Nowotnik A., Kubiak K., Sieniawski J., Rokicki P., Pędrak P., and Mrówka-Nowotnik G., 2014 "Development of nickel based superalloys for advanced turbine engines," in *Materials Science Forum,* 2014, pp. 2491-2496.
[10] Lu Y. L., Chen Z., Wang Y. X., Zhang J., and Yang K., 2011 "Phase-field study the effect of elastic strain energy on the incubation period of Ni–Cr–Al alloys," *Computational Materials Science,* vol. 50, pp. 1925-1931, 2011/04/01/ 2011.
[11] Babalola B. J., Shongwe M. B., Rominiyi A. L., Jeje S. O., and Olubambi P. A., 2019 "A study of nanocrystalline nickel powders developed via high-energy ball milling," *The International Journal of Advanced Manufacturing Technology,* pp. 1-9, 2019.

[12] Diouf S. and Molinari A., 2012 "Densification mechanisms in spark plasma sintering: Effect of particle size and pressure," *Powder technology,* vol. 221, pp. 220-227, 2012.

[13] Kekana N., Shongwe M. B., Johnson O. T., and Babalola B. J., 2019 "Densification behaviour and the effect of heat treatment on microstructure, and mechanical properties of sintered nickel-based alloys," *The International Journal of Advanced Manufacturing Technology,* pp. 1-7, 2019.

[14] Zhou Y., Yang Y., Yang G., Yin D., Qin Y., and Liu J., 2015 "Effects of sintering temperature on the densification of WC-6Co cemented carbides sintered by coupled multi-physical-fields activated technology," *Manufacturing Review,* vol. 2, p. 18, 2015.

[15] Mitra R., 2017 *Intermetallic Matrix Composites: Properties and Applications*: Elsevier, 2017.

[16] Dey G. K., 2003 "Physical metallurgy of nickel aluminides," *Sadhana,* vol. 28, pp. 247-262, 2003.

[17] Shongwe M. B., Diouf S., Durowoju M. O., and Olubambi P. A., 2015 "Effect of sintering temperature on the microstructure and mechanical properties of Fe–30%Ni alloys produced by spark plasma sintering," *Journal of Alloys and Compounds,* vol. 649, pp. 824-832, 2015/11/15/ 2015.

[18] Zhou Y., Yang Y., Yang G., Yin D., Qin Y., and Liu J., 2015 "Effects of sintering temperature on the densification of WC-6Co cemented carbides sintered by coupled multi-physical-fields activated technology," *Manufacturing Rev.,* vol. 2, p. 18, 2015.

IOP Conf. Series: Materials Science and Engineering **655** (2019) 012033 doi:10.1088/1757-899X/655/1/012033

Metal injection moulding of 17-4PH stainless steel: Effects of porosity on the mechanical properties of the sintered products

M N Seerane[1, 2, *], and R Machaka[1]

[1]Light Metals, Manufacturing Cluster, Council for Scientific and Industrial Research, Meiring Naudé Road, Brummeria, Pretoria 0185, South Africa

[2]School of Chemical and Metallurgical Engineering, University of the Witwatersrand, Johannesburg, South Africa

*E-mail: mseerane@csir.co.za

Abstract. 17-4PH stainless steel (SS) is a martensitic precipitation hardenable alloy which normally contains approximately 15.5-18.5 wt.% Cr and 2.5-4.5 wt.% Ni and Cu. It is well known for its high impact strength and fracture toughness, and excellent corrosion resistance at service temperatures below 300 °C promoted by its Cu-rich precipitates. 17-4PH SS is widely used as a structural material in various sectors such as aerospace, medical, petrochemical, etc. The high strength of the alloy makes it difficult to machine, especially when making complex shaped products. Metal injection moulding (MIM) uses the shaping advantage to produce small-to-medium intricate products in high volumes at a relatively low cost. However, due to the nature of MIM process, the mechanical properties of the sintered products depend strongly on various metallurgical MIM variables with porosity included. Porosity is largely dependent on the starting powder characteristics, powder loading, and sintering conditions. In this study, the mechanical properties of the sintered 17-4PH SS MIM samples fabricated at different powder loadings, viz. 55, 60 and 65 vol.%, will be evaluated. The starting powder particle size distribution is -5, -15 and -45 μm. The sintered density, microstructure and hardness response of the samples are presented. These properties are influenced by the nature of the pores, i.e. size, density and distribution.

1. Introduction

17-4PH stainless steel (SS) is a precipitation hardenable martensitic alloy widely used in areas requiring high strength, excellent corrosion resistance and good biocompatibility such as automotive, aerospace, chemical and power plants, and medical industries [1], [2]. The alloy typically contains 15.5 – 18.5 wt.% Cr and 2.5 – 4.5 wt.% Ni and Cu elements which give the alloy its unique properties [3]. This alloy is, however, difficult to machine especially when making intricately shaped products. Metal injection moulding (MIM) uses the shape advantage to produce small-to-medium complex metal parts at a relatively low cost in high volumes. MIM combines powder metallurgy (P/M) and plastic injection processes to fabricate these metal products [4]. Four main stages in MIM include mixing, injection moulding, debinding and sintering. The mechanical properties of the sintered products strongly depend on the porosity, inclusions, oxygen content and other metallurgical variables [5]. By varying the sintering conditions, MIM process can allow injection moulded parts to have a wide range of porosity level which can provide a very useful scope to explore and investigate the mechanical properties of the sintered products. Furthermore, the starting powder loading and particle size were also investigated on how they affect the porosity level and final mechanical properties. Sung et al. [2] studied the influence of sintering temperature on the porosity level of 17-4PH SS and discovered that sintering temperature influences the shape and distribution of the pores. Mukund et al. [6] investigated the effect of bimodal powder particle size distribution on the rheological properties of the feedstocks, and defects and distortion evolution during sintering. Other studies examined the powder size variation at fixed powder loading against the rheological properties [7] and final mechanical properties [8].

Porosity remains a critical topic for MIM practitioners and end-users of MIM products. This work examines a combination of powder loading and powder particle size on the porosity level and hardness response of the final sintered parts.

2. Materials and experimental methodology

Precipitation hardenable martensitic stainless steel (17-4PH SS grade) powders of different sizes viz. -45 μm (supplied by Praxiar Surface Technologies), -15 μm and -5 μm (supplied by Epson Atmix Corporation) were used as starting materials in this study. The typical chemical composition of the 17-4PH SS powders is given in table 1. The powders were analyse for size and morphology using a scanning electron microscope (SEM, Jeol). For feedstock preparation, the powder materials were homogenously mixed with a polymeric binder at a ratio of 55, 60 and 65 vol.% with the binder reducing, and the feedstocks were labelled 55a, 55b, 55c, 60a, 60b, 60c, 65a, 65b and 65c; whereby a, b and c resemble the powder particle size -45, -15 and -5 μm respectively. The feedstocks were injection moulded into a 'dog-bone' shape.

Table 1: Typical chemical composition of the starting 17-4PH stainless steel powders

Elements (wt.%)	C	Si	Mn	Ni	Cr	Cu	Nb	P	S	Fe
17-4PH SS (Praxiar)	-	1.0	-	4.0	16.0	3.0	0.25	-	-	Bal.
17-4PH SS (Epson)	<0.07	<0.5	<0.3	3.0-5.0	15.5-17.5	3.0-5.0	0.15-0.45	<0.04	<0.03	Bal.

The injection moulded green parts were chemically and thermally debound to remove the polymeric binder components prior to sintering. The samples were pre-sintered at 1000 °C for 2 hours to remove the residual binder. Sintering was done under a vacuum furnace at 1300 °C for 3 hours. Sintered samples were cut-off for metallographic analysis, density and hardness evaluation. Sintered density was done through Archimedes' principle using OHAUS explorer precision electronic balance. Polished samples were observed under optical microscope for porosity analysis, and the samples were further taken for micro-Vickers hardness testing at a load of 0.5 kgf.

3. Results and discussion

The as-received 17-4PH stainless steel powders were analysed for size and morphology, as depicted in figure 1. All the powders were spherical, however, the finer powders viz. figure 1b and 1c had agglomerated and this could be attributed to their larger surface area to volume ratio which is proportional to the strength of the cohesive forces [9]. Agglomeration leads to poor packing and flowability of the powder particles [7]. During injection moulding, higher shear rates between 10^2 –

10^5 s^{-1} [6] are introduced to break-off the agglomerates and improve homogeneity and feedstock stability [7], [10]. It is expected that finer particle size will improve sinterability and hence reduce porosity in the final sintered products.

a) -45 μm **b) -15 μm** **c) -5 μm**

Figure 1: SEM images showing morphologies of the as-received 17-4PH stainless steel powders

Figure 2 gives the optical micrographs of the as-sintered parts. It can be seen that porosity in MIM products is inevitable. The size of the pores reduces with particle size (see 55a to 55c, 60a to 60c, etc. for example). However, there is no significant change in terms of porosity size between feedstock 55a and 55b. A decrease in pore size from lower powder loading to higher loading is attributed to lesser binder content leading to closely packed arrangement of the particles and more contact surface areas between the powder particles. In addition, closely packed arrangement of the particles (i.e. less binder content) and finer powder size contributed to finer microstructure and pore size (55c, 60c and 65c), and this result gives a good indication of better mechanical properties.

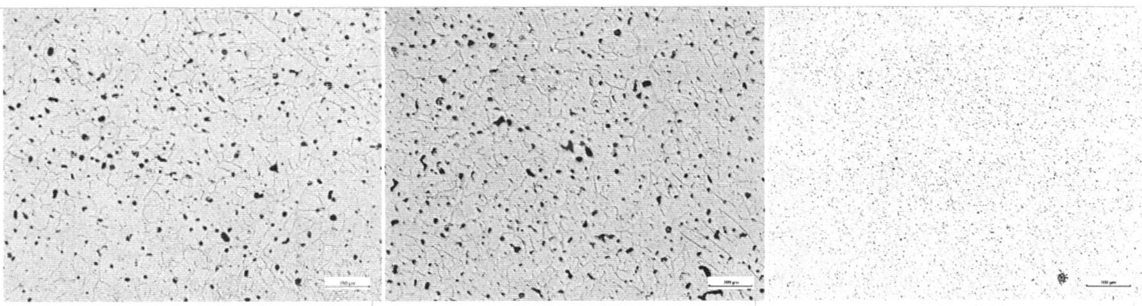

a) 55a **b) 55b** **c) 55c**

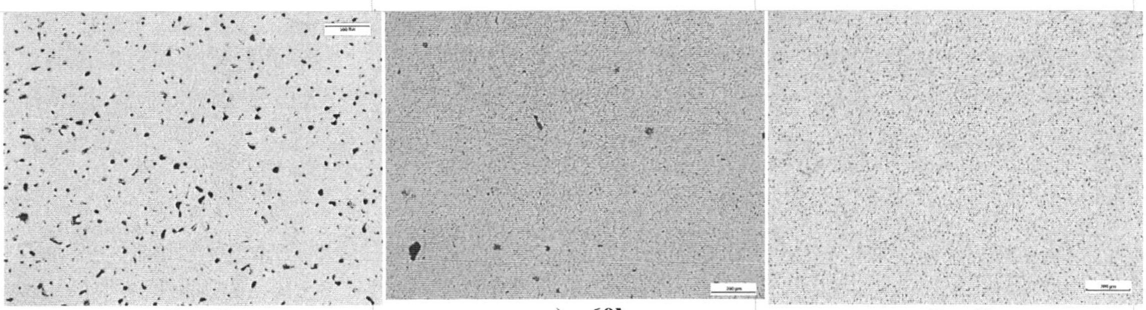

d) 60a **e) 60b** **f) 60c**

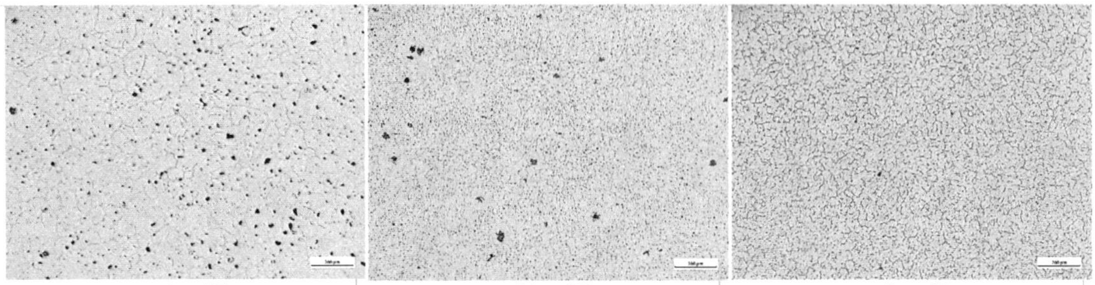

g) 65a	h) 65b	i) 65c

Figure 2: Optical micrographs of the as-sintered specimens

Figure 3 show the density variation of the as-sintered products. Generally, feedstocks -45 µm gave the least sintered density. The -15 µm feedstocks give good densification with 60b and 65b possessing the highest sintered densities. It can be deduced that densification improves with smaller particle size. The improved sintered density is attributed to better sinterability of the finer powder particles.

The porosity level of the as-sintered products can be estimated by using the following equation [11]:

$$Porosity = \left(1 - \frac{\rho_S}{\rho_H}\right) x\ 100\% \qquad (1)$$

Where ρ_S is the apparent density of the as-sintered samples and ρ_H is the apparent bulk or pycnometer density. The pycnometer density (ρ_H) of 17-4PH SS is 7.75 g/cm^3. Using equation (1), the lowest porosity levels recorded is 2.97% and 2.94% for 60b and 65b respectivley. Feedstocks 55a and 65a had the most porosity levels at 5.74% and 5.63% respectively.

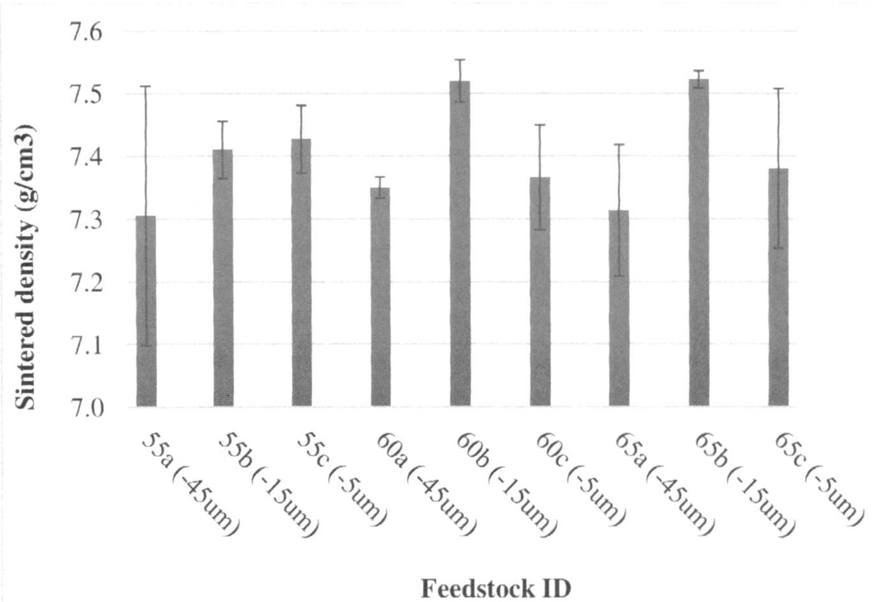

Figure 3: Density profile of the as-sintered parts

In figure 4, the hardness profile of the as-sintered feedstocks is presented. The hardness improves with decreasing particle size from all the feedstocks. Powder loading does not influence, linearly, the hardness variation in the feedstocks. The 60c feedstock, viz. 60 vol.% -5 μm, has the highest hardness. Porosity does not have a significant effect on the hardness of the material.

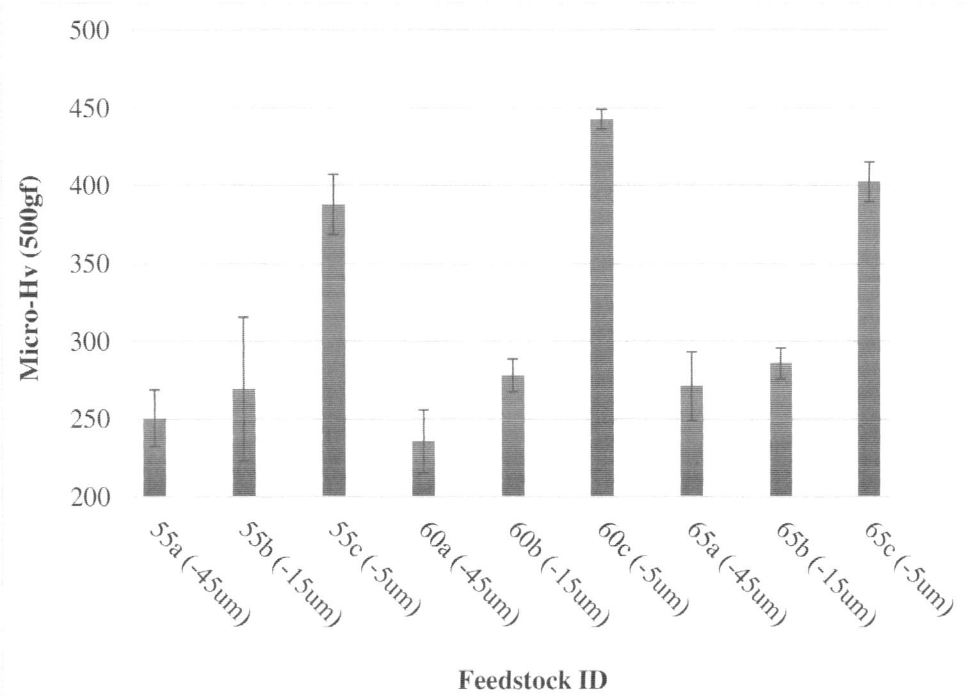

Figure 4: Hardness data of the different feedstocks after sintering

Following the presentation and discussion of results, we can make the following general remarks:

- Porosity is important in some areas, particularly medical sector when building orthopaedic implants for good osseontergration [12]. Therefore, for high porosity level and size, larger powder particle size and lower powder loading (55a and 55b, figure 2) is important.
- For better mechanical properties such as increased hardness, higher powder loading and finer particle size is recommended (55c, 60c and 65c, figure 4). This is critical for applications where high strength is required such as automotive and aerospace industries.

4. Conclusions

From the work presented, the following conclusions can be drawn:

- Finer metal powders tend to agglomerate due to large contact surface area which promote strong cohesive forces between the particles.
- Higher powder loading contributed to the finer pore size after sintering and this was attributed to lesser binder content in the feedstock matrix which led to more contact surfaces between the powder particles.

- Finer powder particle size contributed to finer microstructure and pore size whereby feedstocks 55c, 60c and 65c gave the finest porosities. This effect resulted from better sinterability of fine powders.

- Porosity range obtained from the feedstocks lied between 2.94% - 5.74% whereby feedstock 65b gave the least porosity level and feedstock 55a gave the most porosity level.

- Powder loading influences the size of the pores only, whereas powder particle size significantly influences pore size and pore density, sintered density, and also hardness. Thus for high strength and hardness fine powder is recommended, and for porosity size and control (medical applications) powder loading variation is recommended. The hardness response of the material is not influenced significantly by the porosity level and size.

- Future work will consider the effect of porosity on tensile and fatigue behaviour as literature states that pores are often crack initiation sites and reduce fatigue life. Tensile and cyclic fatigue properties are known to be monotonic, hence reduction in fatigue life influences tensile strength. In addition, analysis of these pores using CT scans will also be considered.

Acknowledgements

The authors would like to acknowledge the Council for Scientific and Industrial Research (CSIR) for providing facilities to conduct this work, and the Department of Science and Technology (DST) for their full support.

References

[1] Murr L E, Martinez E, Hernandez J, Collins S, Amato K N, Gaytan S M and Shindo P W 2012 *J. Mater. Res. Technol.* **1** 167–177
[2] Sung H, Kwon T, Ahn S and Won Y 2002 *J. Mater. Process. Technol.* **131** 321–327
[3] Yeli G, Auger M A, Wilford K, Smith G D W, Bagot P A J and Moody M P 2017 *Acta Mater.* **125** 38–49
[4] Cheng J, Wan L, Cai Y, Zhu J, Song P and Dong J 2010 *J. Mater. Process. Technol.* **210** 137–142
[5] Haynes R 1981 *The Mechanical Behaviour of Sintered of Sintered Metals*. London: Freund Publishing House
[6] Mukund B N, Hausnerova B and Shivashankar T S 2015 *Powder Technol.* **283** 24–31
[7] Machaka R, Ndlangamandla P and Seerane M 2018 *Powder Technol.* **326** 37–43
[8] Seerane M, Ndlangamandla P and Machaka R 2016 *J. South. African Inst. Min. Metall.* **116** 935–940
[9] Ku N, Hare C, Ghadiri M, Murtagh M, Oram P and Haber R A 2015 *Procedia Eng.* **102** 72–80
[10] Amin S Y M, Jamaludin K R and Muhamad N 2009 *Inst. Eng. Malaysia* **71** 59–63
[11] Zhao D, Chang K, Ebel T, Qian M, Willumeit R, Yan M and Pyczak F 2013 *J. Mech. Behav. Biomed. Mater.* **28** 171–182
[12] Geetha M, Singh A K, Asokamani R and Gogia A K 2009 *Prog. Mater. Sci.* **54** 397–425

IOP Conf. Series: Materials Science and Engineering **655** (2019) 012034 doi:10.1088/1757-899X/655/1/012034

Fretting Biocorrosion Behaviour of Titanium-Zirconia composites in Foetal Bovine Serum

L Semetse[1,*], B A Obadele[1], L Raganya[1,2], J Geringer[3] and P A Olubambi[1]

[1]Centre for Nanoengineering and Tribocorrosion, School of Mining, Metallurgy and Chemical Engineering, University of Johannesburg, Doornfontein Campus, Johannesburg, South Africa

[2]Light Metals, Materials Science & Manufacturing, Council for Scientific and Industrial Research, Meiring Naudé Road, Brummeria, Pretoria 0184, South Africa

[3]Univ Lyon, IMT Mines Saint-Etienne, Centre CIS, [STBio] Univ Jean Monnet, INSERM, SainBioSE, F-42023 Saint-Etienne France

E-mail: lerato.semetse@gmail.com

Abstract. This work aims at studying the fretting biocorrosion response of newly developed Ti-6Al-4V/ZrO_2 in simulated body fluid. Ti-6Al-4V alloy with different volume fractions of ZrO_2 produced via powder metallurgy techniques were spark plasma sintered to produce Ti-6Al-4V composites with improved properties. The microstructures of the resulting spark plasma sintered composites were examined using a scanning electron microscope (SEM). Fretting corrosion tests were conducted for each material composition with a special device used for fretting corrosion investigations with a cylinder-on-flat configuration. The tests were done in foetal bovine serum maintained at ambient temperature. Open circuit potential, dissipated energy and coefficient of friction were monitored throughout the experiments. The results show that the microstructures produced after zirconia additions were very different from those observed in pure Ti-6Al-4V. The presence of zirconia promoted the formation of globular zirconia-rich agglomerates throughout the matrix, leading to more improved fretting biocorrosion properties of Ti-6Al-4V.

1. Introduction

Metallic materials account for the widest application in biomedical devices (Jayaraj and Pius, 2018); however, the use of metals also poses some shortcomings and challenges due to their corrosion and degradation in the human body. Body fluids in contact with the implant device act as a corrosive media, which accelerates the degradation process in the presence of mechanical action such as walking, running or jumping (Sivakumar et al., 2018). Tribocorrosion processes result in accelerated material loss due to the combined effect of both wear and corrosion; leading to deterioration, poor performance and ultimate failure of the implant device (Mathew et al., 2009, Cruz et al., 2011). Tribocorrosion effects can thus translate to serious health implications, originating from the potential toxicity of degradation products like metal ions and wear debris when introduced into the human body (Bidhendi and Pouranvari, 2018, Zhang et al., 2018, Matusiewicz, 2014, Leopold et al., 2000).

Austenitic stainless steels, titanium and titanium alloys as well as cobalt-chromium based alloys fall amongst the most common metallic biomaterials due to their excellent corrosion properties and acceptable biocompatibility (Anderson, 2001).

These metals owe their corrosion resistance to the stable oxide layer that forms on the surface, which shields them from most corrosive environments (Cruz et al., 2011). Load-bearing orthopaedic implants often fail due to the breakdown of this protective layer as a result of exposure to high loads and corrosive body fluids (Duisabeau et al., 2004). The breakdown of the oxide layer leads to generation of wear debris and corrosion particles. This interaction between electrochemical and mechanical effects promotes increased material loss (Mathew et al., 2009). Because of this, the goal with most biomaterials used for implants has been to develop materials with excellent biocompatibility and corrosion resistance as well as adequate mechanical properties for the intended function (Gilbert and Mali, 2012). The drawback with titanium and its alloys in load-bearing orthopaedic implant applications is the poor wear resistance (Mindivan, 2016, Obadele et al., 2015a).

Over the years, some attempts have been made into enhancing the properties of titanium by introducing various alloying elements such as Zr, Nb, Ta, and Mo to improve its mechanical and biocompatibility properties (Li et al., 2014, Hua et al., 2017). More recently, advanced manufacturing techniques have made it possible for a broad range of materials like composites, ceramics and new alloys to be investigated for biomedical applications (Bose et al., 2017, He et al., 2015, Buciumeanu et al., 2017, Karanjai et al., 2008). It has been observed that the introduction of inert ceramics like zirconia into the matrices of titanium-based materials through powder metallurgy techniques to produce metal matrix composites, is effective in improving the wear and tribocorrosion properties of titanium-based materials (Obadele et al., 2017). Obadele et al. (Obadele et al., 2015c) studied the wear behaviour of Ti-6Al-4V and composite coatings of laser cladded TiNi with different compositions of zirconia and reported that addition of zirconia produced a more refined microstructure and also led to a considerable improvement in the hardness and wear behaviour of the titanium composite.

This work aims to study the fretting corrosion behaviour of spark plasma sintered Ti6Al4V-ZrO_2 composites in simulated body fluid, for potential use in biomedical applications. The effect of zirconia content on the tribological properties of Ti6Al4V will be established.

2. Materials and method

The materials investigated were spark plasma sintered Ti-6Al-4V, Ti6Al4V+5%ZrO2 and Ti6Al4V+10%ZrO2. The Ti6Al4V and Ti6Al4V composites were cut to 9 × 9 × 15 mm parallelepipeds. Commercially available Ti-6Al-4V was used as the counter material from a forged process. The samples were ground down using 3 µm diamond suspension, thereafter polished with colloidal silica until a mirror surface was achieved. Afterward, the composites were washed with deionized water and dried with an automatic drier.

The fretting corrosion experiments were conducted on a special set-up device used for fretting corrosion investigations (see Figure 1). It is a derivative of the Tribomines® device that has been reported in fretting corrosion investigations by (Geringer et al., 2005, Kim and Geringer, 2012, Geringer et al., 2013). Similar experimental procedure was followed in this work. Linear voltage displacement transducer (LVDT) sensors on the device are in charge of controlling displacement, which exists within a range of ±10-100 µm. The flat and cylindrical samples were fixed onto the device's sample holders by means of insulated metallic screws. Fretting corrosion tests were conducted Foetal bovine serum (FBS), 30g/L of protein, Biowest®). Unlike other simulated body fluids which contain only inorganic species with compositions equal to those in human body fluids, FBS also contains proteins and organic molecules such as serum albumin; which have been found to have a significant effect on the corrosion and tribocorrosion behaviour of a material (Hang et al., 2010) (Geringer et al., 2013). The test was conducted at a frequency of 1 Hz. This was done to simulate the assumed average human gait as reported by Geringer et al., 2005. A normal load F_n of 85 N was used for each composite. A displacement of 0.08 mm sufficient to induce gross slip was imposed on the contact during testing. A relative displacement between the stem and bone of a hip implant is reported to be between 0.02-0.06 mm (Kim and Geringer, 2012). Fretting corrosion experiments were carried out at room temperature and pressure. Each experiment was repeated three times for each composition. Due to the long test durations, and complex test set-up; room temperature was used to ensure reproducibility. Other studies with similar equipment and set-up have reported fretting corrosion tests for medical applications at room temperature (Geringer et al., 2013, Geringer et al., 2005, Kim et al., 2013). A three-electrode configuration was used in order to measure and control the electrochemical measurements of the metal couple; using a Parstat 2263 Potentiostat. A metallic wire was welded onto the sample, which served as the working electrode. For controlling and monitoring the potential, a platinum and saturated calomel electrode (SCE) served as the counter and reference electrodes respectively. Mechanical and electrochemical data acquisition software was employed to record data generated throughout the test such as friction coefficient, tangential force-displacement cycles ($F_T - D$ cycles) and open circuit potential. After each test, the samples were removed from solution and rinsed with deionized water. The wear track area and contact zones were characterized using a non-contact three-dimensional optical profilometry and a scanning electron microscope.

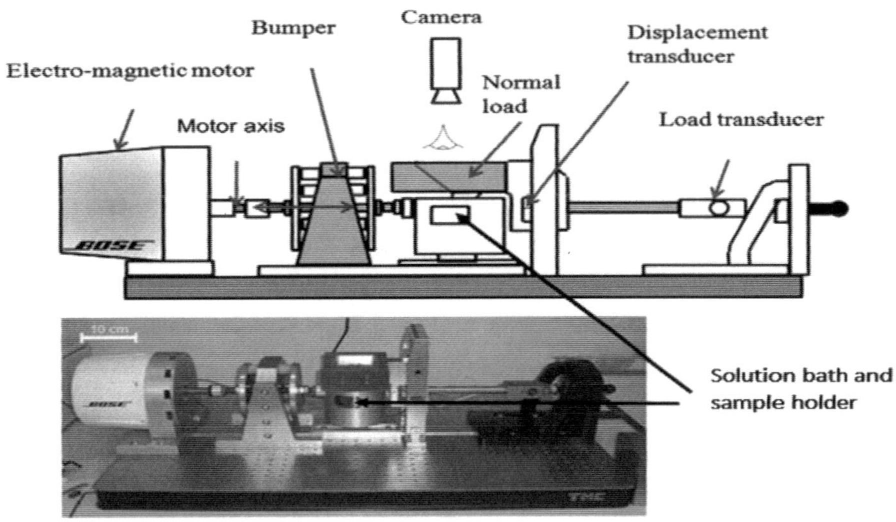

Figure 1: A schematic and image of the fretting corrosion test set-up used in the experiments; reported by (Kim et al., 2013).

3. Results and Discussion

3.1. Microstructural analysis of the as-received samples

The SEM micrographs of the sintered composites without and with ZrO_2 (Ti-Al-4V, Ti-Al-4V /+5/+10%ZrO_2) are shown in Figure 2. The microstructure of Ti-6Al-4V as shown in Figure 2(a) comprises a lamellae structure of alpha and beta (light) phases. The darker is alpha phase while the light is beta phase. This type of microstructures could be related to the alloy fabricated by Obadele et al. (Obadele, 2014, Obadele et al., 2017). This illustrates that unreinforced Ti-6Al-4V alloy sintered at temperatures of 1000 °C and above using spark plasma sintering method, produced a fully lamellae microstructure.

The microstructures shown in Figure 2(b) and (c) are visibly different from those of sintered Ti-6Al-4V (in Figure 2(a)). The presence of zirconia resulted in the formation of agglomerates which are rich in zirconia, as shown on the EDS spectra in Figure 2(d). The presence and amount of the zirconia-rich agglomerates is better illustrated by the smaller OM images taken at a low magnification of 10X. The microstructure of Ti-6Al-4V with 5% ZrO_2 presented a smaller distribution of the zirconia-rich agglomerates, together with some platelet phases. The agglomerates are fewer and not uniformly distributed. Furthermore, the microstructure obtained for Ti-6Al-4V with 10% ZrO_2 showed a higher region rich in zirconia agglomerates within the matrix as a result of the amount of zirconia present in the matrix. In another study by Obadele et al., 2017, SEM micrographs showed flower-like microstructures where the light areas were the zirconia-rich phases.

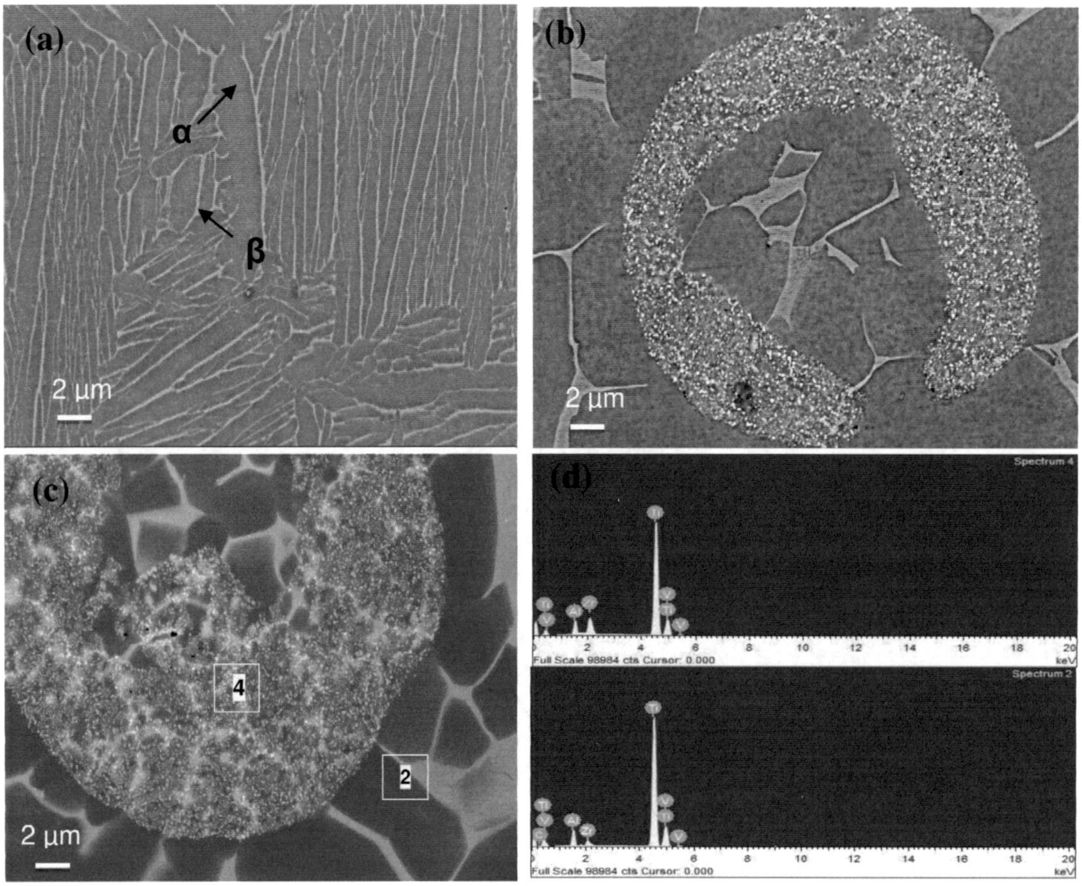

Figure 2. SEM micrographs of the as-received (a) Ti-6Al-4V, (b) Ti-6Al-4V+5%ZrO_2 and (c) Ti-6Al-4V+10%ZrO_2; (d) EDS

3.2. Open circuit potential before, during and after fretting Biocorrosion

The OCP evolution of Ti-6Al-4V without and with 5% and 10% ZrO_2 in foetal bovine serum (FBS) with no applied load and friction is given in Figure 3 (a). The most stable potential is displayed by Ti-6Al-4V with 5% ZrO_2 addition without applied load. The graph showing OCP evolution when fretting corrosion is taking place is given in Figure 3 (b). Generally, open circuit reading during fretting corrosion tests comprises three distinct zones, namely; the initial stabilization stage, fretting region and final stabilization stage. The OCP before, during and after fretting in FBS is presented in Figure 3 (b) for Ti-6Al-4V without and with 5% and 10% ZrO_2 under an applied load of 85 N respectively. It could be seen that stable potentials of -0.5, -0.2 and -0.2 V were recorded for Ti-6Al-4V, 5% and 10% ZrO_2 addition respectively before fretting. At the onset of fretting, there is a significant drop in corrosion potential towards the negative values. The OCP values obtained are -1.0 ±0.03, -0.7 ±0.03 and -0.8 ±0.03 V for Ti-6Al-4V with 5% and 10% ZrO_2 additions. Nevertheless, the highest OCP drop is obtained for sintered Ti-6Al-4V with Ti-6Al-4V counterface, suggesting a higher tendency to metallic dissolution under friction conditions in FBS. This could be due to disruption of the naturally formed oxide layer exposing a fresh underlying metal surface to the electrochemical solutions (Gilbert and Mali, 2012). The low OCP values during fretting are an illustration of the electrochemical activity taking place between the underlying material and solution where the passive film has been destroyed.

The fluctuations observed during fretting indicate repassivation and depassivation, a phenomenon that results in continuous destruction and formation of the metal's surface oxide (McTighe et al., 2015). According to Ponthiaux et al. (Ponthiaux et al., 2004), the OCP recorded during fretting is usually a mixed potential emanating from the worn and unworn areas of the materials in contact, particularly in case whereby the material being tested and the counterface material are the same. When the OCP values of the sintered composites are compared with respect to the ZrO_2 content, Ti-6Al-4V with 10 wt.% ZrO_2 shows the lowest OCP decline, indicating the most stable passive film formation and high resistance to plastic deformation. In contrast, the composite having 5 wt.% ZrO_2 is showing less stability during the combined effect of wear and corrosion, as opposed to when there was no load applied. As a conclusion, the ZrO_2 content has an influence in improving the OCP behaviour. Increasing zirconia from 5 wt.% to 10 wt.% in Ti-6Al-4V increases the stability of the passive layer. This observation was reported by Obadele et al., 2015b, where TiNi+10%ZrO_2 were found to have better tribocorrosion response in acidic media compared to TiNi+5%ZrO_2 under similar conditions.

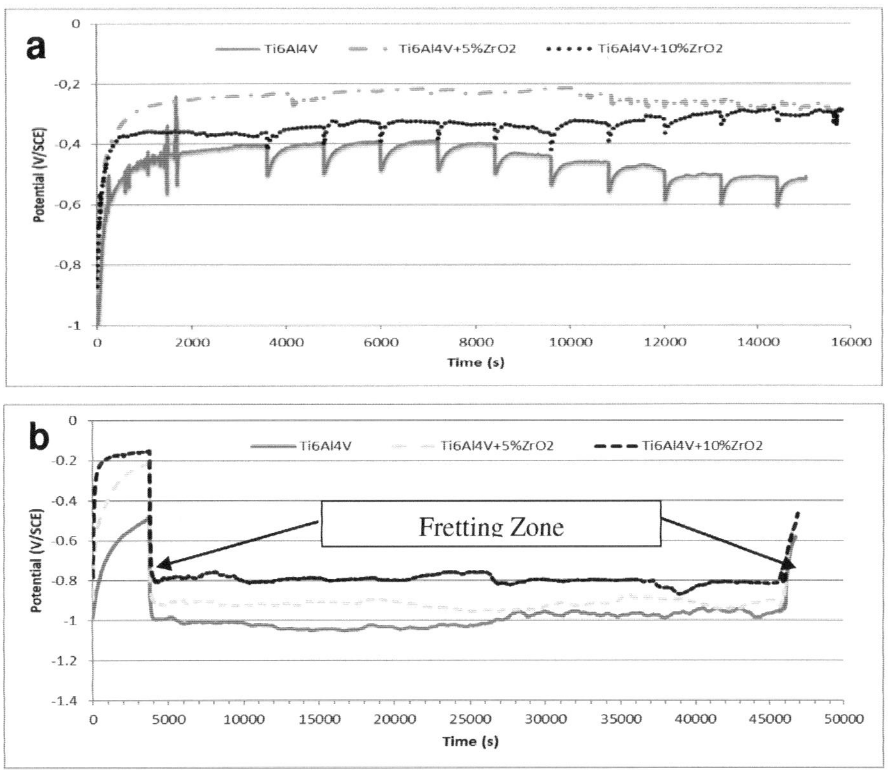

Figure 3. OCP evolution with (a) without an applied load and (b) at applied load of 85 N for Ti-6Al-4V, Ti-6Al-4V+5/+10% ZrO_2 sintered composites.

Generally, a decrease in the electrode potential of a material is indicative of an increased tendency of the material surface to corrosion (Mathew et al., 2011). Mischler stated that the measured potential during the fretting regime is a reflection of the galvanic coupling between the worn areas (underlying metal exposed to solution by destruction of the passive film) and unworn areas (passive metal) (Mischler, 2008). The open circuit potential therefore serves as a method to monitor the corrosion response of a material under different test conditions, and to assess the stability of the oxide film under such conditions. From comparison, Ti-6Al-4V with 10% ZrO_2 showed the best fretting corrosion stability due to the higher potential values maintained throughout the test duration. This could be as a result of higher ZrO_2 content

IOP Conf. Series: Materials Science and Engineering **655** (2019) 012034 doi:10.1088/1757-899X/655/1/012034

within the matrix of this material, which produced a refined microstructure with a high distribution of ZrO_2 rich agglomerates.

3.3. Fretting loops during biocorrosion

The fretting regimes in this study were obtained from the plots of tangential force versus displacement loops obtained during fretting corrosion experiments at 85 N, and these are given in Figure 4. It could be seen that the tests show a gross slip regime for all compositions. Additionally, the development of sharp peaks on the curve can be observed for Ti-6Al-4V with 5% ZrO_2 addition in Figure 4(b) and (c). This effect has been reported to occur due to the ploughing effect or the interlocking of geometric features in the contact surfaces. Flatter parallelogram shapes have been found to produce low coefficient of friction (COF) values (Tobi et al., 2017). Fretting mechanisms maps have demonstrated that low loads together with high displacements lead to gross slip conditions which is associated with wear, whereas high loads with low displacements promote partial slip conditions (Tobi et al., 2017). These observations are in agreement with the current study in terms of the resulting fretting loops, even though the experiments were conducted under the same displacement.

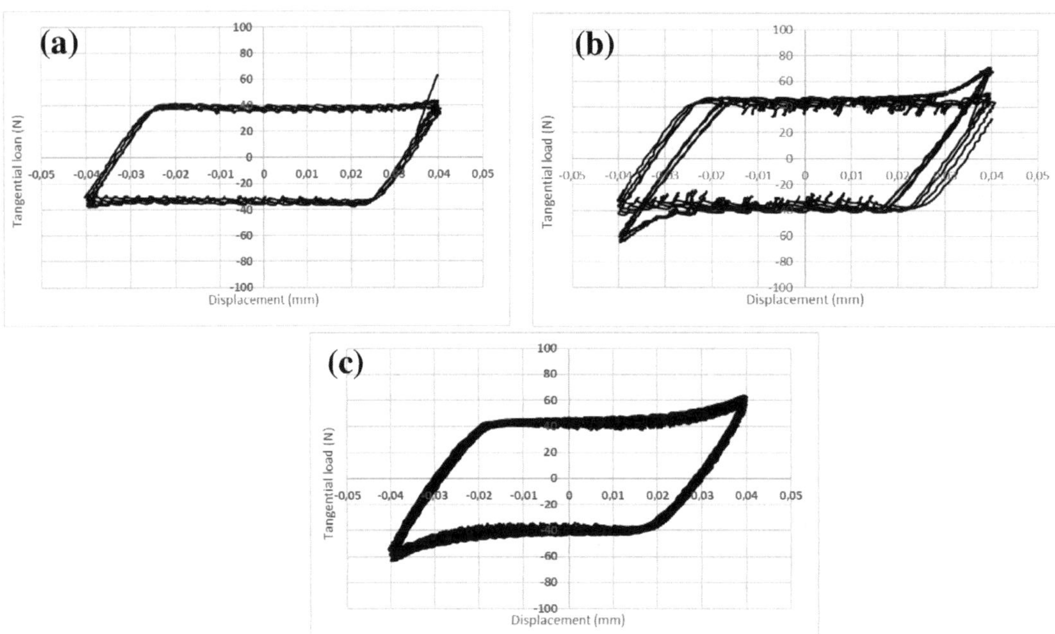

Figure 4. Fretting loops based on tangential force versus displacement curves established during a fretting corrosion tests for (a) Ti-6Al-4V, (b) Ti-6Al-4V+5% ZrO_2, and (c) Ti-6Al-4V+10% ZrO_2 at 85 N.

3.4. Friction coefficient during fretting

Friction coefficient evolution with time during fretting corrosion experiments for Ti-6Al-4V with 5 and 10% ZrO_2 are shown in Figure 5 at an applied load of 85 N and test duration of 16 h. Within the first few seconds of the test, the COF increases from about 0.4 and 0.6 to 0.8 and 0.9, where it stabilises for some time and decreases slightly towards the end of the experiment. The initial increase in COF at the beginning of the test is referred to as the running-in-period and has been reported in literature (Kim et al., 2013, Ding et al., 2018). Johnson (Johnson, 1995) defines the 'running-in' stage as the change in the surface profile that occurs on a smooth or rough surface during repeated sliding. During this running-in period, the COF increases due high friction resulting from surface asperities that may be present on the surfaces of contacting materials in the initial stages of fretting (Kim et al., 2013).

IOP Conf. Series: Materials Science and Engineering **655** (2019) 012034 doi:10.1088/1757-899X/655/1/012034

At around 20000 s and 35000 s in Figure 5(a) and 5(b), fluctuations could be observed for Ti-6Al-4V and 5% ZrO_2 respectively; it is also around this time where the curves tends to decrease slightly. This fluctuations could be as a result of hard ceramic particles of ZrO_2 that may be present on the surfaces of the fretting bodies (Obadele et al., 2015a). Mindivan (Mindivan, 2016) reported a similar behaviour where the fluctuations were attributed to the abrasive effect of the generated wear particles, originating from the ceramic counterface. After the initial running-in, and stabilization of COF, a decrease in COF values could be observed over time for all compositions, to values close to 0.4 at the end of the test. These observations correspond to results obtained by Geringer and Macdonald (Geringer and Macdonald, 2014) were it was observed that during fretting corrosion tests, the friction coefficient of a material couple decreased over time. That is, friction coefficient values were lower at the end of experiments which had a higher number of cycles. This behaviour was attributed to reduced friction due to smoother surfaces forming over long duration experiments. Looking at all OCP curves, it can be observed that there are less fluctuations for Ti-6Al-4V+10% ZrO_2 indicating the stability of this composition under friction.

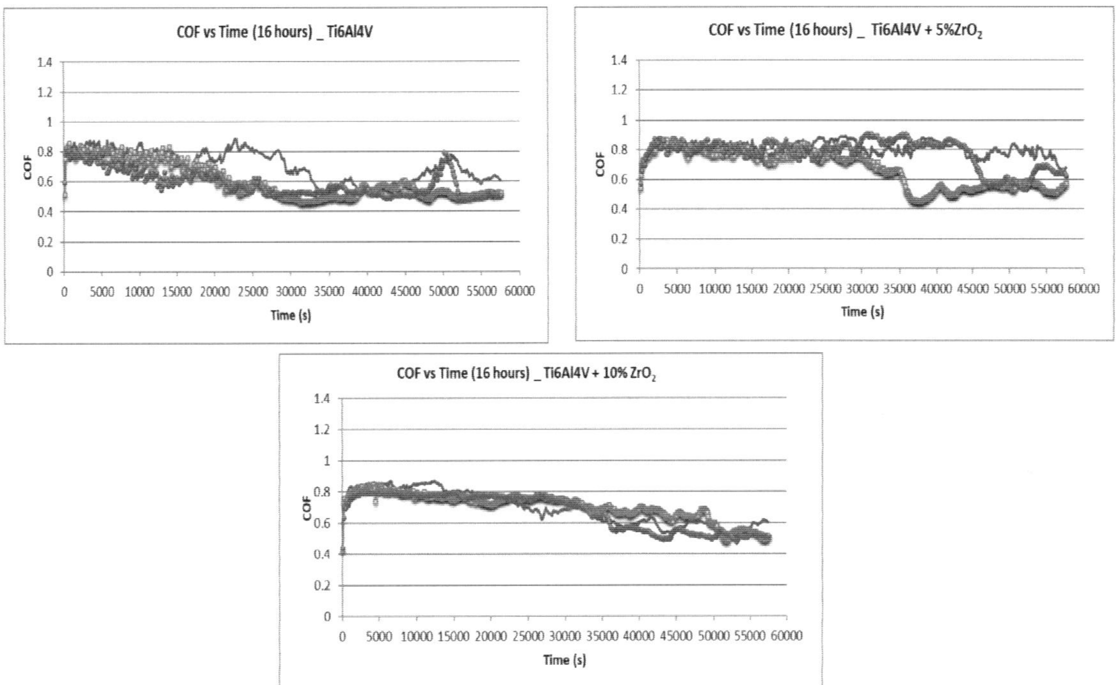

Figure 5. Friction coefficient for (a) Ti-6Al-4V, (b) Ti-6Al-4V+5% ZrO_2, and (c) Ti-6Al-4V+10% ZrO_2 under an applied load of 85 N.

4. Conclusion

The fretting corrosion behaviour of Ti-6Al-4V without and with 5 wt.% and 10 wt.% ZrO_2 were investigated in foetal bovine serum at an applied load of 85 N. Open circuit potential was measured before, during and after fretting and the corresponding friction coefficient was reported. The results obtained showed that higher amount of zirconia (10 wt.%) had a significant effect in improving the overall fretting biocorrosion behaviour of Ti-6Al-4V. This was attributed to the refined microstructure obtained, having uniformly distributed zirconia-rich agglomerates throughout the matrix. Ti-6Al-4V+ 10% ZrO_2 composite exhibited the best tribological properties and better corrosion resistance during fretting, indicating that a higher content of zirconia is more effective in improving the tribological and electrochemical properties of Ti-6Al-4V.

5. Acknowledgement

The authors acknowledge the Centre for Nanoengineering and Tribocorrosion at the University of Johannesburg and the National Research Foundation for financial support. Special thanks to École des Mines de Saint-Étienne in France and Centre Ingénierie et Santé (CIS) laboratory for the training and use of facilities for fretting corrosion experiments.

References

[1] ANDERSON, J. M. 2001. Biological responses to materials. *Annual review of materials research,* 31**,** 81-110.

[2] BIDHENDI, H. R. A. & POURANVARI, M. 2018. Corrosion study of metallic biomaterials in simulated body fluid. *Metallurgical and Materials Engineering*.

[3] BOSE, S., KE, D., SAHASRABUDHE, H. & BANDYOPADHYAY, A. 2017. Additive manufacturing of biomaterials. *Progress in Materials Science*.

[4] BUCIUMEANU, M., ARAUJO, A., CARVALHO, O., MIRANDA, G., SOUZA, J., SILVA, F. & HENRIQUES, B. 2017. Study of the tribocorrosion behaviour of Ti6Al4V–HA biocomposites. *Tribology International,* 107**,** 77-84.

[5] CRUZ, H., SOUZA, J., HENRIQUES, M. & ROCHA, L. 2011. Tribocorrosion and bio-tribocorrosion in the oral environment: the case of dental implants. *Biomedical tribology***,** 1-33.

[6] DING, H., FRIDRICI, V., GERINGER, J., FONTAINE, J. & KAPSA, P. 2018. Influence of diamond-like carbon coatings and roughness on fretting behaviors of Ti–6Al–4V for neck adapter–femoral stem contact. *Wear,* 406**,** 53-67.

[7] DUISABEAU, L., COMBRADE, P. & FOREST, B. 2004. Environmental effect on fretting of metallic materials for orthopaedic implants. *Wear,* 256**,** 805-816.

[8] FERNÁNDEZ-GARCÍA, E., GUTIERREZ-GONZALEZ, C., FERNANDEZ, A., TORRECILLAS, R. & LÓPEZ-ESTEBAN, S. 2013. Processing and spark plasma sintering of zirconia/titanium cermets. *Ceramics International,* 39**,** 6931-6936.

[9] GERINGER, J., FOREST, B. & COMBRADE, P. 2005. Fretting-corrosion of materials used as orthopaedic implants. *Wear,* 259**,** 943-951.

[10] GERINGER, J. & MACDONALD, D. D. 2014. Friction/fretting-corrosion mechanisms: Current trends and outlooks for implants. *Materials Letters,* 134**,** 152-157.

[11] GERINGER, J., PELLIER, J., TAYLOR, M. L. & MACDONALD, D. D. 2013. Fretting corrosion with proteins: The role of organic coating on the synergistic mechanisms. *Thin solid films,* 528**,** 123-129.

[12] GILBERT, J. L. & MALI, S. A. 2012. Medical implant corrosion: electrochemistry at metallic biomaterial surfaces. *Degradation of implant materials.* New York: Springer.

[13] HANG, R., MA, S., JI, V. & CHU, P. K. 2010. Corrosion behavior of NiTi alloy in fetal bovine serum. *Electrochimica Acta,* 55**,** 5551-5560.

[14] HE, D., ZHENG, S., PU, J., ZHANG, G. & HU, L. 2015. Improving tribological properties of titanium alloys by combining laser surface texturing and diamond-like carbon film. *Tribology international,* 82**,** 20-27.

[15] HUA, N., CHEN, W., ZHANG, L., LI, G., LIAO, Z. & LIN, Y. 2017. Mechanical properties and bio-tribological behaviors of novel beta-Zr-type Zr-Al-Fe-Nb alloys for biomedical applications. *Materials Science and Engineering: A,* 76**,** 1154-1165.

[16] IVKOVIC, B., DJURDJANOVIC, M. & STAMENKOVIC, D. 2000. The influence of the contact surface roughness on the static friction coefficient. *Tribology in industry,* 22**,** 41-44.

[17] JAYARAJ, K. & PIUS, A. 2018. Biocompatible coatings for metallic biomaterials. *Fundamental Biomaterials: Metals.* Elsevier.

[18] JOHNSON, K. L. 1995. Contact mechanics and the wear of metals. *Wear,* 190**,** 162-170.

[19] KARANJAI, M., SUNDARESAN, R., MOHAN, T. R. R. & KASHYAP, B. P. 2008. Evaluation of growth of calcium phosphate ceramics on sintered Ti–Ca–P composites. *Materials Science and Engineering: C,* 28**,** 1401-1407.

[20] KIM, K. & GERINGER, J. 2012. Analysis of energy dissipation in fretting corrosion experiments with materials used as hip prosthesis. *Wear,* 296**,** 497-503.

[21] KIM, K., GERINGER, J., PELLIER, J. & MACDONALD, D. D. 2013. Fretting corrosion damage of total hip prosthesis: Friction coefficient and damage rate constant approach. *Tribology International,* 60**,** 10-18.

[22] LEOPOLD, S. S., BERGER, R. A., PATTERSON, L., SKIPOR, A. K., URBAN, R. M. & JACOBS, J. J. 2000. Serum titanium level for diagnosis of a failed, metal-backed patellar component. *The Journal of arthroplasty,* 15**,** 938-943.

[23] LI, Y., YANG, C., ZHAO, H., QU, S., LI, X. & LI, Y. 2014. New developments of Ti-based alloys for biomedical applications. *Materials,* 7**,** 1709-1800.

[24] MATHEW, M., SRINIVASA PAI, P., POURZAL, R., FISCHER, A. & WIMMER, M. 2009. Significance of tribocorrosion in biomedical applications: overview and current status. *Advances in tribology,* 2009.

[25] MATHEW, M. T., UTH, T., HALLAB, N. J., POURZAL, R., FISCHER, A. & WIMMER, M. 2011. Construction of a tribocorrosion test apparatus for the hip joint: Validation, test methodology and analysis. *Wear,* 271**,** 2651-2659.

[26] MATUSIEWICZ, H. 2014. Potential release of in vivo trace metals from metallic medical implants in the human body: from ions to nanoparticles–a systematic analytical review. *Acta biomaterialia,* 10**,** 2379-2403.

[27] MCTIGHE, T., BRAZIL, D., KEPPLER, L., KEGGI, J. & MCPHERSON, E. 2015. Metallic Modular Taper Junctions in Total Hip Arthroplasty. *Reconstructive Review,* 5.

[28] MINDIVAN 2016. Comparative Study of Tribocorrosion Properties of Some Bio-Based Materials

[29] in Simulated Artificial Saliva. *Machines, Technologies, Materials***,** 58-60.

[30] MISCHLER, S. 2008. Triboelectrochemical techniques and interpretation methods in tribocorrosion: A comparative evaluation. *Tribology International,* 41**,** 573-583.

[31] NACHANE, R. P., HUSSAIN, G. F. S. & IYER, K. R. 1998. Theory of stick-slip effect in friction.

[32] OBADELE, B., ANDREWS, A., OLUBAMBI, P., MATHEW, M. & PITYANA, S. 2015a. Tribocorrosion behaviour of laser cladded biomedical grade titanium alloy. *Materials and Corrosion,* 66**,** 1133-1139.

[33] OBADELE, B. A. 2014. Tribocorrosion mechanisms in laser deposited titanium-based smart composite coatings.

[34] OBADELE, B. A., ANDREWS, A., MATHEW, M. T., OLUBAMBI, P. A. & PITYANA, S. 2015b. Improving the tribocorrosion resistance of Ti6Al4V surface by laser surface cladding with TiNiZrO2 composite coating. *Applied Surface Science,* 345**,** 99-108.

[35] OBADELE, B. A., ANDREWS, A., OLUBAMBI, P. A., MATHEW, M. T. & PITYANA, S. 2015c. Effect of ZrO2 addition on the dry sliding wear behavior of laser clad Ti6Al4V alloy. *Wear,* 328–329**,** 295-300.

[36] OBADELE, B. A., IGE, O. O. & OLUBAMBI, P. A. 2017. Fabrication and characterization of titanium-nickel-zirconia matrix composites prepared by spark plasma sintering. *Journal of Alloys and Compounds,* 710**,** 825-830.

[37] PONTHIAUX, P., WENGER, F., DREES, D. & CELIS, J.-P. 2004. Electrochemical techniques for studying tribocorrosion processes. *Wear,* 256**,** 459-468.

[38] ROYHMAN, D., PATEL, M., RUNA, M., JACOBS, J., HALLAB, N., WIMMER, M. & MATHEW, M. 2015. Fretting-corrosion in hip implant modular junctions: new experimental set-up and initial outcome. *Tribology international,* 91**,** 235-245.

[39] SIVAKUMAR, B., PATHAK, L. C. & SINGH, R. 2018. Fretting corrosion response of boride coated titanium in Ringer's solution for bio-implant use: Elucidation of degradation mechanism. *Tribology International.*

[40] TOBI, A. M., SUN, W. & SHIPWAY, P. 2017. Evolution of plasticity-based wear damage in gross sliding fretting of a Ti-6Al-4V non-conforming contact. *Tribology International,* 113**,** 474-486.

[41] ZHANG, Y., XIU, P., JIA, Z., ZHANG, T., YIN, C., CHENG, Y., CAI, H., ZHANG, K., SONG, C. & LENG, H. 2018. Effect of vanadium released from micro-arc oxidized porous

Ti6Al4V on biocompatibility in orthopedic applications. *Colloids and Surfaces B: Biointerfaces,* 169**,** 366-374.

Prospects of graphite - polypropylene/epoxy blend composite for high performance bipolar plate in polymer electrolyte membrane fuel cell

O A Alo[1,2], I O Otunniyi[1] and HCVZ Pienaar[2]

[1] Department of Metallurgical Engineering, Vaal University of Technology, Vanderbijlpark, South Africa

[2] Center for Alternative Energy, Vaal University of Technology, Vanderbijlpark, South Africa

E-mail: 216074185@edu.vut.ac.za

Abstract. Conductive polymer composites (CPCs) are very promising candidate materials for bipolar plates (BPs) in polymer electrolyte membrane fuel cells (PEMFCs). However, a major challenge facing application of CPCs in PEMFC BPs is the difficulty in achieving high electrical conductivity while maintaining adequate mechanical strength. Therefore, a good balance between electrical conductivity and mechanical strength is critical in the development of high-performance CPC BPs. In this study, CPCs consisting of polypropylene (PP)/epoxy blend filled with graphite were investigated for application as BP material, which will combine electrical conductivity and mechanical strength that meet the performance requirements for PEMFC BPs. The CPCs, with 30 – 80 wt% graphite powder, were produced by melt mixing followed by compression molding. The morphology, in-plane electrical conductivity, and flexural strength of the composites were investigated using the scanning electron microscopy, four-point probe method, and three-point bending test, respectively. The results obtained were compared with the technical targets for PEMFC BPs by the United States Department of Energy (DOE). All the PP/epoxy/graphite composites exhibited flexural strength that satisfies the DOE target of > 25 MPa while the composite with 80 wt% graphite content showed the best in-plane electrical conductivity.

1. Introduction

Proton exchange membrane fuel cells (PEMFCs) have a huge potential for addressing the world's energy needs and the environmental issues associated with the current fossil-dependent energy systems. Out of the currently available fuel cell technologies, the PEMFC is preferable, particularly for transportation and other mobile applications, due to its beneficial features such as convenient fuel supply, high energy efficiency, desirable power density, low temperature operation, and long lifetime [1-2]. The bipolar plate (BP) is a key component of the PEMFC, and its functionality and performance greatly impact on the overall performance and durability of the PEMFC stack. Key among the required properties for the BP to perform its functions effectively is good electrical performance.

For PEMFC BP application, conductive polymer composites (CPCs) have the potential to outperform other material options such as graphite and metals. This is due to their attractive properties such as low cost, good corrosion resistance, low density, and ease of forming gas flow channels [3-4]. However, a major drawback of CPC BPs is their low electrical conductivity, and the conductive filler loading in most commonly used CPCs is below 10 vol% [5], which is usually not enough to impart sufficient electrical conductivity for BP application. Despite their generally low conductivity, CPCs with remarkable electrical conductivity values have been reported in the literature [5-10]. However, the conductivities of several of the developed CPCs are still lower than the target set by the United States Department of Energy (DOE). In cases where the DOE target has been met, high filler content (70 – 80 wt%) was required [5, 11]. At such high filler concentrations, issues associated with difficulty in processing as a result of increased melt viscosity set in. This leads to degradation of the mechanical strength of the produced composites. To address this problem, some authors have employed the use of different filler combinations with the goal of achieving a good balance between electrical conductivity and strength [6, 10, 12]. However, the carbon black, carbon fiber, carbon nanotubes, and graphene, which are usually used as secondary fillers in addition to graphite in such multi-filler CPCs, often pose problems such as agglomeration and non-uniform dispersion. This results in detrimental effects on the electrical and mechanical properties of the composites [13]. Another possible approach to achieve CPC BPs with a good electrical conductivity and strength balance is the use of polymer matrices with inherently high strengths since they have the potential to yield composites with high electrical conductivity and satisfactory mechanical strength even at high filler loadings.

Epoxy resins are attractive matrix materials for producing CPC BPs due to their excellent mechanical properties and chemical resistance. However, they are brittle and extremely prone to impact-induced failure due to their highly crosslinked structure. If applied in PEMFC BPs, especially in transportation applications, cracking of the plates and leaking of reactant gases may occur as a result of poor resistance to mechanical shocks and vibrations. Polypropylene (PP) has lower mechanical strength compared to epoxy but it has good toughness and impact resistance [14]. A blend of PP and epoxy have been reported to exhibit improved combination of strength, toughness, and impact resistance [15]. Therefore, the potential exists for obtaining CPC PEMFC BPs with electrical conductivity and mechanical strength that meet the DOE targets, using PP/epoxy as the matrix. In this study, the prospect of PP/epoxy/graphite composite to yield high performance and durable PEMFC BPs was investigated. PP/epoxy blends filled with varying contents of graphite were produced by melt mixing and followed by compression moulding. After various characterization steps, the performances of the produced composites were compared to the DOE targets.

2. Materials and Methods

2.1. Materials

The PP (density 0.9 g/cm^3 and melt index 4) and maleic anhydride-grafted polypropylene (MAH-g-PP) used in this study were purchased from Sigma-Aldrich. Due to immiscibility of PP and epoxy, the MAH-g-PP (with melting point of 156 °C, density of 0.934 g/cm^3 at 25 °C, and MAH content of 8-10 wt%) was used as a compatibilizer. The bisphenol-A epoxy resin (LR 30) and the amine-type hardener (LH 30) were purchased from AMT Composites, South Africa. The densities of the epoxy resin and harderner are 1.13 g/cm^3 (at 25 °C) and 0.94 g/cm^3 (at 20 °C), respectively. The graphite powder used was donated by Graftech South Africa (Pty) Ltd.

2.2. Composite Preparation

Before mixing, the PP and MAH-g-PP were dried under vacuum at 70 °C for 8 hours to remove any moisture that may be present. Different PP/epoxy/graphite composite samples with 30-80 wt% graphite were then prepared by melt mixing in a Haake Rheomix OS mixer at 190 °C and 60 rpm. The graphite powder, PP and MAH-g-PP were first manually mixed. The mixture was then transferred into the Rheomixer and mixed for

3 minutes. The epoxy resin was then added, and mixing was continued for another 4 minutes. The hardener was then added, and the mixing was continued. An epoxy-hardener weight ratio of 5:1 was used as recommended by the supplier. Also, epoxy-MAH-g-PP weight ratio of 3:1 was used for all the prepared composites. The total mixing time was 10 minutes. After mixing, the samples were compression moulded at 200 °C at 2 metric tonnes for 10 minutes.

2.3. *Characterizations*

The morphology of the graphite powder and the prepared composites was observed using a TESCAN VEGA3 scanning electron microscope (SEM). To analyze the morphology of the composites clearly, prior to SEM observation, some freeze-fractured samples were etched in dichloromethane for 30 min to remove the epoxy phase on the surfaces. Composite samples with 37 mm × 13 mm × 1.7 mm dimensions were used for the in-plane electrical conductivity measurements by means of the four-point probe method according to ASTM D4496. The electrical resistance (R) was measured by passing a constant current through the sample and measuring the voltage drop. The electrical resistivity (ρ) was calculated using equation (1):

$$\rho = \frac{\pi}{ln2} \times t \times R \times f_1 f_2 \quad (1)$$

Where f_1 and f_2 are finite thickness and width correction factors, respectively, and t is the thickness of the sample. Based on the dimensions of the test specimens, the values f_1 and f_2 were obtained from data provided in literature [16-17]. Values of the in-plane electrical conductivity (σ) were obtained from the reciprocal of the resistivity values. The flexural properties of the composite samples were measured at room temperature by the three-point bending test using the Instron 5966 universal testing machining according to ASTM D790-03. The flexural specimens with dimensions 100 mm × 12.7 mm × 1.76 mm were cut from the compression moulded plaques using a thin diamond cutting disc. The support span was 30 mm, and the crosshead speed was 1 mm/min.

3. Results and discussion

3.1. *Microstructure of the PP/epoxy/graphite composites*

The morphology of the graphite filler used in the study shows a layered structure (figure 1). This type of structure has the potentials for achieving higher conductivity-CPCs compared to graphite particles with granular structure [18]. However, the larger surface area can result into lower degree of filling and processability. Figure 2 shows the SEM micrographs of the fractured surfaces of the neat polymer blend and composite. The microstructure of the PP/epoxy blend, as shown in figure 2a, shows a co-continuous morphology. Possible morphologies in binary immiscible polymer blends include lamellar structure, matrix-dispersed particle structure, matrix-fiber structure, and co-continuous structure. The formation of co-continuous structure is favoured by a symmetric or close to symmetric blend composition [19]. Therefore, the co-continuous morphology observed for the PP/epoxy blend can be attributed to the close to symmetric composition of the blend.

Figure 2b shows the morphology of the PP/epoxy/80 wt% graphite composite. At such high filler loading, the microstructure of the composite is largely influenced by the morphology of the graphite filler. This is as a result of the larger concentration and volume fraction of the graphite filler compared to the polymer matrix. The composition of such highly filled composite is more likely to represent that of graphite particles with a polymer binder rather than that of graphite filler dispersed in a polymer matrix. Therefore, the morphology of the PP/epoxy/80 wt% graphite composite revealed several micrometers size flakes of graphite protruding from the polymer matrix. The graphite flakes are present throughout, and no large, noticeable distinct polymeric features can be seen. Also, the SEM image (figure 2c) revealed presence of some voids (red arrows) within the material. This can be due to insufficient amount of resin to completely wet the graphite flakes.

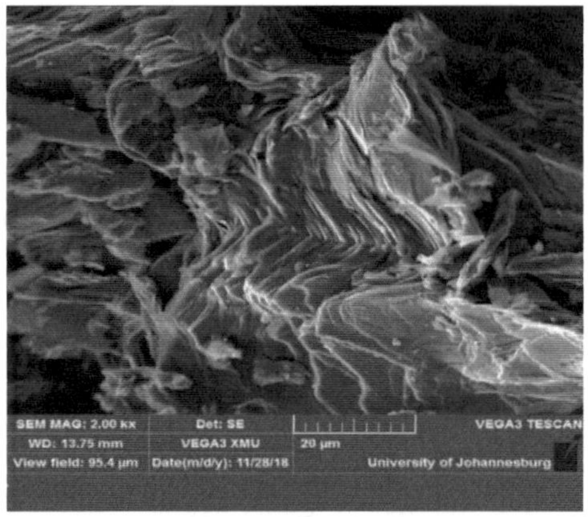

Figure 1. Morphology of the graphite filler used in the study.

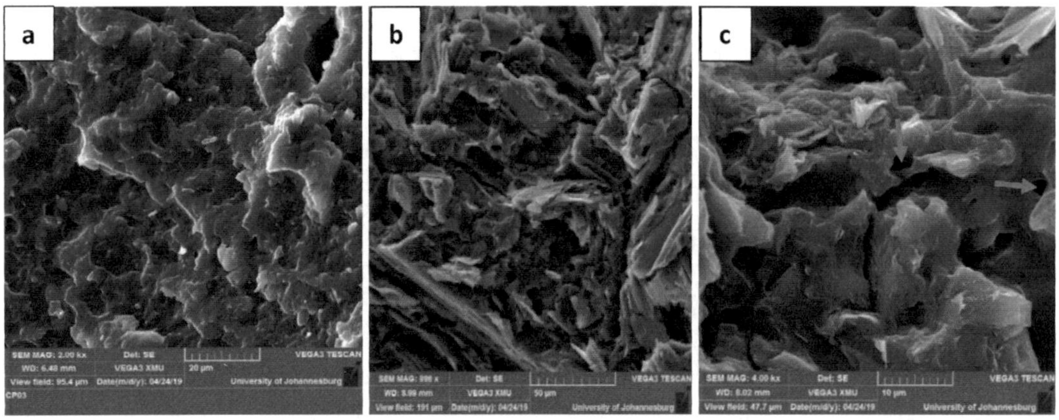

Figure 2. SEM micrograph of (a) neat PP/epoxy blend, (b) PP/epoxy/80 wt% graphite composite, and (c) PP/epoxy/80 wt% graphite composite at higher magnification showing presence of voids (red arrows).

3.2. Electrical properties of the PP/epoxy/graphite composites

The variation of in-plane electrical conductivity of the composite samples with graphite contents is shown in figure 3. The in-plane electrical conductivity of the composites increased as the graphite content increased from 30 to 80 wt%. The electrical conductivity of polymer/graphite composites depends on the electrical conductivity and concentration of the graphite filler and can be explained by the percolation theory. According to classical percolation theory, the electrical conductivity (σ) of polymer/graphite composites above the percolation threshold ($\emptyset_c$) is related to the filler concentration according to the following equation [20]:

$$\sigma = \sigma_0(\emptyset - \emptyset_c)^t \qquad (2)$$

Where σ_0 is the electrical conductivity of the filler, $\emptyset$ is the weight fraction of the filler, and t is a critical exponent related to the dimension of the filler. Therefore, at $\emptyset > \emptyset_c$, electrical conductivity of the composite will increase with increase in filler content. Also, The electrical conductivity of CPCs depends on mutual contact of the conductive filler materials to generate conductive paths throughout the composites [21-22]. Therefore, increasing the graphite filler content would result in more electrically conductive paths for flow of current carrier within the composites. The composite with 80 wt% graphite exhibited the highest in-plane electrical conductivity (68.03 Scm^{-1}) as a result of having the highest graphite content for formation of conducting networks. Although, the conductivity values obtained in this study are still below the DOE target of >100 Scm^{-1}, the range of in-plane electrical conductivity obtained in this study is higher than the values reported for several filled single-polymer composites in previous studies [23-24]. The relatively high conductivities obtained for the PP/epoxy/graphite composites can be attributed to the formation of a co-continuous morphology, which would favor the dispersion of the graphite filler in one of the polymer phases and promote formation of more continuous electrical conducting networks.

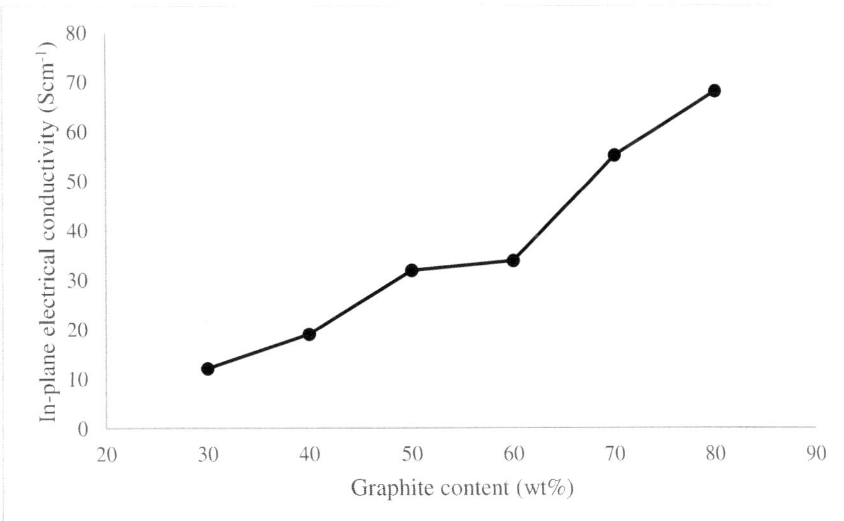

Figure 3. Variation of in-plane electrical conductivity of the composites with graphite content.

3.3. *Flexural properties of the PP/epoxy/graphite composites*

The flexural test results for the neat epoxy, neat PP/epoxy blend, epoxy/50 wt% graphite composite, PP/50 wt% graphite composite, and PP/epoxy/50 wt% graphite composite are shown in figure 4. The PP/epoxy blend exhibited a flexural strength of 58.43 MPa, which is lower than that of the neat epoxy resin with flexural strength of 67.11 MPa. This can be attributed to the inherent lower strength of PP compared to cured epoxy resin. However, since the flexural strength of the PP/epoxy blend is still much higher than the DOE target (>25 MPa), the blend is a promising matrix material for CPC BP for PEMFC. Also, the epoxy/50 wt% graphite composite exhibited a higher flexural strength of 61.78 MPa compared to 39.27 MPa for the PP/50 wt% graphite composite. This can be due to the effects of the lower strength of PP as well as the lower viscosity of the epoxy resin during processing at the elevated temperature, compared to PP, which would result into better wetting of the filler and improve the filler-matrix bonding strength [25].

The PP/epoxy/50 wt% graphite composite exhibited flexural strength of 50.43 MPa, which is lower than that of epoxy/50 wt% graphite composite but is approximately 22% higher than that of the PP/50 wt% graphite composite. For immiscible polymer blend such as PP/epoxy, low strength would be expected due to high interfacial tension. However, the compatibilizing effect of the MAH-g-PP led to good flexural strength in the PP/epoxy blend. The interaction between the carbonyl groups of the maleic anhydride (MAH) in the MAH-g-PP and the hydroxyl groups of the epoxy resin enables the penetration of the MAH-g-PP into the polymer blend constituents. The interfacial adhesion that results from the entangled compatibilizer between the constitutive chains within the blend would offer effective stress transfer from one polymer phase to another and hinders crack initiation at the interface. Therefore, CPC BPs for PEMFC with better mechanical performance can be achieved by using MAH-g-PP compatibilized PP/epoxy blend as the matrix compared to using PP as the matrix.

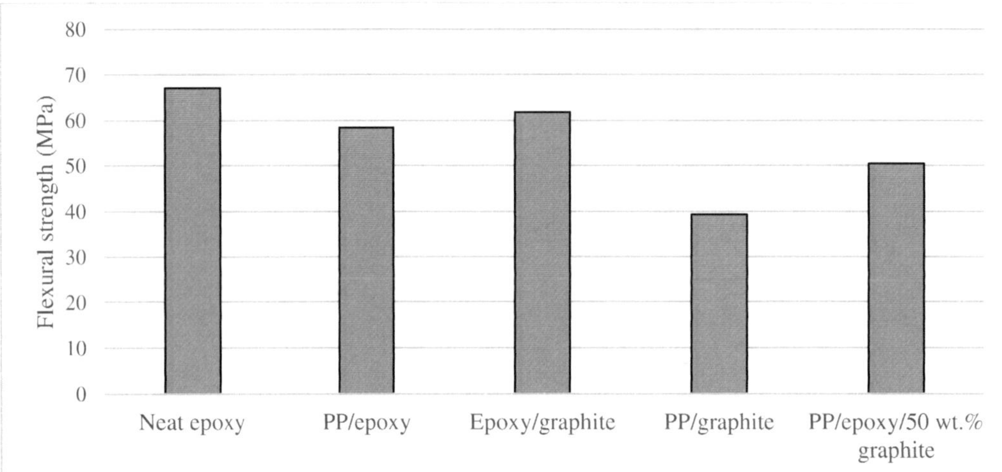

Figure 4. Flexural strength of the neat epoxy resin and polymer blend filled single polymers, and 50 wt% graphite filled polymer blend.

High mechanical strength is required by BPs to be able to provide rigidity to support the membrane electrode assemblies within the PEMFC and withstand the clamping forces within the stack assembly. The effect of graphite content on the flexural properties of the PP/epoxy/graphite composites is shown in figure 5. The flexural strength of the composites increased with increasing graphite content from 46.5 MPa at 30 wt% graphite to a maximum value of 54.36 MPa at 70 wt% graphite. Increase in the graphite content from 70 to 80 wt% led to a decrease in the flexural strength of the composite to 40.16 MPa. This can be due to reduced filler-matrix adhesion and presence of voids (figure 2c) as a result of insufficient amount of resin to wet and bind the graphite powders within the composite. Voids act as potential points of failure within composite materials and their detrimental effect on the mechanical strength of the material becomes significant when their content reaches a certain level [25]. All the PP/epoxy/graphite composites exhibited good flexural strength even at high filler loadings. This can be attributed to the good interfacial bonding between the polymer phases as a result of the compatibilization effect of the MAH-g-PP. Also, increase in the load-bearing capacity of graphite filled CPCs with increasing graphite content as a result of good interfacial bonding and filler-matrix adhesion within the composites has been reported by some authors [26-27]. Increase in graphite content from 30 to 40 wt% resulted in decrease in the flexural modulus from 4177.4 MPa to 3385.9 MPa. Further increase in the graphite content led to an increase in the flexural modulus up to 11,472.8 MPa at a graphite content of 80 wt%. The increase in flexural modulus is due to the higher stiffness of graphite compared to the polymer matrix. All the flexural strength values obtained in this study meet the DOE target of >25 MPa. Therefore, all the PP/epoxy/graphite composites exhibited good flexural properties for PEMFC BP application.

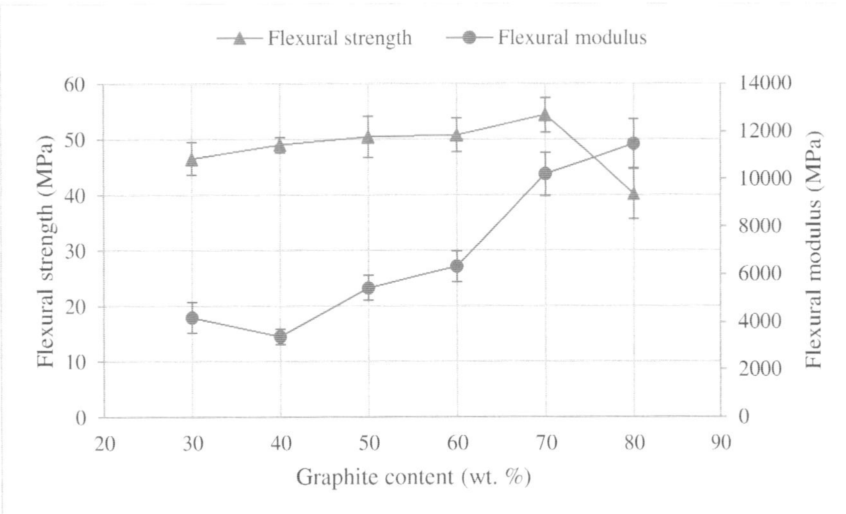

Figure 5. Variation of flexural properties of the composites with graphite content.

4. Conclusion

Different PP/epoxy/graphite composites with 30-80 wt% graphite were prepared by melt mixing followed by compression moulding, and the morphology, electrical conductivity, and flexural properties were investigated. SEM micrograph of the PP/epoxy blend revealed a co-continuous morphology. This could be due to the immiscible nature of PP and epoxy resin, and the close to symmetric composition of the blend. The results also showed that the in-plane electrical conductivity of the composites increased with increase in graphite content from 12.09 Scm^{-1} at 30 wt% to 68.03 Scm^{-1} at 80 wt% graphite. The increase in conductivity with graphite content can be attributed to formation of more conducting networks as the graphite concentration increased. Also, the composite exhibited relatively high electrical conductivities due to formation of a co-continuous morphology, which would favor the distribution of the filler in one of the polymer phases and promote formation of more continuous conducting networks. The in-plane conductivity range obtained in this study is higher than that for several graphite filled-single polymer composites reported in literature. Achieving this level of conductivity shows the potentials of graphite-PP/epoxy composites for meeting the US DOE target for in-plane electrical conductivity for PEMFC BP. The flexural modulus of the composites initially decreased with increase in graphite content and then increased with increase in graphite content from 40 wt% to 80 wt%, while the flexural strength increased with increase in graphite content up to 70 wt% and then decreased with further increase in graphite content. However, all the PP/epoxy/graphite composites exhibited flexural strength that meet the DOE target of >25 MPa. This can be attributed to good interfacial bonding between the polymer phases as a result of the compatibilizing effect of the MAH-g-PP, and good filler-matrix adhesion. Among the produced composites, the PP/epoxy/80 wt% graphite composite has the best combination of in-plane electrical conductivity and flexural strength with respect to the DOE targets. The results obtained in this study showed that graphite-PP/epoxy blend composite has a good potential for application in CPC BPs for PEMFC.

Acknowledgments

The authors will like to thank the National Research Foundation of South Africa and Telkom Centre of Excellence for the financial support for the work. Mr. A. Akinwunmi, Mr. O. Oluokun, and Mr. O. Ogunniyi

(Vaal University of Technology) are greatly appreciated for their assistance during the electrical conductivity measurements.

References

[1] Lee J H, Lee J S, Kuila T, Kim N H and Jung D 2013 Effects of Hybrid Carbon Fillers of Polymer Composite Bipolar Plates on the Performance of Direct Methanol Fuel Cells *Compos Part B: Eng* ***51*** 98-105

[2] Alaswad A, Baroutaji A, Achour H, Carton J, Al Makky A and Olabi A G 2016 Developments in Fuel Cell Technologies in the Transport Sector *Int. J. Hydrogen Energy* ***41*** (37) 16499-16508

[3] Naji A, Pötschke P and Ameli A Melt Processed Conductive Polycarbonate Composites With Ternary Fillers Towards Bipolar Plate Applications, ASME 2018 Conference on Smart Materials, Adaptive Structures and Intelligent Systems, American Society of Mechanical Engineers, **2018,** V001T01A012-V001T01A012.

[4] Lee H E, Chung Y S and Kim S S 2017 Feasibility Study on Carbon-felt-reinforced Thermoplastic Composite Materials for PEMFC Bipolar Plates *Compos Struct.* ***180*** 378-385

[5] Suherman H, Sahari J and Sulong A B 2013 Effect of small-sized conductive filler on the properties of an epoxy composite for a bipolar plate in a PEMFC *Ceram Int.* ***39*** (6) 7159-7166

[6] Radzuan N A M, Sulong A B, Husaini T, Majlan E H, Rosli M I and Aman M F 2019 Fabrication of multi-filler MCF/MWCNT/SG-based bipolar plates *Ceram Int.* ***45*** (6) 7413-7418

[7] Mohd Radzuan N A, Sulong A B, Somalu M R, Abdullah A T, Husaini T, Rosli R E, Majlan E H and Rosli M I 2019 Fibre orientation effect on polypropylene/milled carbon fiber composites in the presence of carbon nanotubes or graphene as a secondary filler: Application on PEM fuel cell bipolar plate *Int. J. Hydrogen Energy*

[8] Naji A, Krause B, Pötschke P and Ameli A 2019 Hybrid conductive filler/polycarbonate composites with enhanced electrical and thermal conductivities for bipolar plate applications *Polym. Compos.*

[9] Bairan A, Selamat M Z, Sahadan S N, Malingam S and Mohamad N 2018 Effect of CNTs on the electrical and mechanical properties of polymeric composite as PEM fuel cell bipolar plate *Jurnal Teknologi* ***80*** (6) 115-122

[10] Zakaria M Y, Sulong A B, Sahari J and Suherman H 2015 Effect of the Addition of Milled Carbon Fiber as a Secondary Filler on the Electrical Conductivity of Graphite/Epoxy Composites for Electrical Conductive Material *Compos Part B: Eng.* ***83*** 75-80

[11] Planes E, Flandin L and Alberola N 2012 Polymer Composites Bipolar Plates for PEMFCs *Energy Procedia* ***20*** 311-323

[12] Liu Z, Cai X, Ke X, You F, Zhang Q and Gao X 2018 Influence of $CaCO_3$/glass fiber hybrid fillers on the mechanical and thermal properties of polytetrafluoroethylene *Adv in Polym Tech.*

[13] Wen B and Zheng X 2019 Effect of the selective distribution of graphite nanoplatelets on the electrical and thermal conductivities of a polybutylene terephthalate/polycarbonate blend *Compos Sci Technol.* ***174*** 68-75

[14] Gulrez S K H, Ali Mohsin M E, Shaikh H, Anis A, Pulose A M, Yadav M K, Qua E H P and Al-Zahrani S M 2014 A review on electrically conductive polypropylene and polyethylene *Polym Compos.* ***35*** (5) 900-914

[15] Francis B and Thomas S Polyolefins/epoxy resin blends, in: Polyolefin blends Nwabunma, D.; Kyu, T., Eds. Wiley & Sons Inc. Publication: New Jersey, 2008; pp 623-624.

[16] Runyan W R and Shaffner T J *Semiconductor measurements and instrumentation*. McGraw-Hill New York: 1975; Vol. 112.

[17] Schroder D K *Semiconductor material and device characterization*. John Wiley & Sons: 2006.

[18] Király A and Ronkay F 2013 Effect of Graphite and Carbon Black Fillers on the Processability, Electrical Conductivity and Mechanical Properties of Polypropylenebased Bipolar Plates *Polym Polym Compos.* ***21*** (2) 93-100
[19] Díez-Pascual A M Mechanical Properties of Epoxy/Thermoplastic Blends. In *Handbook of Epoxy Blends*, Parameswaranpillai J., H. N., Pionteck J., Woo E., Ed. Springer: Cham, 2017; pp 743-774.
[20] Pan Y, Liu X, Hao X, Starý Z and Schubert D W 2016 Enhancing the electrical conductivity of carbon black-filled immiscible polymer blends by tuning the morphology *Eur. Polym. J.* ***78*** 106-115
[21] Taherian R 2016 Experimental and analytical model for the electrical conductivity of polymer-based nanocomposites *Compos Sci Technol.* ***123*** 17-31
[22] Aneli J, Zaikov G and Mukbaniani O 2012 Physical Principles of the Conductivity of Electrically Conductive Polymer Composites (Review) *Mol Cryst Liq Cryst.* ***554*** (1) 166-187
[23] Derieth T, Bandlamudi G, Beckhaus P, Kreuz C, Mahlendorf F and Heinzel A 2008 Development of highly filled graphite compounds as bipolar plate materials for low and high temperature PEM fuel cells *J New Mater Electrochem Syst.* ***11*** (1) 21
[24] Johnson B A. Thermally and Electrically Conductive Polypropylene Based Resins for Fuel Cell Bipolar Plates. Michigan Technological University, Michigan, USA, 2009.
[25] Reddy C V, Raju J S, Babu P R and Ramnarayan R 2017 Study of Voids Effect on Tensile Strength of Carbon Fiber Reinforced Composites for Structural Applications *Int. J. Res. Appl. Sci. Eng. Technol.* ***5*** (8) 1311-1320
[26] Baptista R, Mendao A, Guedes M and Marat-Mendes R 2016 An experimental study on mechanical properties of epoxy-matrix composites containing graphite filler *Procedia Structural Integrity* ***1*** 74-81
[27] Suherman H, Mahyoedin Y, Septe E and Rizade R 2019 Properties of graphite/epoxy composites: the in-plane conductivity, tensile strength and shore hardness *AIMS Mater Sci.*

IOP Conf. Series: Materials Science and Engineering **655** (2019) 012036 doi:10.1088/1757-899X/655/1/012036

Development and characterization of porous Ti-Sn-SiC composite intended for biomedical application

C R Mathebula[1], W Matizambuka[2], and A S Bolokang[1]

[1]Council of Scientific Industrial Research, Material Science and Manufacturing, Light Metals, Meiring Naude Road, P O Box 395, Pretoria, South Africa

[2]Department of Metallurgical Engineering, Vaal University of Technology, Andries Potgieter Blvd, Vanderbijlpark, 1900, South Africa

E-mail address: CMathebula@csir.co.za

Abstract. Titanium-based materials have received great research interest this is mainly due to their promising specific strength, thermal stability, corrosion resistance, and biocompatibility. Porous titanium has been shown to exhibit desirable properties as biomedical materials. Ti-Sn-SiC composites of different compositions were synthesized through mixing Ti, Sn and SiC powders. The study was carried out by blending and cold-pressing followed by sintering at a 1400°C. Microstructures of the materials were studied using an optical microscope (OP) and scanning electron microscopy (SEM). To investigate structural development and phase transformation, x-ray diffraction (XRD) was performed. A porous Ti-Sn-SiC composite was achieved which is favorable for biomedical application.

1. Introduction

Titanium-based materials have attracted research interest across engineering disciples in aerospace, automotive and biomedical applications. This is due to their outstanding properties such as specific strength, low thermal expansion, corrosion resistance and biocompatibility [1]. Even though titanium has relatively good properties, Ti and its alloys are considered expensive owing to its problematic metal extraction and manufacturing processes [2]. Its corrosion resistance and biocompatibility are among the reasons why the material has gained increased applications in the biomedical industry. Powder metallurgy routes have gained considerable acceptance in providing an economical route for processing titanium and its alloys. Moreover, porous titanium implant produced by powder metallurgy has increased proliferation and adhesion activity of osteoblasts unlike the dense titanium implants [3]. Mechanical properties of the porous material can, therefore, be improved by developing the metal matrix composites (MMCs) [4]. In the determination of enhancing the mechanical properties, one approach which has been considered is the development of metal matrix composites (MMC) whereby two dissimilar materials with different mechanical properties and physical properties are combined to attain at a product whose final properties are superior to the individual constituents [4]. MMCs are a class of materials that when combined result in high strength and stiffness. They offer high-temperature competence with good oxidation resistance of titanium from increased stiffness provided by the ceramic reinforcements [5].

A well-applied biomedical Ti-6Al-4V alloy has the limitations of low wear resistance [6], high elastic modulus [7] and low strength [8] that could weaken the usage as implant and as in screw form (stress shielding effect). Aluminum (Al) and Vanadium (V) have recently been found to be detrimental and might cause certain long term health challenges. In particular, vanadium exhibits high cytotoxicity while aluminum induces senile dementia [9]. Replacement of these elements with nobler non-toxic elements is highly desired for the generation of more biocompatible alternatives [10, 11, 13]. The use of Sn and

Nb as non-toxic alloying elements for Ti-based alloys is being investigated. Tin (Sn) is known to be a safe alloying element and often used to strengthen Ti alloys [12]. The incompatibility of mechanical properties between the bone tissue and implant materials lead to tissue loss due to the stress-shielding [14]. To avoid shield stress, an alternative is to use porous material. The porosity of the implant permits bone ingrowth into the pores and enables the stress redistributions to the adjacent bone tissues, therefore, resulting in minimized stress shielding and eventually prolonged device lifetime [15]. Many researchers have indicated that the stress shielding retards bone modeling and healing, which results in increased porosity in the surrounding bone [16, 17]. Therefore there is an increasing interest in altering the elastic modulus of metallic materials to overcome stress-shielding. Additionally, porous materials yield lower modulus of elasticity combined with good strength provided by the SiC reinforcement in Ti-Sn alloy. The main aim of this study is to develop and characterize the porous Ti-Sn-SiC composite. The microstructure, hardness, structural development and phase transformation of the material were studied to evaluate the behavior of the material.

2. Experimental work

Powders of titanium (Ti), tin (Sn) silicon carbide (SiC) were used as starting material. The description of the materials is shown in **Table 1**. A series of $Ti_{\chi}SnSiC$ alloys with weight were prepared by adding Ti, Sn and SiC powder of different weight percentages. The powders were weighed and then blended using a Tubular mixer to provide a homogeneous mixture. The blended powders were pressed using the Hydraulic cold press machine to produce green compact samples with a diameter of 17.5cm. The green compacts samples were then sintered at 1400°C for 6 hours in a tube furnace. After sintering, the samples were subsequently sectioned and mounted. The surface was initially ground using 500, 800, 1000, 2400 and 4000 grit paper, followed by polishing with 9 and 3µm diamond suspension then finally with colloidal silica suspension. The Joel JSM 6510 scanning electron microscope (SEM) equipped with the energy-dispersive X-ray spectroscopy (EDS) were used to analyze the morphology of the powders, the microstructure of the sintered samples as well as to identify the sample chemistry. The microstructure was also characterized using the Leica DMI5000 M optical microscope (OM). The automatic FM-700 microhardness Vickers tester was used to study the hardness profile of the material. The X-ray diffraction (XRD) was performed to investigate the structural development and phase transformation of the material. The chemical composition of the sample alloys is shown in **Table 2.**

Table 1: Description of materials

Powder Blend	Titanium	Tin	Silicon Carbide
Supplier	TLS	Aldrich	-
Particle Size	<45µm	<45µm	-
Purity	99.8%	99.8%	-

Table 2: Composition of Ti-Sn-SiC in wt %

Powder Blend	Titanium	Tin	Silicon Carbide
Ti-Sn-SiC	54	38	8
Ti-Sn-SiC	56	38	6
Ti-Sn-SiC	58	38	4
Ti-Sn-SiC	60	38	2

3. Results and discussion

3.1 Microstructural analysis

Morphology of the as received powders of Ti, Si and SiC are shown on **Figure 1**. Ti powder shows the spherical particles of varying particle sizes while the Sn powder displays a mixture of the spherical and irregular shaped particles. The SiC particles were irregularly shaped with some sharp edges.

The optical microstructures of the Ti-Sn-SiC composites are shown in **Figure 2a-d**. All samples were sintered at 1400 °C in Argon in atmosphere. In all the samples, two phases are evident, the Ti-Sn and

IOP Conf. Series: Materials Science and Engineering **655** (2019) 012036 doi:10.1088/1757-899X/655/1/012036

SiC particles, respectively. Additionally, pores of different size and shapes are present in the composite materials. The $Ti_{5.6}$-$Sn_{3.8}$-SiC in **Figure 2b** reveal more of the spherical pores than all the composite and **2c** confirm the porous structure of the composites alloy. It is roughly estimated that the average porosity was ± 6µm. The porous microstructure are due to the compaction and sintering process. As a preliminary study, this can be improved. Pure Sn melts at 232 °C and Ti at 1668 °C, the two starts to react at lower temperature until the intermetallic phases are formed. At the sintering temperature of 1400 °C, it seems that Sn was unable to vaporize from the Ti-Sn-SiC compact during sintering. However, pre-melting of Sn occurs which is responsible for void and results in the porous material due to the Kirkendall effect. When two distinct materials are placed next to each other and diffusion is allowed to take place between them, that is known as the Kirkendall effect. The diffusion coefficients of the two materials in each other are not the same. They form voids which acts as sinks for vacancies. Porous titanium has shown to exhibit desirable properties as biomedical materials. This is due to the interconnected pores which permit bone ingrowth into the pores and provide not only anchorage for fixation but also a system which enhances stress distributions to the adjacent bone tissues therefore minimises stress shielding and eventually prolong device lifetime [15]. The SEM images shown in **Figure 3a-d** confirm a porous structure and the two phases evident from the microstructure.

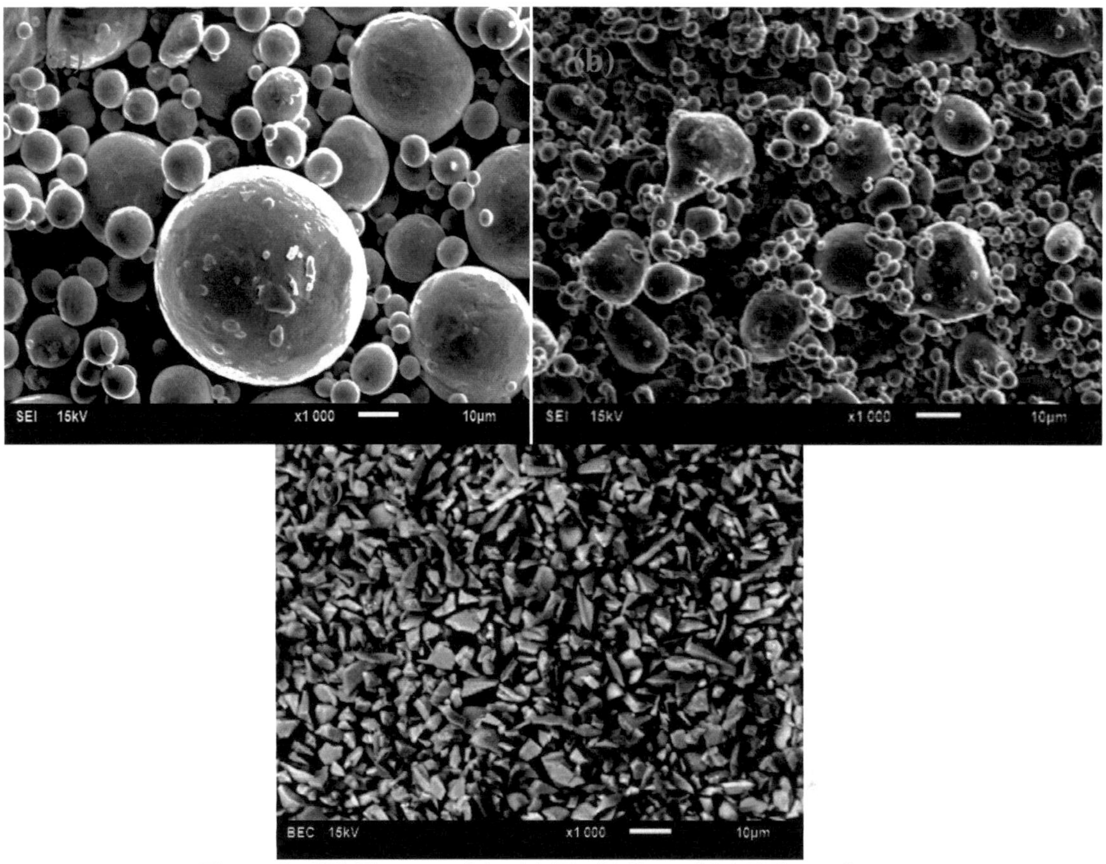

Figure 1: SEM images of the (a) Ti, (b) Sn and (c) SiC powders

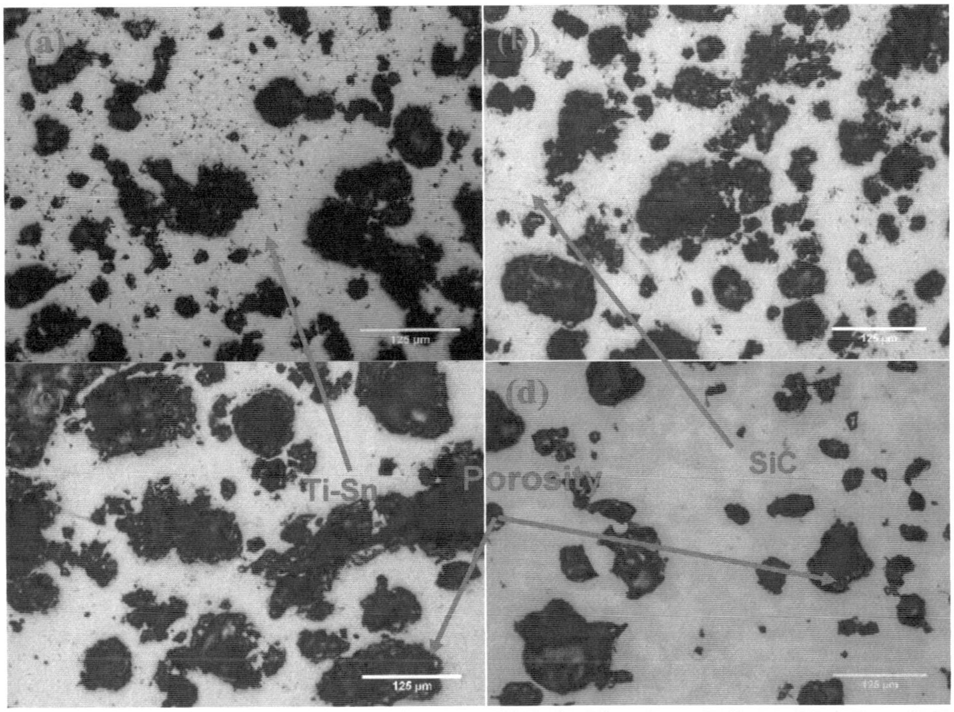

Figure 2: Optical micrographs of the (a), $Ti_{5.4}$-$Sn_{3.8}$-SiC, (b) $Ti_{5.6}$-$Sn_{3.8}$-SiC, (c) $Ti_{5.8}$-$Sn_{3.8}$-SiC and (d) $Ti_{6.0}$-$Sn_{3.8}$-SiC composite materials.

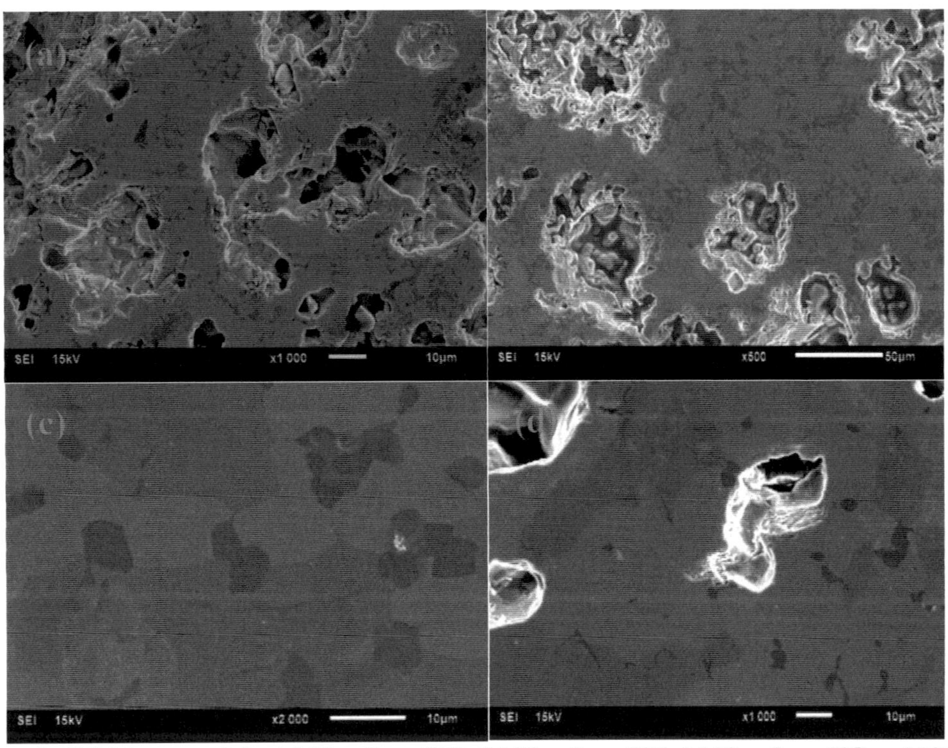

Figure 3: SEM images of the (a) $Ti_{5.4}$-$Sn_{3.8}$-SiC, (b) $Ti_{5.6}$-$Sn_{3.8}$-SiC, (c) $Ti_{5.8}$-$Sn_{3.8}$-SiC and (d) $Ti_{6.0}$-$Sn_{3.8}$-SiC composite materials.

Figure 4d shows the EDS chemical analysis of the composites. **Table 3** indicates the chemical compositions of the material of **Figure 4b**. **Figure 4a** below shows a Ti-rich phase at point 1, 3 and 4 whereas point 2 shows a homogenous phase of Ti-Sn-SiC. **Figure 4b** shows a homogenous phase of Ti-Sn-SiC at point 1 and 5. Point 2 and 4 are Ti-rich phase; whereas point 3 and 6 are Sn-rich phase. **Figure 4c** shows point 1 and 3 are Ti-rich phases, point 2 and 4 shows a homogenous phase of Ti-Sn-SiC; whereas point 5 and 6 are Ti-Sn phases. Point 4 and 5 show a grained formed that consist of Ti-Sn-SiC and Ti-Sn phases. **Figure 4d** shows a Ti-Sn phase at point 1 and 7. Point 2, 4 and 6 are Ti-rich phases; whereas point 3, 5 and 8 shows a homogenous phase of Ti-Sn-SiC. Figure 4c-d have appear to have larger grains compared to **Figure 4a-b**.

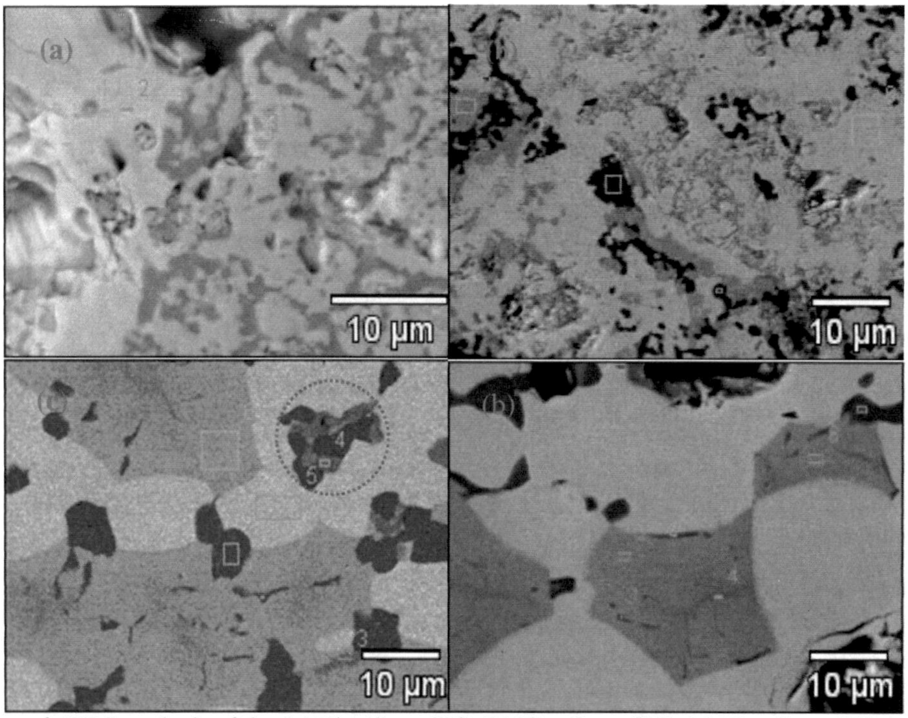

Figure 4: EDS analysis of the (a) $Ti_{5.4}$-$Sn_{3.8}$-SiC, (b) $Ti_{5.6}$-$Sn_{3.8}$-SiC, (c) $Ti_{5.8}$-$Sn_{3.8}$-SiC and (d) $Ti_{6.0}$-$Sn_{3.8}$-SiC composite materials.

Table 3: EDS data of $Ti_{5.6}$-$Sn_{3.8}$-SiC composite

	C-K	Si-K	Ti-K	Sn-L
Point 1	**3.69**		**52.64**	**43.67**
Point 2	**3.11**	**16.15**	**61.73**	**19.01**
Point 3	**1.28**	**3.48**	**44.27**	**50.98**
Point 4	**2.12**	**18.48**	**68.51**	**10.89**
Point 5	**3.81**		**52.67**	**43.51**
Point 6	**1.60**	**3.10**	**43.31**	**52.00**

3.2 X-ray diffraction

XRD analysis of the Ti-Sn-SiC composite of various compositions is shown in **Figure 5**. The HCP crystal structure of lattice parameters a= 2.902 Å; c=4.734 Å with space group and number P63/mmc #194 and c/a ratio of 1.63 was detected for the $Ti_{5.4}$-$Sn_{3.8}$-SiC shown in **Figure 5a**. This lattice parameter is similar to the lattice parameter previously obtained for the TiSn intermetallic alloy induced by mechanical milling [18]. Similar results were obtained for the $Ti_{5.6}$-$Sn_{3.8}$-SiC alloy composition in **Figure 5b**. **Figure 5c** shows the XRD pattern of the $Ti_{5.8}$-Sn_{3}.8-SiC revealing the HCP and FCC crystal

IOP Conf. Series: Materials Science and Engineering **655** (2019) 012036 doi:10.1088/1757-899X/655/1/012036

structures, respectively. However, the HCP phase has lattice parameters a= 2.920 Å; c=4.620 Å with smaller c/a ratio of 1.589. Additionally, lattice parameter a=5.620 Å Fm-3m # 225 was obtained both for **Figure 5c** and **d** XRD patterns. The XRD analysis confirmed the EDS results.

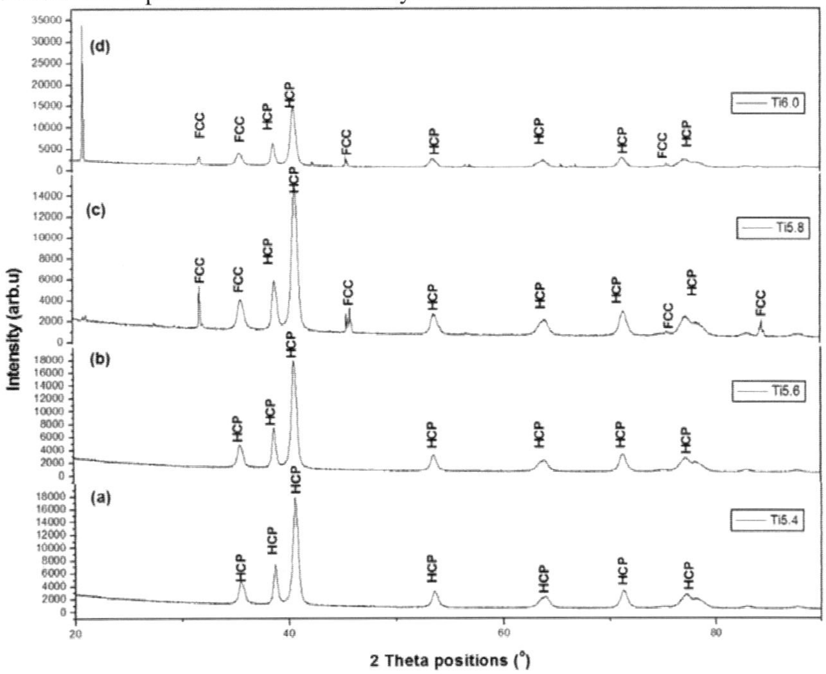

Figure 5: XRD analysis of the Ti-Sn-SiC composite for (a) Ti 5.4, (b) Ti 5.6, (c) Ti 5.8 and (d) Ti 6.0 compositions

3.3 Hardness profile

The hardness profile of the Ti-Sn-SiC composites is shown on Fig. 6. Compared to all three alloys, the $Ti_{5.4}Sn_{3.8}SiC$ alloy resulted in lower hardness with an average of 257.54 HV due to the lower Ti content while $Ti_{6.0}Sn_{3.8}SiC$ alloy showed the highest hardness profile with an average of 487.87 HV. The average hardness of the $Ti_{5.6}Sn_{3.8}$ SiC was 322.351 HV while $Ti_{5.8}Sn_{3.8}SiC$ alloy was 343.03 HV. The hardness matches the densification response of the samples. The hardness variation is due to the presence of different phases in the microstructure as shown by optical micrographs in **Figure 2a-d** and the SEM images in **Figure 3a-d.** The Ti-rich phases contributed to the high hardness while the Ti-Sn phases contributed to the lower hardness.

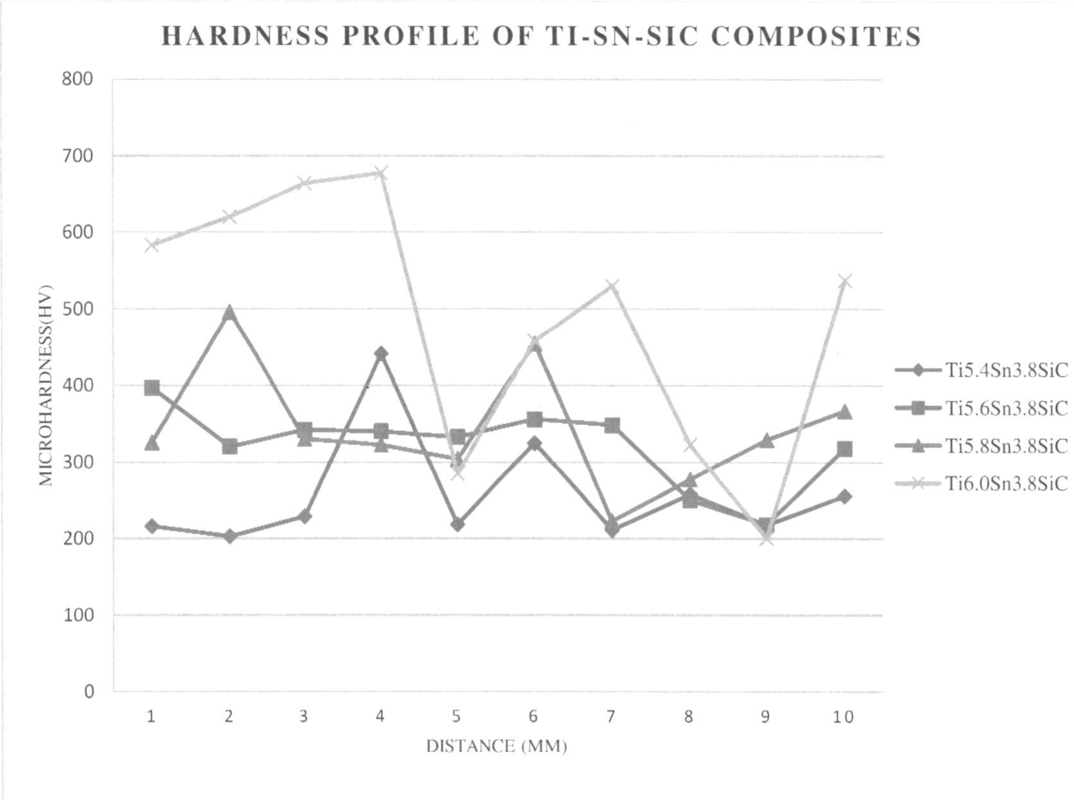

Figure 6: Hardness profile of the Ti-Sn-SiC composites

4. Conclusions

SiC was employed as reinforcement to prepare Ti-Sn-SiC alloy via powder metallurgy. The Ti-Sn-SiC composite was successfully synthesised. The pore structures were evenly distributed throughout the materials. The optical microstructures and SEM images, as well as the hardness profile indicated a porous material with pores of different size and shapes with an average porosity of ±6μm. The $Ti_{5.6}$-$Sn_{3.8}$-SiC reveal more of the spherical pores than all the composites. EDS results showed different phases in the microstructure. There is a homogenous phase of Ti-Sn-SiC present as well as Ti-rich phase, Sn-rich phase and a Ti-Sn phase. XRD analysis detected the HCP crystal lattice structure for the $Ti_{5.4}Sn_{3.8}SiC$ and $Ti_{5.6}$-$Sn_{3.8}$-SiC composites. XRD pattern of the $Ti_{5.8}$-$Sn_{3.8}$-SiC revealing the HCP and FCC crystal structures, respectively. The HCP phase has lattice parameters a= 2.920 Å; c=4.620 Å with smaller c/a ratio of 1.589. Additionally, lattice parameter a=5.620 Å Fm-3m # 225 was obtained both for $Ti_{5.8}Sn_{3.8}SiC$ and $Ti_{6.0}Sn_{3.8}SiC$ XRD patterns. The XRD analysis confirmed the EDS results. The different phases also contribute to the variation of hardness. The results showed that an improvement, in terms of processing and homogenisation, is required for future work.

5. Acknowledgement

The authors acknowledge the technical and financial support of the CSIR. The contributions of P.Daswa, M.Seerane, N.Mnguni and Maria Mathabathe are greatly acknowledged.

References

[1] Jayaprakah M, Ping D.H, Yambe-Mitarai Y 2015 Enhanced yielding strength of near –α Ti- Al-Zr-Sn high temperature alloys *Materials Science & Engineering* **625** 131-139
[2] Liu H.W, Paul Bishop D, Plucknett K.P 2015 A comparison of Ti-Ni and Ti-Sn alloys processed using powder metallurgy *Materials Science Engineering* **644** 392-404

[3] Civantos A et al 2019 Designing bioactive porous titanium interfaces to balance mechanical properties and in vitro cells behavior towards increased osseointegration. *Surface & Coatings Technology* **368** 162-174
[4] Smith P.R, Froes F.H 1984 Developments in Titanium metal matrix composites *Journal of metals* **36** 19-26
[5] Godfrey T.M.T, Goodwin P.S, Ward-close C.M 2000 Titanium particulate metal matrix composites reinforcement, production methods, and mechanical properties *Advanced Engineering Materials* **2** 85-92
[6] Dong H, Bell T 2000 Enhanced wear resistance of titanium surfaces by a new thermal oxidation treatment **238** 131-7
[7] Li X, Wang C.T, Zhang W.G et al 2009 Properties of porous Ti-6Al-4V implant with a low stiffness for biomedical application *Proc Institute Mechanical Eng.* **223** 173-8
[8] Kong F, Chen Y, Zhang D 2011 Interfacial microstructure and shear strength of Ti-6Al-4V/TiAl laminate composite sheet fabricated by hot packed rolling *Materials and Design* **32** 3167-72
[9] Davidson J.A, Mishra A.K, Kovacs P et al 1994 New surface hardened, low modulus, corrosion resistant Ti-13Nb-13Zr alloy for total hip arthroplasty *Biomedical Materials Engineering.* **4** 231-43
[10] Niinomi M 2002 Recent metallic materials for biomedical applications *Metal Materials Transactions* **33** 477-486
[11] Niiomi M, Nakai M 2011 Titanium-based biomaterials for preventing stress shielding between implant devices and bone *International Journal Biomaterials*
[12] Niinomi M 2003 Recent research and development in titanium alloys for biomedical applications and healthcare goods *Science Technology Advance Master* **4** 445-454
[13] Eisenbarth D et al 2004 Bicompatibility of β-stabilising elements of titanium alloys *Biomaterials* **25** 5705-5713
[14] Long M, Rack H.J 1998 Titanium alloys in total joint replacements-a materials science perspective *Biomaterials* **19** 1621-39
[15] Xiao-hua WANG, et al 2013 Mechanical properties of porous titanium with different distributions of pore size *Trans. Nonferrous Met. Soc. China* **23** 2317–2322
[16] Rack H.J, Qazi J.I 2006 Titanium alloys for biomedical applications *Materials Science and Engineering C* **26** 1269-1277
[17] Bahraminasab M, Sahari B.B, Edwards K.L, Farahmand F, Arumugam M 2013 Aseptic loosening of femoral components - Materials engineering and design considerations *Materials & Design* **44** 155-163
[18] Bolokang A.S, Motaung D.E, Arendse C.J, Muller T.F.G 2015 Production of titanium–tin alloy powder by ball milling: Formation of titanium–tin oxynitride composite powder produced by annealing in air *Journal of Alloys and Compounds* **622** 824-830

Influence of warm temperature (0.3-0.5T_m) uniaxial compression on the distribution of reinforcing particles in a 2124-Al/10%SiC MMC

Z Gxowa-Penxa [1, 2, 3], L H Chown [2, 3], G Govender [1]

[1] Materials Science and Manufacturing, Council for Scientific and Industrial Research (CSIR)

[2] School of Chemical and Metallurgical Engineering, University of the Witwatersrand (Wits)

[3] DST-NRF Centre of Excellence in Strong Materials, hosted by Wits

*ZGxowa@csir.co.za

Abstract. Experimental work was carried out to study the influence of uniaxial compression tests, performed on a Gleeble 3500, on the distribution of reinforcing particles in SiC reinforced Al Metal Matrix Composites (MMCs). Al MMC powder (2124-Al with 10 vol. %SiC) was produced through blending in a high energy ball mill. Unreinforced 2124-Al and Al MMC powders were compacted at ambient temperature (*i.e.* cold compaction) and the green compacts were then sintered at 490°C for 1 hr. Warm temperature (280°C) uniaxial compression tests were performed on a Gleeble 3500 thermomechanical simulator at a strain rate of 5 s^{-1}, strain of 0.3 and a soaking time 20 minutes.

Microstructures of the uniaxially compressed 2124-Al with 10% SiC MMC revealed that deformation had an influence on the distribution of reinforcing particles. Reinforcing particles were only distributed inside 2124-Al grains that had undergone deformation (evident through change in shape of grains). This finding supports statements made in literature regarding the positive influence of deformation on the distribution of reinforcing particles. Engineering stress-strain curves showed that reinforcing 2124-Al with 10 vol. % SiC had a positive effect on flow stress as the 10%SiC MMC had a flow stress of 153 MPa while the flow stress of the unreinforced 2124-Al was 95 MPa. An improvement in hardness was also observed where the MMC had a higher hardness than the unreinforced Al alloy; further indicating that reinforcing the Al alloy improved properties.

1. Introduction

Metal matrix composites (MMCs) are still seen as viable alternatives to conventional metals and alloys for applications which require high temperature performance as well as high stiffness and strength [1]. MMCs may be continuously reinforced with fibres, or discontinuously reinforced by dispersing particles or whiskers with higher hardness in the matrix material [2]. Discontinuously reinforced MMCs are favoured over continuously reinforced MMCs due to lower cost [3]. The focus of this work is particle reinforced MMCs. One of the determining factors for producing particle reinforced MMCs that are of good quality is achieving a uniform distribution of reinforcing particles in the matrix [4].

Literature states that the fabrication process selected significantly influences the distribution of reinforcing particles in MMCs [5]. Powder metallurgy (PM) processes have been said to surpass liquid state processes in their ability to achieve a uniform distribution [6]. However, many studies have reported the formation of a *necklace microstructure* after the blending and compaction steps in PM processes[5][7][8]. In this necklace microstructure, the reinforcing particles cluster on the periphery of the matrix grains along the grain boundaries as illustrated in Figure 1 [5]. This is undesirable because an MMC with this necklace microstructure is susceptible to intergranular cracking which may result in failure [9]. As a result, PM processes consist of a secondary processing step where a deformation process (*e.g.* uniaxial compression, extrusion or drawing) is used to induce plastic flow in the MMC in an attempt to improve reinforcing particle distribution [10].

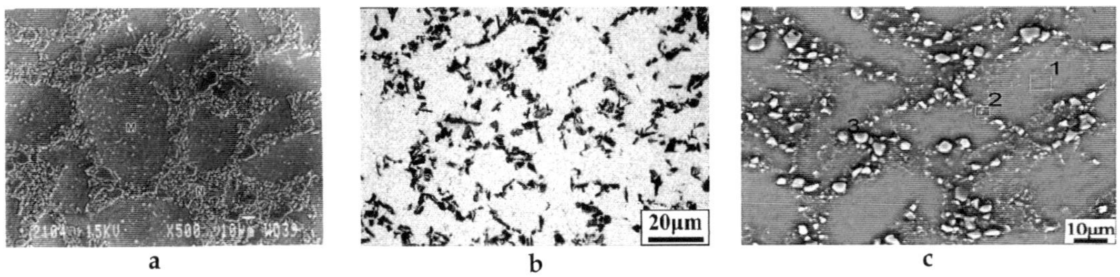

Figure 1. Typical necklace microstructures observed in literature by a) Prasad et al. (2002), marker = 10μm, b) Liu et al. (2010), marker = 20μm and c) Chen et al. (2015), marker = 10μm.

Deformation temperature is a critical factor. Hot deformation (*i.e.* temperatures higher than 350°C) is associated with inferior quality due to poor dimensional accuracy and defects [11][12]. The presence of reinforcing particles increases resistance to deformation which makes deforming MMCs at ambient temperatures cumbersome and impractical for industrial applications; especially where mass production is required since the service life of equipment and die sets is adversely affected and their frequent replacement increases production costs [12]. Most deformation processes are carried out at temperatures above 350°C (hot deformation). There is therefore a knowledge gap to be filled regarding deformation at intermediate or warm working temperatures.

This work aims to investigate the influence of warm temperature uniaxial compression on the distribution of reinforcing particles in powder metallurgically produced 2124-Al/10%SiC MMCs. A part of the work is to study microstructures (evolution from compaction to uniaxial compression) and assess if statements about the positive influence of deformation on distribution of reinforcing particles, made in literature, can be corroborated.

2. Experimental procedure

An Enerpac VLP® 100 tonne hydraulic press and 8 mm diameter die set made from high speed steel (HSS) were used to compact unreinforced 2124-Al alloy and blended 2124-Al with 10 vol. %SiC powders to form green compacts. Ambient temperature compaction was carried out because the limitation of this Enerpac press is that powders can only be compacted at ambient temperature. The green compacts were sintered at 490°C for 1 hr prior to uniaxial compression. To determine the warm working temperature range, a differential scanning calorimetric (DSC) study was done on the unreinforced 2124-Al and MMC. The DSC results showed that the materials started melting at ~560°C (solidus temperature). This meant that the warm working temperature range (30 and 60% of the matrix phase melting point) was between 168°C and 336°C. In a previous study, it was observed that at temperatures below 280°C the green compacts fractured; indicating that deformation was poor. Uniaxial compression tests were performed on a Gleeble 3500® thermomechanical simulator at 280°C, strain rate of $5s^{-1}$, strain of 0.3 and soaking time of 20 minutes. To evaluate the influence of this uniaxial compression on the distribution of the reinforcing particles, the sample preparation technique was critical. Sample preparation involved only one grinding step with a 1000 grit paper, at a force of 20N for 2 minutes. Only one grinding step was used because SiC papers are used for grinding and it has been observed that SiC particles from the grinding paper tend to adhere to the MMC samples and may be mistaken for the SiC particles that were added. The focus of this work is evaluating the influence of processing steps, particularly uniaxial compression, on the distribution of reinforcing particles. Therefore, the possibility of mistaking SiC particles from the grinding paper for those that were added had to be eliminated so as to acquire a good representation of the influence of the processing steps on reinforcing particle distribution. The grinding step was followed by polishing using 15 μm, 3 μm and 1 μm polycrystalline diamond paste using a force of 20 N for 5 minutes at each step. The samples were then polished with colloidal silica at 15 N for 15 minutes and water for 10 minutes. A Leica DM15000M® optical microscope and Jeol-6510® scanning electron microscope (SEM) were used to analyse the resultant microstructures. Microhardness tests with a 100g load were done on the compacts after uniaxial compression tests using a Future-Tech Vickers microhardness tester FM-700®. Fifteen indentations were made on each sample across the diameter of the compacts using a distance of 0.200 mm between indents. An average of the fifteen hardness values was calculated to determine hardness for each sample. Hardness tests were carried out according to ASTM standard E92 (ASTM E92-17, 2017).

3. Results and discussion

3.1. Blending & compaction

3.1.1. Unreinforced 2124-Al

Figure 2 shows that the unreinforced 2124-Al alloy powder particles were spherical, the powder compacted well and no defects (*e.g.* surface cracks) were observed on the green compacts. Microstructural analysis revealed that the microstructure of the 2124-Al green compacts was inconsistent from the surface to the mid-height (also referred to as the core). The darker phase in the microstructure from the core represents the pores and it can be seen that there is more porosity (as well as larger pore size) at the core of the green compacts. This inconsistent microstructure illustrates that there was a density variation in the green compacts due to pressure gradients that are caused by the friction between the powder and the die walls. It can also be observed that the Al alloy particles, under the compressive load, adhered to one another, bonded relatively well and showed good consolidation overall.

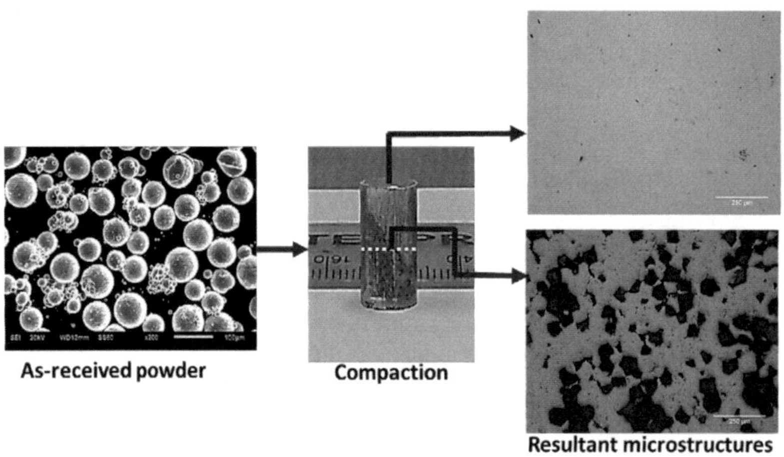

Figure 2. Images showing as-received unreinforced 2124-Al alloy powder (marker =100μm), 2124-Al green compact as well as resultant microstructures at the surface and core of green compact (marker = 250μm).

3.1.2. 2124-Al with 10%SiC

The microstructure showing results from the blending step in Figure 3 shows that there were some SiC particles which adhered to the Al alloy particle surfaces and a considerable amount of loose SiC particles situated around the Al alloy particles. These loose SiC particles indicate that the SiC powder did not blend fully with the Al alloy powder. Compaction of the blended powder was successful and green compacts produced were of good quality and comparable to those of the unreinforced Al alloy. Microstructures show that the typical necklace microstructure (Figure 1 compared to Figure 3) was formed after cold compaction. When one analyses microstructures after blending and compaction one can infer that, in this context, this necklace microstructure formed as result of poor blending and type of compaction used. Figure 4 illustrates that as the load applied during compaction forces the Al alloy particles to move closer together, the loose SiC particles (together with those attached to Al alloy particle surfaces) were caught in between the Al alloy particles. As a result, as the Al alloy particles consolidated the SiC particles agglomerated in between them. The SiC particles were then stuck at the periphery of the Al alloy grains once they had consolidated because plastic flow was limited (due to cold compaction) and there was no mechanism for the SiC particles to move from the grain boundaries to the inside of the grains. This result highlights that, where process costs are not a major concern, hot compaction should be considered to increase plastic flow and possibly improve distribution of reinforcing particles in compacts. It was also observed that sintering had no influence on the distribution of reinforcing particles and this makes sense because conventional sintering, which does not involve the use of pressure, was used. The white phase inside the Al alloy grains represents the eutectic.

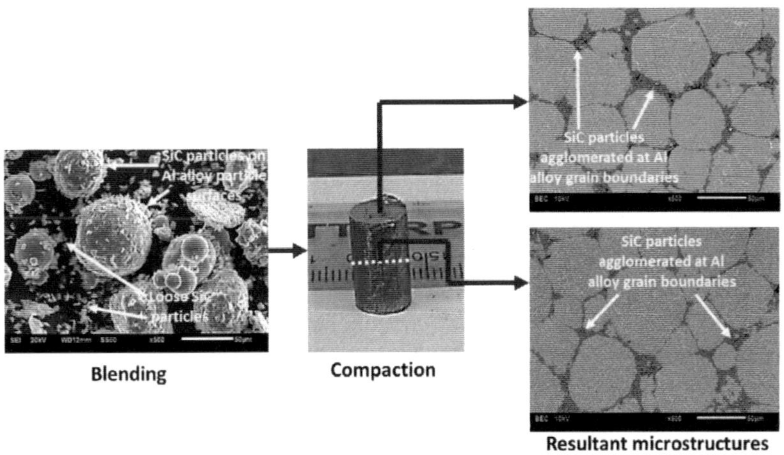

Figure 3. Images showing blended 2124-Al with 10%SiC powder (marker=50μm), 10%SiC green compact as well as resultant microstructures at the surface and core of green compact (marker=50μm).

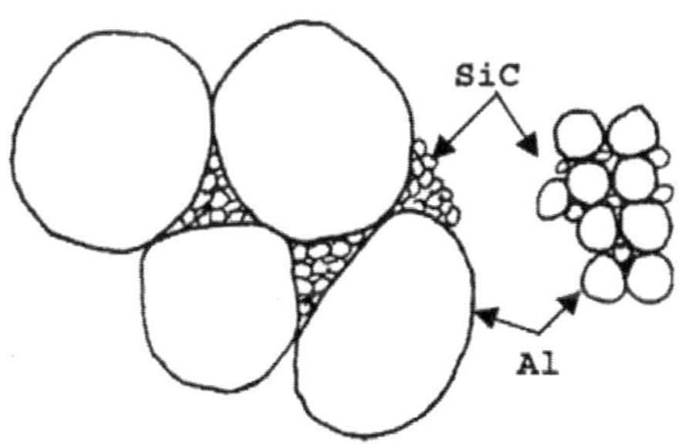

Figure 4. Image showing mechanism by which SiC particles agglomerated on Al alloy grain boundaries during cold compaction taken from Prasad et al. (2002).

3.2. Uniaxial compression

Figure 5 illustrates that reinforcing 2124-Al alloy with 10%SiC improved flow stress significantly because the 2124-Al with 10%SiC MMC had a higher flow stress of 153 MPa when compared to the unreinforced Al alloy whose flow stress was 95 MPa. It can also be observed that the MMC was able to deform at the maximum flow stress of 153 MPa up to the maximum strain of 0.3. This result is good considering that MMCs have been reported to be more difficult to deform than their unreinforced counterparts [14]. Moreover, this good deformation was achieved within the warm working temperature range at a temperature of 280°C.

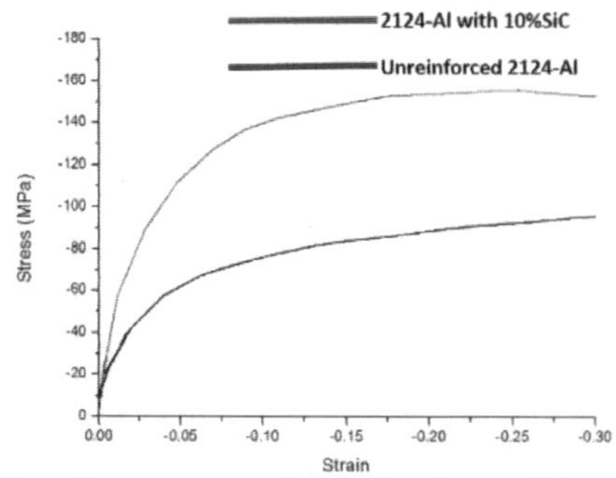

Figure 5. Stress-strain curves showing compressive behaviour of the unreinforced 2124-Al alloy and 2124-Al with 10% SiC MMC at 280°C, 5 s^{-1}, strain of 0.3 as well as sintering at 490°C for 1 hr and 20 minutes soaking time prior to uniaxial compression.

Figure 6 shows microstructures of the 2124-Al with 10%SiC MMC after uniaxial compression. The microstructures clearly show that the distribution of reinforcing particles was only influenced and slightly improved in areas where deformation had occurred. Deformation is identified by the Al alloy grains' change in shape. It can be observed that where the 2124-Al alloy grains deformed, as a result of uniaxial compression, some of the SiC particles migrated from the grain boundaries to the inside of the Al alloy grains and these areas are shown with white circles in Figure 6a. This result, without a doubt, corroborates statements made in literature regarding the positive influence of deformation on the distribution of reinforcing particles [5]. This finding also illustrates that there is potential to improve the distribution of reinforcing using a deformation process such as uniaxial compression; however in this case it seems that more deformation was required. It can be inferred that plastic deformation of the Al alloy grains is the mechanism by which distribution is improved by deformation processes since, from the microstructures shown in Figure 6, it is clear that in cases where the Al alloy grains did not plastically deform the SiC particles remained on the grain boundaries. It can further be deduced that the necklace microstructure that is often acquired after compaction in powder metallurgical processes can be improved with the aid of a deformation process that induces sufficient deformation to the MMC. In Figure 7 the SiC particles can be clearly seen inside the deformed Al alloy grains. The white square in Figure 7a shows the spheroidised eutectic (as a result of sintering) which should not be mistaken for smaller SiC particles [15].

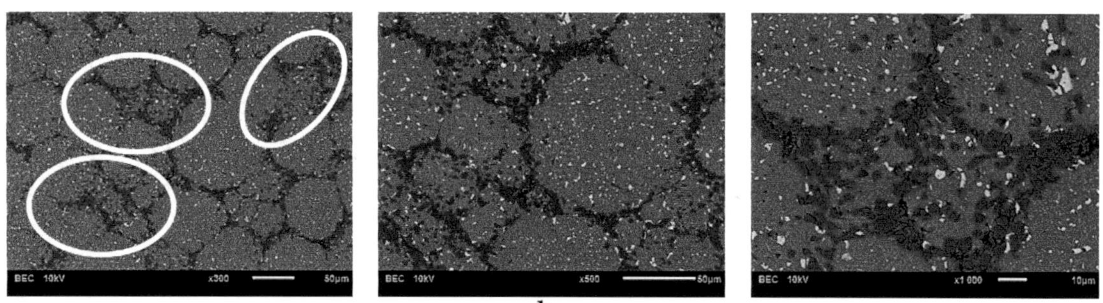

a b c

Figure 6. SEM BEI images showing microstructure of the uniaxially compressed 2124-Al with 10%SiC at a) 300X magnification (marker = 50μm), b) 500X magnification (marker = 50μm) and c) 1000X magnification (10μm).

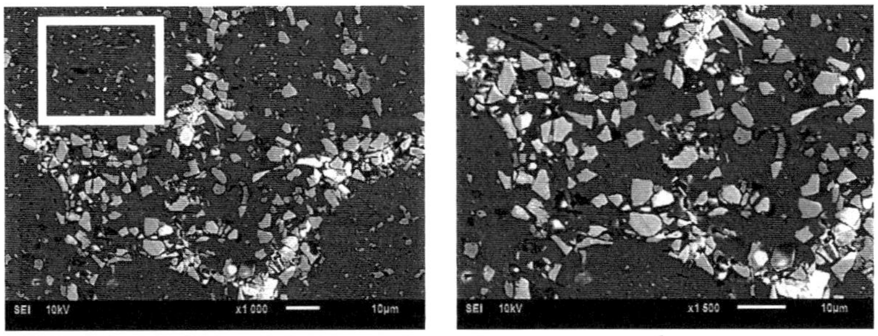

a b

Figure 7. SEM SEI images showing more detailed microstructures of uniaxially compressed 2124-Al with 10%SiC a) 1000X magnification (marker = 10µm) and b) 1500X magnification (marker = 10µm).

3.3. Hardness

Hardness results after uniaxial compression (Figure 8) show that the MMC had higher hardness when compared to the unreinforced Al alloy. The higher hardness in the MMC was attributed to the presence of reinforcing particles. Reinforcing particles are believed to improve hardness of materials by improving load bearing capacity through the load transfer effect. The load transfer effect explains why MMCs (and composite materials in general) are able to bear larger loads than unreinforced materials. According to the load transfer effect, any load that is applied to composite materials is shared between the matrix material and the reinforcement. The reinforcement is stiffer and stronger and is able to bear larger loads than the matrix material. This therefore explains why composites tend to have higher hardness values when compared to the unreinforced material[14].

It can also be observed that there is a variation in hardness from the surface to the core in both the materials. The hardness of the unreinforced 2124-Al was found to be 51.60 $HV_{0.1}$ and 59.16 $HV_{0.1}$ at the surface and core respectively. Hardness values of the MMC at the surface and core were 85.73 $HV_{0.1}$ and 97.96 $HV_{0.1}$ respectively. A higher hardness was observed at the core of the samples and it was thought that this could be due to better densification at the core of the samples during uniaxial compression. There seems to be a relationship between density and hardness since density measurements showed that the 10%SiC MMC had a higher relative density (99.431%) than the unreinforced 2124-Al (95.072%).

Hardness was measured at the surface and core of the samples to evaluate if hardness was consistent from the surface to the core. This inconsistency in hardness implies that there are inconsistencies in densification from the surface to the core, which was observed during the compaction stage and is a well-known phenomenon in compaction. This result is significant because it highlights how complex the deformation process is during uniaxial compression and also shows that deformation can be inconsistent across the length of deforming compacts. It can then be inferred that this variation in hardness implies that resistance to applied load is not uniform. This means that for optimal performance of these materials, this variation in hardness should be taken into account. Furthermore, this result illustrates that surface hardness measurements may not necessarily give a true reflection of the hardness of a sample or component.

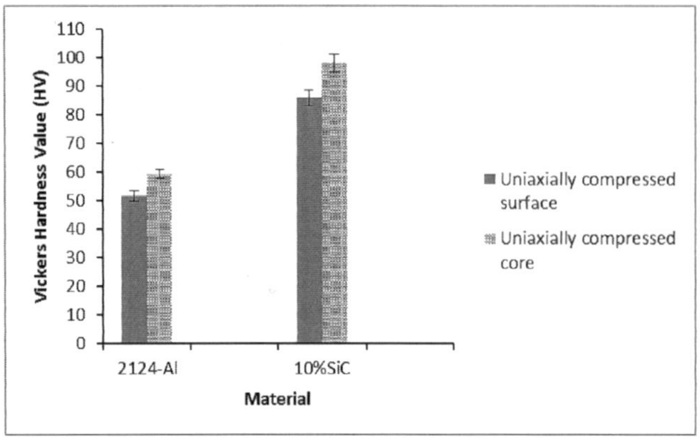

Figure 8. Hardness of surface and core of uniaxially compressed unreinforced 2124-Al and 2124-Al with 10%SiC MMC samples.

4. Conclusions

The influence of warm temperature (0.3-0.5T_m) uniaxial compression on the distribution of reinforcing particles in a 2124-Al/10%SiC MMC was studied and the following conclusions were drawn:

1. Deformation, as a result of uniaxial compression, had a positive influence on the distribution of reinforcing particles since an improvement in distribution of reinforcing particles was only observed in grains that had undergone deformation. This finding corroborated statements made in literature regarding the positive influence of deformation on distribution of reinforcing particles.

2. Reinforcing the 2124-Al alloy with 10 vol. % SiC particles improved flow stress since the 10%SiC MMC had a flow stress of 153 MPa while flow stress of the unreinforced 2124-Al was 95 MPa.

References

[1] Shin K, Chung D and Lee S 2009 The Effect of Consolidation Temperature on Microstructure and Mechanical Properties in Powder Metallurgy – Processed 2XXX Aluminum Alloy Composites Reinforced with SiC Particulates *Metall. Mater. Trans. A* **28**

[2] Tumanov A . and Portnoi K . 1972 Composite materials *Met. Powder Rep.* **57** 35

[3] Delannay F, Colin C, Marchal Y, Tao L, Boland F, Cobzaru P, Lips B, Dellis M, Delannay F, Colin C, Marchal Y, Tao L and Boland F 1993 Processing and properties of metal matrix composites reinforced with continuous fibres for the control of thermal expansion , creep resistance and fracture toughness To cite this version : HAL Id : jpa-00251903 *J. Phys. IV Colloq.* **3** 1675–84

[4] Karnezis P ., Durrant G and Cantor B 1998 Cast Al-Alloy / SiC p Composites C haracterization of Reinforcement Distribution in *Mater. Charact.* **5803** 97–109

[5] Prasad V V B, Mahajan Y R, Ramakrishnan P and Bhat B V . 2002 Structure Á property correlation in discontinuously reinforced aluminium matrix composites as a function of relati v e particle size ratio *Mater. Sci. Eng.* **337** 179–86

[6] Behera R, Majumdar G and Oraon B 2010 Experimental Investigation on the Effect of Reinforcement Particles on the Forgeability and the Mechanical Properties of Aluminum Metal Matrix Composites *Mater. Sci. Appl.* **1** 310–6

[7] Liu Z Y, Wang Q Z, Xiao B L, Ma Z Y and Liu Y 2010 Experimental and modeling investigation on SiC p distribution in powder metallurgy processed SiC p / 2024 Al composites *Mater. Sci. Eng. A* **527** 5582–91

[8] Chen J, Yong C B, Liu W J and Suryanarayana C 2015 Microstructure and Lattice Parameters of AlN Particle-Reinforced Magnesium Matrix Composites Fabricated by Powder Metallurgy *Acta Metall. Sin. (English Lett.* **28** 1354–63

[9] Dutta B, Samajdar I and Surappa M 2013 Particle redistribution and matrix microstructure evolution during hot extrusion of cast SiCp reinforced aluminium alloy matrix composites *Mater. Sci. Technol.* **14** 36–46
[10] Hirianiah C S R A, Harishanad K S and Noronha N P 2012 A review on hot extrusion of Metal Matrix Composites **1** 30–5
[11] Fard R R and Akhlaghi F 2007 Effect of extrusion temperature on the microstructure and porosity of A356-SiCp composites *J. Mater. Process. Technol.* **187–188** 433–6
[12] Saravanan L and Senthilvelan T 2015 Investigations on the Hot Workability Characteristics and Deformation Mechanisms of Aluminium alloy -Al2O3 Nanocomposite *Mater. Des.* **79** 6–14
[13] E92-17 A 2017 *Standard Test Methods for Vickers Hardness and Knoop Hardness of Metallic Materials* (West Conshohocken,PA)
[14] McWilliams B, Ramesh K . and Yen C 2012 Effect of Particle Size Distribution on the Response of Metal Matrix Composites *Dynamic Behaviour of Materials* ed V Chalivendra, B Song and D Casem (New York: Springer Berlin Heidelberg) p 197
[15] Falticeanu C,Chang I T, Kim J and Cook R 2006 *Sintering behaviour of Al-Cu-Mg-Si blends* (Busan,Korea)

IOP Conf. Series: Materials Science and Engineering **655** (2019) 012038 doi:10.1088/1757-899X/655/1/012038

Preparation and characterization of glass-ceramic composites from South African coal fly ash

A Modiga[1,*], N Sosibo[1] and N Singh[1]

[1] Mintek, 200 Malibongwe Drive, Randburg, 2194, South Africa

E-mail: agnesm@mintek.co.za

Abstract. Though fly ash in South Africa has mainly been utilized in the construction industry, a large portion of it is still discarded. Production of glass-ceramic composites through fly ash devitrification has gradually become one of the applications being examined internationally. This study reports the investigation of producing glass-ceramic composites from fly ash and varying quantities of beverage waste glass and magnesium oxide as main raw materials. Glasses were obtained by melting fly ash and its additives using an arc furnace. Based on their behaviour when subjected to differential thermal analysis, the glasses were then crystallised through a double-staged thermal treatment to form glass-ceramic composites. The composition and behaviour of the glass-ceramic composites were investigated using X-ray diffraction, Scanning Electron Microscopy, mechanical and chemical assessments. The crystal content in the glass-ceramic composites increased as the magnesium oxide content was increased. The desired diopside phase decreased with an increased magnesium oxide content and ultimately resulted in the formation of forsterite and anorthite as the major and second major phases respectively. The glass-ceramic composites with lesser amounts of magnesium oxide were chemically and thermally resistant and they also demonstrated high cold compressive strength.

1. Introduction

South Africa has an abundance of coal deposits, and this has led to the country's reliance on coal for energy generation. Eskom, the energy generation utility depends on the combustion of ±120 million tonnes of pulverised coal per annum for producing 90% of its electricity and yielding 30 million tonnes of fly ash [1]. During the year 2014-15, Eskom reported that 34.4 million tonnes of ash were produced and only 7 volume% of it was sold to the construction industry for cement and bricks manufacturing [2]. The remaining quantity was stored in ash dumps or dams which require: lining to prevent seepage of toxic elements contained in the ash into the groundwater, monitoring of groundwater and rehabilitation of the topsoil. These dumps are unsightly and environmentally undesirable resulting in an unprofitable use of land resources as well as a continued financial liability through their long term maintenance. The main commercial utilization of fly ash has been in the construction industry for cement manufacturing due to its cementitious properties. It has also been used as a raw material in brick manufacturing, though the success of new businesses has been a challenge due to its classification as a hazardous waste [2]. Whoever intends to use the fly ash is required by legislation to go through a lengthy process of applying for a Waste Management License, which also is costly. Consequently, the current rate of fly ash production far outweighs its consumption.

Besides its hazardousness, owing to the existence of the trace quantities of toxic heavy metals in its chemical composition, generally, fly ash has substantial amounts of oxides that are key in the ceramic industry. In its molten state, the oxides are categorised as the acidic component as well as the basic component [3]. Silicon dioxide (SiO_2), aluminium oxide (Al_2O_3) and titanium dioxide (TiO_2) form the acidic component and are considered to behave as receptors (network formers). The basic component is formed by calcium oxide (CaO), iron (III) oxide (Fe_2O_3), magnesium oxide (MgO), sodium oxide (Na_2O) and potassium oxide (K_2O). These basic oxides behave as oxygen donators (network modifiers). Additionally, fly ash is partially vitrified with few crystal phases [4] and it can be completely vitrified by melting it with other glass-forming oxides. Consequently, it has been explored as starting material in the production of glass-ceramics [4], [5], [6].

Glass-ceramics are finely structured and multi-crystalline materials (also referred to as advanced or technical ceramics), that are produced by the controlled crystallization (devitrification) of the parent glass. They are partially crystalline. Usually, the microstructure is 50-95 volume% crystalline, with the remainder being an amorphous content [7]. They present an option of combining superior properties of conventional sintered ceramics with the unique characteristics of glasses resulting in physical and mechanical properties such as: high bending strength and hardness, good corrosion as well as chemical, abrasion and high-temperature resistance. They've found use in electrical, magnetic, electronic and optical applications. They are appealing as construction and architectural material or for other specialised technical applications requiring integration of suitable biological and thermo-mechanical properties. Fly ash, if melted at high temperatures with the proper amount of glass-forming additives results in relatively inert glass products. When these glasses are subjected to thermal treatments they are then converted to glass-ceramics with improved mechanical and thermal properties. Moreover, the process is said to yield fly ash reductions of a third to half of the original quantity [8].

Melting fly ash without any additives gave melts that were difficult to handle. Barbieri et al mentioned that melting fly ash with low a SiO_2/Al_2O_3 ratio without any additives have been found to yield glasses with high viscosities at melting temperatures [9]. High amounts of Al^{3+} ions in the ash, act as network formers by replacing the Si^{4+} ions and filling tetrahedrally matched spots with charge-balancing alkali or alkaline earth cations [10]. As a result, the degree of polymerization increases causing the viscosity to increase as well. Nevertheless, an increase of the basic oxides counteracts this behaviour by lowering the connectivity of the glassy network and the liquidus temperature. The weakening of the network results in low viscosity. Beverage waste glass was then added to the fly ash as a source of additional SiO_2 and basic oxides like Na_2O and CaO. Their presence increase the chemical and mechanical strength of the final glass-ceramic product. Additionally, dolomite ($CaCO_3.MgCO_3$) was also added to provide MgO, which facilitates crystallization and improves the thermal resistance of both the parent glasses and the glass-ceramic composites [10]. Moreover, the presence of MgO favours the formation of Diopside ($CaMgSi_2O_6$) as the major crystal phase [10], which have excellent thermal and mechanical properties [11].

This paper reports the conversion of South African fly ash to glass-ceramics. The effects of varying the quantities of beverage waste glass and analytically pure MgO on the microstructure and properties of the glass-ceramics were also studied.

2. Experimental

2.1. Preparation of materials

Fly ash from Kriel power station, beverage waste bottles and analytically pure MgO were used as raw materials. The chemical composition of fly ash and beverage waste glass were determined through inductively coupled plasma optical emission spectrometry (ICP-OES) and atomic absorption spectroscopy (AAS). Beverage waste bottles were added as a source of more SiO_2, Na_2O and CaO.

For the production of four parent glasses, the fly ash quantity was kept constant while the MgO and the beverage waste glass varied as indicated in Table 1. The glasses were labelled as per proportions listed in table 1. The crushed beverage waste glass, fly ash and the MgO were mixed and milled in a laboratory ring pulverizer to increase homogeneity. The raw materials were then packed in alumina crucibles and melted in a large induction furnace at 1550°C for two hours. Thereafter melts were annealed at 550°C for two hours to relieve internal stresses that can potentially cause spontaneous breakage of the glass and subsequently cooled at room temperature.

Table 1. Proportions of raw material used.

	G1	G2	G3	G4
Fly ash (wt %)	60	60	60	60
Beverage waste glass (wt %)	40	30	20	10
MgO (wt %)	0	10	20	30

In order to convert the parent glasses to glass-ceramic composites, the nucleation and crystallization temperature must be determined. Thus differential thermal analysis (DTA) was used and the glass samples were analysed with a Stanton Redcroft STA thermal analyser to 1000°C in the air with a heating rate of 10°C/minute. The glasses were milled to sub 75 microns, and 34 mg of each sample was lightly compacted in a platinum test crucible. Based on the DTA results, a two-staged thermal treatment was employed. The glasses were heated at a rate of 100°C/minute to a nucleation temperature of 750°C and held for 5 hours, thereafter heated to a crystallization temperature of 950°C at a rate of 150°C/minute and held for 20 minutes. Annealing of the samples was at a cooling rate of 450°C/minute to 600°C and the samples were held at the temperature for 2 hours and thereafter cooled to room temperature.

2.2. Characterization of the glass-ceramics

To determine the glass and the crystal content in the glass-ceramic composites, X-ray powder diffraction (XRD) was employed. The analysis was done using a Bruker D8 Advance powder diffractometer, with LynxEye detector and Co-Kα radiation from 3-80° (2θ) with a 0.02° step size. The observed peaks were matched to known minerals or compounds using DIFFRACplus EVA software. Only crystalline materials were detectable and the detection limit for this technique was approximately 3 mass percent. The method is dependent on the diffraction behaviour of the individual phases.

The microstructure of the final products was studied by Scanning Electron microscopy (SEM). A Zeiss EVO® MA15 SEM equipped with Bruker Energy Dispersive X-ray Spectrometer (EDS) and Quantax software were used. The analysis mainly focused on determining the chemical composition of both crystalline and non-crystalline materials by EDS elemental analysis. The backscattered electron (BSE) images were taken to illustrate the phases identified.

2.3. Properties

Numerous South African National Accreditation System (SANAS) accredited methods were used to assess the mechanical and physical characteristics of the glass-ceramic composites. The samples were subjected to a compressive load at room temperature to determine their cold crushing strength based on works procedure WP-308 (Rev 04). Thermal expansions were carried out according to works procedure WP-317 (Rev 02) to determine the fractional change in the size of the samples per degree change in temperature at a constant pressure. A low coefficient of thermal expansion at a high temperature is desired in this instance, which will imply a high thermal shock resistance for the glass-ceramic products. Bulk density by Archimedes was carried out according to works procedure WP-300 (Rev 04). The phase fraction of porosities was quantified using a DSX510 optical microscope. Chemical resistance tests were conducted by immersing unpolished glass-ceramic samples in 10% nitric acid (HNO_3) solution at 25°C for 1 week, 10% sodium hydroxide (NaOH) solution at 25°C for 48 hours and in the water at 25°C for 48 hours.

3. Results and discussions

3.1. Characterization of the glass-ceramics

Table 2 shows the chemical composition of fly ash and beverage waste glass. The elements found in fly ash are commonly reported in their oxide form because the oxide anion is superior in fly ash [3], hence the representation of the results below. The fly ash was classified as "class F" fly ash, with CaO less than 15 wt% and a combination of SiO_2+Al_2O_3+Fe_2O_3 greater than 70 wt%. A low SiO_2/Al_2O_3 ratio of 1.98 was obtained, which is said to result in melts that are too viscous to pour at the melting temperature [12]. Alkaline earth metal oxides, CaO and MgO were also detected in the fly ash at 7.62 wt% and 1.84 wt% respectively. The beverage waste glass had CaO at 11.87 wt% and MgO at 0.64 wt%. Having dual roles in silicate glasses, the metal oxides loosen the silica network by lengthening a bond between two silica tetrahedra and also breaking the Si-O bonds to reduce the viscosity caused by high polymerization [12]. The beverage waste glass had a significant amount of SiO_2, which was also necessary to increase the SiO_2/Al_2O_3 ratio and to ensure high mechanical strength of the glass-ceramic material.

Alkali metal oxides, N_2O and K_2O, which break the Si-O bonds were at 12.96 wt% and 0.39 wt% respectively. Na_2O is the most effective in reducing the liquidus temperature. However, its addition decreases the chemical durability and the hardness of the obtained parent glass, while increasing the coefficient of thermal expansion [12]. To stabilize the glass melt, small quantities of Al_2O_3 (1–5 wt %) are required, thus improving the durability of the parent glass and the glass-ceramics [13].

Table 2. Chemical composition of fly ash and beverage waste glass.

	Composition (wt %)								
	SiO_2	CaO	Al_2O_3	MgO	Na_2O	K_2O	Fe_2O_3	TiO_2	Cr_2O_3
Fly ash	55.95	7.62	28.40	1.84	-	-	4.44	1.75	-
Beverage waste bottles	71.41	11.87	1.59	0.64	12.96	0.39	0.85	-	0.29

Nucleating agents, Fe_2O_3 and TiO_2 were detected at 4.44 wt% and 1.75% respectively in the fly ash, whereas the beverage waste glass had Fe_2O_3 at 0.85 wt% and Cr_2O_3 at 0.29 wt%. These nucleating agents facilitate the development of small crystals within the bulk glass structure, which will result as nucleation sites for the main crystal phases in the glass-ceramic structure [14].

3.2. Thermal behaviour of the glasses

For effective control of nucleation and crystallization processes in glass-ceramics, the behaviour of the parent glasses under different temperatures must be determined. DTA is used to determine glass transition, crystallization, and melting temperatures of the glass as the temperature is gradually increased. During a physical change such as melting or dehydration of the sample, the temperature is absorbed. This will be denoted by an endothermic peak on the DTA curve. Contrarily when a chemical reaction occurs and the microstructure changes, heat will be released resulting in an exothermic peak on the DTA curve.

DTA results in figure 1 showed that G2 sample had a small endothermic event (shallow dip) at ~740°C and a large exothermic event (sharp peak) at ~930°C on heat up. G3 and G4 did not show the endothermic events but showed small exothermic events at ~950°C on heat up. Due to the limited heating range of up to 1000°C, G1 sample had no endothermic or exothermic events during heat up, just a shallow hump at 900°C. The additional SiO_2 from the beverage waste glass made G1 to be stable under the heating range. All the glasses showed the same small endothermic event at ~680°C on cooldown.

At 10 wt% MgO, the exothermic event of the G2 was sharp at 930°C but as the MgO content increased from G3 to G4, the peak became shallow and almost disappeared in glass sample 4. This can be attributed to the increased melting temperature of the samples by the addition of MgO. The thermal treatments for all the glasses were then conducted at a nucleation temperature of 750°C which was 10°C higher than the glass transition temperature for G2 and crystallization temperature of 950°C.

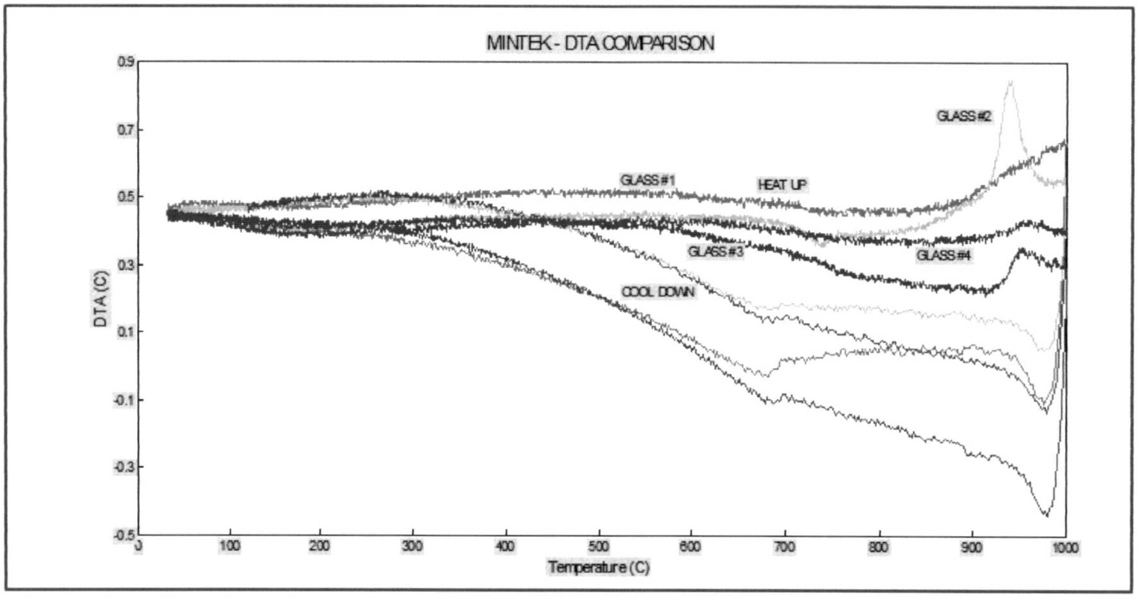

Figure 1. DTA comparison of the glass samples.

3.3. Characterization of the glass-ceramic

XRD analysis in table 3, indicated that after thermal treatments, the glass phase decreased as the MgO content increased. This can be attributed to the fact that increasing the addition of MgO content while decreasing the beverage waste glass content favoured crystallization. The silicate network was broken by MgO which facilitated the formation of crystalline phases. Initially, single silicate chain, pyroxene crystal phases were formed. Diopside ($CaMgSi_2O_6$) was formed from G1, composed of only fly ash and beverage waste glass. An introduction of MgO resulted in a slight decrease in diopside phase and a formation of another pyroxene crystal phase, augite ($(Ca,Na)(Mg,Fe,Al)(Al,Si)_2O_6$). The pyroxene phase was formed by the dissolution-precipitation process, wherein the dissolution of MgO in the viscous glassy phase initially occurred, and thereafter precipitated as pyroxene [15]. As the MgO content increased to 20 wt%, the diopside decreased with augite transforming to forsterite (Mg_2SiO_4), which is a mineral of the isolated silicate group called olivines.

At 30 wt% MgO content, the diopside quantity remained constant while forsterite dropped slightly and the glass phase having a notable decrease with the formation of very strong 3-dimensional framework silicates of the plagioclase feldspar group. The crystal phases formed were anorthite ($CaAl_2Si_2O_8$), dmisteinbergite ($CaAl_2Si_2O_8$) and bytownite ($(Ca,Na)(Si,Al)_4O_8$), indicating limited dissolution of MgO in the glass phase.

Table 3. Main phase compositions of the glass-ceramic composites with different MgO contents.

Phases	MgO content (wt %)			
(wt %)	0	10	20	30
Glass phase	90	77	55	24
Pyroxene (diopside)	10	8	5	5
Pyroxene (Augite)	-	12	-	-
Olivine (Forsterite)	-	-	39.6	31
Anorthite	-	-	-	16
Dmisteinbergite	-	-	-	9
Bytownite	-	-	-	14

Figure 2 shows SEM images of the glass-ceramic components. By increasing the MgO content from 0-30 wt%, the crystal phases also increased. The phases progressed from the "worm-like" textures of Diopside phase at 40 wt % and 0 wt% of beverage waste glass and MgO respectively, to a dendritic pattern when 10 wt% MgO was introduced. As the MgO content increased to 20 wt%, spinifex texture of acicular olivine phenocrysts (dark grey lines) was observed in the glass groundmass. Ultimately at 30 wt% MgO and 10 wt% beverage waste glass, large euhedral crystals were formed. The plagioclase feldspar group minerals (light grey), anorthite, bytownite and dmisteinbergite were the major phases formed. Forsterite (dark grey) was also observed. As the MgO content was increased, the beverage waste glass content was reduced to make a total mixture of 100 wt %.

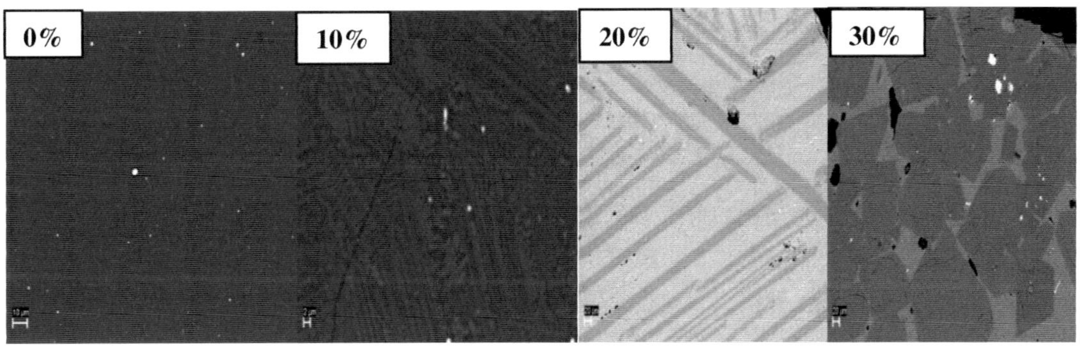

Figure 2. SEM images of the glass-ceramic composites formed.

3.4. Properties

Figure 3 shows the correlation between the bulk density and the MgO content of the glass-ceramic samples. Addition of MgO increased the bulk density, which stabilized at 10 wt% MgO at 2.72 g/m^3. More additions of the MgO resulted in a drop to 2.70 g/m^3. As more crystals were formed due to the increased MgO and the reduced glass contents, porosity increased as well from 0.21% at 0 wt% MgO to 5.38% at 30 wt% MgO resulting in less compaction of the structure, hence the decline in the densities. This can be attributed to the decline in the glass-network interconnectivity due to increased crystallisation of more than 70%.

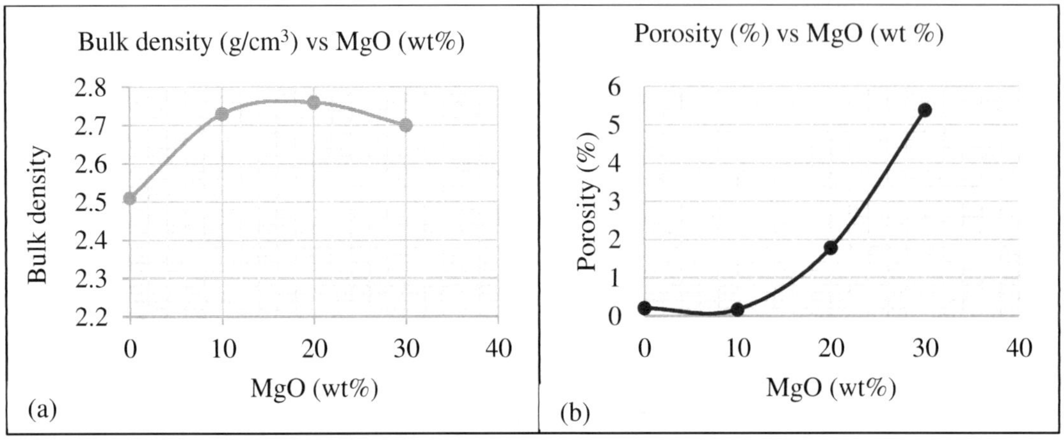

Figure 3. (a) Bulk density and (b) porosity as functions of MgO content.

The cold crushing strength (CCS) of the glass-ceramic in figure 4 increased to a maximum at 10 wt% MgO but dropped afterwards. This can be attributed to increased crystal content and porosity in the glass-ceramic composites that promoted cracking. The maximum CCS achieved was 118.4 Megapascal (Mpa) at 10 wt% MgO content.

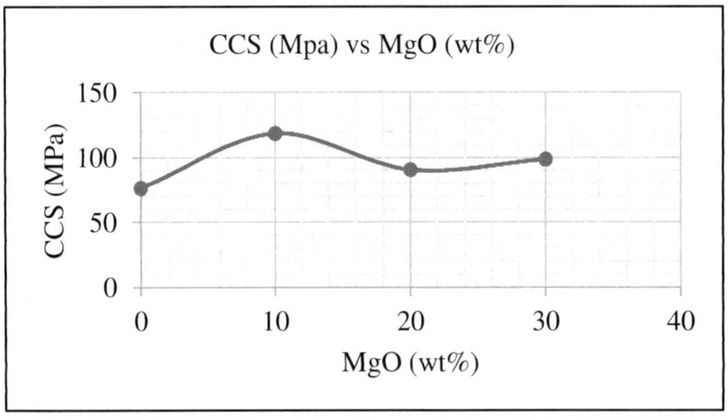

Figure 4. Cold crushing strength as a function of MgO content.

The coefficient of thermal expansion (CTE) of the glass-ceramic composites in table 4 increased with the MgO content from 6.8 x10^{-6}/°C at 0 wt% MgO to 10.8x10^{-6}/°C at 30 wt%. Increasing of the basic oxides, while reducing the silica content causes the CTE of the resulting glass to increase.

Table 4. Coefficient of thermal expansion (CTE) of the glass-ceramic composites.

MgO wt%	Melting Temperature	CTE(/°C)
0	724°C	6.8 x 10^{-6}
10	977°C	7.5 x 10^{-6}
20	>999°C	8.9 x 10^{-6}
30	>999°C	10.8 x 10^{-6}

IOP Conf. Series: Materials Science and Engineering **655** (2019) 012038 doi:10.1088/1757-899X/655/1/012038

The addition of oxides results in breakage of the connectivity of the glass network, which then softens the structure of glass. The increased CTE indicates that the glass-ceramic composites are less stable to temperatures above 999°C.

Figure 5. Glass-ceramic samples after soakage in distilled water.

After soakage in distilled water and being air-dried, there were no changes in all the glass-ceramic samples shown in figure 5. The traces of rust accumulation on glass-ceramic samples 3 and 4 were identified before its immersion.

Figure 6. Glass-ceramic samples after soakage in HNO_3 solution.

Figure 6 displays the glass-ceramic samples after soakage in an acidic solution. The glass-ceramic samples with 0-20 wt% MgO did not undergo any mass loss, but there were noticeable rust accumulations on the samples after they were air-dried. Glass ceramic sample with 20 wt% MgO had more accumulations than the others. At 30 wt% MgO, the glass-ceramic sample turned greyish and there was a 25% mass loss.

Figure 7. Glass-ceramic samples after soakage in NaOH solution.

In an alkali solution, all glass-ceramic samples were stable as shown in figure 7, except for the glass-ceramic sample with 30 wt% MgO which turned whitish with no mass loss though. The stability of glass-ceramic samples 1, 2 and 3 respectively can also be attributed to the glass phase content which was between 55-90 wt%. The glass network connectivity was stable enough to withstand the alkali attack. However, the glass network connectivity for glass-ceramic sample 4 was weaker due to increased crystallization.

4. Conclusion

The formation of glasses from South African fly ash with the addition of beverage waste glass and magnesium oxide was achieved by mixing and melting them at 1550°C. The parent glasses were easily crystallized through thermal treatment to form glass-ceramic composites with a diopside crystal phase. The conclusions are as follows:

- Addition of 10 wt% MgO seems ideal with less crystallinity and improved mechanical properties.
- The glass-ceramic sample with 10% MgO behaved excellently, displaying the best mechanical properties.
- Increasing the MgO content, while reducing the beverage waste glass resulted in a decline of the glass network connectivity leading to increased crystallization and poor mechanical properties of the glass-ceramic composites.
- The glass-ceramic composite with 10 wt% MgO had a cold crushing strength of 118.4 Mpa, which is comparable to the ones for bricks (50-140 Mpa) and stone (34-248 Mpa).
- The coefficient of thermal expansion at 10 wt% MgO was 7.5 x10-6/°C at a melting temperature of 977°C which is comparable to the ones for granite (7.92 x10-6/°C) at melting temperatures 1215-1226°C.

5. Acknowledgments

The authors would like to thank Mintek for the financial and technical support as well as Cermalab for their guidance and mechanical testing of the samples.

6. References

[1] Mashau S, Gitari W and Akinyemi A 2018. *Int. J. Environ. Res. Public Health* **15** (12) p 2841
[2] Reynolds-Clausen K and Singh N 2019. *CCGP Journal* **11** pp 10-7
[3] Xuan W, Zhang J and Xia D 2018. *Fuel* **222** pp 523-8
[4] Goodarzi F 2006. *Fuel* **85** (10-11) pp 1418-27
[5] Wang S, Zhang C and Chen J 2014. *J. Mater. Sci. Technol.* **30** (12) pp 1208-12
[6] Barbieri L, Lancellotti I, Manfredini T, Pellacani C, Rincón M and Romero M 2001. *J. Am. Ceram. Soc.* **84** (8) pp.1851-8
[7] Rawlings D, Wu P and Boccaccini R 2006. *J. Mater. Sci.* **41** (3) pp 733-61
[8] Ghouleh Z, Isac M, Guthrie R, Carabin P, and Kozinski J 2009. *8th Int. Conf. on Molten Slags, Fluxes, and Salts* pp 18-21
[9] Barbieri L, Manfredini T, Queralt I, Rincón M and Romero M 1997. *Glass Technol.* **38** (5) pp 165-70
[10] Xuan W, Whitty K, Guan Q, Bi D, Zhan Z and Zhang J 2015. *Fuel* **144** pp 103-10
[11] Park J and Heo J 2002. *Ceram. Int.* **28** (6) pp 689-94
[12] Vargas S, Frandsen F and Dam-Johansen K 2001. *PECS* **27** (3) pp 237-429
[13] Hülsenberg D, Harnisch A and Bismarck A 2008. *Microstructuring of glasses* Vol. 87 R Hull (Berlin: Springer) p 326
[14] Rezvani M, Eftekhari-Yekta B, Solati-Hashjin M and Marghussian V 2005. *Ceram. Int.* **31**(1) pp 75-80
[15] Partyka J 2015. *Ceram. Int.* **41** (10) pp 14013-20
[16] Lu Z, Lu J, Li X and Shao G 2016. *Ceram. Int.* **42** (2) pp 3452-9

IOP Conf. Series: Materials Science and Engineering **655** (2019) 012039 doi:10.1088/1757-899X/655/1/012039

Development of $MgAl_2O_4$ grain refiner in Al in-situ composite through H_3BO_3 addition

[*]P Chandramohan[1,2], P A Olubambi[2], R Raghu [1]and B A Obadele[2]

[1]Department of Mechanical Engineering, Sri Ramakrishna Engineering College, Coimbatore 641 022, India

[2]School of Mining, Metallurgy and Chemical Engineering, University of Johannesburg, South Africa 2028

[*]E-mail: pcmohu@yahoo.co.in

Abstract. In the present work, a novel approach has been made for synthesis of in-situ Magnesium Aluminate ($MgAl_2O_4$) particles in the Al-4Mg alloy by Boric Acid (H_3BO_3) precursor addition (1 wt %, 1.5 wt % and 2 wt %) during the casting process. The developed composite has been investigated for its microstructural characteristics and corrosion performance. Scanning Electron Microscopy and Energy Dispersive Spectroscopy examination revealed the formation of $MgAl_2O_4$ particles in the composite. Potentiodynamic polarization corrosion experiments were performed on the Al-4Mg/H_3BO_3 composite specimens (1 wt %, 1.5 wt % and 2 wt %) in three different medium (3.5 % Sodium Chloride-NaCl, 1 M Sulphuric Acid-H_2SO_4 and 1 M Hydrochloric Acid-HCl). Corrosion results showed that Al-4Mg/1.5 wt % H_3BO_3 composite specimen exhibited better corrosion resistance in 3.5 % NaCl, 1 M H_2SO_4 and 1 M HCl medium due to the significant grain refinement produced by $MgAl_2O_4$ particles. The developed composite with better corrosion properties can be utilized for marine and naval application.

1. Introduction

Aluminium 5xxx series alloys (Al-Mg) are widely used in the transportation field, marine industry and other industries due to their better corrosion resistance and weldability [1,2]. Among 5xxx series alloys, the 5083 alloy (Al-4Mg) contains Mg as the solute alloying element, which aids in improvement of corrosion performance. However, the 5083 alloy containing more than 3 % Mg suffers from intergranular and pitting corrosion because of β phase (Al_3Mg_2) precipitation at the grain boundaries [3,4]. Corrosion attack in this alloy reduces the lifetime of the component and safety of the application [5, 6]. It is therefore highly essential to control the corrosion resistance in these aluminium alloys.

The corrosion behaviour of aluminium relies on several factors such as the condition of the material, microstructural features of the material and the electrolyte. Among these, microstructure affects the corrosion of the alloy significantly [7-9]. Various researchers have attempted to increase the corrosion resistance of aluminium alloys through the addition of elements such as Strontium (Sr), Neodymium (Nd) and Zinc (Zn) [10-12]. It is stated that grain refinement in the aluminium alloy improves the corrosion resistance [13]. Traditionally, addition of inoculants is the commonly approach for grain refinement and also distribution of the second phase in the aluminium alloy [14-17]. Various methods have been attempted by several researchers to produce grain refinement in the 5083 aluminium alloy for better corrosion performance.

Ali Torkan et al. (2018) [18] investigated the microstructural and corrosion characteristics of the accumulative roll bonded (2, 4 and 6 passes) aluminium alloy (5083) in 3.5% NaCl solution. Microstructural examination of the aluminium alloy revealed that accumulative roll bonding aids in development of nanostructured (up to 10 nm) microstructure. Polarization tests were carried out on the aluminium specimens and it was reported that the corrosion rate decreases up to 4 passes and then increases. Shuiqing Liu et al. (2019) [19] attempted to improve the corrosion performance of 5083 aluminium alloy through the addition of nano-Cerium Hexaboride (CeB_6)/Al inoculant. Potentiodynamic polarization tests were carried out on the inoculated alloy, in a 3.5 wt% NaCl solution, and its corrosion performance was compared to a cold-rolled alloy and as-cast alloy. It was observed that grain refinement (up to 100 μm) occurred with the addition of nano-CeB_6/Al inoculant, which resulted in high corrosion resistance better than the other two alloys.

Sadegh Rasouli et al. (2014) [20] used a Friction Stir Processing (FSP) technique to improve the microstructural and mechanical properties of the AA5083 aluminum alloy. It is observed that dynamic recrystallization resulted due to intense plastic deformation produced during FSP. Polarization tests performed on the FSPed alloy resulted in higher corrosion resistance than base metal which is due to the microstructural refinement caused by FSP.

Various techniques are available for grain refinement of the Al5083 alloy. Among those, recently developed technique is the generation of the Al5083 composites through the addition of reinforcements. Prabhakar et al. (2018) [21] fabricated the Al5083 aluminum/fly ash surface composites through friction stir processing (FSP) under different speeds and feed rates, and investigated its mechanical properties. Results revealed that the FSPed composite (97.2 HV) has higher hardness than the FSPed alloy (83.5 HV) and unprocessed alloy (68.3 HV) which is attributed to the grain refinement due to the addition of flyash.

Meijuan Li et al. (2016) [22] developed a nanostructured Al5083/nano-Titanium Diboride (TiB_2) composite through cryomilling and spark plasma sintering (SPS). Microstructural observation revealed greater refinement of the Al matrix having an average grain size of ~74 nm and a homogeneous distribution of (n-TiB_2) in the Al matrix. Ranjit Bauri et al. (2014) [23] incorporated tungsten particles in the aluminium 5083 matrix through FSP and studied its microstructural and mechanical properties. Uniform dispersion and greater interfacial bonding was observed between the tungsten and aluminium matrix. Results revealed that Ultimate Tensile Strength (UTS) of the composite increased more than 100 MPa and ductility increased up to 30 % which is due to substantial grain refinement of the matrix because of dynamic recrystallization.

From the literature review, it is observed that various research has concentrated on the microstructural and mechanical properties of Al5083 alloy and ex-situ 5083 composites. It is observed that in these studies, the improvement of properties was attained through grain refinement. However, the development of in-situ Al composite has rarely been reported. In addition, corrosion behaviour of such composites has not been investigated. Hence in the present study, Al5083 (Al-4Mg) alloy was prepared in the laboratory and in-situ Al-4Mg/$MgAl_2O_4$ composite was produced with the addition of H_3BO_3 (1 wt %, 1.5 wt % and 2 wt %). It is anticipated that the addition of $MgAl_2O_4$ in the alloy will aid in grain refinement and improvement of corrosion resistance.

2. Experimental procedure

Al-4Mg alloy was cast using Electrical Conductor (EC) grade pure aluminium and commercial grade pure magnesium, melted using an electric resistance furnace. H_3BO_3 of varying wt.% (1, 1.5 and 2) with an average grain size of 20 μm was taken as the oxide source and incorporated into the Al-4Mg melt. H_3BO_3 is an oxide source which tends to react with the Al and forms $MgAl_2O_4$ reinforcement. The melt is kept at 750 °C for 15 min to aid the reaction of Al-4Mg melt with H_3BO_3. The melt is then poured into the die. The final product is a composite that was examined through Scanning Electron Microscope (SEM) with Energy Dispersive Spectrometer (EDS), to observe the morphology and confirm the presence of $MgAl_2O_4$.

IOP Conf. Series: Materials Science and Engineering **655** (2019) 012039 doi:10.1088/1757-899X/655/1/012039

Al-4Mg/$MgAl_2O_4$ in-situ composite specimens were taken for potentiodynamic polarization experiments and the experiments were conducted (Potentiostat - Autolab 302 NFRA2, Metrohm). The electrochem system is equipped with Nova 2.1 software which is used for Tafel extrapolation. For the corrosion experiments, a platinum rod is taken as the counter electrode, saturated silver/silver chloride (Ag/AgCl) as the reference electrode and the Al-4Mg/ H_3BO_3 composite specimens were taken as the working electrode. Polarization experiments were conducted from -0.3 V to 2 V at a scan rate of 2 mV s^{-1} using three different solutions 3.5 % NaCl, 1 M H_2SO_4 and 1 M HCl at room temperature.

3. Results and discussion

The $MgAl_2O_4$ products are produced due to various reactions through the reactive wetting of the H_3BO_3 and the Al-4Mg melt [24-26]

$$4H_3BO_{3(s)} \rightarrow 2B_2O_{3(s)} + 6H_2O_{(g)}, \quad T = 200°C \quad (1)$$

$$B_2O_{3(s)} \rightarrow B_2O_{3(l)}, \quad T = 450°C \quad (2)$$

$$B_2O_{3(l)} + 3Mg \rightarrow 3MgO + 2B, \quad T = 650°C \quad (3)$$

$$B_2O_{3(l)} + 2Al + MgO \rightarrow MgAl_2O_4 + 2B, \quad T = 650°C \quad (4)$$

Reaction products formed tend to increase the wettability between the reinforcement and the melt. The interfacial energy is reduced due to the reaction between the substrate and the reinforcement. The displacement reactions (3) and (4) take place consequently and the reaction products such as MgO or $MgAl_2O_4$ are formed at molten metal/particle interfaces. The presence of Al and Mg at the reaction zone aids the MgO and $MgAl_2O_4$ formation. It is anticipated that MgO formation takes place easier than $MgAl_2O_4$ due to greater reactivity of Mg than Al. MgO transforms into $MgAl_2O_4$ particles upon diffusion of Al atom.

3.1. SEM and EDS analysis

SEM examination has been carried out in the Al-4Mg/1, 1.5 and 2 wt% H_3BO_3 composite specimens to observe the morphology of the $MgAl_2O_4$. EDS was done to confirm the formation of $MgAl_2O_4$ crystals. The polygonal faceted type $MgAl_2O_4$ crystals are observed in the microstructures of the Al-4Mg/1, 1.5 and 2 wt% H_3BO_3 composite specimens and successfully distributed in the Al-Mg matrix through the addition of H_3BO_3 as shown in Figure 1a-c. This was also confirmed by the EDS spectra that show the presence of Mg, Al and O in the $MgAl_2O_4$ crystals (Figure 2a-c).

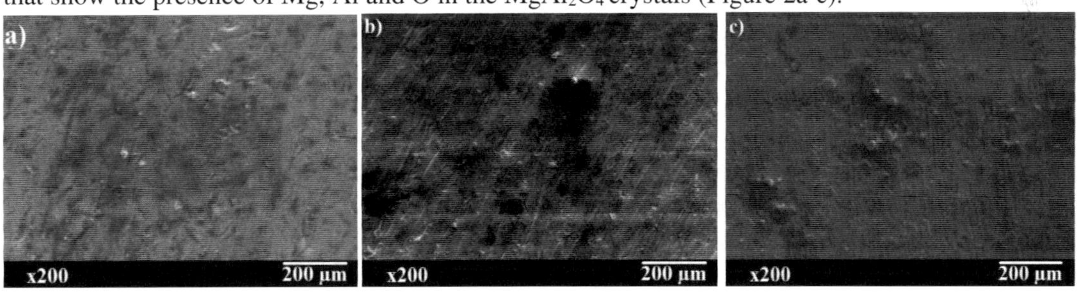

Figure 1.SEM micrograph of the Al-4Mg/H_3BO_3 composite specimen at lower magnification (a) 1 wt % (b) 1.5 wt % and (c) 2 wt %.

Nevertheless, the size and quantity of the $MgAl_2O_4$ crystals formed in the composite specimens are different. The average size and fraction of the reinforcement particles has been determined through the Image analysis software Image J. In the Al-4Mg-1 wt % H_3BO_3, the size of the $MgAl_2O_4$ crystals found very small (4 µm) and number of particles formed is also less (1.5 %). In the Al-4Mg-1.5 wt % H_3BO_3, an increased amount (2.3 %) of $MgAl_2O_4$ with a medium size (6.36 µm) is formed and upon further increase in H_3BO_3 (2 wt %), the amount (3.3 %) and size (7.5 µm) of $MgAl_2O_4$ crystals increases with few agglomerations.

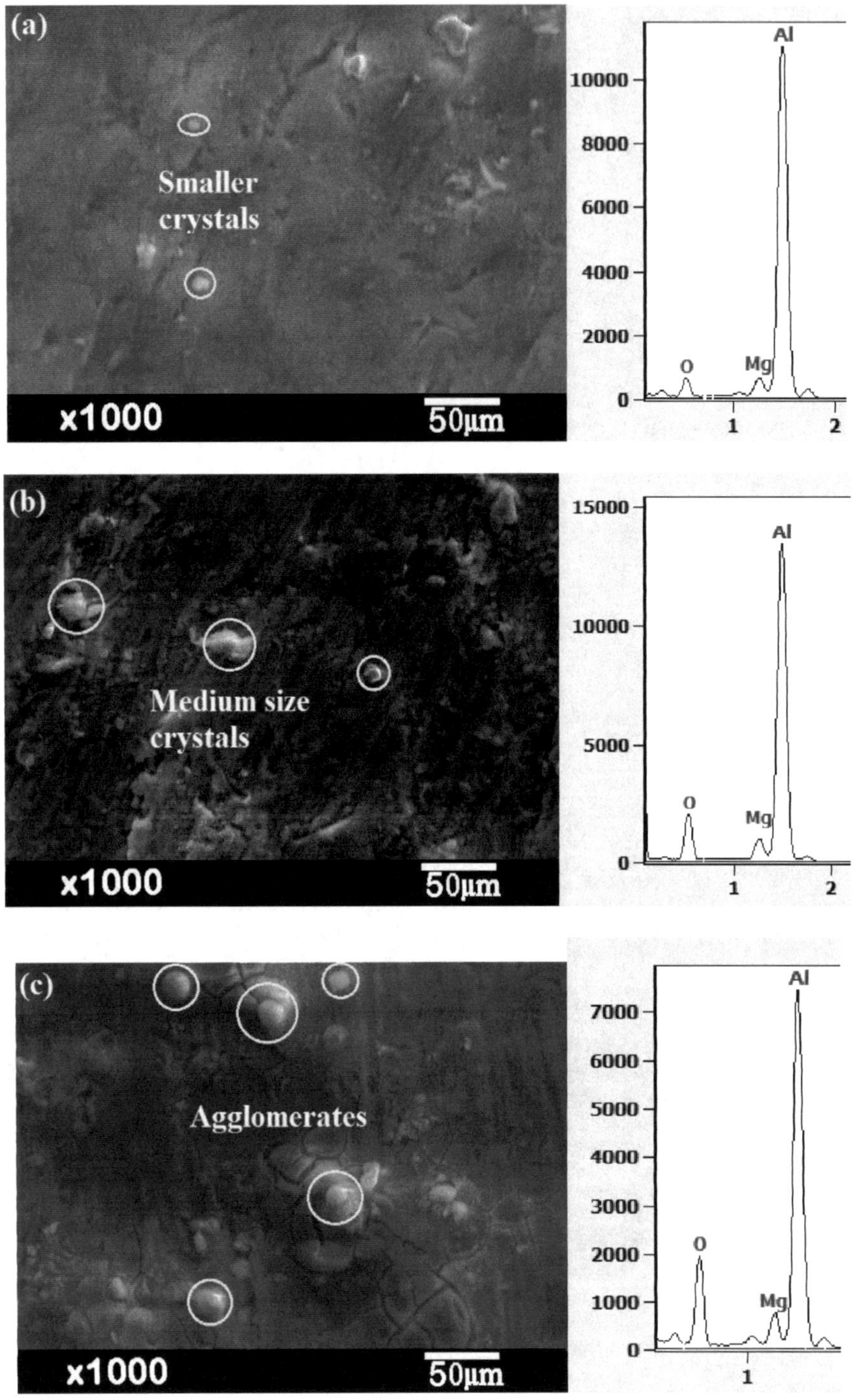

Figure 2.SEM micrograph and EDS spectra of the Al-4Mg/H_3BO_3 composite specimen (a) 1 wt % (b) 1.5 wt % and (c) 2 wt %.

Thus, it is confirmed that the $MgAl_2O_4$ formed in the composite is due to the addition of H_3BO_3 oxide source. The $MgAl_2O_4$ allows for heterogeneous nucleation in the molten melt and refines the structure (Figure 3a-c). $MgAl_2O_4$ has been reported as a stable oxide and potent nucleant of Al grains owing to its cube-on cube orientation relationship [27]. These potent nucleants tend to decrease the activation barrier for nucleation. As per the classical nucleation theory, this activation barrier greatly depends on the interfacial energy between the nucleus metal and the nucleant for the heterogeneous nucleation. The decreasing level of activation barrier required for heterogeneous nucleation relies on the contact angle between the nucleus metal and the nucleant. The significant factor that affects the heterogeneous nucleation is the contact angle and it must be low as possible (lesser than 30°) for the nucleant to be highly potent. The lattice misfit between nucleating phase and nucleant must be low to reduce the interfacial energy to attain a low contact angle. The $MgAl_2O_4$ particles formed in the matrix has a low lattice mismatch and possess a similar cubic structure as that of aluminium hence it is a potent nucleant for the aluminium. A low misfit of 1.41 % between $MgAl_2O_4$ and aluminium is reported by Li et al. [27] on the (111) plane along the direction of [110]. $MgAl_2O_4$ possess a low lattice misfit to act as a potent heterogeneous nucleant. Nevertheless, its potency level relies on various parameters such as number density, particle size and narrow size distribution. A high amount and smaller size $MgAl_2O_4$ particles will result in better heterogeneous nucleation.

3.2. Corrosion behaviour

This study is performed to determine the corrosion performance of the composite specimens in three different solutions. 3.5 % NaCl solution is preferred since the sodium and chloride ions are present at greater levels in the marine atmosphere (seawater). Acidic and alkaline media are generally selected for study and hence the standard solutions of H_2SO_4 and HCl are used. Corrosion rate obtained for the Al-4Mg-H_3BO_3 (1, 1.5 and 2 wt %) composite specimens tested in three different solutions i.e., 3.5 % NaCl, 1 M H_2SO_4 and 1 M HCl at room temperature are displayed in Table 1.

Table 1. Corrosion test results of Al-4Mg/ H_3BO_3 composite

SOLUTION	Parameters	1 wt% H_3BO_3	1.5 wt% H_3BO_3	2 wt% H_3BO_3
3.5 % NaCl	E_{corr} (V)	-0.97	-0.93	-0.97
	i_{corr}(A/cm^2)	1.35E^{-05}	8.89E^{-07}	1.03E^{-05}
	Corrosion rate(mm/yr)	0.44	0.02	0.33
1 M H_2SO_4	E_{corr} (V)	-0.63	-0.64	-0.61
	i_{corr}(A/cm^2)	0.00011	3.82E^{-05}	9.09E^{-05}
	Corrosion rate(mm/yr)	3.79	1.248	2.96
1 M HCl	E_{corr} (V)	-0.71	-0.75	-0.76
	i_{corr}(A/cm^2)	0.0002	0.0001	0.0001
	Corrosion rate(mm/yr)	7.27	3.67	3.30

3.2.1. In the case of the 3.5 % NaCl solution, the corrosion rate of the Al-4Mg/1.5 wt% H_3BO_3 composite (0.029043 mm/yr) specimen is found lower than that of the Al-4Mg/1 wt % H_3BO_3 (0.44169 mm/yr) and Al-4Mg/2 wt % H_3BO_3 composite specimen (0.33594 mm/yr). The lower corrosion rate is due to lower corrosion current density of 8.89E^{-07} compared to other two composite specimens. The corrosion rate is higher for the low reinforcement fraction (1 wt % H_3BO_3) as there is no appreciable formation of $MgAl_2O_4$ and hence no grain refinement as observed in Figure 3a. The low corrosion rate obtained at 1.5 wt % H_3BO_3 is attributed to the grain refinement due to sufficient amount and reinforcement of $MgAl_2O_4$ without agglomeration. Increasing the H_3BO_3 causes agglomeration which might be increasing the corrosion rate.

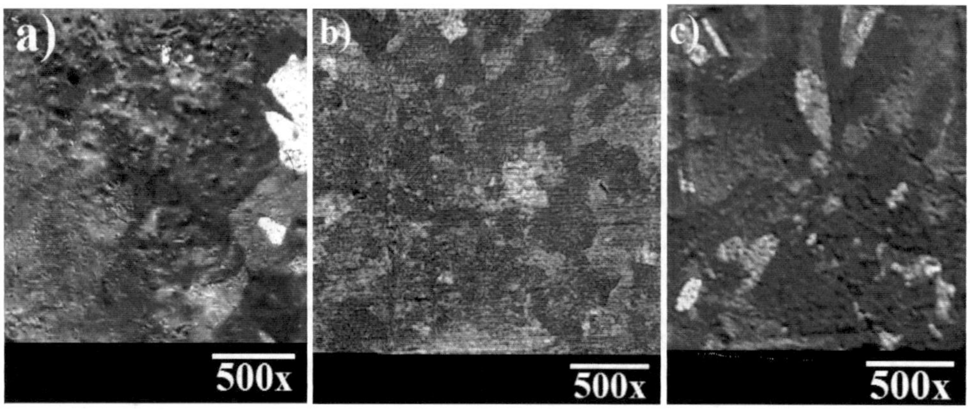

Figure 3. Grain refinement of Al-4Mg/H_3BO_3 composite a) 1 wt % H_3BO_3 b) 1.5 wt % H_3BO_3 and c) 2 wt % H_3BO_3

The SEM analysis results of the surface of the corroded specimen tested in 3.5 % NaCl is shown in Figures 4a-c. The pit density is quantified with the aid of Image analysis software Image J. The corroded surface of the specimens showed pits with the pit density increasing (15 %, 29 % and 35 %) as the H_3BO_3 level (1 wt %, 2 wt % and 3 wt %) increased respectively.

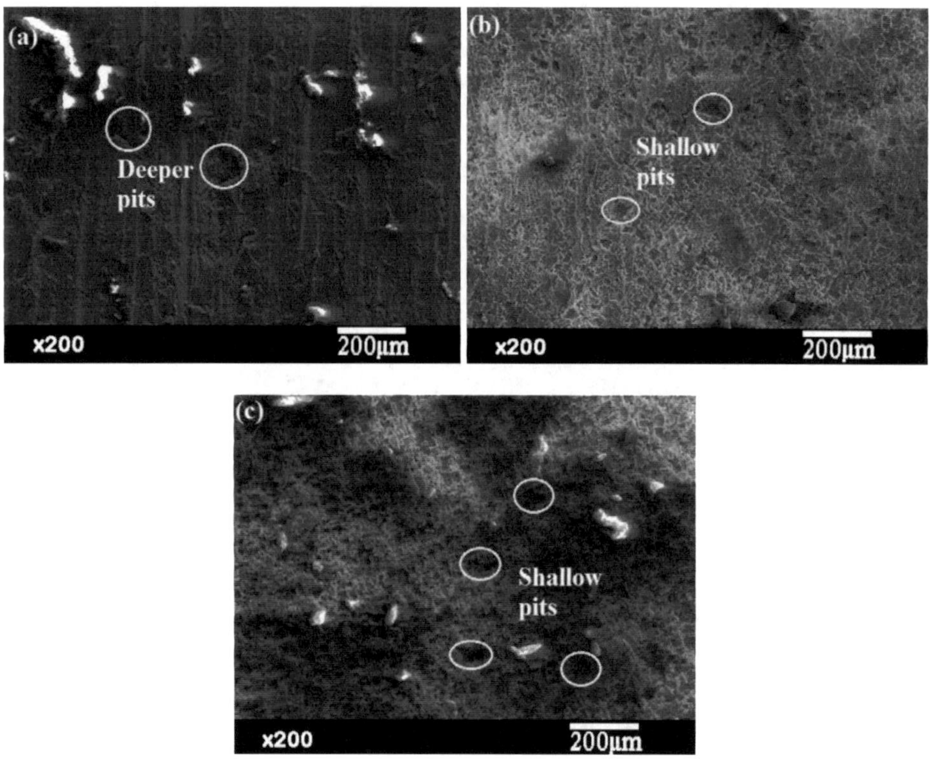

Figure 4. Corroded specimens tested in 3.5 % NaCl solution (a) 1 wt % (b) 1.5 wt % and (c) 2 wt%.

3.2.2. Similarly to 3.5 % NaCl solution, in 1 M H_2SO_4 the Al-4Mg/1.5 wt% H_3BO_3 composite specimen exhibited lower corrosion rate (1.248 mm/yr) than that of Al-4Mg/1 wt % H_3BO_3 (3.7973 mm/yr) and Al-4Mg/2 wt % H_3BO_3 composite specimens (2.9693 mm/yr). This is due to the lowest corrosion current density of $3.82E^{-05}$ obtained for the Al-4Mg/1.5 wt% H_3BO_3 composite specimen compared to other two specimens. The corrosion trend is decreases initially upto 1.5 wt % and further increases when reaching 2 wt%. This is also attributed to the appreciable grain refinement produced by the $MgAl_2O_4$ crystals at the fraction of 1.5 wt% H_3BO_3. The SEM analysis on the surface of the corroded specimen tested in 1 M H_2SO_4 is shown in Figure 5a-c which is also agreeing with the trend of corrosion rate obtained for different fraction. The amount of corrosion products is determined through the Image analysis software Image J. The surface with 1 wt % H_3BO_3 (Figure 5a) revealed high amount of corrosion products (38 %) whereas 1.5 wt % reinforced composite surface (Figure 5b) showed mild pitting with lesser corrosion product (1.3 %).

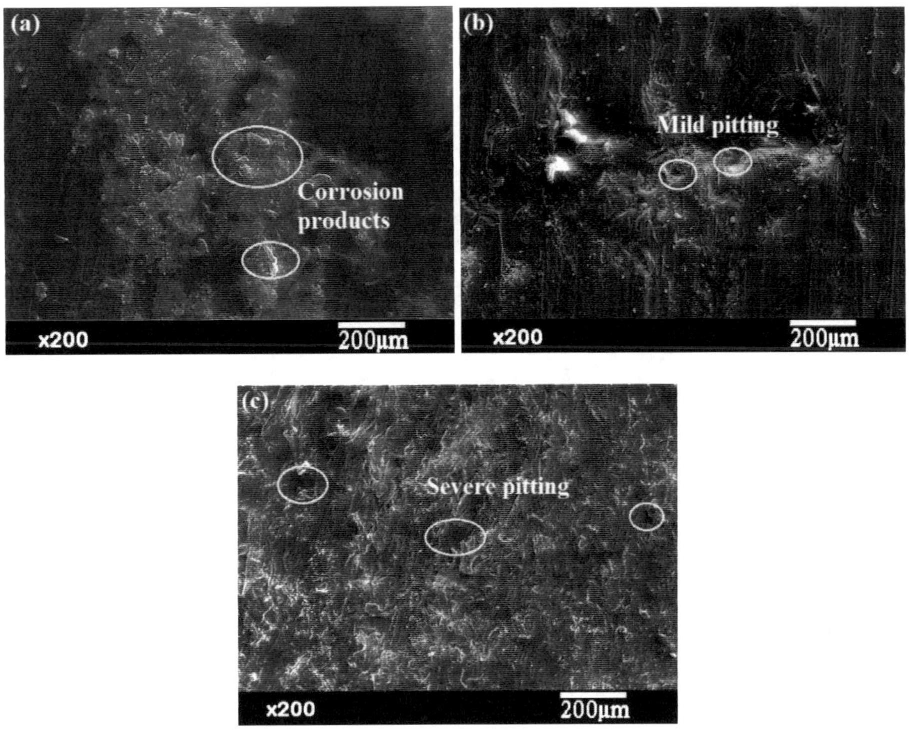

Figure 5. Corroded specimens tested in 1 M H_2SO_4solution (a) 1 wt % (b) 1.5 wt % and (c) 2 wt%.

3.2.3. In 1M HCl solution, the corrosion rate is found lowest in the case of the Al-4Mg/2 wt% H_3BO_3 composite specimen (3.3045 mm/yr) than that of the Al-4Mg/1 wt % H_3BO_3 (7.2793 mm/yr) and Al-4Mg/1.5 wt % H_3BO_3 composite specimens (3.6761 mm/yr). The corrosion rate is found initially higher at 1 wt H_3BO_3 and decreases as the reinforcement fraction increases. There is no appreciable difference in the corrosion rate of 1.5 wt % and 2 wt % H_3BO_3. In HCl, both the composites (1.5 wt % H_3BO_3 and 2 wt % H_3BO_3) possess better corrosion resistance. The SEM analysis on the surface of the corroded specimen tested in 1 M HCl is shown in Figure 6a-c. The corroded surface of the 1 wt % H_3BO_3 specimen (Figure 6a) revealed severe granular corrosion showing cracks on the surface evident the high corrosion rate (7.2793 mm/yr). The corroded surface of 1.5 wt % H_3BO_3 (Figure 6b) and 2 wt % H_3BO_3 (Figure 6c) composites indicates the moderate granular corrosion attack.

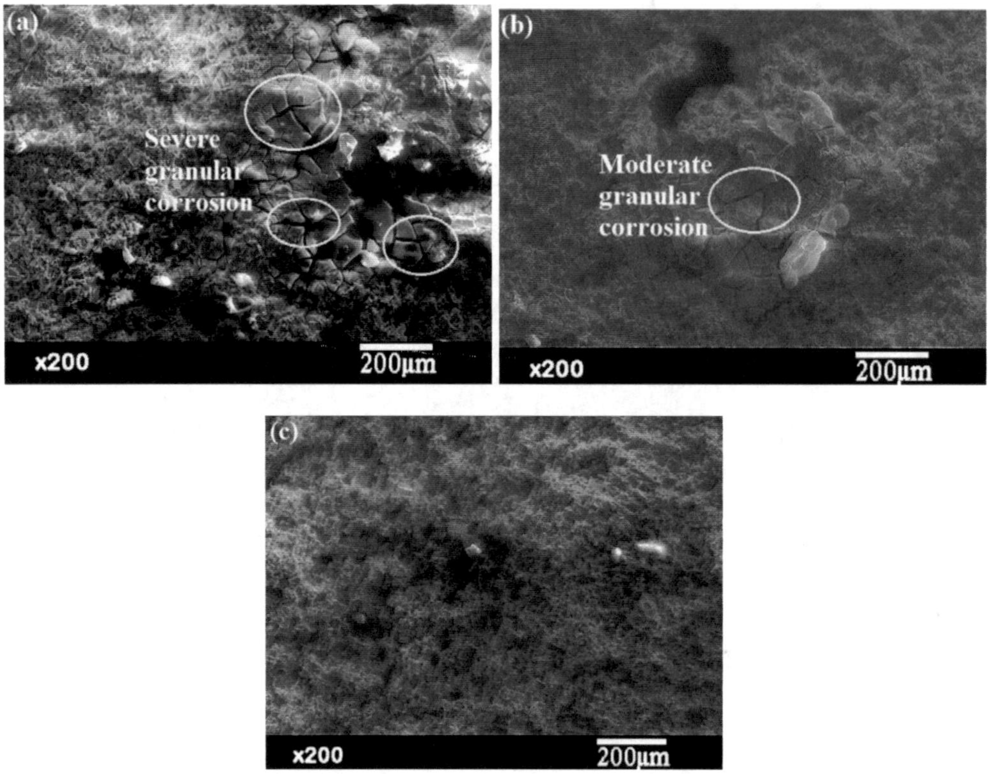

Figure 6. Corroded specimens tested in 1M HCl solution (a) 1 wt % (b) 1.5 wt % and (c) 2 wt%.

In 3.5 % NaCl and 1 M H_2SO_4, Al-4Mg/1.5 wt% H_3BO_3 composite specimen performed well relatively compared to the other two composite specimens (Table 1). Hence, Al-4Mg/1.5 wt% H_3BO_3 composite specimen can be utilized for the development of marine and naval applications exposed to NaCl environment. In 1M HCl, Al-4Mg/2 wt% H_3BO_3 composite specimen has better corrosion resistance in 1 M HCl compared to other two composite specimens (Table 1).

4. Conclusion

- In-situ Al-4Mg/$MgAl_2O_4$ composite specimens were fabricated successfully through the addition of H_3BO_3 (1 wt %, 1.5 wt%, 2 wt %) precursor.
- $MgAl_2O_4$ crystals were formed in all the composites upon addition of H_3BO_3 and slight agglomeration of $MgAl_2O_4$ crystals is observed in case of Al-4Mg/2 wt% H_3BO_3 composite.
- $MgAl_2O_4$ particles acted as heterogeneous nucleant in the Al matrix due to cube-on-cube orientation relationship and similar structure with aluminium.
- Corrosion rate is found to be lowest in case of Al-4Mg/1.5 wt % H_3BO_3 composite in 3.5 % NaCl and 1 M H_2SO_4 due to its lower corrosion current density and significant grain refinement by the $MgAl_2O_4$ particles.

Acknowledgement
The author gratefully acknowledge the technical and financial support from Indoshell Cast Private Limited, Coimbatore, India to carry out this research work.

5. References

[1] Mills J Robert, Lattimer Y Brian and Scott W 2018 *Corros. Sci.* **143** 1
[2] Holroyd N J H, Burnett T L, Seifi M and Lewandowski J J 2017 *Mater. Sci. Eng.* A **682** 613
[3] Yi G, Zeng W, Poplawsky J D, Cullen D A, Wang Z and Free M L 2017 *J. Mater. Sci. Technol.* 33 991
[4] Goswami R, Spanos G and Pao P S 2010 *Mater. Sci. Eng.* A **527** 1089
[5] Searles J L, Gouma P I and Buchheit R G 2001 *Metall. Mater. Trans.* A **32** 2859
[6] Scotto D, Antuono D, Gaies J and Golumbfskie W 2014 *Scr. Mater.* **76** 81
[7] Zhang R, Gupta R K, Davies C H J, Hodge A M, Tort M, Xia K and Birbilis N 2015 *Corros.* **72** 160
[8] Zhang R, Qiu Y and Qi Y 2018 *Corros. Sci.* **133** 1
[9] Li, Zhang Z and Vogt R 2011 *Acta. Mater.* **59** 7206
[10] Gupta R K, Zhang R, Davies C H J and Birbilis N 2014 *Corros.* **70** 402
[11] Wang Y, Gupta R K, Sukiman N L, Zhang R, Davies C H J and Birbilis N 2013 *Corros. Sci.***73** 181
[12] Carroll M C, Gouma P I, Mills M J, Daehn G S and Dunbar R 2000 *Scr. Mater.* **42** 335
[13] Zhang R, Gupta R K, Davies C H J, Hodge A M, Tort M, Xia K and Birbilis N 2015*Corros.***72** 160
[14] Liu S Q, Wang X, Cui C X, Zhao L C, Liu S J and Chen C 2015 *Mater. Des.* **65** 432
[15] Liu S Q, Wang X and Cui C X 2017 *J. Alloys Compd.* **701** 926
[16] Liu S Q, Cui C X and Wang X 2018 *Appl. Surf. Sci* .**431** 202
[17] Liu S Q, Cui C X and Wang X 2017 *Met.* 72 04
[18] Ali Torkan, Amin Rabiei Baboukani and Iman Khakpour 2018 *Nanosci. Nanometro.* **4** 34
[19] Liu S, Wang X and Tao Y 2019 2019 *Appl. Surf. Sci.* **484** 403
[20] Sadegh Rasouli, Reza A Behnagh, Abdolrahman Dadvand and Noureyeh Saleki-Haselghoubi 2014 *J Mater. Des.Appl.* 1
[21] Prabhakar G V N B, Ravi Kumar N and Ratna Sunil B 2018 *Mater. Today: Proceed.* **5** 8391
[22] Meijuan Li, Kaka Ma, Lin Jiang, Hanry Yang, Enrique J. Lavernia, Lianmeng Zhang and Julie Schoenung 2016 *Mater. Sci. Engg.* A **656** 241
[23] Ranjit Bauri, DevinderYadav, Shyam Kumar C N and Balaji B 2014 *Mater.Sci. Engg.* A **620** 67
[24] ChengGuo, TianchunZou, Chunsheng Shi, Xudongyang, Naiqin Zhao, Enzuo Liu and Chunnian He 2015 *Mater. Sci. Engg.* A **645** 1
[25] Zhenyang Yu, Naiqin Zhao, Enzuo Liu, Chunsheng Shi, Xiwen Du and Jian Wang 2012 *Compos. Part A* **43** 631
[26] Li Xing, Yaxuan Zhang, Chunsheng Shi, Yang Zhou, Naiqin Zhao, Enzuo Liu and Chunnian He 2014 *Mater. Sci. Engg.* A **617** 235
[27] Li H T, Wang Y and Fan Z 2012 *Acta. Mater.* **60** 1528

IOP Conf. Series: Materials Science and Engineering **655** (2019) 012040 doi:10.1088/1757-899X/655/1/012040

Plasma dissociation of monazite

H Bissett[1], M M Makhofane[1], R van der Merwe[1], S J Lotter[1]

[1] The South African Nuclear Energy Corporation SOC Ltd (Necsa),Church Street West Ext, Pelindaba, Pretoria 0001, South Africa

E-mail: hertzog.bissett@necsa.co.za

Abstract. Monazite is a phosphate mineral which contains a significant amount of rare earth elements (REE) with thorium (Th) and a low concentration of uranium (U). The radioactive components of monazite, Th and U, make it an unattractive source of REE due to the added cost associated with radioactive waste handling. The mineral is also chemically inert requiring the use of corrosive reagents at elevated temperatures to effectively extract REEs. According to thermodynamic calculations, monazite dissociation into oxides of REE and phosphorus (P) is favourable at temperatures greater than 1000°C, while the reverse reaction occurs well below a 1000°C. Rapid cooling is required to prevent the reverse reaction. Achieving these high temperatures, with rapid heating and cooling, is possible by means of thermal plasma treatment such as a radio frequency plasma. This type of plasma system offers superior control of plasma/particle interaction, quenching rates and plasma gas composition. The objective of this study was to determine the parameters for dissociation of the monazite under various conditions to reduce its chemical inertness. Dark and grey powders were obtained under reducing conditions, while lighter powder formed in the presence of an oxygen plasma with increase particle circularity observed after plasma treatment. While SEM-EDS analysis indicated no notable change in powder composition after plasma treatment, dissolution tests proved that monazite was not only more reactive, but also more susceptible chemical reaction after treatment.

1. Introduction

Monazite is a valuable source of rare-earth elements (REE) such as cerium (Ce), lanthanum (La), praseodymium (Pr), neodymium (Nd) and yttrium (Y) while also containing the radioactive elements Th and U [1]. REE are increasingly being used in the development of advanced materials in clean energy technology applications such as wind turbines, electric cars and permanent magnets [2].
The processing of monazite is difficult and expensive due to the fact that highly corrosive chemicals such as sodium hydroxide and sulfuric acid at elevated temperatures are required to dissolve this phosphate based mineral [3]. New and innovative technologies have been suggested to improve extraction efficiencies by changing the monazite structure in such a way that it makes it more susceptible to chemical attack. One such method is thermal cracking of the monazite crystal lattice [4].

The use of thermal plasma technology has been successfully applied at the South African Nuclear Energy Corporation SOC Ltd. (Necsa) to dissociate zircon ($ZrSiO_4$), which made use of a direct current non-transferred arc plasma. In past investigations, harsh alkaline treatment was required to extract zirconium (Zr) from the chemically inert crystal structure of zircon. With the assistance of thermal plasma technology, the zircon was converted into plasma-dissociated zircon (PDZ – $ZrO_2.SiO_2$) which was more chemically reactive and could be processed making use of less aggressive chemicals at lower temperatures [5, 6]. Necsa has purchased a 15 kW RF thermal plasma system from TEKNA Plasma

Systems Inc., by making use of the National Equipment Programme managed by the National Research Foundation of South Africa. This system operates with a versatile, inductively coupled radio-frequency (RF) plasma torch, which allows for the use of various plasma gas compositions (H_2/Ar, He/Ar, O_2/Ar) in a range of 9 -15 kW plasma plate powers.

The purpose of this paper is to provide details on how the monazite was treated under reducing, inert and oxidizing plasma conditions. The untreated monazite was compared with the plasma treated monazite making use of optical microscopy and Scanning Electron Microscopy - Energy Dispersive X-Ray Spectroscopy (SEM-EDS) to observe morphological and compositional particle changes. Subsequently, all untreated and plasma treated monazite samples were evaluated by hydrochloric acid dissolution tests.

2. Thermal decomposition of monazite

In the 1990s a direct current (DC) non-transferred arc plasma process was developed by Necsa to dissociate zircon [5]. Analogous to monazite, zircon sand is chemically inert and requires aggressive chemical attack to dissolve the mineral structure for beneficiation of valuable zirconium metal. Although an arc plasma pilot plant was used to dissociate the zircon, followed by rapid quenching to prevent the re-association of the ZrO_2 and SiO_2, the same principle can be applied for thermal "cracking" of the monazite in a RF thermal plasma.

In a paper by Kemp and Cilliers [4] the theoretical decomposition of monazite was discussed. Inthis study it was implied from a physical strength perspective that the monazite crystal is weaker and could therefore be dissociated more easily than zircon [7]. As the zircon crystal lattice was dissociated by means of the thermal plasma, it was expected that the monazite crystal could also be dissociated in a similar manner. Although monazite contains various rare earth phosphates, the most important rare earths (RE) in the monazite structure are Ce, La, Nd, Pr and Y. Calculating the Gibbs free energy to estimate the temperature at which the individual rare earth phosphates dissociate into the given oxides and phosphorus-oxygen complexes, it can be predicted whether plasma dissociation of the monazite can be accomplished using Equation 1:

$$2\ REPO_4 \rightarrow RE_2O_3 + 2\ PO_2\ (g) + \tfrac{1}{2}O_2\ (g) \tag{1}$$

According to literature zircon dissociates into PDZ at 1673 ± 10°C, while from the Gibbs free energy calculations, the dissociation of $CePO_4$ occurs at 1700 °C and the remaining rare earth phosphates dissociate between 2500 and 3000 °C. Therefore, it is expected that a higher plasma temperature will be required to dissociate the monazite into the various rare earth oxides compared to zircon dissociation into ZrO_2 and SiO_2. The conversion of monazite into rare earth oxides would improve the leachability when using diluted mineral acid when compared to leaching of rare earth phosphates [8].

Various studies have investigated the dephosphorisation behavior of rare earth phosphates. Converting the $REPO_4$ to the oxides should make the mineral more susceptible to chemical attack required for enhanced dissolution rates. The high temperature conversion of the rare earth phosphate to obtain rare earth oxides was examined. The studies indicated that the reaction temperature was the main factor influencing the dephosphorisation efficiency. According to thermodynamic calculations and real initial reaction temperature conditions, all rare earth phosphates were converted to the oxides at temperatures higher than 1400 ^{0}C. [9, 10].

3. RF thermal plasma for dissociation or "cracking" of the monazite

Thermal plasmas, characterised by their extremely high temperatures (3000 - 10 000 ^{0}C) and rapid heating and cooling rates (~ 10^6 ^{0}C/s) under oxidizing, reducing or inert conditions. This makes theme particularly suitable for thermal treating of powders at high temperatures, followed by rapid quenching once they pass through the plasma. Residence times of particles in the plasma region ranges from 5 to 20 ms and is dependent on the gas flow rates used during the treatment. Although various plasma methods can be used, radio-frequency (RF) induction plasmas are the preferred method, based on the longer residence times possible. For an induction torch, the energy coupling between the electric generator and the plasma itself is done by a cylindrical coil [11]. A typical induction set-up is illustrated in Figure 1.

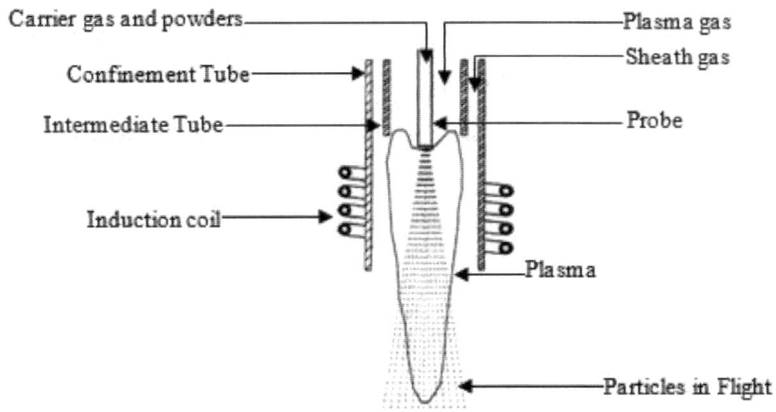

Figure 1. RF induction plasma within the confinement tubes with an induction coil

The plasma "flame" properties are dependent on various parameters such as input power, frequency from the RF power supply, pressure inside the reactor and the plasma gas composition. Less power is required at higher frequencies and lower gas pressures to maintain a stable plasma. For instance; when comparing argon and hydrogen, less power is required to maintain an argon plasma since argon ionizes more easily than hydrogen. A higher plasma temperature can however be obtained when making use of gases that ionize less readily, while higher plasma gas temperatures can also be obtained at higher operating pressures. In most instances Ar was used as plasma gas (central gas) and carrier gas, while the sheath gas was a mixture of Ar and an optional gas such as H_2, He and O_2 depending on the application [11].

4. Experimental

Mineral sand from Richard's Bay was upgraded / floated to a monazite concentrate (> 90 %) by Mintek and used for all experimental work. The monazite was plasma treated at various conditions and characterised by optical microscopy and SEM-EDS. Dissolution tests on all samples were conducted in hydrochloric acid.

4.1. Thermal decomposition of monazite

TEKNA's 15 kW induction plasma system makes use of a PL-35M induction torch which can operate between 2 to 5 MHz. For the purpose of this study, the PL-35 torch was mounted on a reactor chamber equipped with a "catch pot" for the collection of the plasma treated monazite powder. Experiments were conducted in either a reducing, inert and/or oxidizing environment. In Table 1 the plasma treatment conditions are indicated.

Table 1. Thermal plasma conditions for the treatment of monazite. For all experiments the powder feed rate was ± 0.5 kg/h and the carrier gas (Ar) flow rate 2 standard litres per minute (slpm)

Plasma gas mixture	Optional gas concentration (% v/v)	Plasma plate power (kW)
H_2/Ar	7.5	15
He/Ar	50	15
O_2/Ar	60	12

A mass balance of the powder before and after plasma treatment was conducted in order to determine if there were any loss in material during operation. Optical images of the plasma treated powders collected in the "catch pot" below the reactor were acquired using a Nikon SMZ-T2 optical microscope equipped with a Nikon D70s camera.

4.2. SEM-EDS and image processing

Four samples of monazite, one raw (untreated) and three plasma treated, were mounted and materialographically prepared (ground and polished down to 1 μm) beforehand, with the surface carbon coated prior to produce a conductive surface for required for SEM imaging of the powders.
SEM and EDS analyses were performed using a Quanta FEI 200 D SEM with an EDAX Octane Elect SD detector running the TEAM™ EDS Analysis System. Elemental and phase mapping of each sample surface (cross-sections) was conducted at 155 x magnification.
Image processing was applied on the Back-scattered Electron (BSE) images when applying elemental maps. The Carl Zeiss Zen 2 Core software package was used to determine if there was a quantifiable change in the particle morphology under different conditions. The average circularity was determined by evaluating a number of particles and making use of Equation 2.
The shape factor f_{circ} is the circularity, a function of the perimeter P and the area A:

$$f_{circ} = \frac{4\pi A}{P^2} \quad (2)$$

In this instance the circularity varies between 1 and approaching 0, with the circularity of a circle equal to 1.

4.3. Dissolution study

The raw (untreated) monazite and the three plasma treated samples were subjected to acid dissolution tests in order to compare the chemical amenability of the feed monazite and the variously treated plasma products by comparing the degree of digestion observed during each test. For each sample a mass of between 0.1 and 0.3 g was weighed off accurately to 0.0001 g into a round bottomed flask. A total of 5 ml of 32 % analytical grade hydrochloric acid was added to the flask was connected to a reflux condenser and placed in a heating mantle. The sample was heated and allowed to boil for a period of 4 h in order to ensure that as much sample material dissolved into the acid as possible. The acid selected was chosen based on commonly used reagent with the reflux conditions selected to ensure a constant concentration throughout a dissolution test. Hydrochloric acid forms a constant boiling azeotrope at 108.6 °C with a concentration of 20.2% [12]. Upon completion, the sample was allowed to cool, then diluted slightly with deionised water during filtration and transferred quantitatively to a 50 ml volumetric flask, after which it was filled to the mark with more deionised water. These samples were submitted to Pelindaba Analytical Laboratories (PAL) for inductively coupled plasma - optical emission spectrometry (ICP-OES) analysis of the elemental content of the dissolved component. Though the dissolution rate was not determined for the purpose of this portion of the study, the ease with which plasma treated monazite did dissolve in comparison to untreated powders was investigated.

5. Results and discussion

5.1. Thermal decomposition of monazite

For the powder treated making use of H_2/Ar and O_2/Ar a loss in mass of 15 % in both these instances was observed, while a mass loss of 14 % was observed in the He/Ar plasma. Although a negligible quantity of powder deposited on the reactor wall, this loss in mass might indicate the evaporation of light elements from the monazite. It was apparent that a deposit formed on the top-section inside the plasma reactor. In the presence of H_2 a yellow deposit was observed while a "brownish" and a white deposit was formed when He and O_2 was used, respectively. Due to the small quantity (<1 g) of deposit that formed, accurate analysis of this powder was not possible. It is however realistic to assume that all these deposits were some kind of phosphorus containing compounds caused by dephosphorisation of the monazite at high temperatures, based on the colouration of the deposits observed [9,10].

In Figure 2 the optical images of all samples are shown. It is clear that a colour change of the monazite had occur in all instances, with mostly dark particles forming after treatment in the H_2/Ar plasma and lighter particles formed using a O_2/Ar plasma. The use of the He/Ar plasma resulted in a mixture of dark, grey and white particles with predominantly grey particles present. From this visual comparison it was apparent that the respective reducing and oxidizing environments tested had a physical effect on the plasma treated monazite.

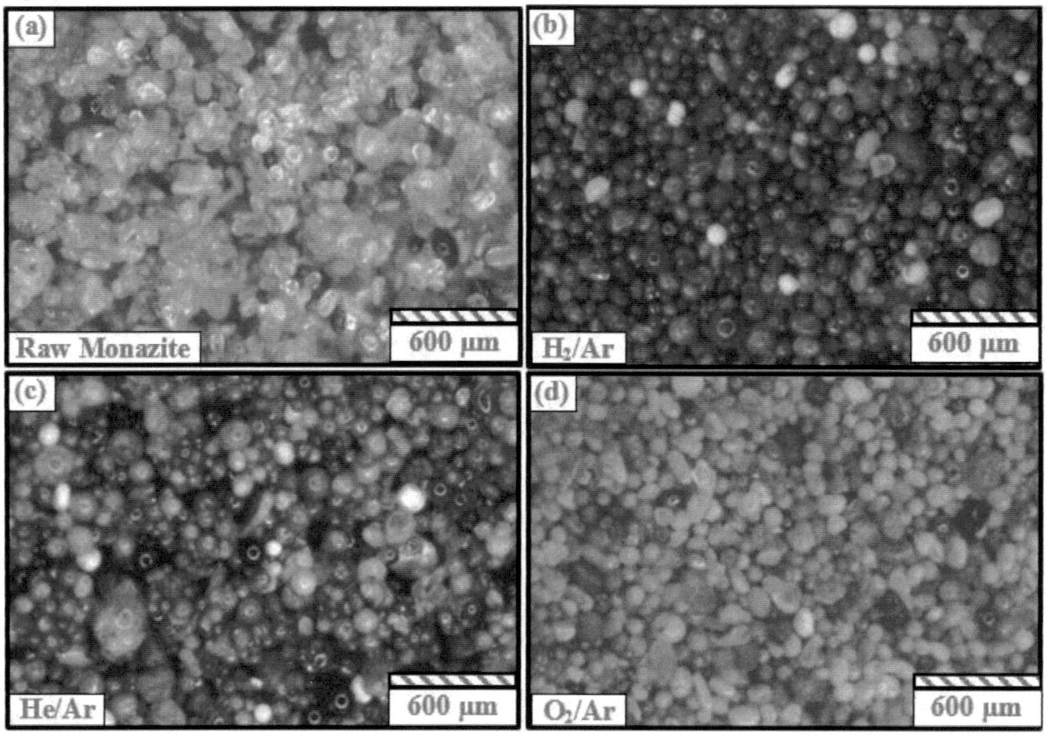

Figure 2. Optical images of the (a) raw untreated monazite, (b) plasma treated monazite in H_2/Ar, (c) plasma treated monazite in He/Ar and (d) plasma treated monazite in O_2/Ar.

5.2. SEM-EDS and Optical image processing

In Figure 3 the SEM-EDS images of the raw monazite cross-sections are shown. From the BSE image it is clear that the monazite particles were irregularly shaped. According to the element / phase mapping conducted, most of the particles consisted of REEs with some contamination of $Zr_xSi_yO_z$ (typically zircon) and some oxide-containing beach sand.

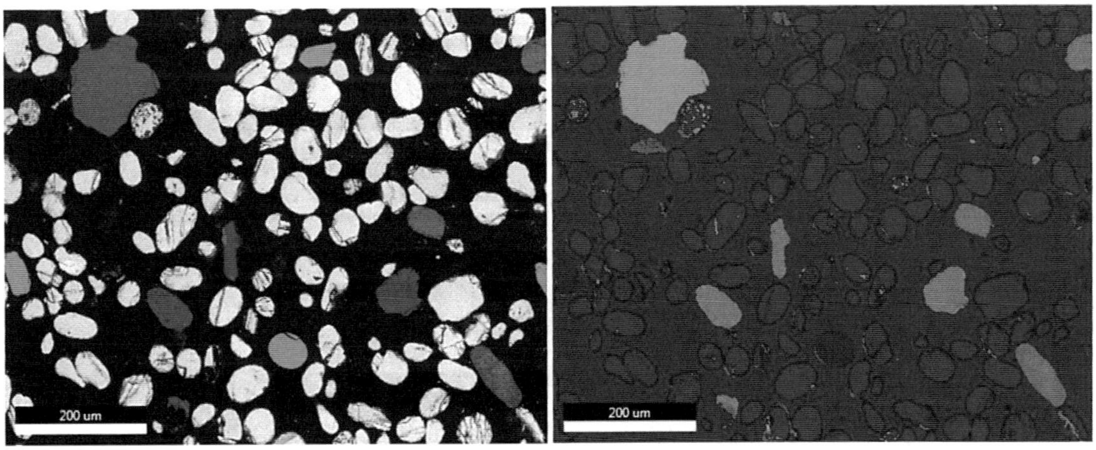

Figure 3. SEM-EDS images of the floated monazite. The BSE image is shown on the left and the image overlaid with elemental/phase mapping on the right. The colour coding indicates the following elements/phases: ■ - O/Ce/Nd/P/La/C /Zr; ■ - C/O; ■ - O/Si/Fe; ■ - Si/Zr/O/P; ■ - Ti/O/La/Ce; ■ - Ca/Si/O/C/Fe.

In Figure 4 to 6 the samples plasma treated in H_2/Ar, He/Ar and O_2/Ar are shown. Elements, in phases as identified on the backscatter images (BSE), were successfully quantified. Still, no noteworthy compositional changes were recorded for raw monazite samples after plasma treatment. Some elements were more abundant in samples after treatment, which could be attributed to the inhomogeneity of the sample. EDS should not be used to assume the elemental composition of the entire bulk sample as it takes into account elements in a very small spot (300 μm maximum) on the sample surface, also carrying an error of ± 1 % (wt. %). For the above mentioned reasons it was also difficult to quantify the difference in phosphorus composition before and after plasma treatment which might have been attributed to the dephosphorisation of the monazite.

Some morphological changes were observed after plasma treatment of the monazite powder. The most apparent were the increase in particle circularity and voids present after plasma treatment of the raw monazite powder. Though some voids were distinct on the BSE images, they were more evident on the secondary SEM images (taken with the Everhart-Thornley detector) not presented in the current work.

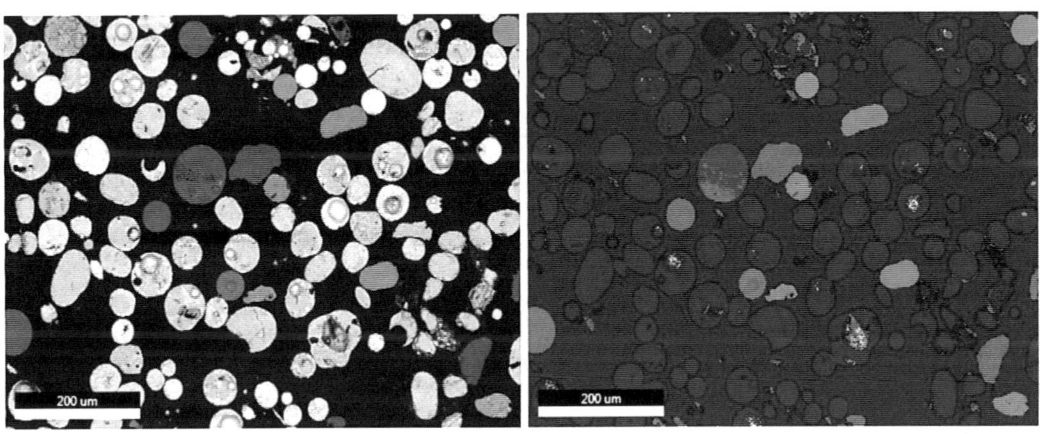

Figure 4. SEM-EDS images of the monazite treated in H_2/Ar plasma. The BSE image is shown on the left and the image overlaid with elemental/phase mapping on the right. The colour coding indicates the following elements/phases: ■ - O/Ce/Nd/P/La/C /Zr; ■ - C/O; ■ - O/Si/Fe/Al/Ca; ■ - Ti/O/La/Ce; ■ - Ca/Si/O.

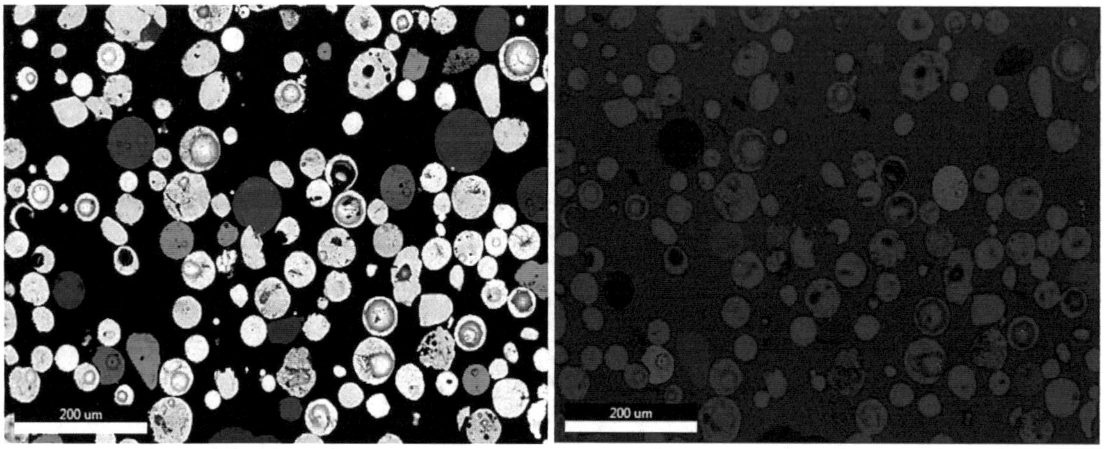

Figure 5. SEM-EDS images of the monazite treated in He/Ar plasma. The BSE image is shown on the left and the image overlaid with elemental/phase mapping on the right. The colour coding indicates the following elements/phases: ■ - O/Ce/Nd/P/La/Zr; ■ - C/O; ■ - O/Si/Fe/Ca; ■ - Si/O/Zr/O/P; ■ - Ca/Si/O.

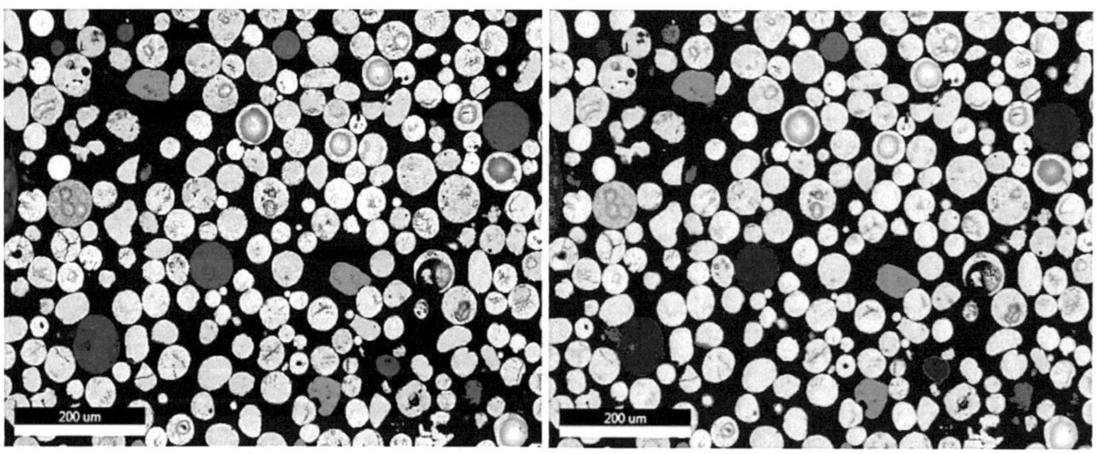

Figure 6. SEM-EDS images of the monazite treated in O_2/Ar plasma. The BSE image is shown on the left and the image overlaid with elemental/phase mapping on the right. The colour coding indicates the following elements/phases: - O/Ce/Nd/P/La/Zr; ■ - C/O; ■ - O/Si/Ca; ■ - Si/O/Zr/O/P.

The SEM images obtained were subjected to optical image processing making use of the Carl Zeiss Zen 2 Core software package, used to quantify changes in the particle morphology under different plasma conditions. The results can be seen in Table 2. All plasma treated samples showed an increase in circularity, a metric which was used as a 2-dimensional equivalent to sphericity.

Table 2. Average circularity determined by image processing software

Plasma gas composition	Number of particles measured	Average circularity (-)
Raw Monazite	130	0.79
Plasma treated in H_2/Ar	133	0.83
Plasma treated in He/Ar	140	0.86
Plasma treated in O_2/Ar	188	0.85

5.3. Dissolution study

Initial results can be seen in Figure 7. The sample labeled "raw monazite" is the recovery of the untreated monazite from the HCl dissolution medium. The other samples were plasma treated with varying power settings and chemical environments. The best extraction from a monazite powder appears to be associated with the 15 kW plasma setting with hydrogen as a gas additive with an extraction of 44 % obtained when compared to the 20% extraction of raw monazite sample. Further tests are required to properly define the degree of improvement from the baseline sample, but an initial increase in extraction of more than double is highly encouraging. An extraction of 35 % was also observed for the sample treated in the He/Ar plasma. The sample treated at 12 kW in the presence of oxygen did not result in an improvement in the extraction during dissolution. The reason for this is not clear at this stage. Only total extraction percentage was measured here. The extraction rate (extraction percentage vs time) and optimum conditions will be determined in future work. This basic dissolution study was intended solely to guide the plasma treatment process.

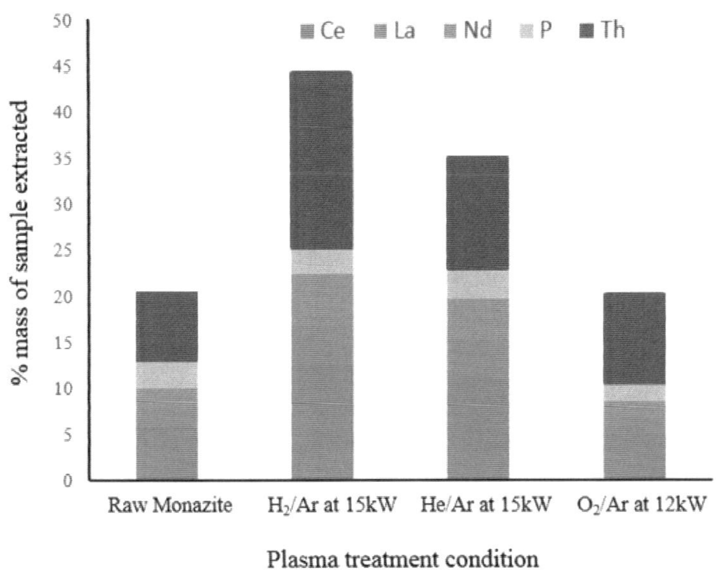

Figure 7. Dissolution results of the raw monazite and the three plasma treated samples in HCl. The colour coding indicates the following main elements determined in solution by ICP-OES: ■ - Ce, ■ - La, ■ - Nd, ■ - P, ■ - Th

6. Conclusions

Raw monazite concentrate samples were treated in the TEKNA 15 kW induction plasma system in reducing, inert and oxidizing environments from 12 to 15 kW plasma plate powers. In all instances a loss in material mass around 15 % were observed indicating possible dephosphorisation of the rare earth phosphate. According to visual observations, the treatment of the monazite powder in H_2/Ar and He/Ar plasma resulted in the formation of mostly dark and grey particles, while lighter particles were produced making use of a O_2/Ar plasma. SEM-EDS analysis of the samples indicated no notable compositional changes when comparing the untreated monazite to all the plasma treated samples. Clear morphological changes of the powder were observed with an increase in particle circularity for all plasma treated samples. The average circularity increased from 0.79 for the untreated monazite to around 0.85 for all plasma treated powders. Dissolution of the raw monazite and the three plasma treated samples indicated an extraction of 44 % and 35 % for the samples treated in H_2/Ar and He/Ar plasma respectively, compared to the 20 % extraction obtained for the raw monazite. The results showed that plasma treatment could make the monazite mineral more reactive and susceptible to reaction with chemicals.

7. Acknowledgements
The authors would like to acknowledge the Department of Science and Technology (DST), through the Advanced Material Initiative (AMI), for the financial support in conducting this study. The South African Nuclear Energy Corporation (Necsa) is acknowledged for their financial support. Mintek is also acknowledged for supplying monazite concentrated.

8. References

[1] Kim W, Bae I, Chae S and Shin H 2009 *J. Alloy. Comp.* **486** 610

[2] Abreu R D and Morias C A 2010 *Miner. Eng.* **23** 536

[3] Kumari A, Panda R, Jha M K, Kumar J R and Lee J Y 2015 *Miner. Eng.* **79** 102

[4] Kemp D and Cilliers A C 2016 *J. S. Afr. Inst. Min. Metall.* **116** 901

[5] Havenga J L and Nel J T 2012 *J. S. Afr. Inst. Min. Metall.* **112** 12

[6] Rentorff N M, Suarez G, Conconi M S, Singh S K and Aglietti E F 2012 *Proc. Mater.Sci.* **1** 337

[7] Cotrelli T L 1958 2nd edn. Butterworth, London

[8] Peelman S, Sun Z H I, Sietsma J and Yang Y 2014 *Proc. ERE2014L 1st Eur.RER Conf.* 446

[9] Xing P, Li F, Guo J and Tu G 2010 *J. Rare Earths* **28** 194

[10] Xing P, Zhuang Y, Tu G and Guo J 2010 *Trans. Nonferrous Met. Soc. China* **20** 2392

[11] Boulos M I 1985 *Pure Applied Chem.* **57** (9) 1321

[12] http://www.jsia.gr.jp/data/handling_02e.pdf

IOP Conf. Series: Materials Science and Engineering **655** (2019) 012041 doi:10.1088/1757-899X/655/1/012041

The effect of molar ratio and flow velocity on the surface area required for pertraction of a Zr/Hf mixture

D Malan[1], DJ van der Westhuizen[1,*], HWJP Neomagus[2], and HM Krieg[1]

[1] Chemical Resource Beneficiation (CRB), Membrane Technology Group, North-West University, Potchefstroom Campus, Private Bag X6001, Potchefstroom, 2520, South Africa

[2] School of Chemical and Minerals Engineering, North-West University, Potchefstroom Campus, Private Bag X6001, Potchefstroom, 2520, South Africa

*Derik.VanDerWesthuizen@nwu.ac.za

Abstract. Zirconium (Zr) and hafnium (Hf) need to be separated before being used in nuclear applications. Equilibria data have shown that Hf can be selectively extracted over Zr from a sulphate medium using organophosphorus extractants. This study investigated a modification of Zr and Hf solvent extraction whereby a hollow-fibre membrane contactor was used to separate the aqueous and organic phases, termed pertraction (PX). The aim of the study was to investigate the effects of the molar D2EHPA:Hf ratio and the volumetric flow rate on the estimated surface area required for an industrial PX extraction unit. Equilibrium experiments showed that the maximum separation factor (8.8) was found at D2EHPA:Hf ratio of 350:1. Mass transfer coefficients and kinetic selectivities of hafnium and zirconium were determined using a hollow fibre PX set-up. Variables investigated were: i) D2EHPA concentration, and ii) volumetric flow rate at a constant organic to aqueous ratio. The mass transfer coefficients of hafnium and zirconium and the separation ratio increased at higher flow rates. The best experimental conditions estimate that an industrial extraction unit capable of producing 500 ton per annum nuclear grade zirconium would require approximately 3 200 m^2 membrane surface area, at a volumetric flow rate of 1.7 $m^3.h^{-1}$ when operated under similar conditions to those in this study.

1. Introduction

The industrial separation of zirconium (Zr) and hafnium (Hf) is most commonly achieved by means of either the methyl isobutyl ketone (MIBK) or tributyl phosphate (TBP) processes [1,2]. These two processes are hydrometallurgical solvent extraction (SX) processes in which one component is selectively extracted into the organic phase using an extractant as carrier. Most commonly these processes utilize mixer-settler units which have a number of drawbacks such as large chemical inventories, product lockup and flooding at high flow rates [3]. In this study, a novel adaptation to traditional SX of Hf and Zr was investigated by means of introducing a porous membrane to separate the phases as follow-up work on the patent by Van der Westhuizen *et al.* [4]. Membrane based solvent extraction, also termed pertraction (PX), counters the drawbacks of traditional mixer-settler units due to two major changes. First, the separated state of the phases negates the need for large settling tanks, reducing chemical inventory and product lockup. Second, the hydraulic conditions of each phase can be adjusted independently, allowing significantly varying phase ratios to be used without flooding or unloading [3]. Flow conditions have a significant impact on the performance of the process and the separate phases can be manipulated individually to suit a particular application. The membrane offers a large, fixed surface area through which mass transfer can occur.

This study aimed to estimate the membrane surface area of an industrial PX extraction unit to selectively extract Hf from a mixed Zr/Hf feed stream. The variables investigated were the molar D2EHPA:Hf ratio and the volumetric flow rate of the fluid phases inside a membrane contactor column.

2. Design Considerations

Previous work has shown that organophosphorus extractants can selectively extract Hf over Zr from a sulphate medium. Wang & Lee have published a number of articles on this [2,5,6] and proposed a chemical reaction for Hf extraction using D2EHPA from H_2SO_4 media [5]:

$$Hf^{4+} + 3(HA)_{2Org} \rightleftharpoons HfA_4(HA)_{2Org} + 4H^+ \quad (1)$$

In a PX process, separation is achieved by a difference in the distribution coefficients and the rates of mass transfer of the components in the system. Allahyari *et al.* describes the mass flux of a PX process by way of a mass balance over the shell side of a membrane contactor [7]. Equation 1 describes the overall mass transfer coefficient.

$$k_{exp} = \frac{F\, \Delta C_{aq}}{A_{eff}\, \Delta C_{lm}} \quad (2)$$

where k_{Exp} is the experimental overall mass transfer coefficient ($m.s^{-1}$), F is the volumetric flow rate ($m^3.s^{-1}$), ΔC_{aq} ($kg.m^3$) is the concentration difference between the shellside inlet and outlet, A_{eff} (m^2) is the effective membrane area available for mass transfer, and ΔC_{lm} is the log mean concentration difference between the aqueous and organic phases ($kg.m^{-3}$).

3. Materials and Methods

3.1. Reagents

The chemicals used in this study along with the purity and supplier are listed in Table 1. All chemicals were used without further purification.

Table 1. Chemicals used in the study.

Reagent	Purity	Supplier
$Zr(Hf)SO_4{\cdot}4H_2O$	98% (2.3wt% Hf)	Hangzhou Dayangchem Co., Ltd.
H_2SO_4	98	Labchem (Pty) Ltd.
Shellsol 2325	N/A	Chemquest (Pty) Ltd.
D2EHPA	98%	Sigma Aldrich
1-Octanol	95%	Sigma Aldrich
Oxalic Acid	96%	Labchem (Pty) Ltd.

The aqueous feed solutions were prepared by dissolving the required amounts of $Zr(Hf)SO_4{\cdot}4H_2O$ salt in deionized water (Evoqua Labostar), the acidity was adjusted using sulphuric acid. The aqueous feed conditions were kept constant throughout all the experiments with [Zr] = 15 g/L, [Hf] = 0.34 g/L, and [H_2SO_4] = 1.0 M. Fresh feed solutions were prepared for every experiment.

The organic phase was prepared by mixing the required amounts of D2EHPA and 1-octanol with Shellsol 2325 as diluent. Pandey *et al.* reported that oxalic acid was found to be a very effective and non-selective stripping agent [8]. Between experiments, the organic phase was stripped and regenerated by contacting overnight with 1.0 M oxalic acid at an O/A ratio of 1.

3.2. Batch shake-out procedure

Baudot *et al.* reported that the overall mass transfer of a species in a PX process is strongly influenced by the distribution coefficient ($HfA_4(HA)_{2Org}/Hf^{4+}_{Aq}$ in Equation 1) of the species [9]. They reported that species with high distribution coefficients exhibit increased mass transfer coefficients. To maximize the distribution coefficient, shake-out experiments were conducted in which the D2EHPA:Hf ratio was varied from 10 to 500 by changing the D2EHPA concentration of the organic phase. Equal volumes (25ml) of the aqueous and organic phases were contacted for 60 minutes at 350 revolutions per minute (Labcon 3081U mechanical agitator). The temperature was kept constant at 25°C. After 60 minutes the phases were allowed to separate and the aqueous phase was sampled for analysis by ICPOES (inductively coupled plasma optical emission spectrometry, Agilent 5110, Chemetrix (Pty) Ltd).

3.3. Pertraction extraction procedure

PX experiments were conducted using the experimental set-up shown in Figure 1. The membrane column consisted of two 3M (Liqui-Cell) 1.7x8.75 mini-modules connected in series and operated under counter-current flow conditions. The specifications for the modules and membrane contained within are listed in Table 2. To regulate the aqueous flow through the shell side of the membrane column a gear pump (Cole-Parmer Instrument Company, LLC.) was used in combination with a return valve to regulate the return flow to the reservoir. Due to the hydrophobic nature of the membrane, organic breakthrough may occur at low aqueous pressures. To prevent this, a 1.8m stand pipe was used to create a minimum pressure head inside the module. A peristaltic pump was used to move the organic phase through the membrane lumen. Chemical incompatibility between the organic phase and the polycarbonate shell material prevented experiments with the organic phase in the shell.

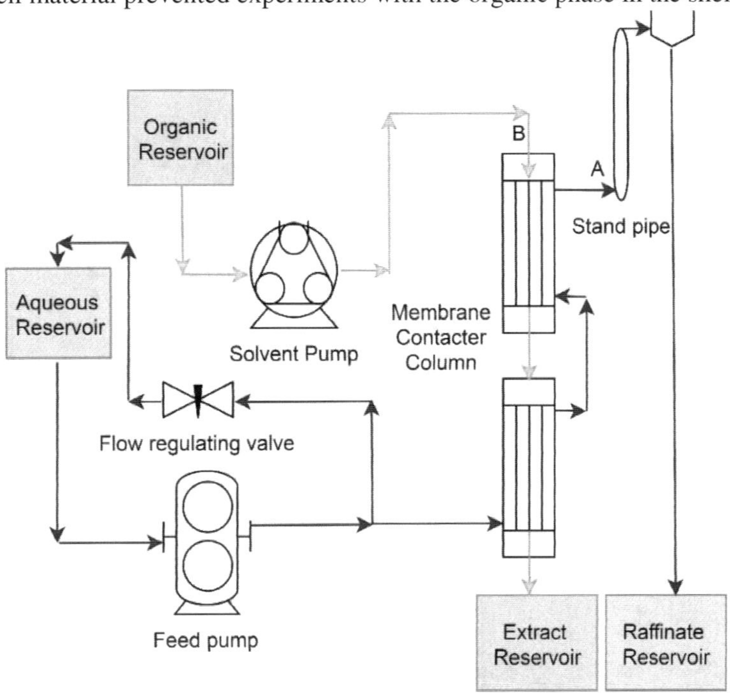

Figure 1. Pertraction column experimental set-up.

Table 2. 3M (Liqui-Cell) 1.7x8.75 mini-module specifications [10,11].

Module specifications			**Membrane specifications**		
Parameter	**Value**	**Unit**	**Parameter**	**Value**	**Units**
Shell OD	42.5	mm	Membrane Material	Polypropylene	
Shell ID	36.5	mm	Wetting phase	Organic	
Overall length	260	mm	Fiber OD	0.3	mm
Membrane length	198	mm	Fiber thickness	40	µm
Shell total volume	270	ml[a]	Pore size	0.04	µm
Shell flow volume	190	ml[a]	Porosity	40	%
Lumen flow volume	45	ml[a]	Approx. fiber count	6500	N/A
Surface Area	0.9	m^2	Membrane volume	50	ml

[a] Approximated value

Before each experiment the flow rates were calibrated using fluids with similar viscosity and density characteristics to the experimental solutions. The aqueous loop was calibrated first and left running to ensure sufficient pressure head inside the column to avoid organic breakthrough. Once both flow rates had been calibrated the column was drained and the calibration fluids replaced by the experimental solutions. To start the experiment the aqueous phase was pumped until fluid exited the top of the column, point A in Figure 1, at which point the organic pump was switched on. The experimental time was started when the organic reached the column inlet, point B in Figure 1. In-line samples of the aqueous phase were taken at fixed intervals for analysis by ICP-OES.

4. Results

4.1. Benchmark

Wang & Lee reported liquid-liquid equilibria data that found a D2EHPA:Hf ratio of 62.5 and 1.0 M H_2SO_4 feed solution resulted in a separation factor of 8.8 [6]. These conditions were tested to determine a benchmark for the rate of mass transfer of a PX system. The aqueous and organic phases were prepared as described in Section3.1, with a D2EHPA concentration of 40.1 g/L in the organic phase. The flow rates of both phases were set to 1.1×10^{-6} $m^3.s^{-1}$. Using Equation 2, the overall mass transfer coefficients were found to be 8.9×10^{-8} $m.s^{-1}$ and 8.3×10^{-8} $m.s^{-1}$ for Hf and Zr, respectively. The separation ratio (k_{ovHf}/k_{ovZr}) of 1.07 resulted in nearly equal extraction of Hf and Zr and very poor separation.

4.2. Distribution enhancement

The shake-out results shown in Figure 2 indicate that the higher D2EHPA:Hf ratio resulted in increased distribution coefficients of both Hf and Zr. Increasing the D2EHPA:Hf ratio from 10 to 350 resulted in the separation factor (D_{Hf}/D_{Zr}) increasing from 1.5 to 7.6 as the extraction is more selective towards Hf. Above a D2EHPA:Hf ratio of 350 a marked decrease in separation was found. This was due to near complete extraction of Hf extraction leaving the excess D2EHPA to co-extract Zr, as seen at a ratio of 500:1.

The D2EHPA:Hf ratio of 350:1 was found to be the best operating point for the purification as the separation factor was at a maximum while high distribution coefficients would facilitate higher mass transfer rates compared to that of the benchmark.

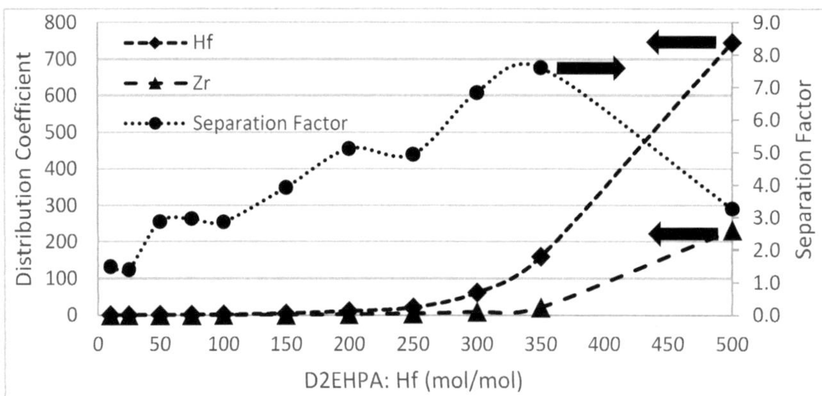

Figure 2. Effect of D2EHPA:Hf ratio on the distribution coefficients of Hf and Zr, and the separation factor. [Zr] = 15g/L, [Hf] = 0.34 g/L, [H2SO4] = 1.0 M. [D2EHPA] = 0.019 – 0.95 M. [1-C8OH] = 5 vol%, diluent: Shellsol 2325. O/A = 1.

4.3. Mass transfer

To determine the effect of the volumetric flow rate on the overall rate of mass transfer the flow rate was varied between 0.17×10^{-6} $m^3.s^{-1}$ (10 $mL.min^{-1}$) and 8.3×10^{-6} $m^3.s^{-1}$ (500 $mL.min^{-1}$). The linear velocities of the aqueous and organic phases varied between 0.03 $cm.s^{-1}$ and 1.5 $cm.s^{-1}$ and 0.07 and 3.4 $cm.s^{-1}$ respectively. Reynolds numbers for the shell and lumen were determined to be 170 and 4.1 respectively at 8.3×10^{-6} $m^3.s^{-1}$, indicating laminar flow ($Re < 2100$) conditions for both phases during all the experiments conducted. The overall mass transfer coefficients of Hf and Zr were determined using Equation 2 and are reported in Figure 3. From Figure 3 it can be seen that the separation ratio and mass transfer coefficients of both Hf and Zr increased with flow rate. Higher flow rates reduced the boundary layer thickness through which species need to diffuse in order to react (aqueous-membrane interface) or move toward the bulk organic solution (membrane to bulk organic). Since Hf and Zr are known for their chemical similarity their diffusion rates are expected to be similar as well. The increased separation ratio at higher flow rates indicates that the extraction of Hf is kinetically favoured over that of Zr.

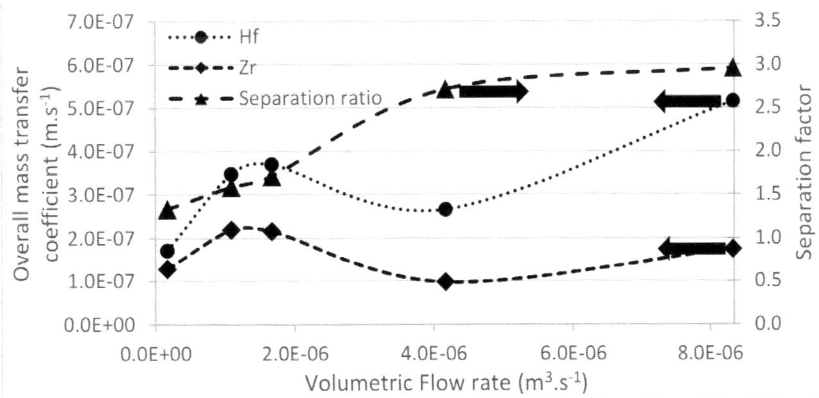

Figure 3. Overall mass transfer coefficient and separation factor as function of volumetric flow rate. [Zr]= 15 g/L, [Hf] = 0.34 g/L, [H2SO4] = 1.0 M, [D2EHPA] = 0.95 M, [1-C8OH] = 5 vol% diluent = Shellsol 2325.

A summary of the increase in overall mass transfer coefficients relative to the benchmark experiment is provided in Table 3. Increasing the molar D2EHPA:Hf ratio from 62.5:1 to 350:1 resulted in significant improvements to the mass transfer rates of Hf and Zr. The overall mass transfer coefficients of Hf and Zr increased by factors of 3.9 and 2.6 respectively. Additionally the separation ratio also showed an increase from 1.07 to 1.6. At the highest flow rate tested, Hf transferred almost six times faster compared to the benchmark while the rate for Zr doubled. This significantly improved the separation ratio from 1.6 to 3.0, as reported in Figure 3 and Table 3.

Table 3. Relative mass transfer rates and separation ratio compared to the benchmark experiment:

D2EHPA:Hf (mol:mol)	**Vol. flow rate ($m^3.s^{-1}$) x10^6**	**Linear velocity ($cm.s^{-1}$)**		**Relative overall mass transfer factor**		**Separation ratio**
		Shell U_S	**Lumen U_L**	**Hf**	**Zr**	**k_{Hf}/k_{Zr}**
62.5:1 (Benchmark)	1.1	0.2	0.4	1.0	1.0	1.1
350:1	0.17	0.03	0.1	1.9	1.5	1.3
350:1	1.1	0.2	0.4	3.9	2.6	1.6
350:1	1.7	0.3	0.7	4.1	2.6	1.7
350:1	4.2	0.7	1.7	3.0	1.2	2.7
350:1	8.3	1.5	3.4	5.8	2.1	3.0

4.4. Industrial estimation

The industrial estimation was constrained by the Zr production rate of 500 tpa, and the product purity of less than 100 ppm Hf. Since the raffinate purity is a function of both the Hf and Zr extraction, both have to be included in the estimation. In order to estimate the required membrane surface area and required volumetric flow rate a steady state mass balance was created for each species.

Table 4. Industrial plant estimation

Shell linear velocity U_S ($cm.s^{-1}$)	**Lumen linear velocity U_L ($cm.s^{-1}$)**	**Mass Flux ($kg.m^{-2}.s^{-1}$)**		**Flow rate ($m^3.h^{-1}$)**	**Industrial Area (m^2)**
		Hf·10^8	**Zr·10^6**		
0.2 (Benchmark)	0.4	2.9	1.2	20.1	214 000
0.03	0.1	2.9	1.2	9.3	100 500
0.2	0.4	9.0	2.7	3.5	12 000
0.3	0.7	11	3.2	3	8 000
0.7	1.7	8.5	1.4	1.8	6 700
1.5	3.4	17	2.6	1.7	3 200

It was assumed that the mass flux of each species, listed in Table 4, were constant throughout the length of the membrane column. These were substituted into the relevant mass balances and the membrane area and volumetric flow rate were solved. The results are listed in Table 4. It was further assumed that the mass flux behaviour inside the membrane column would be similar to that of the experimental if the linear velocities of the phases were similar as the hydraulic conditions would be expected to be similar. Constant linear velocities inside the column are maintained by increasing the cross sectional area of the shell and lumen at higher volumetric flow rates.

5. Conclusions and Recommendations

This study estimated the membrane surface area that would be required by an industrial PX extraction unit to selectively extract Hf from a mixed Zr/Hf feed stream. The equilibrium distribution coefficients and separation factor were optimised as a function of D2EHPA:Hf molar ratio. The effect of volumetric flow rate on the overall mass transfer coefficients were investigated using the optimised conditions and compared to a benchmark based on literature conditions.

It was found that a D2EHPA:Hf ratio of 350:1 provided a balance between increased distribution coefficients and a high separation factor. The mass transfer investigation showed that higher the D2EHPA:Hf ratio and volumetric flow rate both improved the rate of mass transfer for Hf and Zr as well as the selectivity toward Hf. The highest tested flow rate of $8.3x10^{-6}$ $m^3.s^{-1}$ (500 $mL.min^{-1}$) showed the highest rates of mass transfer for both Hf and Zr as well as the greatest separation ratio. Compared to the required membrane surface area and flow rate of the benchmark case the above conditions resulted in the required membrane area and flow rates being reduced by factors of 67 and 12 respectively. The reduction of both surface area and flow rate would have significant cost savings associated as the size of the required equipment is reduced significantly. It is suspected that the membrane modules will comprise a significant portion of capital expenditures. Reduced pumping costs related to the decreased flow rate will impact the operational costs as well.

It is clear that the rate of mass transfer and separation ratio are vital to the performance of a PX extraction unit. As the extraction of Hf seems to be kinetically favoured it is recommended that ongoing work should focus on the effect of the hydraulic conditions of the aqueous and organic phases, which each contribute to the resistance to mass transfer in a PX process. Other work includes the investigation of scrubbing and stripping of the organic phase for the purpose of recycling back to the extraction unit.

Acknowledgements

The authors wish to acknowledge the Advanced Metals Initiative (AMI) of the Department of Science and Technology (DST), managed by NECSA, for funding this research and ChemQuest for supplying the Shellsol 2325 diluent.

References

[1] Lee M S, Banda R and Min S H 2015 *Hydrometallurgy* **152** 84–90
[2] Wang L Y and Lee M S 2016 *J. of Korean Inst. Resources Recycling* **25** 3–9
[3] Gabelman A and Hwang S-T 1999 *J. Memb. Sci.* **159** 61– 106
[4] Van der Westhuizen D J, Lachmann G and Krieg H M 2012 WO2012168915-PAMPH 20121213-9962- NWU.pdf 55
[5] Wang L Y and Lee M S 2016 *Sep. Sci. Technol.* **51** 759–66
[6] Wang L Y and Lee M S 2016 *Hydrometallurgy* **160** 12–7
[7] Allahyari S A, Minuchehr A, Ahmadi S J and Charkhi A 2017 *Prog. Nucl. Energy* **100** 209–20
[8] Pandey G, Mukhopadhyay S, Renjith A U, Joshi J M and Shenoy K T 2016 *Hydrometallurgy* **163** 61–8
[9] Baudot A, Floury J and Smorenburg H E 2001 *AIChE J.* **47** 1780–93
[10] LiquiCel 2011 Product data sheet. Available online: http://www.liquicel.com/product information/data-sheets.cfm (Accessed 2 May 2017)
[11] Membrana Liqui-Cel Design and Operating Procedures. Available online: http://www.liquicel.com/technical-resources/startup-operating-guides.cfm (Accessed 2 May 2017)

Density functional theory study of TiPd alloying with Os as potential high temperature shape memory alloys

R G Diale[1,*], R Modiba[2], P E Ngoepe[1], and H R Chauke[1]

[1] Materials Modelling Centre, University of Limpopo, Private Bag X 1106, Sovenga, 0727, South Africa

[2] Future Production: Manufacturing, CSIR, PO Box 395, Pretoria, 0001, South Africa

E-mail: Ramogohlo.diale@ul.ac.za

Abstract. $Ti_{50}Pd_{50}$ alloy exhibits thermoelastic martensitic phase transformation above 823 K and has the potential for high temperature shape memory material applications. Previous studies showed that this alloy is mechanically unstable with a negative C′ at room temperature. A systematic investigation of the structural, thermodynamic, electronic and elastic properties of TiPdOs is conducted using a first-principles calculation based on density functional theory. The calculated heats of formation show that this compound is thermodynamically stable as Os content is increased. It is found that an increase in Os content stabilizes the TiPd with a positive C′ observed above 18.25 at. %. An increase in Os content results in an increase of the Bulk modulus and Young 's modulus above 18.25 at. %. Anisotropy factor and Vickers hardness are studied and hardness is found to increase with an increase in Os content. The results suggest that the stability of the B2 phase can be significantly enhanced by the addition of Os in TiPd alloy. These findings have important applications for future materials design in aerospace industries.

1. Introduction

Shape memory alloys (SMAs) have the ability to remember their shape after being deformed and play an important role in many industries such as automotive, aerospace and medical. SMAs have drawn attention in commercial applications, due to their shape memory effects (SMEs) and superelasticity; this commercial development has been supported by research studies [1]. Amongst the SMAs, the Ti-Ni has been studied for a long period and is useful because of their unique sufficient ductility and shape memory properties [2]. However, these alloys have limited temperature around 373 K, so their applications are limited. Other SMAs such as Cu–Zn and Cu–Al alloys are also of commercial importance because of their low price, but they suffer instability of the martensitic phase and brittleness [3]. Recently, SMAs are being developed to suit many applications in many fields, especially for engineering properties. For example, TiNi–X SMAs (X=Pt, Pd, Hf, and Zr) have been considered for high temperature [4, 5]. However, their transformation temperature of remained below 830 K. Currently, the new SMA systems have been established in aerospace industry which can be alternative systems to TiNi include TiPd and TiPt as TiPd undertakes a B2-to-martensite phase transformation with Ms between 783 and 836 K.
TiPd is well known as high temperature shape memory alloy (HTSMA) that can be used in engineering and medical industry due to the shape memory effect and superelasticity. This alloy has extraordinary mechanical behaviour and biocompatibility at 823 K [6]. TiPd HTSMA undertakes martensitic

transformation temperature (MTT) from the parent B2 (cubic) high temperature phase to the product B19 (orthorhombic) low temperature phase [7]. However, TiPd displays very poor shape-memory behavior during transformation, which results in poor corrosion resistance [2, 8]. Hence, the binary TiPd alloy has no strength for the use in actuators and aeronautic industry and ternary alloys will need to be established to improve their properties [9]. To enhance the performance of TiPd SMA which includes hardening, forgeability, corrosion resistance, and thermal stability, ternary alloying has been suggested. Osmium is most often used as an alloying agent with other PGMs and found in electrical contacts, styluses and medical devices and in other applications that need great strength and stiffness [10]. The addition of Os may enhance the transformation temperature of TiPd to above 1000 K. In this paper, the structural, thermodynamic, electronic and mechanical properties of the $Ti_{50}Pd_{50-x}Os_x$ alloys are systematically studied by the first-principles approach based on density functional theory (DFT). We found that addition of Os on B2 TiPd resulted in thermodynamically, electronically and mechanically stable system. The calculated results provided useful and practical guidance for further experimental and theoretical investigations on the future ternary shape memory alloys.

2. Methodology

The theoretical calculations in this paper were achieved using the Vienna ab initio simulation package (VASP) code based on the density functional theory (DFT) [11, 12]. In this calculation, the exchange and correlation functional was defined by Perdew, Burke, and Ernzerhof (PBE) [13] version of the generalized gradient approximation (GGA) [14]. A cutoff energy of 500 eV was used to converge the total energy of the structures. A 12 × 12 × 12 k-point was used for TiPd unit cell and equivalent numbers of k-points were used for the supercell, this according to Monkhorst and Pack [15]. The calculations were carried out using a 2x2x2 supercell having total number of 16 atoms. The substitutional search tool in VASP was used to substitute Pd with Os which provided the most stable composition at the desired symmetry. The calculations were determined at room temperature. For all structures, the elastic constants were calculated for a small strain of 0.005.

3. Results and discussions

3.1. Structural and thermodynamic properties

DFT equilibrium properties were obtained by firstly optimizing the crystal structures of $Ti_{50}Pd_{50-x}Os_x$ where the initial atomic positions and lattice constants were fully relaxed. Figure 1 (a) shows the calculated equilibrium lattice parameter of B2 $Ti_{50}Pd_{50-x}Os_x$ alloys. It is noted that the lattice parameter of $Ti_{50}Pd_{50-x}Os_x$ alloys decrease almost linearly as the Os content is increased. This trend is attributed to the fact that the atomic radius of Os is smaller than that of Pd. The stability of $Ti_{50}Pd_{50-x}Os_x$ is discussed through the predicted heats of formation (ΔH_f) and can be approximated by:

$$\Delta H_f = E_C - \sum_i x_i E_i \tag{1}$$

where E_C is the determined total energy of the system and E_i is the determined total energy of element in the system. The lowest negative value of ΔH_f indicate stability of the structure, otherwise a positive value implies instability.

From figure 1 (b), it can be noted that the heats of formation of $Ti_{50}Pd_{50-x}Os_x$ alloys are all negative, which indicates the great possibility of form these compounds in the experiment. From the DFT ternary results, heats of formation decrease with an increase in Os content indicating thermodynamic stability. The result of the heats of formation are plotted in figure 1 (b) for $Ti_{50}Pd_{50-x}Os_x$ alloys.

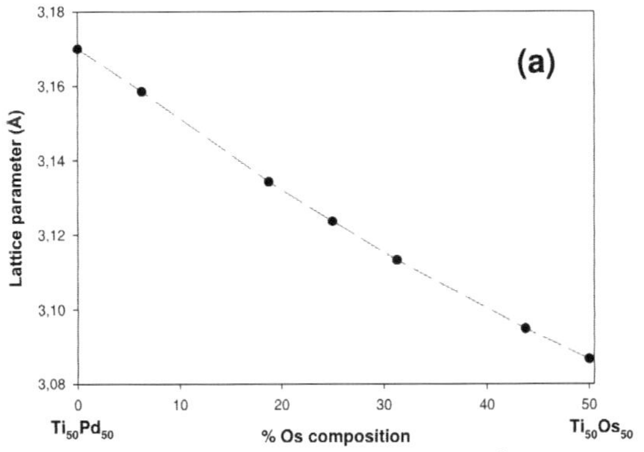

Figure 1(a). Lattice parameter, a (Å) against atomic percent Os (at. % Os) for $0 \leq x \leq 50$.

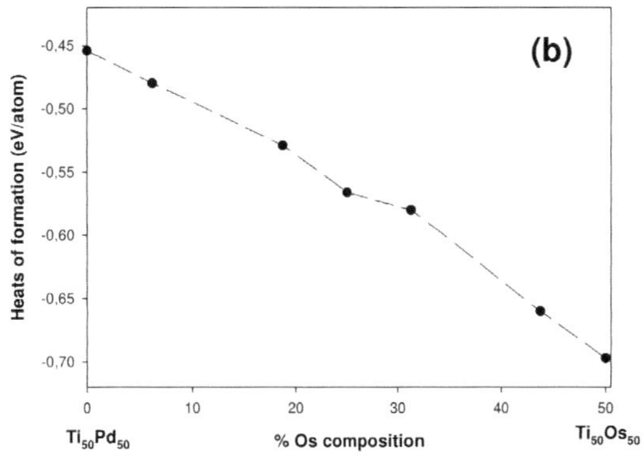

Figure 1(b). Heats of formation, ΔH_f (eV/atom) against atomic percent Os (at. % Os) for $0 \leq x \leq 50$.

3.2. Density of states

Electronic structure calculations have been performed to investigate the stability of TiPd-Os alloys by observing the trend of the total density of states (tDOS) near the Fermi level with respect to the pseudogap. The stability of TiPd when Os is added can be thus attributed to the lower and flatter tDOS near the Fermi level. When the fermi level (E_f) falls on the pseudogap, the structure is said to be the most stable otherwise the least stable structure. Similar analysis has been used by R Mahlangu and D A Pankhurst previously to describe the electronic stability trend for structures [16, 17]. In figure 2, we plot the total DOS for B2 $Ti_{50}Pd_{50-x}Os_x$ ($0 \leq x \leq 50$) alloys where the black line marks the Fermi line (E_F). The $Ti_{50}Pd_{50}$ structure hits the top of the total DOS peak indicating instability system. So adding the element that decreases the number of electrons will stabilize the system [18], hence Os was added. The Os has shown to have fewer electrons than Pd. As the composition of Os is added, the pseudogap moves towards the E_F which may imply that the system becomes electronically stable above 18.75 at. %. It is clearly seen that at 50 at. % Os ($Ti_{50}Os_{50}$) the Fermi level coincides with the pseudogap. This is consistent with the fact that Os has fewer electrons compared to Pd. The stability trend according to the density of states agrees very well with the predicted heats of formation results.

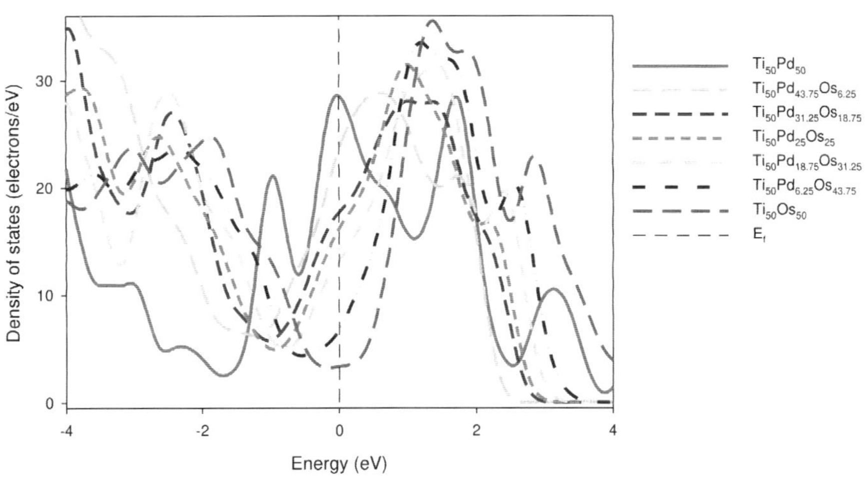

Figure 2. Comparison of the total density of states for TiPdOs structure against Energy.

IOP Conf. Series: Materials Science and Engineering **655** (2019) 012042 doi:10.1088/1757-899X/655/1/012042

3.3. Elastic properties

The elastic constants of a solid are very important parameters for determining the mechanical stability of the compounds. For the cubic symmetry B2 crystal structure of $Ti_{50}Pd_{50-x}Os_x$, there are three independent elastic constants which are c_{11}, c_{12} and c_{44}. The calculated elastic constants and Anisotropy of $Ti_{50}Pd_{50-x}Os_x$ alloys are shown in figure 3. The mechanical stability condition for the cubic system as defined elsewhere [19] are given as follows:

$$c_{44} > 0; c_{11} > c_{12} \text{ and } c_{11} + 2c_{12} > 0, \tag{2}$$

where the shear modulus C' can be calculated as:

$$C' = \frac{1}{2}(c_{11} - c_{12}) \tag{3}$$

whereas the anisotropy can be calculated as:

$$A = \frac{c_{44}}{C'} \tag{4}$$

In figure 3, the calculated elastic properties of the $Ti_{50}Pd_{50-x}Os_x$ alloys ($0 \leq x \leq 50$) are shown. For the structure to be stable, the stability condition for a cubic system must be fulfilled. The positive C' ($(1/2(c_{11}-c_{12})>0$) indicates the mechanical stability of the system, otherwise, unstable. The binary B2 TiPd alloy is mechanically unstable at 0 K due to negative C' (-5.37 GPa). From figure 3 (a), it is observed that c_{11}, c_{12} and c_{44} are greater than zero in all-inclusive ranges of $Ti_{50}Pd_{50-x}Os_x$ alloys ($0 \leq x \leq 50$). At small Os content (below 6.25 at. % Os), the C' is negative suggesting that the structure is elastically unstable at this concentration. The shear modulus (C') is positive above 18.75 at. % Os which indicates the structures are mechanically stable. It is very important to study elastic anisotropy to understand material properties and improve their mechanical strength. When A=1, the crystal is isotropic otherwise show elastic anisotropy ($A \neq 1$). The results showed that the elastic anisotropy values are smaller below 25 at. % Os and larger than the unity above. Anisotropy (A) becomes negative below 6.25 at. % which suggests that the material is anisotropic (figure 3 (b)). The calculated A for 43.75 at. % Os is close to the unit ($A \approx 1$) which indicates that the material is isotropic. Furthermore, the calculated A can be used to check the ductility in metals and the ratio should be greater than 0.8 [20]. The anisotropy is calculated and the values are greater than 0.8 above 18.75 at. % Os which reveal ductility behavior. Interestingly, an anisotropy ratio (A) is also of importance for determining the martensitic transformation of the material. A higher value of A indicates martensitic transformation from B2 to B19 while smaller A implies a good correlation between c_{44} and C' while leading to the transformation from B2 to B19′ [21]. As the composition of 18.75 at. % Os is added to the system, A is higher which confirms the transformation from B2 to B19. There is a strong coupling observed between the c_{44} and C' at 43.75 at. % Os which resulted in smaller anisotropy leading to transformation from B19 to B19′. It can be concluded that $Ti_{50}Pd_{50-x}Os_x$ alloys undergo transformation from B2 to B19 martensite and then B19 transforms into B19′ due to a coupling of the c_{44} and C' at room temperature.

IOP Conf. Series: Materials Science and Engineering **655** (2019) 012042 doi:10.1088/1757-899X/655/1/012042

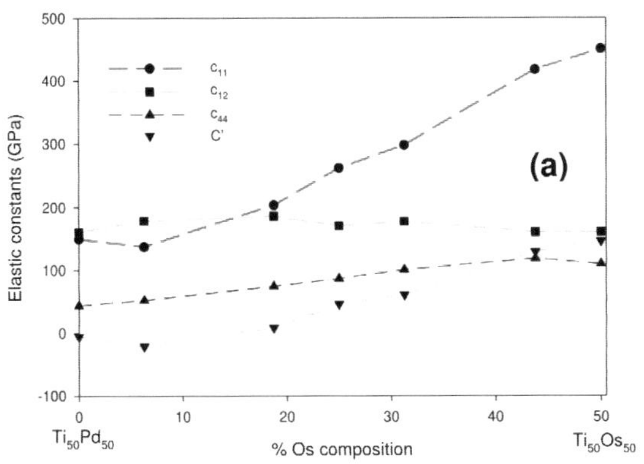

Figure 3(a). The elastic properties (GPa) of $Ti_{50}Pd_{50-x}Os_x$ SMAs.

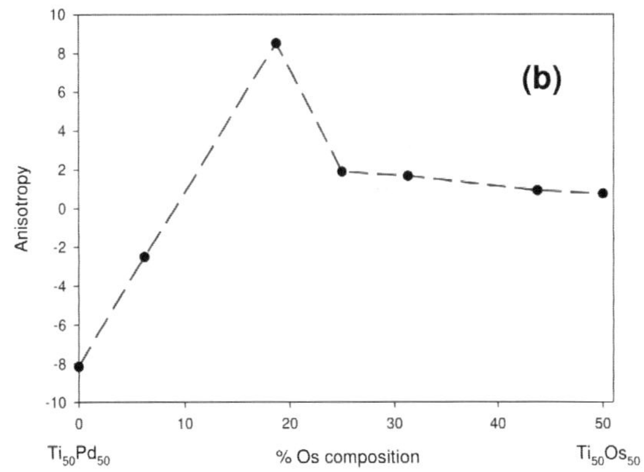

Figure 3(b). The calculated anisotropy of $Ti_{50}Pd_{50-x}Os_x$ SMAs ($0 \leq x \leq 50$).

The calculated Bulk (B), Shear (G) and Young (E) 's moduli for $Ti_{50}Pd_{50-x}Os_x$ alloys are shown in figure 4. Mostly, the Bulk modulus is a degree of the hardness in materials. The calculated Bulk modulus increase with an increase in Os content, indicating that the material becomes harder at high content of Os. At 6.25 at. % Os, the structure appears to be more compressible and easily breakable as revealed by its relatively low negative moduli. The Shear modulus becomes positive and increases above 18 at. % Os, showing less compressibility. The calculated Young 's modulus can approximate the stiffness and the largest value relates to the stiffer material. The findings propose that $Ti_{50}Pd_{6.25}Os_{43.75}$ is the stiffest with the highest value of Young 's modulus.

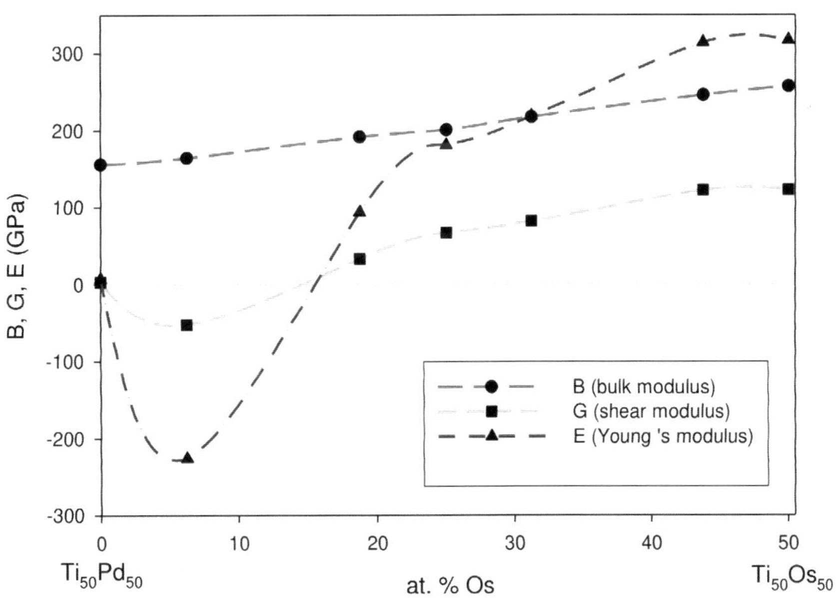

Figure 4. Simulated elastic moduli B, G and E (GPa) for B2 $Ti_{50}Pd_{50-x}Os_x$ SMAs.

Figure 5 shows the calculated B/G ratio, Cauchy pressure, Poisson 's ratio and Vickers hardness (H_V) at different compositions to assess the ductility/brittleness of the $Ti_{50}Pd_{50-x}Os_x$ alloys. The ratio of Bulk to Shear modulus was calculated as suggested by Pugh which predicts ductile/brittle behavior of the materials [22]. The highest value (B/G> 1.75) is related with ductility while low value (B/G< 1.75) relates brittleness. The values of B/G are lower than 1.75 at 6.25 at. % Os and higher above. The results suggest that 18.75 at. % Os is more ductile with the highest value of 5.75 and more brittle at 6.25 at. % Os with the lowest value of -3.13 (see figure 5 (a)). This result further proves the same deduction with the anisotropy ratio which indicates a good correlation between B/G and anisotropy. Metals with a Poisson 's ratio greater than 0.26 are ductile otherwise brittle [23]. The results showed that all compositions are greater than 0.26 which indicates ductile behavior (see figure 5 (b)). For cubic crystal, the Cauchy pressure is well-defined as (c_{12}-c_{44}) for (100) plane. As suggested by Pettifor [24], the material with positive Cauchy pressure reveal their metallic bond character and ductile behaviour while a negative value indicates a weak covalent bond and exhibits brittle behaviour. The calculated Cauchy pressure is positive below 31.25 at. % Os which reveals their ionic character and ductile behaviour while above is negative indicating their weak covalent character and brittle behaviour (see figure 5 (c)). The Vickers hardness (H_V) of $Ti_{50}Pd_{50-x}Os_x$ alloys are also calculated and are shown in figure 5 (d). It was found that the H_V increases as the concentration of Os are increased. This trend indicates that the hardest material can be obtained at the high content of Os, for example, $Ti_{50}Pd_{6.25}Os_{43.75}$ with the hardness of 12.49.

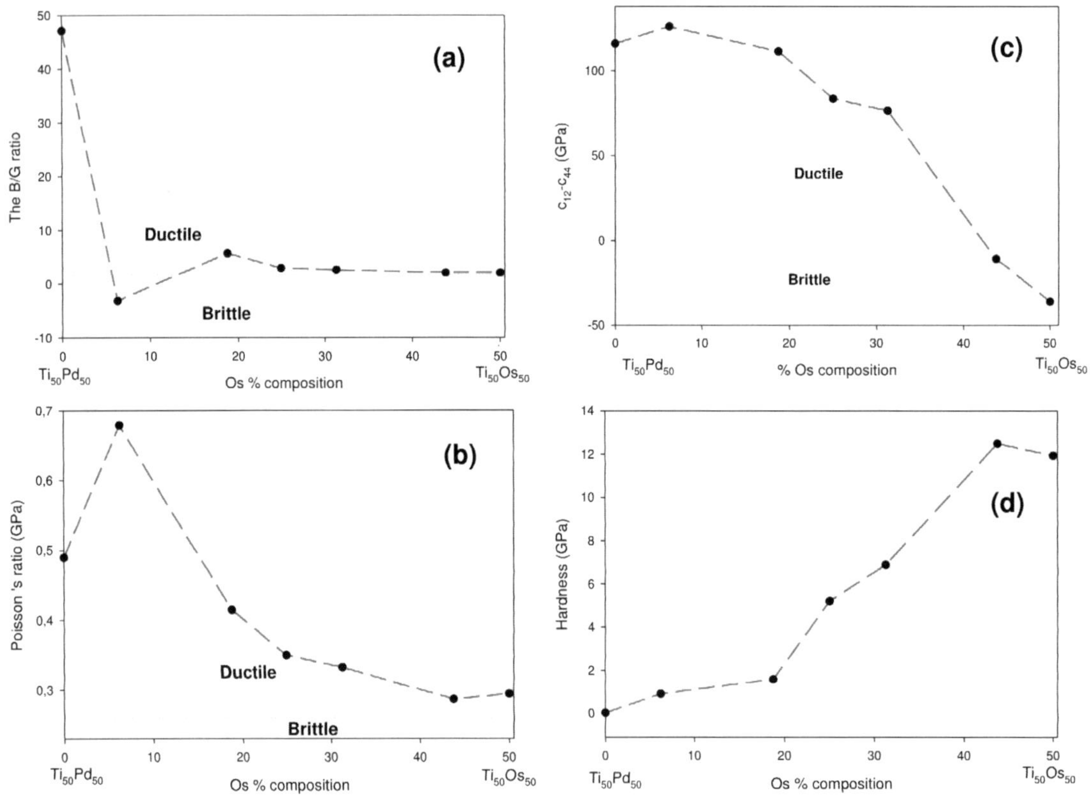

Figure 5. **(a)** The B/G ratio, **(b)** Poisson 's ratio, **(c)** Cauchy pressure (c_{12}-c_{44}) and **(d)** Vickers Hardness (H_V) for B2 $Ti_{50}Pd_{50-x}Os_x$ SMAs.

4. Conclusion

Using the DFT approach, we studied the lattice parameters, heats of formation, electronic structure and elastic properties of $TiPd_{50-x}Os_x$ for potential HTSMAs. The results suggest that $Ti_{50}Pd_{50-x}Os_x$ is thermodynamically stable with the lowest value of the heats of formation. It was found that $Ti_{50}Pd_{50-x}Os_x$ alloys are mechanically stable above 18.75 at. % Os according to the criteria of mechanical stability. The ductile nature of $Ti_{50}Pd_{50-x}Os_x$ alloys was confirmed from the value of the B/G ratio, Poisson 's ratio, Cauchy pressure (c_{12}–c_{44}) and anisotropy. Increasing Os above 6.25 at. % could effectively improve the ductility of the compound. The Bulk modulus (B), Shear modulus (G), Young 's modulus (E), and Vickers hardness (H_V), for the $Ti_{50}Pd_{50-x}Os_x$ alloys, showed an increase tendency with an increase in Os content. The current findings and analysis may provide helpful direction for future design of new $Ti_{50}Pd_{50-x}Os_x$ for potential high temperature applications.

Acknowledgments
This work was funded by National Research Foundation (NRF) and Council for the scientific and Industrial research (CSIR). The authors acknowledge support of the South African Research Chair Initiative of the Department of Science and Technology. The calculations were achieved at the Materials Modelling Centre (MMC), University of Limpopo.

References

[1] Jani J M, Leary M, Subi A, and Gibson M A 2014 *Mater. Des.* **56** 1078-1113.
[2] Otsuka K and Wayman C M 1998 *Cambridge University Press* 220-239.
[3] Dasgupta R 2014 *J. Mater. Res.* **29** 1681.
[4] Bigelow G 2006 MSc. *Thesis Colorado School of Mines Golden CO.*
[5] Choudhary N and Kaur D 2016 *Sens. Actuators* A **242** 162–181.
[6] Huang X, Karin M, and Ackland J 2003 *Phys. Rev.* B **6**7 24101-24107.
[7] Cai W, Tanaka S and Otsuka K 2000 *Mater. Sci. Forum* **279** 327-328.
[8] Frankel D J and Olson G B 2015 *shape memory and superelasticity* **1** 162-179.
[9] Bozzolo G, Mosca H O, and Noebe R D 2007 *Intermetallics* **15** 901-911.
[10] Haynes, William M ed. (2011). *CRC Handbook of Chemistry and Physics (92nd ed.). CRC Press.* ISBN 978-1439855119.
[11] Kresse G and Hafner J 1993 *Phys. Rev.* B **47** 58-561.
[12] Kresse G and Furthmüller J 1996 *Phys. Rev.* B **54** 11169-11186.
[13] Perdew J P, Burke K, and Ernzerhof M 1996 *Phys. Rev. Lett.* **77** 3865.
[14] Kohn W and Sham L J 1965 *Phys. Rev.* **140** 1133-1138.
[15] Monkhorst H J and Pack J D 1996 *Phys. Rev.* B **13** 5188−1592.
[16] Mahlangu R, Phasha M J, Chauke H R and Ngoepe P E 2013 *Intermetallics* **33** 27-32.
[17] Pankhurst D A, Nguyen-Manh D, and Pettifor D G 2004 Phys. *Rev. B* **69** 075113.
[18] Koshi N A and John R 2018 *IJRASET* **6** 415-422.
[19] Mehl M J and Klein B M 1994 *Intermetallic Compd.* **1** 1−26.
[20] Gschneidner K et al 2003 *Nat. Mater.* **2** 587-591.
[21] Ren X, Miura N, Taniwaki L, Otsuka K, Suzuki T, Tanaka K, Chumlyakov Y I and Asai M 1999 *Mater. Sci. Eng. A* **273-275** 190-194.
[22] Pugh S F 1954 *Phil. Mag.* **45** 823-843.
[23] Frantsevich I N and Voronov S A 1983 *Naukova Dumka, Kiev* 60-180.
[24] Pettifor D G 1992 *Mater. Sci. Technol.* **8** 345-349.

IOP Conf. Series: Materials Science and Engineering **655** (2019) 012043 doi:10.1088/1757-899X/655/1/012043

First principle study of HF molecule adsorption on TiO_2 (110) surface

D M Tshwane[1,2*], R Modiba[1,2], H R Chauke[2], G Govender[1] and P E Ngoepe[2]

[1] Materials Science and Manufacturing, CSIR, PO Box 395, Pretoria, 0001, South Africa

[2] Materials Modelling Centre, University of Limpopo, Private Bag X1106, Sovenga, South Africa

*DTshwane@csir.co.za

Abstract. Titanium and its alloy components are one of the most important technological materials, which has found extensive application in various industries. However, surface defects play a key role in the mechanical properties of these components. Currently, wet chemical etching is one of the most important procedure for surface processing due to the presence of HF since it can etch metal oxide. Therefore, there still a need to investigate the etching mechanism. In this work, adsorption of HF on TiO_2 (110) surface has been studied using density functional theory to investigate the fundamental process of etching. HF molecule is adsorbed on the TiO_2 surface by dissociation to form Ti-F and O-H species. The interaction between HF and TiO_2 surface become more favorable at high HF coverage. The pre-adsorbed of water molecule is favorable for HF adsorption process, which is in good agreement with experimental results. Fluorination processes show the formation and desorption of water intermediate at 0.50 ML coverage. We also investigated the relation between the work function and Mulliken charge for HF adsorption. The results indicate that the interaction of F on the surface attracts electrons due to its higher electronegativity than oxygen. Our results suggest that adsorption of HF is considered chemisorption process.

1. Introduction

Titanium and its alloys are one of the most important materials used in various industries such as aerospace, biomedical and automotive [1]. During the investment casting, oxygen is absorbed from the ceramic mould, which results in the formation of an alpha case layer and is very brittle [2]. Formation of this alpha case layer causes deterioration of mechanical properties that result in material micro-failure [3]. Surface modification is required due to spontaneously growth of a native oxide layer. Recently, wet chemical/ion etching process has been widely employed as a surface modification using hydrogen fluoride (HF) as an etchant, to remove alpha case layer and improve the surface quality [4, 5]. Metal surface etching by HF is one of the most important material tailoring techniques in the manufacturing of metal-based alloyed components. The main purpose of chemical etching is to remove alpha case layer by breaking down titanium oxide layer [6]. The presence of fluorine ion significantly influences the electrochemical behaviour of a metal surface, which is a major relevance for etching and corrosion [7]. There are many theoretical and experimental studies on surface etching but the mechanism of hydrogen fluoride etching process is still not well-understood [8]. In order to get a better understanding of atomic

etching mechanism, a density functional theory (DFT) approach has become an attractive theoretical method to provide chemical bonding, atomic interaction and hybridisation [9].
Recently, research interest has emerged driven by the demand of a new method that is capable of etching metal surface oxides. However, previous studies focused mainly on the Al_2O_3 surface [10] and the adsorption behaviour on a single atom or probe molecules on metal oxides [11]. Wang *et al*, [12] investigated the selective etching phenomenon on anatase-TiO_2 surfaces. Their results demonstrated that the etching processes is energetically preferable and merely occur on the (100) surface under high HF concentrations. Recently, work done by Yang *et al* [13] demonstrated that the adsorption of fluorine (F) ion on the surfaces of anatase-TiO_2 makes (100) surface energetically preferable to (101) surface.

To investigate deeply on the fluorination process of titanium or metal oxides surface, in this work DFT is employed to study the adsorption behavior of HF on rutile TiO_2 (110) surface. These present work greatly improve our understanding on electronic properties of titanium dioxide and reaction behavior of HF. The structure of HF on TiO_2 (110) surface with and without pre-adsorbed water at different coverage were systematically investigated. The adsorption energies, the relative energy, Mulliken charge analysis and work function of the systems were calculated. The main goal of the study is to identify the key steps to remove oxygen atoms from the metal oxide surface.

2. Methodology

The study employed DFT to investigate adsorption behavior of HF on TiO_2 surface using plane-wave/pseudopotential approach as implemented in Cambridge Serial Total Energy Package (CASTEP) [14]. The electron-exchange and correlation were described by the Perdew-Burke-Ernzerh functional of the generalized gradient approximation (GGA) [15]. The GGA-PBE functional underestimate the band gap, which reduces photo-induced electron transfer during reactions [16]. In order to correct the strong electronic correlation problem, DFT+U method is utilized with a U parameter of 5.0 eV for Ti in TiO_2 that was first proposed by Dudarev *et al* [17]. The Brillouin-zone integrations are using the Monkhorst-Pack approach [18] with K-point sampling grids 4x4x1 for geometry calculation with cut-off energy of 400 eV. The adsorption energy is defined as:

$$E_{\mathrm{ads}} = E_{\mathrm{HF/slab}} - [E_{slab} + E_{HF}] \tag{1}$$

Where $E_{HF/slab}$ is the total energy of the molecule-surface system, E_{slab} and E_{HF} represent the energy of the pure surface and free HF molecule, respectively. A E_{ads} is negative for exothermic adsorption process and positive for endothermic.

3. Results and discussion

3.1 Structural geometry and adsorption energy

Surface model is an essential basis precision for adsorption calculations. In figure 1, the optimized structure of TiO_2 (110) surface is presented wherein figure 1 (a) shows the side-view and figure 1 (b) top-view. The TiO_2 (110) surface consist of three-fold oxygen atoms (O3c), five- and six-fold titanium atoms (Ti5c and Ti6c, respectively). The resulting of geometry of the surface slabs are relaxed to obtain energetic minima. The surface energies of the two slabs with four and six layers relaxed are 1.39 and 1.93 Jm^{-2},which were found to be in good agreement with the results obtained Idriss *et al* [19]. The surface energy depends on the atomic number of layers; 7-layer slab was found to be more stable with lower surface energy than 13-layer. Therefore, 7-layer slab model was selected to describe the adsorption of HF molecule on TiO_2 (110) surface.

IOP Conf. Series: Materials Science and Engineering **655** (2019) 012043 doi:10.1088/1757-899X/655/1/012043

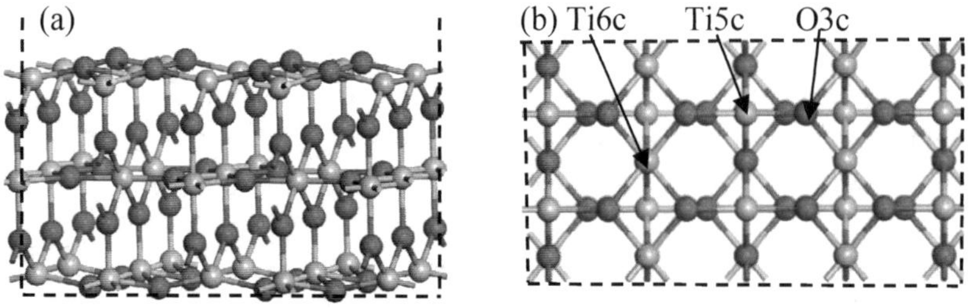

Figure 1. Optimized structure of TiO_2 (110) surface slab: (a) side view and (b) top view with atomic coordination number.

Table 1 presents the adsorption energies and geometric parameters of HF interacting with TiO_2 (110) surface. Firstly, HF molecule was placed on Ti5c and Ti6c site as shown in figure 1 (b) and the systems were relaxed to determine their adsorption energy. Comparing their adsorption energies, it was found that HF molecule thermodynamically favors adsorption site of Ti5c (-4.36 eV) interaction than that of Ti6c (-4.16 eV) on the TiO_2 (110) surface. This clearly reveals that bonding is crucial for HF to react with TiO_2 surface and that could drive HF dissociation. The adsorption energy values are negative, entailing that initial fluorination process of TiO_2 (110) surface is energetically favorable.

Table 1. Adsorption energy of HF on different surface site.

Surface site	E_{ads} (eV)	d_{Ti-F} (Å)	O-H (Å)
Ti5c	-4.36	1.94	1.03
Ti6c	-4.16	1.97	1.01

In order to understand adsorption energy trends, the HF molecules are adsorbed at different coverages and their corresponding adsorption energies, geometric parameters and Mulliken charges are shown in table 2. For all the HF coverage considered, their adsorption energy become more stable with an increasing HF coverage. According to Mulliken charge analysis, Ti atoms gained some positive charge from F-ion surrounding. This suggest that Ti surface atoms loses electrons to F atoms during the interaction.

Table 2. Adsorption energies, Geometrical parameters and Mulliken charge of HF adsorption at different coverage.

Coverage (ML)	E_{ads} (eV)	Mulliken charge |e|	Bond length (Å)			Angle H-O-H (^{0})
			Ti5c-F	Ti6c-F	H-O-H	
0.25	-4.36	1.13	1.94	-	1.01	-
0.50	-5.40	1.15	1.93	1.97	1.04,1.02	104.1
0.75	-8.16	1.22	1.95	2.01	1.11,1.10	107.9
1.0	-9.19	1.34	1.95	2.05	1.00,1.13	107.2,110.4

Coverage: number of molecules (ML) adsorbed per adsorption site.

It was found that the Mulliken charge increases with HF coverage. This suggest that exposing TiO_2 surface to various HF molecules results in surface saturation. Figure 2(a-d) shows the optimized structure of HF adsorption at different coverage. When placing single HF molecule on the surface slab corresponding to 0.25 ML coverage as shown in figure 2(a), the H-F bond raptures to form Ti-F bond with bond length of 1.94 Å. Hydrogen ion chemically bonds to a neighbour oxygen to form hydroxyl with bond length of 1.01 Å. However, Ti-O bond elongate from the initial bond of 2.01 Å to 2.32 Å. When the second HF molecule is introduced, as shown in figure 2(b), the HF molecule dissociate to form new bond of Ti-F2 with bond length of 1.97 Å. Interestingly, for 0.25 to 0.75 ML coverage, the oxygen (O3c) atom is pulled out of the surface forming H_2O molecule as shown in figure 2 (b) and (d). This suggests a stronger chemical interaction of HF and TiO_2 (110) surface. The maximum number of HF molecule coverage was four. Similar adsorption mechanism was observed for all HF coverage and this suggests that HF molecules dissociate completely to the TiO_2 (110) surface.

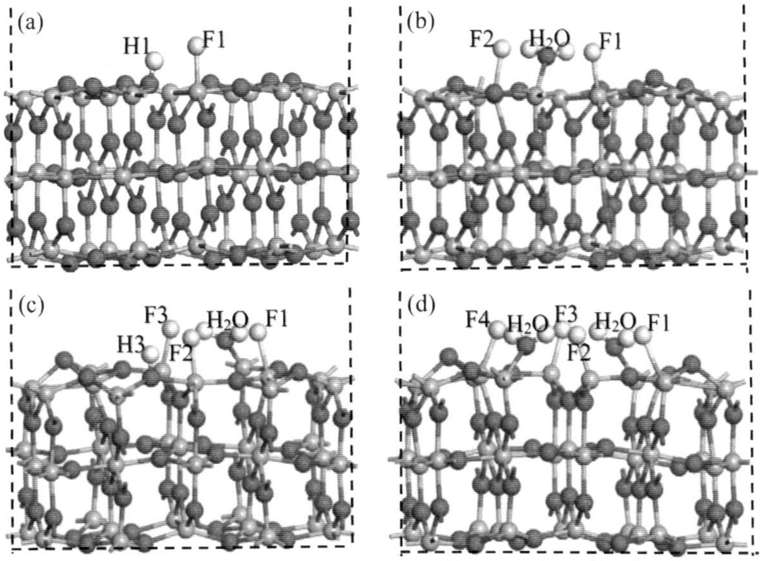

Figure 2. (a)-(d) Adsorption structures of HF on the TiO_2 (110) surface at different coverage.

3.2 Relative Energy

To further understand the fluorination process of titanium dioxide, the successive relative energy of two dissociated HF molecules in the TiO_2 (110) surface are also calculated and their results are shown in figure 3. The total relative energies of the systems for the water formation and desorption were found to be 2.371 eV and 0.291 eV, respectively; which suggest that the reaction is not spontaneous. It appears that two HF molecules dissociate completely to the TiO_2 (110) surface and the two hydrogen atoms are transferred to the neighbouring oxygen to form intermediate water molecule. This shows a stronger chemical interaction of O-H bonding. Therefore, the intermediate water desorbed from the surface to become a free molecule suggest that the oxide layer can be removed as the water intermediate from the surface.

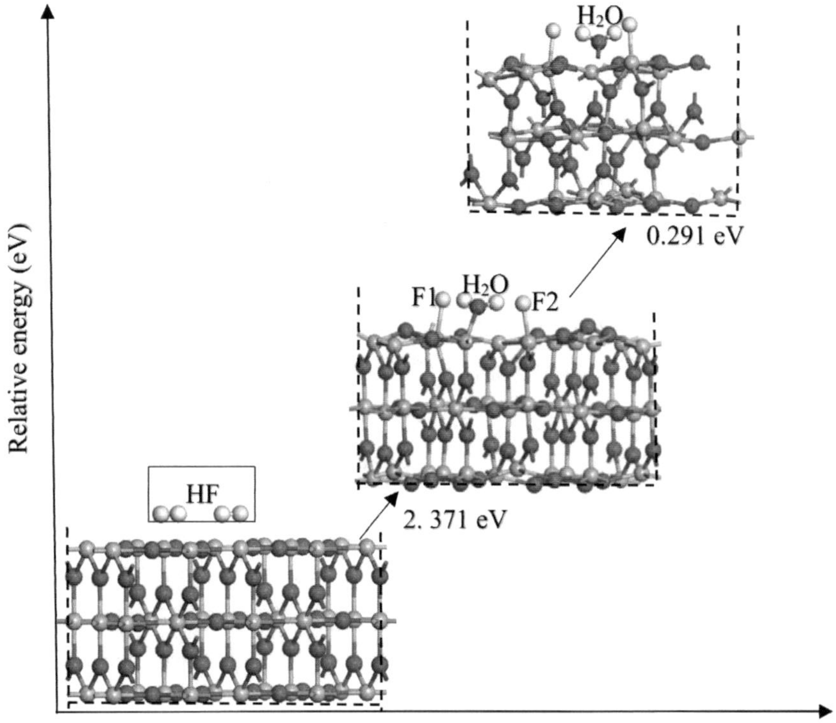

Figure 3. Relative energy of HF on TiO_2 (110) surface.

3.3 Co-adsorption of H_2O and HF on the TiO_2 (110) surface

Previous result revealed that etching titanium oxide surfaces through a wet fluorination process is more favourable than dry etching [20], whereby the existence of water plays a crucial role. One-way to establish this theory, co-adsorption behaviour of H_2O and HF on the TiO_2 (110) surface was also investigated. The same technique was done by Quan *et al* [10] on aluminium oxide surfaces. Figure 4 present the co-adsorption of H_2O-HF molecules on TiO_2 (110) surface. When the water molecule is adsorbed on the TiO_2 (110) surface, a Ti-O bond formed with the bond length of 1.906 Å. The calculated adsorption energy of -4.83 eV was found, this is in good agreement with the value of -4.796 eV found by Dai *et al* [21]. When HF molecule is adsorbed on the TiO_2 (110) surface with the present of H_2O molecule, HF dissociate completely to form Ti-F and O-H with the bond length of 1.989 Å and 1.001 Å, respectively. Adsorption energy of HF with pre-adsorbed H_2O (-6.02 eV) is more stable than the adsorption energy (-4.36 eV) without pre-adsorbed water molecule. When the second HF molecule is introduced on the surface, the adsorption energy is (-7.28 eV) more favourable than the adsorption energy without H_2O molecule. This suggest that the pre-adsorption of water or wet HF is favourable for fluorination of TiO_2 surface than adsorption of HF gas. This correspond with the experimental observation of dry and wet etching on TiO_2 and SiO_2 surfaces [22].

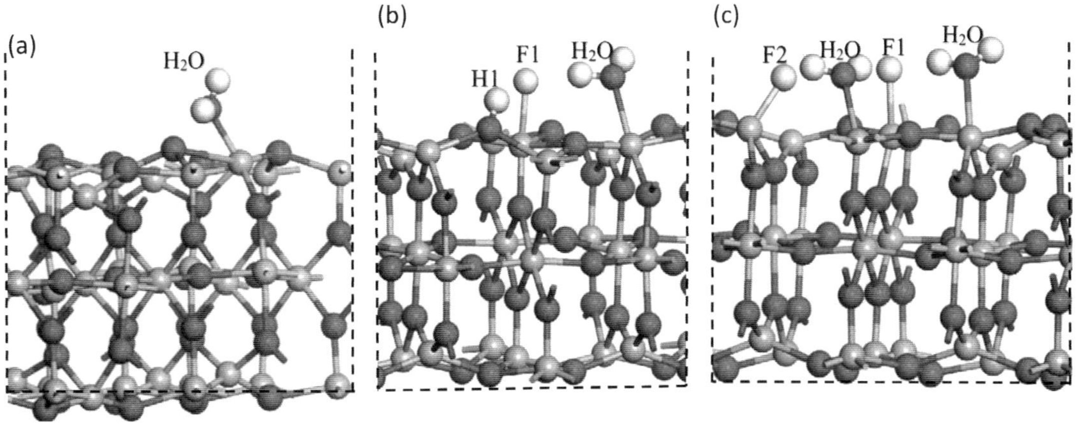

Figure 4. (a)-(c) Adsorption structures of H_2O and HF on the TiO_2 (110) surface.

3.4 Work function and Mulliken charge

Etching mechanism is often electrochemical in nature and this concept arises from corrosion [23]. The process involves electron charge transfer hence Mulliken charge analysis were calculated and summarized in table 3. The trend seen is that the Mulliken charge increases with an increasing HF molecules which implies that more charges are depleted from the surface upon the Ti-F interaction. The changes in surface charge shown to affect surface work function. When a single HF molecule is adsorbed, the work function slightly increases from 2.651-2.815 eV. It was also found that the work function is coverage depended as shown in table 3. The work function offers a good measure of how prone is the corrosion of a certain material [24].

Table 3. The work function and Mulliken charge analysis for different coverage.

Coverage (ML)	Work function	Mulliken charge |e|
0.25	2.815	1.13
0.50	3.392	1.15
0.75	3.237	1.22
1.00	3.682	1.34

In this study change in the work function has been found to be coverage-depended see figure 5. The large relative increase in work function with increasing HF coverage is due to the high electronegativity of fluorine atoms. However, for 0.75 ML coverage the work function decreases as shown in figure 5 due to the repulsive interaction between closed adsorbed fluorine-ion. Different changes in the work function are cause by charge transferred associated with fluorine ion.

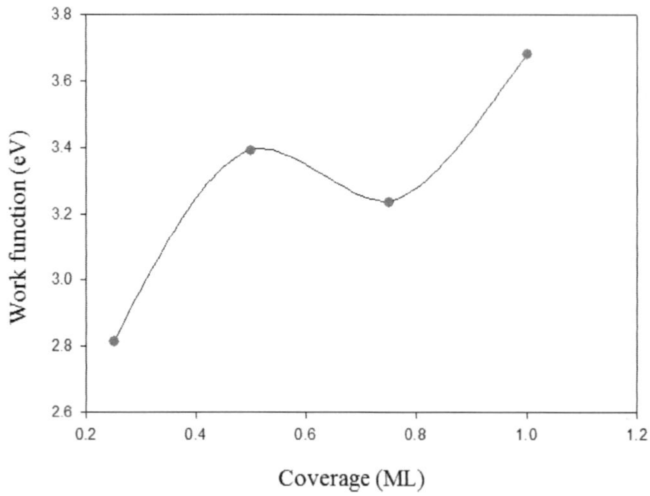

Figure 5. Change in the work function as a function of coverage.

In figure 6 the electrostatic potential is shown where (a) is pure surface slab and (b) the highest coverage (1 ML). The electrostatic potential of the pure surface tends to be flat in the vacuum region, while after the adsorption of HF the is a potential step in the vacuum region for all coverage. The potential step in the middle of the vacuum is due to the dipole introduced by the fluorine-ion, which result in two surfaces upper and lower. The upper surface (Φ_1) consist of Ti-F with the higher value of the work function while lower surface (Φ_2) consist of Ti-O layer with the lowest value of work function. The decrease of lower work function (Φ_2) shows that the surface lost electrons during adsorption, which implies the manifestation of surface corrosion.

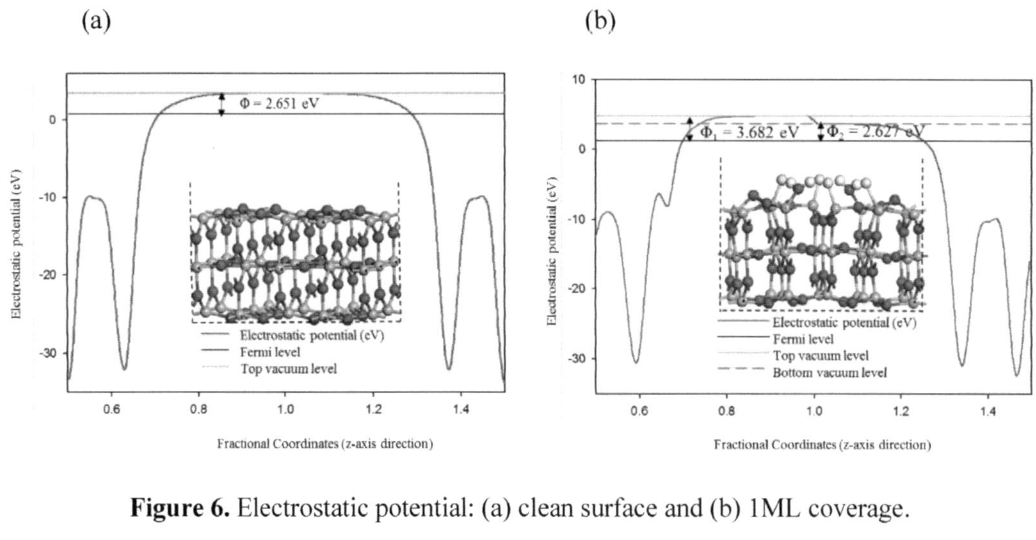

Figure 6. Electrostatic potential: (a) clean surface and (b) 1ML coverage.

4. Conclusion

The adsorption characteristic of HF molecule on TiO_2 (110) surface at different coverage was investigated using DFT approach. The explicit surface slab calculations were conducted to study the adsorption of HF molecules per cell on a rutile TiO_2 (110) surface. HF molecule dissociate completely upon adsorption on the TiO_2 surface, to form Ti-F bond while the hydrogen ion is chemically bonded to the nearest oxygen to form hydroxyl. The adsorption stability of HF molecule increases with an increasing HF coverage. Interestingly for all coverages their adsorption energies are negative, this implies that adsorption process is exothermic and etching mechanism is energetically permitted at high coverage. In addition, the pre-adsorption of water molecule was investigated on TiO_2 surface and it was found to be more favourable for HF adsorption. The total energy of the two dissociated HF molecules on the TiO_2 surface were found to be 2.371 and 0.291 eV for the formation and desorption of water intermediate. This preliminary reaction suggests that etching of TiO_2 (110) surface occurs at high temperature. Furthermore, reaction showed intermediate water desorbed from the surface, which mimic one of several possible cases when the oxide layer is suddenly destroyed by etching mechanisms. The study revealed a significant increase on the work function at different coverage due to the high electronegativity of the fluorine atom. Furthermore, Mulliken charge analysis showed a positive charge increase with an increase in HF adsorption molecule. Since during adsorption, fluorine atom withdraws electrons from Ti atoms

Acknowledgements

The authors acknowledge Department of Science and Technology (DST) and Council of Scientific Industrial Research (CSIR) for their financial support. Centre for High Performance Computing (CHPC) for computing resources.

References

[1] Trevisan F, Calignano F, Aversa A, Marchese G, Lombardi M, Biamino S, Ugues D and Manfredi D, 2017 Appl Biomaterials and Functional Material **16** 57-67
[2] Sung S Y, Choi B J, Han B S, Oh H J and Kim Y J 2008 J.Mater.Sci.Technol. **24** 70-74
[3] Lin K F and Lin C C, 1999 J.Mater.Sci. **34** 5899-5906
[4] David M, Muhida R, Roman T, Nakanishi H, Diño W, Kasai H, Takano F, Shima H and Akinaga H, 2008 Vacuum **83** 599-601
[5] Natarajan S K, and Elloitt S D, 2018 Chem.Mater. **30** 5912-5922
[6] Ezaki T and Ohno T, 2000 Surf. Sci. **444** 79-86
[7] Tereshchenko O E, Paget D, Toropetsky K V, Alperovich V L and Eremeev S V, 2012 J. Phys.Chem C **116** 8535-8540
[8] Czekala P T, Panosetti C, Lin H and Hofer W A, 2014 Surf.Sci. **621** 152-161
[9] Obi-Egbedi N O, Obot I B and El-Khaiary M I, 2011 Journal of Molecular Structure **1002** 86-96
[10] Quan J L, Teng B T, Wen X D, Zhao Y, Liu R and Luo M F, 2012 J.Chem.Phys. **136** 11470-11477
[11] Bouzoubaa A, Diawara B, Maurice V, Minot C and Marcus P, 2009 Corr.Sci. **51** 2174-2182
[12] Wang Y, Zhang H, Han Y, Liu P, Yao X and Zhao H, 2011 Chem.Comm. **47** 2829-2831
[13] Yang H G, Sun C H, Qian S Z, Zou J, Liu G, Smith S C, Cheng H M and Lu G Q, 2008 Nature, **453** 638-641
[14] Segall M D, Philip J D L, Probert M J, Pickard C J, Hasnip P J, Clark S J and Payne M C, 2002 J.Phys: Condens.Matter. **14** 2717-2744
[15] Perdew J P and Wang Y, 1992 J.Phys.Rev. B **45** 13244-13249
[16] Di Valentin C, Pacchioni G and Selloni A, 2016 Phys.Rev.Lett. **97** 1-9
[17] Dudarev S, Botton G, Savrasov S, Humphreys C and Sutton A, 1998 Phys.Rev. B **57** 1505-1509
[18] Monkhorst H J and Pack J D, 1976 Phys.Rev. B **13** 5188-5192
[19] Waterhouse G I N and Idriss H 2009 Asia, John Wiley and Sons 78-89

[20] Park H and Choi W, 2004 J.Phys.Chem. B **108** 4086-4093
[21] Dai J and Song Y, 2016 Materials **09** 1-14
[22] Danga G, Chob H, Ip K P, Pearton S J, Chu S N G, Lopata J, Hobson W S, Chirovsky L M F and Rena F 2001 J.Electrochem. Soc. **148** G25-G28
[23] Taylor C D and Marcus P, 2015 Molecular Modelling of Corrosion Processing, Dublin, John Wiley and Son Inc
[24] Li W and Li D Y, 2005 Appl.Surf.Sci. **240** 388-395

IOP Conf. Series: Materials Science and Engineering **655** (2019) 012044 doi:10.1088/1757-899X/655/1/012044

First-principles studies on the structural, electronic and mechanical properties of $L1_0$ and $L1_2$ Fe_xPt_{1-x} alloys

N L Lethole[1], H R Chauke[1] and P E Ngoepe[1]

[1] Materials Modelling Centre, School of Physical and Mineral Sciences, University of Limpopo, Sovenga 0727, South Africa

E-mail: Corresponding Author: Lesley.lethole@ul.ac.za

Abstract. Bimetallic Fe_xPt_{1-x} alloys with the $L1_0$ and $L1_2$ structures have recently gained a lot of consideration in practical applications for solid-state devices, storage of ultra-high density magnetic data and biomedicine. This is due to their high magnetic and magnetocrystalline anisotropy, density, and coercivity. In order to gain knowledge on the structural, electronic and mechanical properties of the cubic and tetragonal Fe_xPt_{1-x} alloys, we have calculated their equilibrium lattice constants, density of states, and elastic constants at 0 K, employing first-principles calculations. The calculated equilibrium lattice constants were found to be in good agreement with the experimental data to within 3 %. All independent elastic constants satisfy the necessary stability conditions for both cubic and tetragonal systems, suggesting mechanical stability. The shear anisotropic factors predict that the tetragonal Fe_xPt_{1-x} crystals are highly anisotropic along the {001} plane than {100}. Moreover, the percentage of bulk (A_B) and shear (A_G) anisotropies revealed completely isotropic systems in the bulk and slightly anisotropic in shear modulus.

1. **Introduction**

Nanocomposites of ferromagnetic alloys made of hard and soft nanoparticles have attracted a considerable amount of attention for the past few decades. This is due to their potential application as permanent magnets and new storage of ultra-high density magnetic data and in magnetic recording [1, 2, 3]. However, several experimental findings have reported that the density of magnetic recording is restricted by the "superparamagnetic limit" i.e. loss of data due to activated fluctuations of magnetisation. Amongst other possibilities, Wang *et al.* have suggested that superparamagnetic limit can be overcome by designing new magnetic storage, in which magnetic nanoparticle composite forms a patterned media and each bit (0 or 1) is stored in a single magnetic nanoparticle [4]. Hence, it is of great significance to understand how material's properties such as electronic, elastic etc. relate with magnetic stability of magnetic nanoparticle.

Amongst other ferromagnetic alloys, the binary Fe_xPt_{1-x} alloys are perceived to overcome the superparamagnetic limit, due to their high magnetocrystalline anisotropy ($\sim 6 \times 10^6\ J/m^3$) [5, 6] along {001} direction in their chemically ordered states. Fe_xPt_{1-x} alloys exist in four chemically ordered phases categorised into two main crystal symmetry families ($L1_0$ and $L1_2$) when in thermodynamic equilibrium (see Figure 1) [7, 8]. The $L1_0$ symmetry crystallises in the tetragonal distorted FePt (space group P4/mmm) in the equiatomic (50:50) composition. Antoniak *et al.* showed that the orbital magnetic moment of FePt nanoparticles is significantly influenced at the Fe site, whereas the spin magnetic moment remains largely unaffected at the Pt site [3]. Moreover, FePt possesses high magnetocrystalline

anisotropy and high coercivity. On the other hand, the $L1_2$, existing between 25 % -75 % Pt composition crystallises in three different space groups, namely; cubic Fe_3Pt (space group Pm-3m), tetragonal Fe_3Pt (space group I4/mmm) and cubic $FePt_3$ (space group Pm-3m).
In this work, we report preliminary studies on the structural, electronic and elastic properties of the bulk Fe_xPt_{1-x} alloys; in particular, we have calculated the lattice constants, density of states, elastic constants, moduli and anisotropy at 0 K. Moreover, we have determined thermodynamic stability by calculating heats of formation.

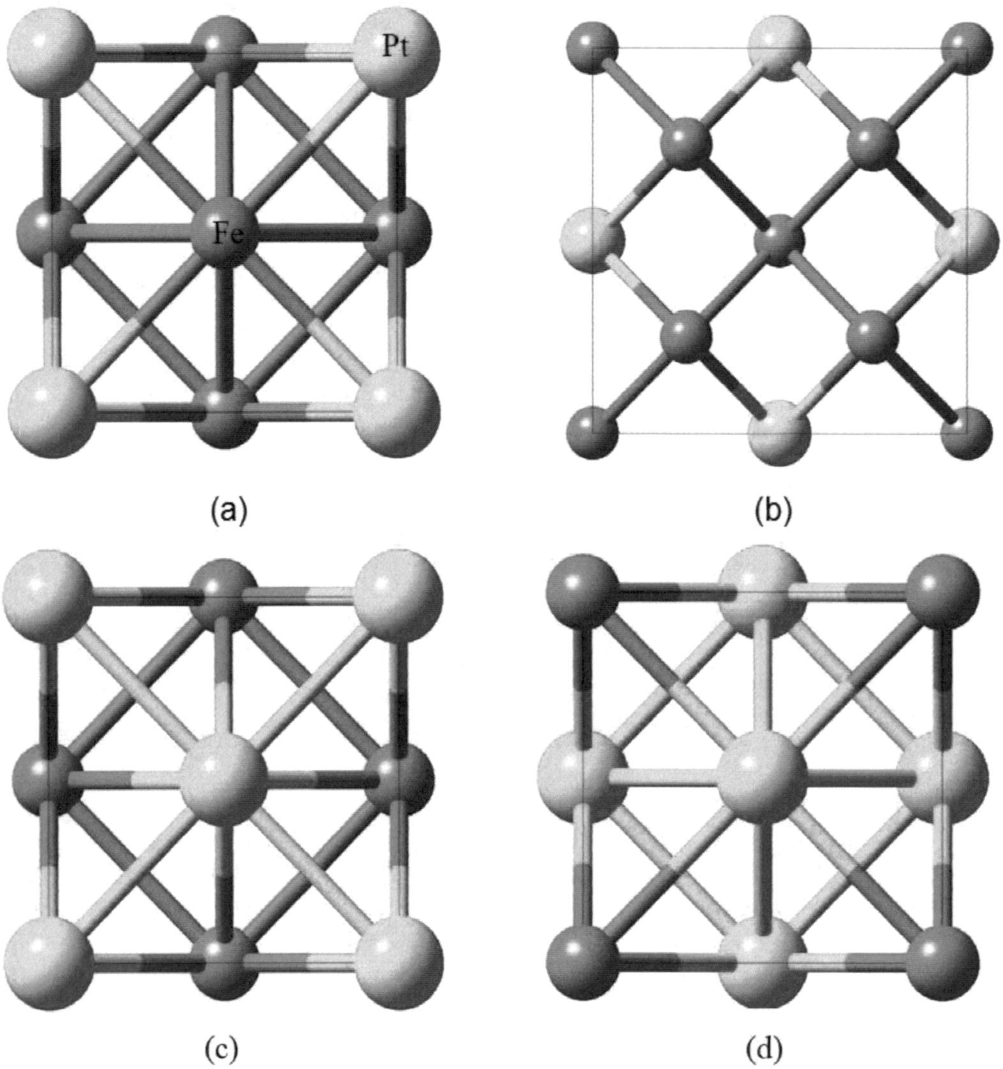

Figure 1. Schematic representation of (a) Pm-3m Fe_3Pt, (b) I4/mmm Fe_3Pt, (c) P4/mmm FePt and (d) Pm-3m $FePt_3$.

2. **Methodology**
First-principles DFT+*U* calculations were performed using the Vienna *Ab Initio* Simulation Package (VASP) [9] to determine the structural, electronic and elastic properties of bimetallic Fe_xPt_{1-x} alloys. We applied the spin-polarised local density approximation (LDA) exchange correlation energy functional [10]. Moreover, the Hubbard correction (U) necessary to treat strongly correlated transition metals was employed on the *d* orbital of Fe; U=1 eV and J= 1 eV were found to correctly predict the lattice constants to within 3 % in agreement with the experimental data. A plane-wave cut-off of 450 eV and suitable k-points spacing of 0.2 1/Ang obtained by performing numerical convergence test to less than 0.01

eV/atom were employed for structure optimisation. The equilibrium values were achieved by performing full relaxation of cell volumes and atomic positions within the plane-wave pseudopotential method. The elastic properties were calculated employing a strain of 0.005, while the densities of states were calculated in the spin-polarised magnetism with the Methfessel–Paxton smearing width of 0.02 eV [11].

3. Results

3.1. Structural properties and thermodynamic stability

The equilibrium lattice constants and heats of formation for the considered Fe_xPt_{1-x} alloys are shown in Table 1 and Figure 2, respectively. Our calculated DFT+*U* lattice constants are in good agreement with the experimental data to within 3 %, indicating the validity of the approach employed. The thermodynamic heats of formation were calculated according to equation 1.

$$\Delta H_f(eV) = \left[E_{Fe_xPt_{1-x}} - (xE_{Fe} + 1 - xE_{Pt})\right] \quad 1.$$

where $E_{Fe_xPt_{1-x}}$ is the total energy of Fe_xPt_{1-x}, while E_{Fe} and E_{Pt} are the total elemental energies of Fe and Pt in their respective ground-states. We note that the calculated heats of formation decrease with the Fe concentration (see Figure 2), indicating an increase in thermodynamic stability. This suggests that Pm-3m $FePt_3$ alloy is more thermodynamically favourable over P4/mmm FePt, I4/mmm Fe_3Pt and Pm-3m Fe_3Pt, respectively. Moreover, the heats of formation values of I4/mmm Fe_3Pt, P4/mmm FePt and Pm-3m $FePt_3$ are negative, suggesting that these structures can be formed experimentally.

Table 1. Lattice constants and heats of formation for Fe_xPt_{1-x} alloys.

Structure	a (Å)	c (Å)	V (Å^3)	ΔH_f (eV)
Fe_3Pt (Pm-3m)	3.722	-	51.58	0.060
Exp.	3.73		51.90 [12]	
Fe_3Pt (I4/mmm)	5.282	7.416	206.85	-0.080
Exp.	5.73	6.34	208.16 [13]	
FePt (P4/mmm)	3.841	3.766	55.56	-0.350
Exp.	3.85	3.70	54.81 [14]	
$FePt_3$ (Pm-3m)	3.902	-	59.41	-0.679
Exp.	3.86		57.78 [15]	

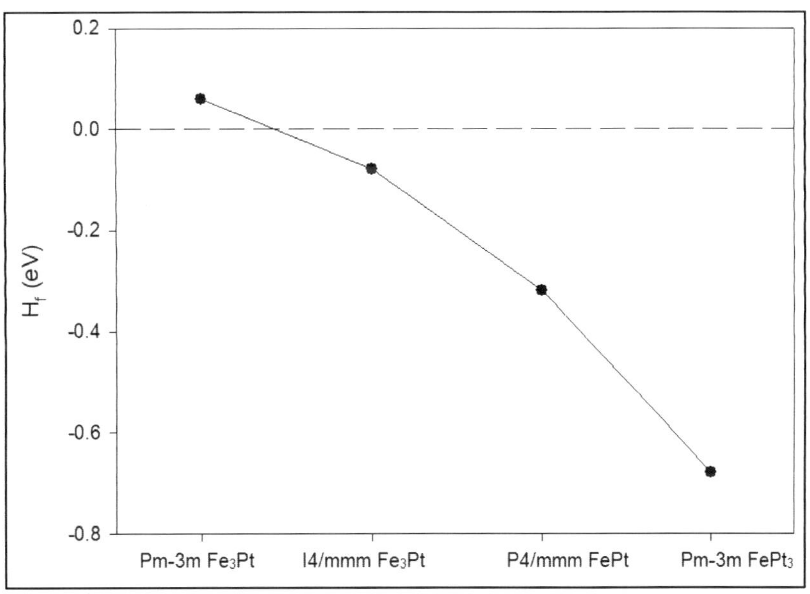

Figure 2. Plots of heats of formation at different Fe and Pt concentrations.

3.2. Electronic properties

In order to determine the electronic properties of Fe_xPt_{1-x} alloys, we have calculated their densities of states (DOS) as presented in Figure 3. We note that the states in the up-spin bands are located below the Fermi level, while the down-spins overlap above the Fermi level. The analysis of electronic structure shows that the cubic Pm-3m Fe_3Pt and $FePt_3$ systems are magnetic metals, characterised by the absence of energy band gaps (a continuous overlap of states) around the Fermi level. The tetragonal I4/mmm Fe_3Pt and P4/mmm FePt systems are predicted to be strong half-metallic ferromagnets. This is in good agreement with the first principles calculations reported by MacLaren *et al.* [16]. Bands for spin up are semiconducting, while bands for spin down are metallic. Moreover, we note that the Pm-3m $FePt_3$ structure shows a trivial shallow pseudo gap around the Fermi level, suggesting stability, corresponding well with the heats of formation. The other Fe_xPt_{1-x} alloys show much broader and flat peaks around the Fermi level.

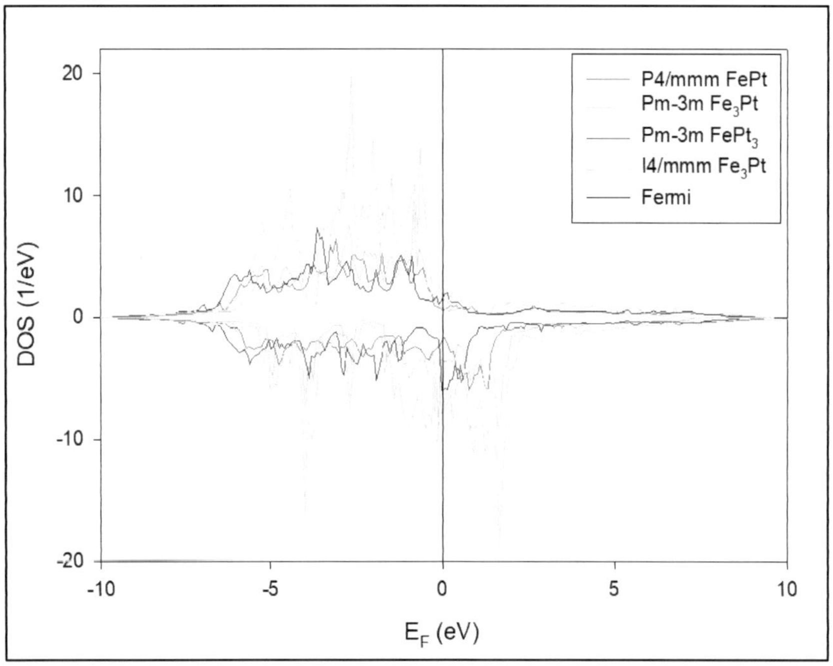

Figure 3. Total density of states for FePt (P4/mmm), Fe_3Pt (Pm-3m), Fe_3Pt (I4/mmm) and $FePt_3$ (Pm-3m) alloys. The Fermi energy is used as the zero of the energy scale.

3.3. Mechanical properties

In Table 2, we present the calculated elastic constants, moduli, anisotropy factors and Pugh ration for the Fe_xPt_{1-x} alloys under study. Calculation of elastic constants is essential in gaining insights of mechanical stability and elastic properties of solids. The Taylor expansion of the total energy of a strained system [17, 18] (see equation 2) was used in the calculations of stress tensors on strains applied to the equilibrium structure to obtain the elastic constants C_{ij}.

$$U(V,\varepsilon) = U(V_0,0) + V_0\left[\sum_i \tau_i\varepsilon_i\xi_i + \frac{1}{2}\sum_{ij} C_{ij}\varepsilon_i\xi_i\varepsilon_j\xi_j\right], \qquad 2.$$

where $U(V_0, 0)$ is the energy of the unstrained system with equilibrium volume V_0; τ_i and ξ_i are elements in the stress tensor and a factor taking care of the Voigt index, respectively. Cubic and tetragonal crystal systems contain three (c_{11}, c_{12}, c_{44}) and six $(c_{11}, c_{33}, c_{44}, c_{66}, c_{12}, c_{13})$ independent elastic constants, respectively. For cubic crystals (Pm-3m Fe_3Pt and $FePt_3$) to be considered mechanically stable, the following necessary Born stability criterion must be satisfied [19, 20].

IOP Conf. Series: Materials Science and Engineering **655** (2019) 012044 doi:10.1088/1757-899X/655/1/012044

$$C_{11} + 2C_{12} > 0,\ C_{11} > |C_{12}|\ and\ C_{44} > 0 \quad 3.$$

We note that our LDA+U elastic constants values satisfy the necessary stability conditions, suggesting mechanical stability on the Pm-3m Fe_3Pt and $FePt_3$ cubic systems. On the other hand, the mechanical stability criterion of tetragonal P4/mmm FePt and I4/mmm Fe_3Pt are outlined according to equation 4 [21]. Similarly to cubic crystals, we note that the stability criterion are satisfied for tetragonal systems, suggesting mechanical stability. Furthermore, we observe a positive C' value, also indicating the mechanical stability for all the crystal systems.

$$C_{11} - C_{12} > 0;\ \ C_{11} + C_{33} - 2C_{13} > 0; \quad 4.$$

$$C_{ii} > 0\ (i = 1, 3, 4, 6);\ \ 2C_{11} + C_{33} + 2C_{12} + 4C_{13} > 0$$

Table 2. Elastic constants (C_{ij}), moduli, anisotropy factors (A) and Pugh ration (K) for FePt (P4/mmm), Fe_3Pt (Pm-3m), Fe_3Pt (I4/mmm) and $FePt_3$ (Pm-3m) alloys.

C_{ij} (GPa)	Pm-3m Fe_3Pt	P4/mmm FePt	I4/mmm Fe_3Pt	Pm-3m $FePt_3$
C_{11}	207.97	297.67	277.26	316.01
C_{12}	174.75	170.90	95.43	180.81
C_{13}	-	158.95	172.27	-
C_{33}	-	308.99	204.53	-
C_{44}	97.25	117.08	94.91	106.53
C_{66}	-	162.74	19.59	-
C'	16.61	63.39	90.91	67.60
B	185.82	209.11	182.09	225.87
G	49.02	101.33	48.48	88.77
B/G	3.790	2.063	3.756	2.5445
A_1	-	1.6218	2.766	-
A_2	-	2.567	0.21	-
A	5.855	-	-	1.576
A_B (%)	0	0	0	0
A_G (%)	32.57	5.67	30.25	2.467

From the calculated independent elastic constants for cubic and tetragonal systems, the macroscopic elastic bulk and shear moduli were determined according to the Voigt [22] and Reuss [23] approximations of simple and linear relations between the isotropic bulk and shear moduli of polycrystalline constants. Moreover, Hill proved that the Voigt and Reuss equations represent upper and lower bounds and proposed an arithmetic average moduli value [24]. The bulk modulus is a measure of resistance to volume changes by applied pressure, while the shear defines resistance to plastic deformation. From the obtained values (Table 2), we note that $B > G$ for all the systems, indicating that the shear modulus is the parameter limiting mechanical stability for Fe-Pt alloys. Pugh proposed the B/G ratio to predict the ductility or brittleness [25] of solids: if B/G is greater than the critical value 1.75, the material is ductile, otherwise brittle. We obtained the ratio B/G values greater than 1.75 for all Fe-Pt alloys, indicating ductility.

$$B = \left(\frac{C_{11} + 2C_{12}}{3}\right);\ G_V = \left(\frac{C_{11} - C_{12} + 3C_{44}}{5}\right);\ G_R = \left(\frac{C_{44}(C_{11} - C_{12})}{4C_{44} + 3(C_{11} - C_{12})}\right); \quad 5.$$

$$G_H = \left(\frac{G_V + G_R}{2}\right);\ C' = \left(\frac{C_{11} - C_{12}}{2}\right)$$

$$B_V = \frac{1}{9}[2(C_{11} + C_{12}) + C_{33} + 4C_{13}]; \quad 6.$$

$$G_V = \frac{1}{30}(M + 3C_{11} - 3C_{12} + 12C_{44} + 6C_{66});$$
$$M = (C_{11} + C_{12}) + 2C_{33} - 4C_{13};$$
$$B = B_H = \left(\frac{B_V + B_R}{2}\right)$$

$$B_R = \frac{C^2}{M};\ G_R = 15/(18B_V/C^2 + 6/(C_{11} - C_{12}) + 6/C_{44} + 3/C_{66});$$ 7.
$$C^2 = (C_{11} + C_{12})C_{33} - 2C_{13}^1;$$
$$G = G_H = \left(\frac{G_V + G_R}{2}\right)$$

Moreover, we have calculated the elastic anisotropy factors, which are essential to determine the susceptibility of materials to cracks induced in alloys owing to the anisotropy of the coefficient of thermal expansion as well as elastic anisotropy [26, 27]. These parameters measure the degree of anisotropy in bonding between atoms in different planes: a material is considered completely isotropic if $A = 1$, while any divergence from unity shows a degree of elastic anisotropy. The anisotropy factor for cubic system defined as the ratio of the extreme values of the orientation-dependent shear moduli is given as follows [28]:

$$\mathrm{A} = \frac{2\mathrm{C}_{44}}{\mathrm{C}_{11} - \mathrm{C}_{12}},$$ 8.

while the shear anisotropic factors along {100} and {001} shear planes is defined by [29].

$$A_1 = A_{\{100\}} = \frac{4C_{44}}{C_{11} + C_{33} - 2C_{13}};\ A_2 = A_{\{001\}} = \frac{2C_{66}}{C_{11} - C_{12}}$$ 9.

We note that all Fe_xPt_{1-x} systems show greater anisotropy behaviour, since anisotropy factors A, A_1 and A_2 are far from unity. In addition, the percentage elastic anisotropy proposed by Chung and Buessem for bulk modulus A_B and shear modulus A_G in polycrystalline materials was also calculated as follows [30]:

$$A_B = \frac{B_V - B_R}{B_V + B_R};\ A_G = \frac{B_G - B_G}{B_G + B_G}$$ 10.

where, *B* and *G* are the bulk and shear modulus, whereas subscripts *V* and *R* represents the Voigt and Reuss bounds. For these two expressions, a value of zero represents elastic isotropy and a value of 100 % is the largest possible anisotropy. We note that Fe_xPt_{1-x} alloys are completely isotropic in the bulk ($A_B = 0$) and anisotropic ($A_G > 0$) in shear moduli.

4. **Conclusion**

First-principles calculations were performed to determine the structural, thermodynamic, electronic and mechanical properties of Fe_xPt_{1-x} alloys at 0 K. The structural lattice constants obtained by full structure optimisation were in good agreement with the experimental data to within 3 %. The heats of formation showed that thermodynamic stability improves with decreasing Fe concentration. Electronic structure analysis predicted that the cubic Fe_3Pt and $FePt_3$ are magnetic metals, whereas the tetragonal FePt and Fe_3Pt are half-metallic ferromagnets. Lastly, the elastic properties showed that all Fe_xPt_{1-x} alloys are mechanically stable and behave in a ductile manner.

5. Acknowledgements

This work has been performed at the materials modeling centre, university of Limpopo, supported by the national research foundation.

6. References

[1] Bian B, Laughlin D E, Sato K and Hirotsu Y 2000 *J. Appl. Phys.* **87** 6962
[2] Ulmeanu M, Antoniak C, Wiedwald U, Farle M, Frait Z and Sun S 2004 *Phys. Rev. B* **69** 054417
[3] Antoniak C, Spasova S, Trunova A, Fauth K, Farle M and Wende H 2009 *J. Phys.: Conf. Ser.* **190** 012118
[4] Wang Y, Stocks G M, Rusanu A, Nicholson D M C, Eisenbach M, Zhang Q and Liu J P 2007 *IEEE Trans. Magn.* **43** 3103
[5] Ivanov O A, Solina L V, Demshina V A and Magat L M 1973 *Phys. Met. Metall.* **35** 81
[6] Shima T, Takanashi K, Takahashi Y K and Hono K 2004 *Appl. Phys. Lett.* **85** 2571
[7] Rellinghaus B, Kastner J, Schneider T, Wassermann E F and Mohn P 1995 *Phys. Rev. B* **51** 2983
[8] Whang S H, Feng Q and Gao Y-Q 1998 *Acta. mater.* **46** 6485
[9] Kresse G and Furthmüller J 1996 *Phys. Rev. B* **54** 11169
[10] Perdew J P and Wang Y 1992 *Phys. Rev. B* **45** 13244
[11] Methfessel M and Paxton A 1989 *Phys. Rev. B* **40** 3616
[12] Buschow K H J, van Engen P G and Jongebreur R 1983 *J. Magn. Magn. Mater.* **38** 1
[13] Nakata Y 2003 *Mater. Trans.* **44** 1706
[14] Yuasa S, Miyajima H and Otani Y 1994 *J. Phys. Soc.* **63** 3129
[15] Cabri L J and Feather C E 1975 *Can. Mineral.* **13** 117
[16] MacLaren J M, Duplessis R R, Stern R A and Willoughby S 2005 *IEEE Trans. Magn.* **41** 4374
[17] Chen H S 1996 "Elastic Anisotropy of Metal" Metallurgy Industry Press, Beijing
[18] Fast L, Wills J M, Johansson B and Eriksson O 1995 *Phys. Rev. B* **51** 17431
[19] Karki B B, Ackland G J and Crain J 1997 *J. Phys.: Condens. Matter.* **9** 375
[20] Kittel C 2005 Introduction to Solid State Physics eighth ed. *John Wiley & Sons Inc.*
[21] Born M and Huang K 1956 *Dynamical Theory of Crystal Lattices. Oxford, Clarendon*
[22] Voigt W 1928 Lehrbuch der Kristallphysik Leipzig, Taubner
[23] Reuss A 1929 *Z. Angew. Math. Mech.* **9** 55
[24] Hill R 1952 *Proc. Soc. London A* **65** 350
[25] Pugh S F 1954 *Philos. Mag.* **45** 823.
[26] Wang D, Wu X, Wang Z and Chen L 2005 *J. Power Sources* **140** 125
[27] Goumri-Said S and Kanoun M B 2007 *Comput. Mater. Sci.* **43** 243
[28] Zener C 1948 Elasticity and Anelasticity of Metals *University of Chicago* Chicago
[29] Ravindran P, Fast L, Korzhayi P A and Johansson B 1998 *J. Appl. Phys.* **84** 4891
[30] Chung D H and Buessem W R 1967 *J. Appl. Phys.* **38** 2010

Surface morphology characterisation for parts produced by the high speed selective laser melting

M Shange[1, 2], I Yadroitsava [2], S Pityana[1], I Yadroitsev[2], and D Bester[1]

[1] Laser Enabled Manufacturing, National Laser Centre, CSIR, Pretoria, 0001, South Africa.

[2] Department of Mechanical Engineering, Central University of Technology, Free State 9301, South Africa

E-mail: mshange@csir.co.za

Abstract. Additive Manufacturing (AM) systems are unique in terms of fast production and lead time to market. South Africa has built the AM Laser Powder Bed Fusion (LPBF) machine called Aeroswift with the world's largest build volume and highest speed. Surface roughness of LPBF parts depends on process-parameters. High surface roughness and deformations during processing limit the application of this technology for certain industrial applications that require high precision. This study characterises the surface roughness for self-supported parallelepipeds samples produced using the Aeroswift platform. Test artefact parallelepipeds were built with 0-90 degrees sloping angles with respect to the building direction. Surface roughness for the as-built samples was analysed using Mitutoyo SURFTEST SJ-210 system and Zeiss Smartzoom 5 digital microscope. It was found that roughness values were higher for the bottom surfaces (overhang part) compared to upward surfaces, as expected. This was attributed to the higher amount of attached partially melted powder particles that were observed on the downward surface. Absolute values of *Ra* and *Rz versus* scanning direction and slope angle were found and analysed.

1. Introduction

Additive manufacturing (AM) is described by ASTM F2792 as a method of joining materials to create parts from 3-D model data with a layer-by-layer process as opposed to conventional techniques, also the 3D printing term is often used synonymously with additive manufacturing [[1]]. Laser powder bed fusion (LPBF) or selective laser melting, direct metal laser sintering and LaserCusing are AM processes that use a laser to consolidate powder material to create a 3D object. The melting of consecutive layers of powder is achieved through the laser beam energy. Moreover, standard metallic powders that melt completely are used, thus ensuring the mechanical properties match 100 percent density or even beat conventional techniques [[2]]. LPBF is a layer by layer manufacturing technique therefore complex geometries can be produced [[3]]. Furthermore, the complexity of the product does not increase the cost [[4]]. However, the more complex the part the more difficult it is to carry-out post processing treatment [[5]]. The common types of post processing: include support structure removal, improving surface finish, post heat treatment etc. Bremen *et al.* [[2]] stated that improving the surface finish is the most costly of the post processing treatments of LPBF parts. Due to the growing interest for parts manufactured by LPBF, it is important to define limitations at an earlier phase of design [[6]]. Therefore, designers for AM are encouraged to possess a quality control or inspection tool to predict the level of the resulting surface roughness before parts are manufactured. The proposed tools include fabricating test artefacts with varying angles to characterise different surfaces topographies. Commonly used test artefacts comprise parallelepipeds, truncheons etc. Figure 1 shows one example of a test artefact.

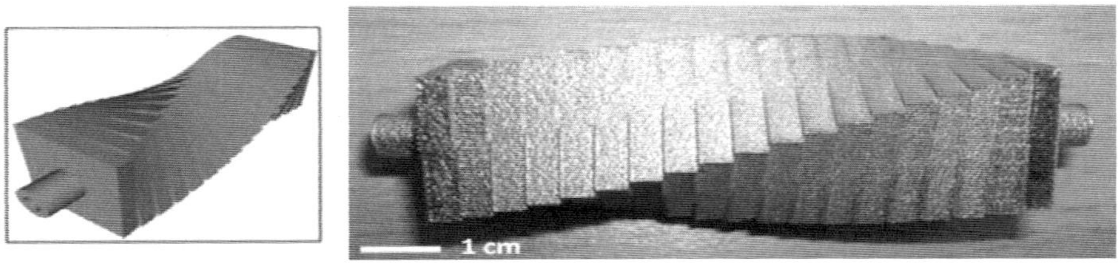

Figure 1: Truncheon test artefacts studied by Strano et al [[7]]

Covarrubias and Eshraghi [[5]] conducted a study on two different types of test artefacts: parallelepipeds and cubes with varying angles. The cubes had inclinations of 45°, 60°, 75° and 90°. The parallelepipeds had of seven rectangular cubes with angles from 0° to 90° with a 15° increment between the angles. Similarly, Strano *et al.* [[7]] investigated surface texture on a truncheon with sloping angles from 0° to 90° with a 5° increment between the angles. Barari *et al.* [[8]] performed a study on cusp geometry with different angles. These studies found that a decrease in angle towards the base-plate is associated with increased surface roughness. Also the downward facing surfaces for overhanging objects have been identified to possess a poor surface quality than upward surfaces. Similarly, it was found that surface quality improves as the sloping angle increases from the base plate.

Authors did not give a distinct difference in the roughness for the two surfaces, downskin and upskin. The downskin is defined as the downward facing surface whereas upskin is defined as the upwards facing surface. However, this study seeks to address the difference between the two surfaces for as built Ti6Al4V samples of parallelepipeds, with angles from 25° to 90° shown in Figure 2. This will be achieved by comparing the arithmetic mean average (*Ra*) and ten-point height (*Rz*) as a function of an angle for both upskin and downskin surfaces.

2. Materials and Methods

2.1. Materials characterization

All samples were fabricated using gas atomised pre-alloyed Ti6Al4V (grade 5) powder. Ti6Al4V amongst all the titanium grades is the widely used alloy due to its good mechanical properties [[9]]. Furthermore, parts that are manufactured with Ti6Al4V are lightweight which makes it ideal for aerospace application. The powder used to build the samples is supplied by TLS Technik and has a powder distribution of 20 to 60 μm. The chemical composition for the Ti6Al4V is shown in Table 1.

Table 1: Chemical composition of employed Ti6Al4V powder.

Al	V	C	Fe	O	N	H	Y	Zn	Sn	Ti
6..8%	3.96%	0.006%	0.161%	0.087%	0.008	0.002	< 0.01	< 0.002	<0.004	balance

2.2. Methods

Before the building process the powder was dried at 120°C in an oven for 3-4 hours to remove moisture. During processing, an inert argon atmosphere was used to avoid oxidation and the amount of oxygen was maintained below 500 ppm. Layer thickness was maintained at 50 μm which is the distance equated to the movement of the z-axis. To build the samples, bi-directional scanning strategy was used. The Aeroswift platform uses the ytterbium laser with a wavelength of 1076 nm and a maximum power output of 5 kW. This machine has a build volume of 2000 mm × 600 mm × 600 mm. After completing the building process, samples were submerged in an ultrasonic-bath for 10 minutes in water. The surface roughness for the samples was measured according to ISO 427:1997 [[10]] using a Mitutoyo Surftest SJ-210. The samples were measured in the as-built condition, meaning no surface finishing was conducted. The stylus, which is the measuring tip of the instrument, was made to traverse perpendicular to the scanning direction to account for the stair-stepping effect. Scanning direction is described in [[7]] as a direction in which the laser traverses, and [[5]] defines that stair effect occurs as a result of additive manufacturing layer by layer fabrication process. Roughness parameters *Ra* and *Rz* in [[11]] can be represented as follows:

$$Ra = \frac{1}{l}\int_0^l |y(x)|\, dx \qquad (1)$$

$$Rz = \frac{1}{n}\left(\sum_{i=1}^{n} p_i - \sum_{i=1}^{n} v_i\right) \qquad (2)$$

Where l is sampling length, p_i is vertical distance from the highest peak and v_i the lowest valley within five sampling lengths. Ten-point height (*Rz*) evaluates the average value of five highest peaks and five lowest valleys within the measured sampling length [[11]]. The sampling length that was used is 2.5 mm based on the standard [[10]]. Subsequently, a non-contact 3D Zeiss Smartzoom 5 digital microscope was used to validate and evaluate surface morphology. The tested samples had angles from 25° to 90° with a 5° increment.

Figure 2: Fabricated test artefact parallelepipeds samples showing angles from 25° to 90° (left to right)

3. Experimental Results and Discussions

3.1. Surface roughness characterization:

A total of 14 upskin surfaces and 14 downskin surfaces were tested and data for these surfaces were measured in three different places for each surface and averaged. The profile of the resulting surface roughness (*Ra*) for both downskin and upskin surfaces as a function of building angle for parts manufactured on the Aeroswift platform are represented in Figures 3 and 4, respectively.

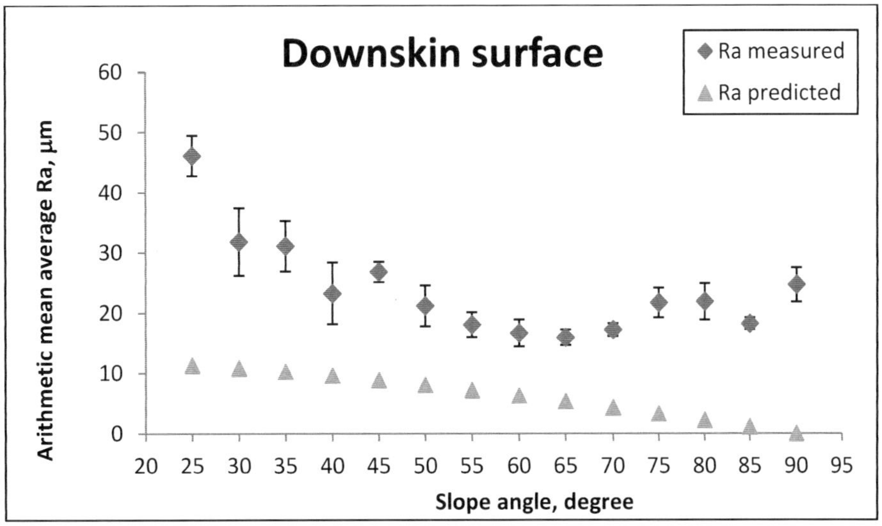

Figure 3: *Ra* values for downward facing surfaces

As expected, it can be seen from Figure 3 that angle 25° had the highest roughness value of all the manufactured angles of 46.4 µm which is a significant amount. This is a result of un-melted powder particles that adhere to the surface of the part during processing especially when the build angle is low (25°). This can be supported by the image in Figure 6(a), which shows high peaks representing unevenness of the surface. According to Triantaphyllou *et al.* [[12]] the increased roughness for downskin surfaces is a result of gravity and capillary forces where the melt pool sags into the un-melted powder particles. As the melt pool solidifies, it causes the surrounding un-melted powder particles to attach on the surface thereby increasing roughness. Furthermore; poor surface roughness at lower angles was found to be a result of the stair stepping effect as previously reported by Covarrubias and Enshragi [5], Barari *et al.* [[8]] and Kaji and Barari [[13]]. As observed from the results a 65° angle generated a better surface quality than all other angles. This can be confirmed by the low peaks as seen in the Figure 6c. This is because in LPBF as the sloping angle increases from the base plate it is associated with better staircase effect. This confirms a similar trend that was observed by Strano *et al.* [[7]] where it was found that better surface quality can be achieved at the angle of 75° as opposed to 90°. On the other hand, Covarrubias and Enshragi [[5]] showed that the *Ra* and *h (Rz)* values in AM can be predicted according to the analytical equation presented by:

$$h = \frac{L_t}{\sin(\propto)} \quad (3)$$

$$Ra\ predicted = \frac{L_t \sin(\alpha)}{4\tan(\alpha)} \quad (4)$$

Where h is the distance between consecutive step edges, L_t is the layer thickness, and ∝ is the inclined angle. This approach was previously used by Reeves and Cobb [[15]] for surface deviation modelling of stereolithography components. The latter equation 4 neglect that LPBF processed parts had round corners, it assumes an ideal environment with perfect sharp edges which make it not ideal for quantifying the degree of surface roughness (Figures 3 and 4). Predicted *Rz* values (Figures 5a and b) were calculated taking into account about 50% shrinkage of powder material and 50 µm distance equated to the movement of the z-axis (see equation 3).

According to the measured roughness values seen in Figure 4, initially the roughness value increase from 25° to 35 °and subsequently formed a zig-zag profile until 90°. From the graph it can be deduced that both 75° and 35° had the highest peaks whereas angle 85° had the lowest *Ra* value of 19,5 µm.

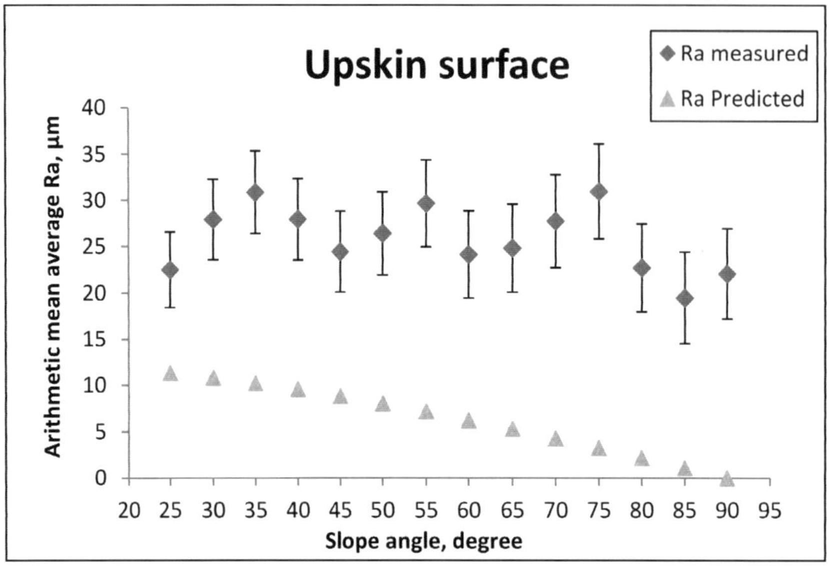

Figure 4: *Ra* values for upward facing surfaces

Since *Ra* is the mean value of the whole profile, *Rz* values also were investigated to analyse individual peaks and valleys (Figure 5a and b). The qualitative analysis for angle 25°, 65° and 90° are represented in Figure 6(b) (d) (f) below for comparison with the previous trend in Figure 4. The melt pool for the top surface does not directly solidify on top of the un-melted powder as in the case of the downskin as previously mentioned. The zig-zag trend might be aggravated by stair-stepping effect especially for smaller sloping angles wherein as the angle increases partially sintered powder particles contributed to the higher roughness. It is well known in literature that angles less than 45 degrees are greatly influenced by stair stepping effects [[5]]. This type of profile has been defined by Strano *et al.* [[7]] and Krol and Tanski [[14]] who stated that the amount of powder concentration of un-melted particles sticking on the edges increases as the inclination angle rises towards the vertical direction. Furthermore, because at low angles the particle size is smaller than the step edges, as the angle increases the edges become smaller thus leading to a higher concentration of partially sintered particles thus resulting in an increased poor surface quality.

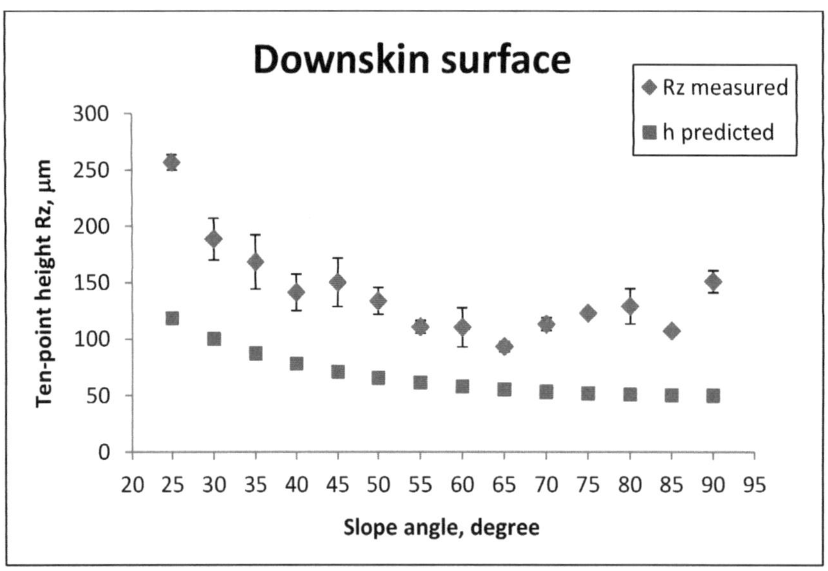

(a)

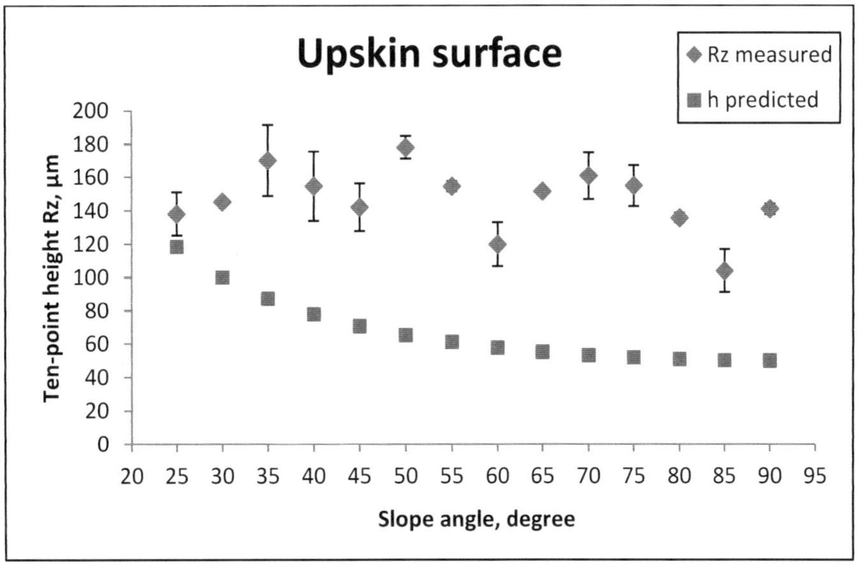

(b)

Figure 5: *Rz* values for downward (a) and upward (b) facing surfaces.

3.1.1. Qualitative analysis of surface roughness

Figure 6 (a) to (f) show the difference between the downward facing surface on the left and the upward facing surface on the right. As it can be seen, the noticeable difference from the 3D images is from the downskin surfaces comprised of large peaks compared to upskin surfaces. Furthermore the samples on the downskin showed greater irregularities or unevenness on to their surface.

a) 25° angle downskin

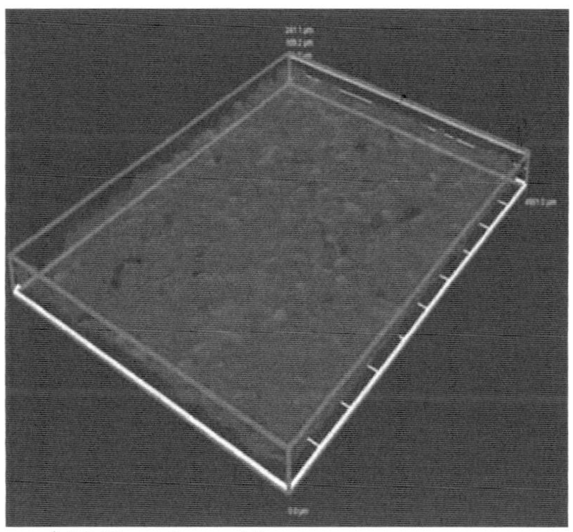

b) 25° angle upskin

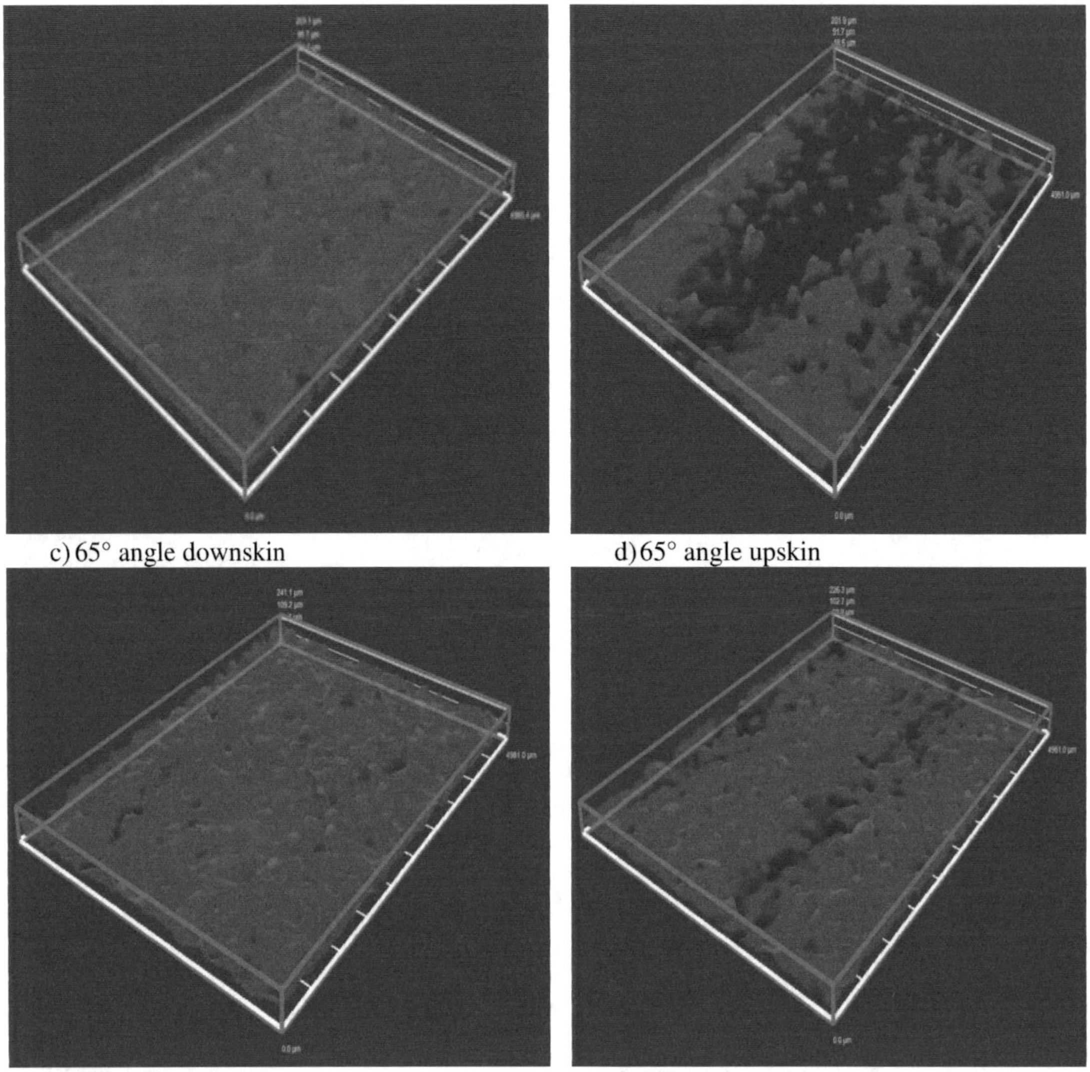

c) 65° angle downskin　d) 65° angle upskin

e) 90°angle downskin　f) 90° angle upskin

Figure 6: 3D surface reconstructions by optical microscope for different angles. Blue colour is valleys, red colour is peaks.

4. Conclusion

The aim of this study was to highlight the fundamentals of the level of surface roughness that can be expected on the as-built components produced by the Aeroswift platform. Based on Figure 2, it was determined that better surface roughness of downskin can be achieved when building at angles between 40° and 90°. There were no significant differences in surface roughness of upskin surfaces *versus* slope angle. Thus, upskin surface morphology depends on process parameters and scanning strategy. According to the obtained results, it can be concluded that powder particles and staircase-effect played a key role in the degree of roughness mostly for downskin surfaces. The collected data of the surface roughness is the first step in diagnosing the as-built manufactured component and will advise the suitable angles to use before parts are manufactured by the Aeroswift platform.

Acknowledgments
The Authors would like to extend our thanks to the CSIR – National Laser Centre and to the South African Research Chairs Initiative of the Department of Science and Technology and National Research Foundation of South Africa (Grant №97994) for all their support to this research.

5. References

[1]. ASTM F2792-12a, Standard Terminology for Additive Manufacturing Technologies, (Withdrawn 2015), ASTM International, West Conshohocken, PA, 2012.
[2]. Bremen, S., Meiners, W. and Diatlov, A., 2012. Selective laser melting: a manufacturing technology for the future. Laser Technik Journal, 9(2), pp.33-38.
[3]. Buchbinder, D., Schleifenbaum, H.B., Heidrich, S., Meiners, W. and Bültmann, J., 2011. High power selective laser melting (HP SLM) of aluminum parts. Physics Procedia, 12, pp.271-278.
[4]. Karme, A., Kallonen, A., Matilainen, V.P., Piili, H. and Salminen, A., 2015. Possibilities of CT scanning as analysis method in laser additive manufacturing. Physics Procedia, 78, pp.347-356.
[5]. Covarrubias, E.E. and Eshraghi, M., 2018. Effect of Build Angle on Surface Properties of Nickel Superalloys Processed by Selective Laser Melting. JOM, 70(3), pp.336-342.
[6]. Kranz, J., Herzog, D. and Emmelmann, C., 2015. Design guidelines for laser additive manufacturing of lightweight structures in TiAl6V4. Journal of Laser Applications, 27(S1), p.S14001.
[7]. Strano, G., Hao, L., Everson, R.M. and Evans, K.E., 2013. Surface roughness analysis, modelling and prediction in selective laser melting. Journal of Materials Processing Technology, 213(4), pp.589-597.
[8]. Barari, A., Kishawy, H.A., Kaji, F. and Elbestawi, M.A., 2017. On the surface quality of additive manufactured parts. The International Journal of Advanced Manufacturing Technology, 89(5-8), pp.1969-1974.
[9]. Mathoho, I., 2017. Enhancing the integrity of machined titanium alloy (Doctoral dissertation, University of Johannesburg).
[10]. ISO, E., 1997. 4287–Geometrical Product Specifications (GPS)–Surface Texture: Profile Method–Terms, Definitions and Surface Texture Parameters. International Organization for Standardization: Geneva, Switzerland.
[11]. Gadelmawla, E.S., Koura, M.M., Maksoud, T.M.A., Elewa, I.M. and Soliman, H.H., 2002. Roughness parameters. Journal of materials processing technology, 123(1), pp.133-145.
[12]. Triantaphyllou, A., Giusca, C.L., Macaulay, G.D., Roerig, F., Hoebel, M., Leach, R.K., Tomita, B. and Milne, K.A., 2015. Surface texture measurement for additive manufacturing. Surface Topography: Metrology and Properties, 3(2), p.024002.
[13]. Kaji, F. and Barari, A., 2015. Evaluation of the surface roughness of additive manufacturing parts based on the modelling of cusp geometry. IFAC-PapersOnLine, 48(3), pp.658-663.
[14]. Krol, M. and Tański, T., 2016. Surface quality research for selective laser melting of Ti-6Al-4V alloy. Archives of Metallurgy and Materials, 61(3), pp.1291-1296.
[15]. Reeves, P. E. and Cobb, R. C., 1997. Reducing the surface deviation of stereolithography using in-process techniques. Rapid Prototyping Journal, 3(1), 20–31.

Thermo-mechanical simulation of steam turbine blade with spark plasma sintering fabricated Inconel 738LC superalloy properties

O F Ogunbiyi [1*], S A Salifu[1], T Jamiru[1], E R Sadiku[2], O T Adesina[1]

[1]Department of Mechanical Engineering, Mechatronics and Industrial Design, Tshwane University of Technology, Pretoria South Africa

[2]Institute for NanoEngineering Research (INER) and Department of Chemical, Metallurgical and Materials Engineering, Tshwane University of Technology, Pretoria South Africa

Corresponding author*: olugbengaogunbiyi@gmail.com

Abstract
The exponential increase in the demand for energy has placed tremendous pressure on the power generation companies and the components used for the power supply. The efficiency of such supply is dependent on the operating parameters of the plants but at the expense of the useful life of the components. Turbine blade is one of the components that contribute immensely to such efficiency and it should be made of high-strength material with an exceptional ability to withstand the harsh operation environment. The properties of spark plasma sintering technique produced Inconel 738 low carbon (IN738LC) superalloy was used in finite element analysis software, Abaqus CAE/2017 to simulate the thermo-mechanical behaviour of steam turbine blade. The maximum thermo-mechanical stress and strain were developed on the root of the turbine blade (along the tailing edge) and the value of the maximum stress developed is far below the yield stress of the superalloy. Hence, spark plasma sintered Inconel 738LC is suitable for turbine blade production as such blade will survive several thermo-mechanical stress and strain cycles prior to failure.

1. Introduction

The exponential growth in human population and industries has placed profuse demands on the power generation companies to provide substantial energy for the growing population. This imply that there is a need for energy capacity expansion and an urgent call to maximize the available power generation plants for better energy output such that minimal break down of the plant's component is recorded. This rising demand for energy has prompt modern turbine engines to operate under extreme conditions with the sole aim of exceeding the manufacturer's recommended conservations, but still maintaining the necessary safety requirements [1].

In order to increase the performance of the turbine, the operating temperature must be raised, thus, making the blade to operate in a harsh environment. Turbine blade are vital components that makes up the turbine assembly of a gas or steam turbine engine and it helps to convert the kinetic energy of the steam or burning gases into mechanical energy [2]. The efficiency of the turbine is conditioned by the vanes of the turbine blades, and the leading edge of the blade is a site where severe thermal damage is experienced [2-5]. In operation, the blades are subjected to stresses due to numerous factors such as very high temperature of steam or gases, centrifugal force creep, thermal gradients developed due to start up and shutdown, vibration fatigue, stresses induced due to complexity in geometry and hot corrosion [2]. Hence, superalloys are used to fabricate turbine blades due to their excellent mechanical properties and good structural stability at high temperature environment [6]. Superalloys differ in their percentage composition and their phases differ with respect to the alloying elements present.

For Inconel 738 low carbon (IN738LC), the secondary intermetallic gamma prime prime (γ''-Nb, Mo,) phase and carbon may sometimes be absent [7]. The principal phases often found in this type of alloy are gamma matrix phase (γ-Ni), precipitate intermetallic gamma prime phase (γ'-Ni_3(Al, Ti)) and some hard-solid solution strengthening phases [7, 8]. These are phases are responsible for the excellent creep and thermal properties exhibited by this alloy at high temperature environments [9, 10]. Also, the presence of chromium improves the corrosion resistance properties of the alloy.

In this paper, the thermo-mechanical behaviour of 60 MW low pressure spark plasma sintered (SPS) Inconel 738LC produced steam turbine blade in operation was determined using finite element analysis (FEA) software, Abaqus CAE/2017.

2. Material Properties

The tables below show the dimensions and material properties of Inconel 738LC used.

Table 1: Weight percentage of IN738LC constituent elements

Material	Ni	Cr	Co	Ti	Aℓ	W	Ta
Percentage (%)	64.58	16	8.3	3.4	3.4	2.6	1.72

Table 2: Dimensions of the blade

Airfoil (m)	Root (m)	Leading Edge (m)	Trailing Edge (m)	Chord Length (m)	Pitch (°/)
0.2	0.05	0.0092	0.00245	0.05	14.2

Table 3: Modulus of elasticity of Inconel 738LC as a function of temperature

Elasticity (GPa)	Poisson Ratio	Temperature (°C)
200.638	0.28	24
195.120	0.27	93
190.295	0.27	204
184.780	0.28	315
179.264	0.28	426
175.127	0.30	538
167.543	0.30	649

Table 4: Specific heat capacity and thermal conductivity of inconel 738LC

Temperature (°C)	Specific Heat Capacity (J/KgK)	Thermal Conductivity (W/mK)
100	460.548	30.243
200	502.416	30.726
300	523.35	30.033
400	544.284	32.924
500	565.218	34.734
600	586.152	37.061

Table 5: Thermal expansion of inconel 738LC

Temperature (°C)	Thermal Expansion ($10^{-6}K^{-1}$)
93	6.45
204	6.75
315	7.15
426	7.55
538	7.75
694	8.05

3. Analysis Methodology

A sequentially coupled procedure was used for the thermo-mechanical simulation in Abaqus CAE/2017 FEA software. The analysis was carried out on a scaled down, 60 MW low pressure turbine blade model shown in figure 1(a). The procedure involves the development of a scaled down turbine blade model which is a representative of a blisk. The material composition of the SPS fabricated Inconel 738LC turbine blade and its dimensions are shown in table 1 and 2 respectively while the material properties are shown in tables (3-5). These material properties are added to the appropriate material properties section of Abaqus CAE/2017. The creation of section and assignment of the relevant section to part was also done, after which a dependent instant was created before assigning the section to a part as shown in figure 1(b).
The blade was subjected to 550 °C sink temperature which represents the typical operating temperature of a steam power plant and the assumed film coefficient (convective heat transfer coefficient) at the operating temperature is 10 000 W/(m^2K) [11]. Also, a sink temperature of 25 °C with an assumed film coefficient of 18 W/(m^2K) [11] representing the convective heat transfer coefficient of air was applied to the blade. This represent the initial condition of the blade prior to operation. For the heat transfer analysis, a 10-node quadratic heat transfer tetrahedron (DC3D10) element type was used while for the thermo-mechanical analysis, a 10-node quadratic tetrahedron (C3D10) element type was used. Text element shape with global seed size of 0.8 mm was used for the analysis as shown in figure 1(c). The temperature distribution profile from the output data base (ODB) file in the heat transfer (thermal) analysis was imported and used as the predefined temperature. This temperature profile was applied in step 1 of the thermo-mechanical stress analysis. In the thermo-mechanical analysis, a pressure of 18 MPa representing the typical operating steam pressure was applied to the blade in the loading section of Abaqus CAE/2017 and the displacement/rotation boundary condition was specified on the trailing edge of the rotor such that the model was constrain in the axial direction [12]. This specified boundary conditions represent the real operating process of the blade.

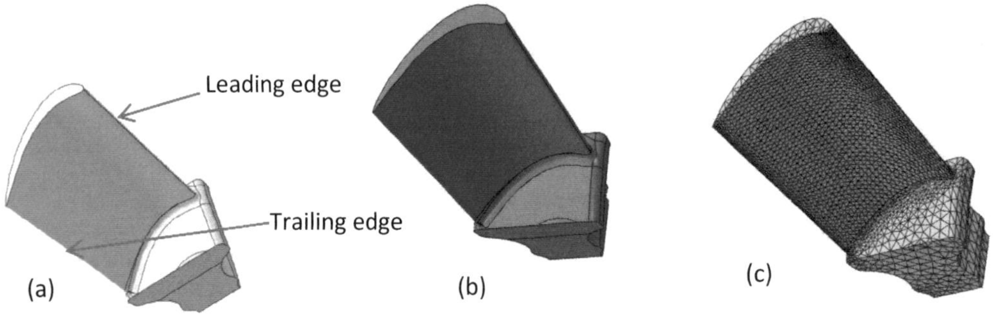

Figure 1: (a) A scaled down model, (b) assembly mode and (c) meshed mode of a power plant turbine blade

4. Results and Discussion

The temperature distribution result and profile in figure 2(a). indicates that the maximum operating temperature of steam was used for the operation of the turbine blade and the 0.91 % drop in temperature in the temperature output result is due to heat loss in the form of conduction and convection. The maximum heat flux was obtained at the root as shown in figure 2(b). This is due to the difference in temperature between the blade and the root which resulted in the transfer of heat from higher temperature region to lower temperature region.

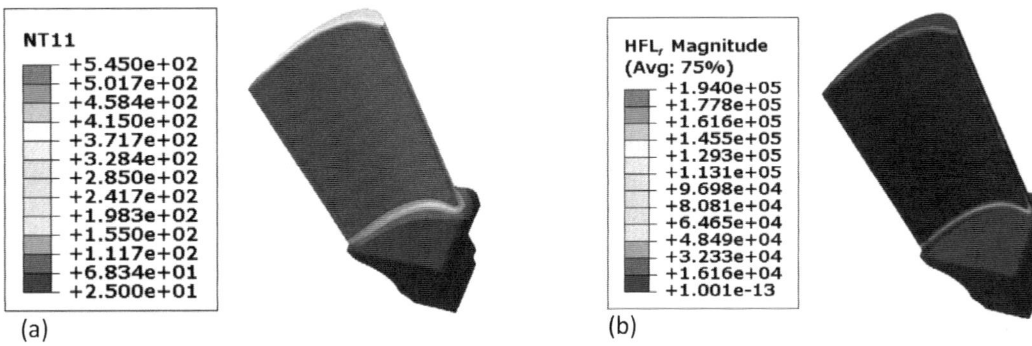

Figure 2: (a) Temperature distribution result and profile and (b) heat flux across the turbine blade in operation

The thermo-mechanical stress distribution profile and output result of the blade is shown in figure 3. The maximum principal stress (443.3 MPa) is located at the root of the blade on the trailing edge. This is the region where the largest stress gradient was experienced. The maximum stress value of the steam turbine at this region under the operating condition of 550 °C and 18 MPa is far below the 896 MPa yield strength of Inconel 738LC [13]. Hence the blade will undergo several cycles prior to failure due to thermo-mechanical stress or thermal fatigue. figure 4 shows the strain distribution profile of the blade and as expected, the maximum strain (2.703 x 10^{-3}) was developed at the root of the blade on the trailing edge. This signifies that the failure of the blade will likely emanate from the root since it is the region with the propensity for crack initiation and propagation.

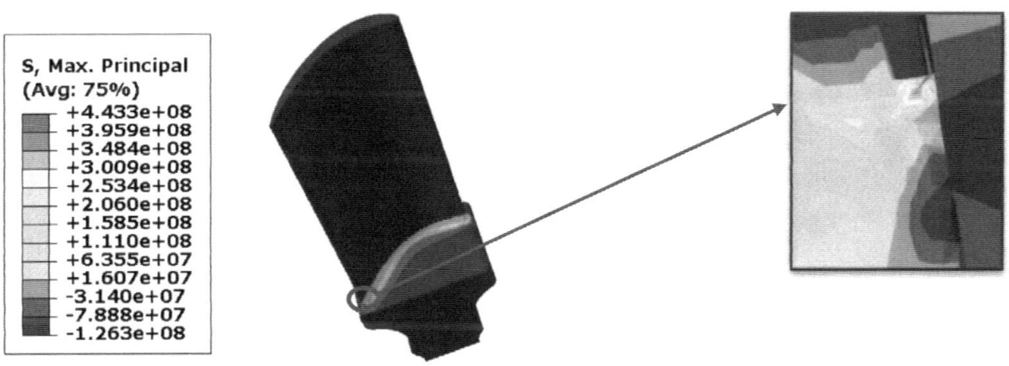

Figure 3: Stress distribution result and profile across the thickness of the turbine blade

Figure 4: Strain distribution deformation profile across the thickness of the turbine blade

The deformation profile across the turbine blade is shown in figure 5. The maximum deformation was obtained at the tip of the trailing edge. This signifies that the blade could also experience failure at this region during operation.

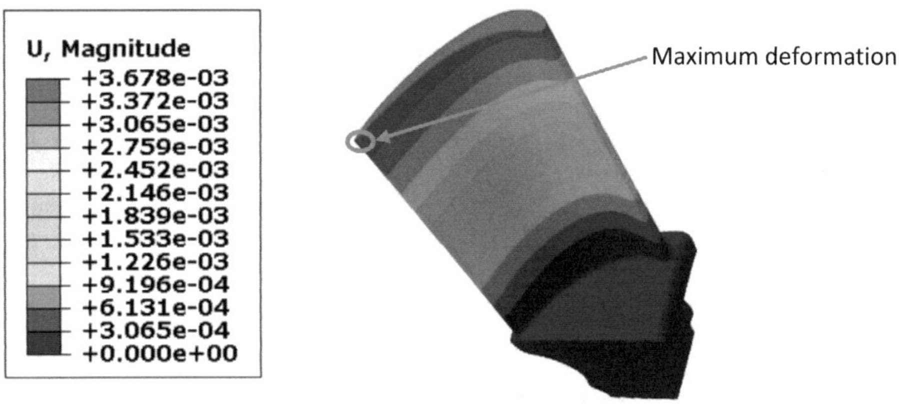

Figure 5: Total deformation across the turbine blade

5. Conclusion

A numerical model for the analysis of thermo-mechanical behaviour of a power plant turbine blade with the material properties of SPS fabricated Inconel 738LC (nickel-based superalloy) was carried out in Abaqus CAE/2017. A sequentially coupled procedure was adopted for the simulation and the improved material properties of IN738LC alloy emanating from the SPS fabrication technique were used during the simulation. The chemical compositions of the alloy were also presented. In the simulated output result, the temperature distribution profile shows that high thermal efficiency was maintained and there was 0.91 % drop in temperature as compared to the applied operating steam temperature. This slight decrease in temperature was due to the heat loss in the form of conduction and convection. The maximum value of stress and strain was developed on the root of the blade (trailing edge). Also, the maximum deformation was experienced at the tip of the trailing edge of the blade. This indicates that crack could initiate at the trailing edge on the root of the blade due to stress developed or the tip of the trailing edge being the location with maximum deformation. The value of the maximum stress developed (443.3 MPa) is far below the yield strength (896 MPa) of Inconel 738LC at the specified operating condition. Hence, failure of the blade due to thermal stress, strain or thermal fatigue will only occur after several cycles of operation. Also, the value of the stress developed by the SPS produced turbine blade compare favourably well with that manufactured through other processes since they all survive several cycles prior to failure under the specified operating condition.

6. Acknowledgments

This work is funded by the Research and Innovation Directorate of Tshwane University of Technology and supported, in part, by the Department of Mechanical Engineering, Mechatronics and Industrial Design, Institute for Nano Engineering Research (INER), Department of Chemical, Metallurgical and Materials Engineering and the Faculty of Engineering and Built Environment of the Tshwane University of Technology, Pretoria, South Africa.

7. References

[1] Brandao P, Infante V and Deus A 2016 Thermo-mechanical modeling of a high pressure turbine blade of an airplane gas turbine engine *Procedia Structural Integrity* **1** pp. 189-196.

[2] Amroune S, Mohamad B, Moussaoui M and Saaidi H 2018 Geometric regeneration and mechanical analysis of a gas turbine blade type Frame 9001 GE *Eng. Solid Mech* **6** pp. 105-112.

[3] Xie Y. J, Wang M. C, Zhang G and Chang M 2006 Analysis of superalloy turbine blade tip cracking during service *Eng.Failure Anly* **13** pp. 1429-1436.

[4] Gollapothu Boyaraju S. R, Sridhar A. V, Hari Narayana Rao 2015 Thermal Analysis Of A Gas Turbine Rotor Blade *Int. J. Scie. Eng. Adv. Tech. IJSEAT* **3** pp. 1181-1187.

[5] Padture N. P, Gell M and Jordan E. H 2002 Thermal barrier coatings for gas-turbine engine applications *Scie.* **296** pp. 280-284.

[6] Maleki V, Omidvar H and Rahimipour M. R 2018 Influences of gap size and cyclic-thermal-shock treatment on mechanical properties of TLP bonded IN-738LC superalloy *Trans. Nonf Metals Soc. China* **28** pp. 920-930.

[7] Rani S, Agrawal A. K and Rastogi V 2017 Failure analysis of a first stage IN738 gas turbine blade tip cracking in a thermal power plant *Case stud. Eng. failure anly.* **8** pp. 1-10.

[8] Golezani A, Bageri M and Samadi R 2016 Microstructural change and impact toughness property of Inconel 738LC after 12years of service *Eng. Failure Anly.* **59** pp. 624-629.

[9] Jalilvand V, Omidvar H, Shakeri H and Rahimipour M 2013 Microstructural evolution during transient liquid phase bonding of Inconel 738LC using AMS 4777 filler alloy *Mats Chart.* **75** pp. 20-28.

[10] Maleki V, Omidvar H and Rahimipour M. R 2016 Effect of gap size on microstructure of transient liquid phase bonded IN-738LC superalloy *Trans. Nonf. Metals Soc. China* **26** pp. 437-447.

[11] TLV Steam Specialist Company, "Overall Heat Transfer Coefficient."

[12] Booysen C 2014 Fatigue Life Prediction of Steam Turbine Blades During Start-Up Operation Using Probabilistic Concepts (University of Pretoria South Africa).

[13] Ikpe A, Efe-Ononeme O and Ariavie G 2018 Thermo-Structural Analysis of First Stage Gas Turbine Rotor Blade Materials for Optimum Service Performance *Int. J. Eng. App Scie.* **10** pp. 118-130.

Exploring the potential of N-acylated chitosan for the removal of toxic pollutants from wastewater

S Naidoo[1,*], N Nomadolo[1], W M R Matshe[1], Z Cele[1] and M Balogun[1]

[1] Chemicals Cluster, Chemicals, Food, Agriculture and Health Division, Council for Scientific and Industrial Research, Meiring Naudé Road, Brummeria, Pretoria 0184, Gauteng, South Africa.

*Corresponding author E-mail: SNaidoo4@csir.ac.za

Abstract. Modified chitosan was prepared by chemical derivatisation using caproyl and palmitoyl fatty acid chains. The performance of the materials on the removal of phosphates in aqueous solutions was evaluated through a series of batch adsorption experiments. The phosphate adsorption studies showed enhanced adsorption efficiencies for caproyl chitosan (63%) and for palmitoyl chitosan (71%) in aqueous solutions containing phosphate at a pH of 6.2. The pristine and modified chitosans showed increases in the removal efficiency at an optimal solution pH of 4. The adsorption kinetics studies performed on palmitoyl-chitosan indicated that a contact time of approximately 30 minutes is required to reach equilibrium at solution concentrations ranging from 10-20 mg/l. Co-existing anions such as nitrates and sulphates do affect the phosphate removal efficiency of modified chitosans in aqueous solutions. This study demonstrates that modified chitosans has the potential of being used as a rigid and degradable material for water treatment.

1. Introduction

Water scarcity, poor water quality and water-related disasters are among the major concerns related to current and future water resource management globally, and in particular South Africa [1]. An ongoing challenge persists in ensuring water security for the country, particularly in poor economically developed areas. It is critical to improve water resource management to ease the burden of water scarcity and availability [1]. Although conventional technologies do work, they are not as effective in treating complex and complicated polluted waters and their capability for the adequate removal of a wide spectrum of toxic chemicals and pathogens in wastewater is still limited [2-5]. This has resulted in a growing emphasis, and consequently opportunities, to develop new water treatment technologies and innovations.

Adsorption has been identified as the most suitable method for wastewater treatment in comparison to other physical and chemical methods owing to its high operation efficiency, limited power requirements, cost efficiency and robustness [5, 6, and 7]. While commercial activated carbon, a widely used adsorbent, has shown excellent adsorption capacity, it is very expensive and derived from non-renewable resources which places serious ecological strains [7].

Biopolymers have become highly attractive alternatives for adsorption owing to their availability, lower cost and lower impact on the environment [6, 7, and 8]. Chitosan is an abundant, hydrophilic biopolymer that has been identified as a promising material for water treatment applications. Its high versatility include good adsorption, chelating and crosslinking properties [4, 9]. Owing to its multivalent backbone, chitosan can be readily chemically modified with hydrophobic moieties to yield N- and O-acylated materials. The hydrophobic modifications to chitosan have been shown to improve bacterial capture and increase its selectivity and reactivity to enhance its wastewater treatment capability [10]. This can render chitosan a good candidate for the treatment of wastewater over a wide pH range.

In this study, N-acylated chitosans were synthesized by chemical derivatisation with caproyl (C6) and palmitoyl (C16) fatty acid chains. The potential of the caproyl and palmitoyl chitosan for the removal of toxic pollutants such as phosphates was investigated by evaluating their performance in a series of batch adsorption experiments. The kinetics of the process, effects of initial concentration and pH, dosage and the effects of co-existing ions, were all used to evaluate the performance of the chemically derived chitosan.

2. Materials and Methods

2.1. Materials

Chitosan (CS) (low viscous, Fluke), caproyl chloride (d=0.976 g/ml, BioChemika Fluka), and palmitoyl chloride (d=0.906 g/ml, BioChemika Fluka) were all purchased from Sigma-Aldrich (South Africa). All other chemicals and reagents were of analytical grade, unless otherwise stated and purchased from Sigma and Merck. Deionized water (dH_2O) was used in the experiments. .

2.2. Synthesis of caproyl and palmitoyl chitosan

The caproyl (CC) and palmitoyl chitosan (PC) were prepared by a method adapted from Le Tien *et al* (2003) and Han *et al* (2008), with modifications as highlighted in Table 1. Briefly, chitosan solutions of varying concentrations were prepared by dissolution in 0.12 M acetic acid solution. The solutions were left to stir for 16 hours, to ensure complete dissolution.

Table 1 List of process parameters for the derivatisation of chitosan with acyl chlorides.

	Processing Parameters						
[1]Sample ID	Acyl chloride	Chitosan concentration (%)	pH of CS solution	Acyl Chloride:CS (w/w)	Temperature (degC)	Time (Hr)	Reaction Mix:Acetone (v/v)
CC1	Caproyl	0.83	6.3	1:2	Ambient	1	2:1
CC2	Caproyl	0.83	7.0	2:5	40	2	1:1
CC3	Caproyl	0.83	7.5	1:4	50	4	1:2
CC4	Caproyl	1.00	6.3	1:2	40	2	1:2
CC5	Caproyl	1.00	7.0	2:5	50	4	2:1
CC6	Caproyl	1.00	7.5	1:4	Ambient	1	1:1
CC7	Caproyl	1.20	6.3	2:5	Ambient	4	1:1
CC8	Caproyl	1.20	7.0	1:4	40	1	1:2
CC9	Caproyl	1.20	7.5	1:2	50	2	2:1
PC1	Palmitoyl	0.83	6.3	1:4	50	2	1:1
PC2	Palmitoyl	0.83	7.0	1:2	Ambient	4	1:2
PC3	Palmitoyl	0.83	7.5	2:5	40	1	2:1
PC4	Palmitoyl	1.00	6.3	2:5	50	1	1:2
PC5	Palmitoyl	1.00	7.0	1:4	Ambient	2	2:1
PC6	Palmitoyl	1.00	7.5	1:2	40	4	1:1
PC7	Palmitoyl	1.20	6.3	1:4	40	4	2:1
PC8	Palmitoyl	1.20	7.0	1:2	50	1	1:1
PC9	Palmitoyl	1.20	7.5	2:5	Ambient	2	1:2

[1]C1-CC9 represent the caproyl chitosan samples and PC1-PC9 represent the palmitoyl chitosan samples.

The pH was then adjusted accordingly by the dropwise addition of 7.5 M sodium hydroxide (NaOH). Caproyl chloride and palmitoyl chloride were added in separate reaction flasks to the different chitosan solutions and were left to react at ambient temperature, 40 °C and 50 °C (Table 1). On completion of each reaction, the volume of the slurry was then adjusted to 800ml using dH_2O.

The reaction media was neutralised to a pH range of 6.8-7.0, before being precipitated using acetone. The precipitate was then collected by filtration. The precipitate was then washed three times using excess methanol at 50°C. The product was then further purified using acetone and dried in a vacuum oven at 60°C.

2.3. Fourier Transform Infrared (FTIR) Spectroscopy
The infrared absorption spectra were measured using the Perkin Elmer Spectrum 100 FTIR spectrophotometer. The pristine, caproyl- and palmitoyl-chitosan were characterised to identify the newly formed functional groups of the materials using attenuated total reflectance fourier transform infrared (ATR-FTIR). The spectra were obtained using the Spectrum version 6.2.0 Software. The spectral range was between 650 cm^{-1} and 4000 cm^{-1} with 4 scans per run and analysed at ambient temperature.

2.4. Contact Angle Measurements
Contact angles were measured using the Krüss Drop Shape Analyser (DSA) 100. The contact angle measurements were performed using the sessile drop methodology to determine the hydrophilic/hydrophobic nature of pristine and modified chitosan. De-ionised water (dH_2O) was used as the wetting medium and the analyses were performed at ambient temperature.

2.5. Batch Adsorption Experiments
A 1000 mg/l phosphate ion (PO_4^{3-}) stock solution was prepared by dissolving potassium dihydrogen phosphate in dH_2O. The solution was then further diluted according to the relevant experimental concentrations. The binary solutions were prepared by diluting the stock solutions of PO_4^{3-} ions with nitrates, sulphates or Rhodamine B dye solutions separately.

The performance of the pristine chitosan and the modified chitosans were assessed for their application in the removal of phosphate ions from wastewater. A batch adsorption screening test was conducted by treating 20 ppm phosphate ions water samples at 50 ml volumes with pristine and modified chitosan. The experiments were conducted at 25 °C, at 200 rpm for 24 hrs. 5 ml of the phosphates solutions were aliquoted and analysed using the Mettler Toledo UltraViolet5Bio (UV5Bio) to determine the amount of phosphate ions adsorbed. The effect of varying parameters including dosage, pH, initial concentration and co-existing ions were evaluated under similar conditions.

The batch absorption kinetic studies were carried out in a beaker on a magnetic stirrer using 100 ml of phosphate solution at initial concentrations varying from 10 mg/l – 20 mg/l. 0.2g of palmitoyl chitosan was added to each flask, and left to stir at 200 rpm, at room temperature for 6 hr. The samples containing the supernatants were taken at predetermined time intervals and filtered to prevent any adsorbent interfering with the subsequent UV analysis.

The PO_4^{3-} removal percentage was determined by using Eq. (1):

$$\%\ \text{Removal} = \left(\frac{C_o - C_e}{C_o}\right) X\ 100 \qquad (1)$$

where C_o and C_e are the initial and equilibrium concentration (mg/l) of PO_4^{3-}, respectively.
The equilibrium absorption capacity was determined using Eq. (2):

$$q_e = \left(\frac{C_o - C_e}{m}\right) V \qquad (2)$$

where q_e is the equilibrium amount of PO_4^{3-}adsorbed per unit mass (m) of adsorbent (mg/g) and V is the sample volume (L).

The adsorption capacity at any time (t) was calculated by Eq. (3):

$$q_t = \left(\frac{C_o - C_t}{m}\right)V \tag{3}$$

where q_t is the amount of PO_4^{3-} adsorbed per unit mass (m) of adsorbent (mg/g) at any time (t), C_t is the concentration of (mg/L) of PO_4^{3-} at time (t) and V is the sample volume (L).

3. Results and Discussion

3.1. Synthesis of caproyl and palmitoyl derivatised chitosan

In this study, synthetic N-acylated chitosan was derivatised using caproyl (C6) and palmitoyl (C16) fatty acids. A series of modified chitosan were produced by varying the processing parameters to obtain the optimal processing conditions. Both the caproyl- and palmitoyl-chitosan were successfully synthesized, with yields of up to 71% and 66% for caproyl-chitosan and palmitoyl-chitosan, respectively.

3.2. FTIR Spectroscopy

The structural composition of the pristine and modified chitosan networks were confirmed using the FTIR spectra (Figure 1). The spectra were normalized for comparison. The characteristic peaks of chitosan were observed between 3400-3000 cm^{-1} (-OH stretch) and 2871 cm^{-1} (-CH- stretching bond). For the pristine chitosan, the carbonyl group from amide I and amide II (C-N stretching) are represented by the bands at 1650 cm^{-1} and 1587 cm^{-1}, respectively. The evaluation of caproyl- and palmitoyl-chitosan, saw shifts in the peaks of the amide I and amide II bands to lower wavenumbers as observed with respect to the pristine chitosan. The band at 1587 cm^{-1} disappear while peaks around 1646cm^{-1} (carbonyl stretching of the amide band I) and 1548cm^{-1} (N-H bending vibrations of the amide band II) become more prominent.

Absorbance peaks at 2850-2950cm^{-1}, ascribed to -CH_2, were also observed for both the caproyl- and palmitoyl-chitosan, suggesting successful modification of chitosan with the acyl functional groups. The intensity of the peaks at 2850-2950cm^{-1} were much higher in the case of palmitoyl-chitosan which is ascribed to its longer chain length (C16) in comparison to the caproyl-chitosan (C6 chain).

3.3. Contact Angle

Discs of pristine chitosan and palmitoyl-chitosan were prepared for the contact angle measurements using the DSA100. A qualitative analysis of the hydrophilic/hydrophobic nature of the chitosan and modified chitosan was performed by evaluating the contact angle using dH_2O droplets. As depicted in Figure 2, contact angles of 52.5 ° and 92.8 ° were observed for the pristine and palmitoyl chitosan, respectively. The increase in the contact angle of the palmitoyl chitosan can be largely attributed to the presence of the hydrophobic groups.

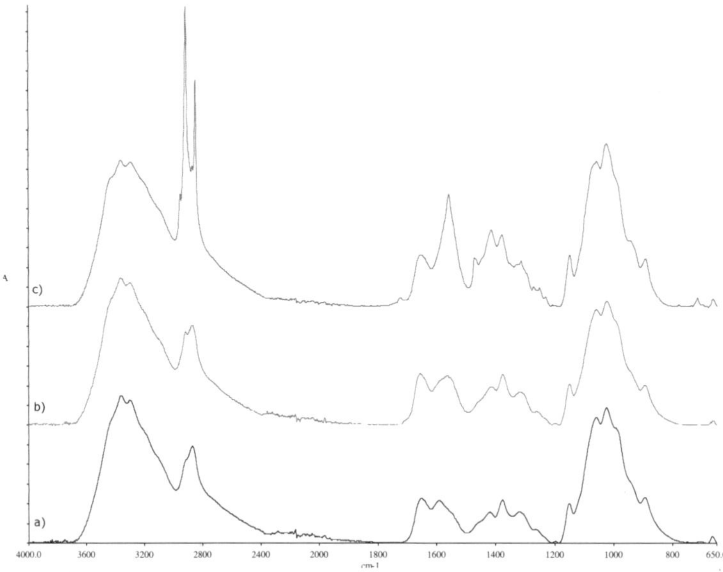

Figure 1: The ATR-FTIR spectra of a) pristine chitosan, b) caproyl chitosan and c) palmitoyl chitosan.

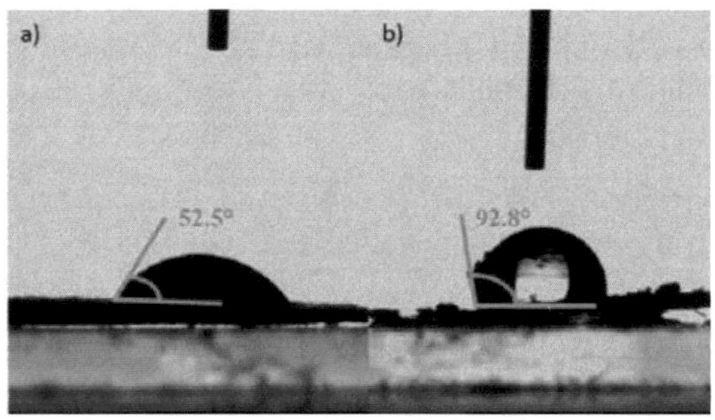

Figure 2: The contact angle measurement of a) pristine chitosan and b) palmitoyl chitosan.

3.4. Batch Adsorption Experiments

Screening for the best performing material. Samples of pristine chitosan and the synthesized caproyl- and palmitoyl-chitosan were assessed for their removal properties of PO_4^{3-} ions from 20 mg/l synthetic phosphate solution. The tests were conducted at an adsorbent dosage of 150 mg at 25 °C. The phosphate removal efficiency of the adsorbents are highlighted in figure 3. There was a significant improvement in removal efficiency of PO_4^{3-} ions from solution using the modified chitosan (40-71%) in comparison to the pristine chitosan (21%).

The best performing caproyl and palmitoyl chitosan (CC4 and PC3) were selected for the studies to evaluate the effects of dosage and pH on the adsorption efficiency of the pristine and modified chitosan.

Effect of solution pH. The adsorption of anions and cations at the solid-liquid interface is largely affected by the pH of the aqueous solution (9). Hence the effect of the solution pH on the adsorption efficiency was investigated. Figure 4 shows the phosphate removal efficiency of the pristine and N-acylated chitosan adsorbents over a solution pH range of 2-10. The study was conducted over 24 hours at 25°C, using a known solution concentration of 20mg/l and an adsorbent dosage of 150mg. An optimal pH of 4 was observed for the pristine and N-acylated chitosan. The pH of the solution directly affects both the speciation of phosphate ions in solution and the protonation of the chitosan.

To better understand the effects of pH on the material, the ionic properties of chitosan need to be considered. It is a cationic polyelectrolyte with a pKa value of approximately 6.3. At pH < pKa, the amine groups become readily protonated, and thus electrostatic interaction takes place between the protonated amine groups and the phosphate molecules. Zero valent H_3PO_4 which is a non-ideal state for adsorption onto active sites due to it being non-ionic, is dominant at pH < 4, and thus low adsorption efficiencies are observed at pH 2 [13]. Monovalent $H_2PO_4^-$ and divalent HPO_4^{2-}, which have high affinity for adsorption sites, are the dominant species between pH 3 and 7 [14]. Thus higher adsorption efficiencies are observed in this range. The increase in pH above 7, showed lowered adsorption affinity for PO_4^{3-} which may be due to competing hydroxyl (^-OH) ions. In general, the N-acylated chitosan performed better than pristine chitosan at the various pH ranges.

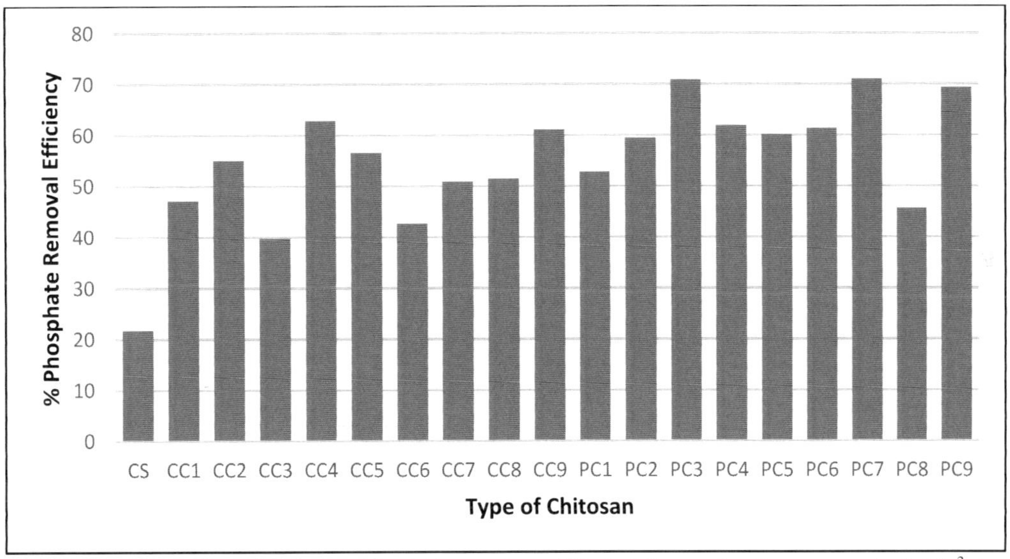

Figure 3: The adsorption efficiency of pristine and modified chitosan for the removal of PO_4^{3-} ions in an aqueous solution.

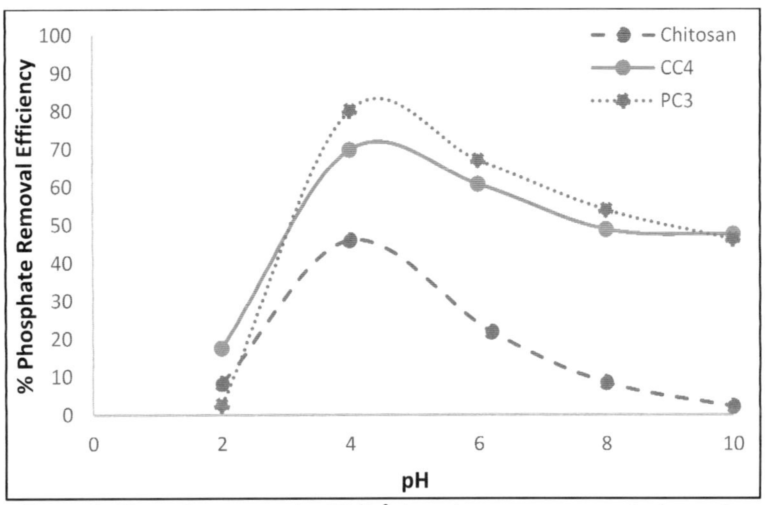

Figure 4: The effect of pH on the removal of PO_4^{3-} ions in an aqueous solution using pristine and N-acylated chitosan.

Effect of adsorbent dosage. To determine the effect of dosage, adsorption tests were conducted using different adsorbent dosages ranging from 50 – 250 mg. The studies were carried out for a period of 24hours using a solution concentration of 20mg/l at its natural pH of approximately 6.2 was used for the evaluation, while the temperature was maintained at 25°C. The phosphate removal efficiency of the pristine and N-acylated chitosan as a function of dosage is represented in Figure 5.

An increase in the adsorption efficiency with increasing dosage for both the pristine (up to 150mg) and modified chitosan (up to 200 mg) was observed (Figure 5). No significant adsorptive effects were observed thereafter. This observed increase and subsequent plateau in adsorption can be attributed to the availability of adsorption sites.

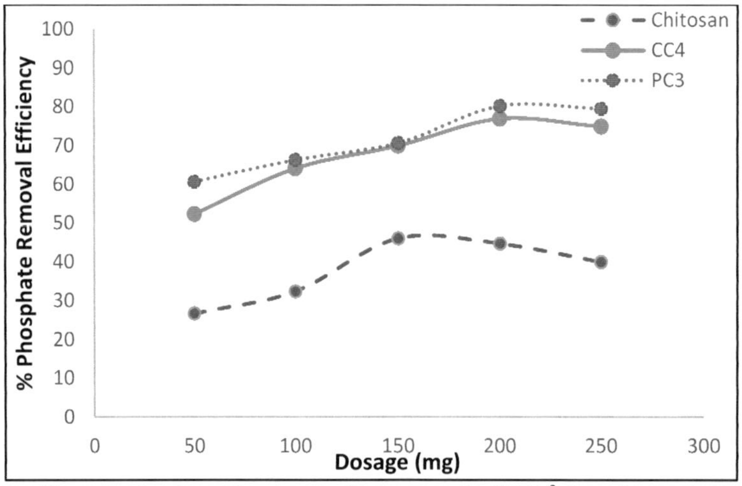

Figure 5: The effect of adsorbent dosage on the removal of PO_4^{3-} ions in an aqueous solutions.

Overall, the performance of the modified chitosan was found to be superior to the pristine chitosan and the longer chain palmitoyl performs better compared to the caproyl-chitosan. Chitosan is mainly characterised by hydrogen bonds, and upon acylation of chitosan the number of hydrogen bonds are reduced, and the modified materials are characterised by hydrophobic interactions due to the acyl chains substituted onto the chitosan [11]. The hydrophobic interactions induce greater stability of the modified chitosan matrix, and therefore enhanced adsorption is observed for the modified chitosan materials. The shorter C6 acyl chain displays weak hydrophobic interactions while the stronger hydrophobic interactions of the longer C16 acyl chain offer greater network stability of the polymer. Thus higher adsorption rates observed for palmitoyl chitosan could be attributed to this. Based on the results obtained, the palmitoyl-chitosan was selected as the adsorbent of choice for the remaining studies.

Effect of contact time. The batch kinetics study for palmitoyl chitosan was carried out at solution concentrations of 10 – 20 mg/l at a pH of 6.2 using the optimum adsorbent dosage of 200 mg. The adsorption capacity as a function of time is represented in Figure 6. Fast adsorption rates are observed in the early stage of the study, and this can be ascribed to the increase in the driving force as a result of the concentration gradient. The adsorption capacity increases with increase in contact time. A contact time of approximately 30 minutes is required for the phosphate solutions at concentrations 10, 15 and 20 mg/l to reach equilibrium, after which the adsorption capacity increases at a minute interval.

Effect of initial concentration. The effect of solution concentration on the removal efficiency and adsorption capacity of palmitoyl-chitosan was studied by varying solution concentration from 15 – 50 mg/l. The test was conducted at the optimum adsorbent dosage of 200 mg at 25°C for 2 Hr.

Figure 7 illustrates the effect of the solution concentration on both the phosphate adsorption capacity and phosphate removal efficiency of palmitoyl chitosan.

The available adsorption sites become saturated as the initial concentration is increased and hence a decrease in the removal efficiency of the palmitoyl-chitosan with increasing concentration is observed. The adsorption capacity increases with the initial solution concentration. According to Leduc et al, this trend could be a result of the effective collision between the phosphate ions and the surface of the palmitoyl chitosan as well as the concentration gradient which provides a greater driving force to lower the mass transfer resistance.

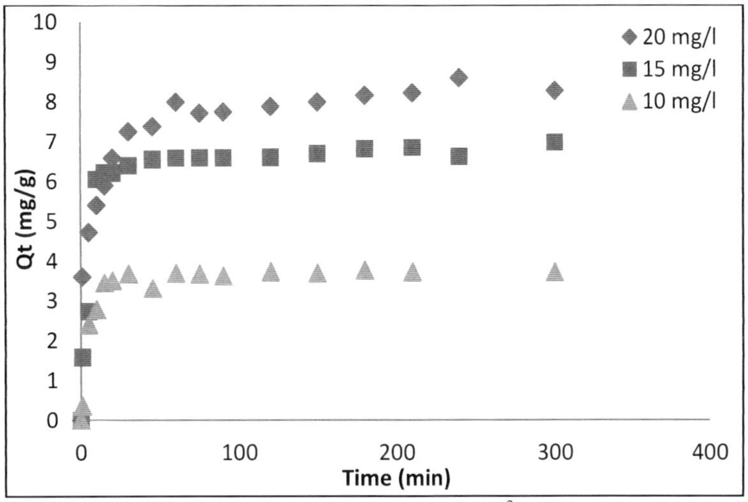

Figure 6: The effect of contact time on the adsorption of PO_4^{3-} at various initial phosphate solution concentrations.

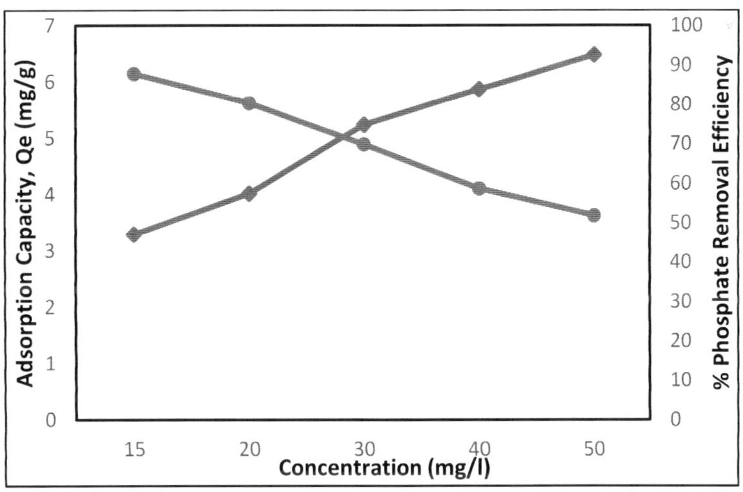

Figure 7: The effect of solution concentration on adsorption capacity and adsorption efficiency of palmitoyl chitosan.

Effect of co-existing ions. Wastewater is often made up of various constituents that could inhibit the overall capability of an adsorbent to remove a particular pollutant, largely due to these constituents competing for the active sites on the modified chitosan. Hence, a study was carried out to investigate the effect co-existing ions have on the removal of phosphates by the palmitoyl-chitosan (Figure 8). The competing ions to be evaluated included anions (nitrates and sulphates) and organics (Rhodamine B).

In general, the phosphate removal efficiency of the palmitoyl-chitosan was reduced in the presence of a competing ion, with a most notable reduction in performance when phosphates and nitrates were in solution. The reduction in the phosphate adsorption in the presence of both nitrates and sulphates can be attributed to the affinity of these anions to the active sites on the modified chitosan. The presence of the Rhodamine B dye did not affect the adsorption of phosphates owing to the electrostatic repulsion between the protonated amine groups on the palmitoyl chitosan and the positively charged Rhodamine B.

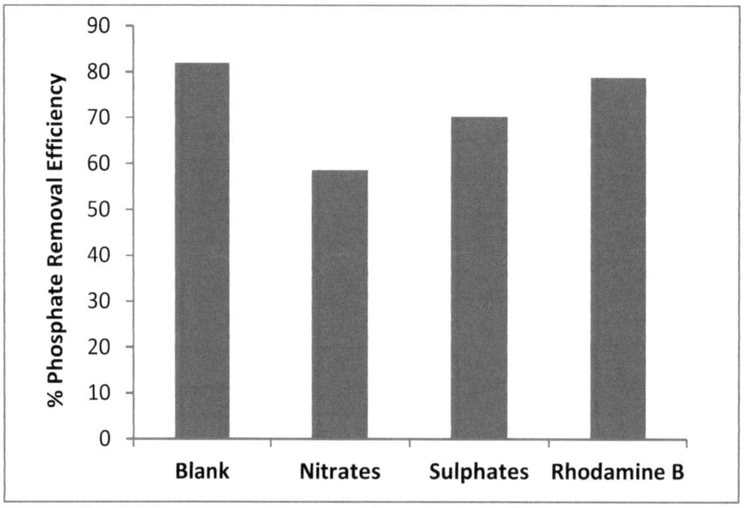

Figure 8: The effect of co-existing ions on the removal efficiency of palmitoyl chitosan.

4. Conclusion

In this study, N-acylated chitosan was synthesized via chemical derivatisation of chitosan with caproyl (C6) and palmitoyl (C16) chains. The functional groups and hydrophobicity of the N-acylated chitosan were confirmed by FTIR spectroscopy and contact angle measurements, respectively. The phosphate adsorption studies indicate an enhancement in the adsorption efficiencies from 20% using pristine chitosan to 63% for caproyl-chitosan and 71% for palmitoyl-chitosan in aqueous solutions containing PO_4^{3-}ions at a pH of 6.2.

However, all the materials show improved performances at the optimal solution pH of 4. The adsorption kinetics studies performed on palmitoyl-chitosan indicated that a contact time of approximately 30 minutes is required to reach equilibrium at solution concentrations ranging from 10-20 mg/l. Co-existing anions such as nitrates and sulphates do have an effect the phosphate removal efficiency of modified chitosan in aqueous solutions. Low adsorption capacities were obtained in the various studies, and hence the material will need to be optimised to increase adsorption capability of the modified chitosan. This study demonstrates that the palmitoyl-chitosan has the potential of being used as a cheap, rigid and degradable material for water treatment. Selectivity of other pollutants will also be investigated in the future.

A life cycle assessment of the N-acylated chitosan will be performed at a later stage. The study will assist in identifying ways to minimize waste generated during the treatment process and determine the environmental load of a single batch of the N-acylated chitosan. The potential use of the phosphate-loaded N-acylated chitosan as fertilizer will be investigated.

5. Acknowledgement
The authors are grateful to the Council for Scientific and Industrial Research for their financial support. Assistance from Dr. Lindani Mdlalose regarding the interpretation of the adsorption data is gratefully acknowledged.

6. References
[1] WWAP (UNESCO World Water Assessment Programme). (2019). *The united nations world water development report 2019: Leaving no one behind.* Paris: UNESCO.
[2] Nechita, P. (2017). Applications of chitosan in wastewater treatment. doi:10.5772/65289
[3] Gehrke, I., Geiser, A., & Somborn-Schulz, A. (2015). Innovations in nanotechnology for water treatment. *Nanotechnology, Science and Applications, 8*, 1-17. doi:10.2147/NSA.S43773
[4] Huang, P., Zhao, X., & Ye, L. (2016). Synthesis of hydrophobic cationic chitosan flocculant and its sludge dewatering property. *Journal of Macromolecular Science, Part B, 55*(3), 299-309. doi:10.1080/00222348.2016.1146975
[5] Rajeswari, A., Amalraj, A., & Pius, A. (2015). *Removal of phosphate using chitosan-polymer composites* doi:10.1016/j.jece.2015.08.022
[6] Leduc, J., Leduc, R., & Cabana, H. (2014). Phosphate adsorption onto chitosan-based hydrogel microspheres. *Adsorption Science and Technology, 32*, 557-569. doi:10.1260/0263-6174.32.7.557
[7] Tan, I. A. W., Ahmad, A. L., & Hameed, B. H. (2008). *Adsorption of basic dye using activated carbon prepared from oil palm shell: Batch and fixed bed studies* doi:https://doi.org/10.1016/j.desal.2007.07.005
[8] Schmuhl, R., Krieg, H., & Keizer, K. (2001). *Adsorption of cu(II) and cr(VI) ions by chitosan: Kinetics and equilibrium studies* doi:10.4314/wsa.v27i1.5002
[9] Pandey, N., Shukla, S. K., & Singh, N. B. (2017). Water purification by polymer nanocomposites: An overview. *Nanocomposites, 3*(2), 47-66. doi:10.1080/20550324.2017.1329983
[10] Muñoz-Bonilla, A., & Fernández-García, M. (2012). Polymeric materials with antimicrobial activity. *Progress in Polymer Science - PROG POLYM SCI, 37*, 281-339. doi:10.1016/j.progpolymsci.2011.08.005
[11] Le Tien, C., Lacroix, M., Ispas-Szabo, P., & Mateescu, M. (2003). N-acylated chitosan: Hydrophobic matrices for controlled drug release. *Journal of Controlled Release, 93*(1), 1-13. doi:DOI: 10.1016/S0168-3659(03)00327-4
[12] Han, J., Guenier, A., Salmieri, S., & Lacroix, M. (2008). Alginate and chitosan functionalization for micronutrient encapsulation. Journal of Agricultural and Food Chemistry, 56(7), 2528-2535. Retrieved from SCOPUS database.
[13] Yue, Q., Wang, W., Gao, B., Xu, X., Zhang, J., & Li, Q. (2010). Phosphate removal from aqueous solution by adsorption on modified giant reed doi:10.2307/25679789
[14] Mdlalose, L., Balogun, M., Setshedi, K., Chimuka, L., & Chetty, A. (2018). Adsorption of phosphates using transition metals-modified bentonite clay. Separation Science and Technology, 1-12. doi:10.1080/01496395.2018.1547315
[15] Liu, Q., Hu, P., Wang, J., Zhang, L., & Huang, R. (2016). Phosphate adsorption from aqueous solutions by zirconium (IV) loaded cross-linked chitosan particles doi:https://doi.org/10.1016/j.jtice.2015.08.012
[16] Liu, X., & Zhang, L. (2015). Removal of phosphate anions using the modified chitosan beads: Adsorption kinetic, isotherm and mechanism studies doi:https://doi.org/10.1016/j.powtec.2015.02.05

IOP Conf. Series: Materials Science and Engineering **655** (2019) 012048 doi:10.1088/1757-899X/655/1/012048

Pillared interlayered clays as catalyst for wastewater treatment: Cost-effective and sustainable materials

S J Baloyi [*], T W P Seadira and T A Ntho

Advanced Materials Division, Mintek, Private Bag X3015, Randburg 2125, South Africa

[*]jeffreyb@mintek.co.za

Abstract. Metal oxide pillared interlayered clays (PILC) represent a novel class of materials that have attracted interests in catalytic oxidative removal of organic pollutants in wastewater. In this work, South African bentonite clay were modified by polymeric Al/Fe, Al/Zr and Al/Cr species, and the obtained pillared solids were characterized and as potential catalysts in catalytic wet air oxidation (CWAO) of phenol in aqueous solution. The placement of metal oxide pillars between bentonite clay sheets modifies their physicochemical properties (surface area and basal spacing) and catalytic activity. The Al/Zr pillared clay exhibited a higher activity compared to Al/Fe and Al/Cr for phenol oxidation. All the pillared clay catalysts allowed the total oxidation of phenol and significant removal for the total organic carbon. The metal leaching was very low in all catalysts studied.

1. Introduction

Environmental problems related to water pollution has been described as one of the major issues and recently attracted much attention [1–5]. One of the major pollution sources comes from wastewater containing phenolic compounds commonly discharged from dyestuffs, pesticide, petroleum, paper and pharmaceuticals industries [6]. Phenolic compounds can be present in industrial wastewaters at different concentration levels depending on the source of pollution such at low concentrations of 10–100 mg/L and high concentrations 1000 mg/L from cleaning containers or spills and wastewater generated in oil refinery, respectively [7]. Phenols are highly toxic, low biodegradability and good water solubility and, so, phenols are in the "Priority Pollutants List" by the US Environmental Protection Agency [8]. The discharge of phenols into water streams may destroy the ecological system and cause disease [6]. Therefore, more and more attention is paid to its treatment before being discharged to water bodies.

Significant efforts have been done by many researchers to find appropriate treatment methods in order to remove phenols pollutants from wastewaters [1,3,9]. Numerous conventional chemical and physical methods of phenols removal including chemical precipitation [10], electro-coagulation [11] membranes [12] and adsorption techniques [13] are currently used for this purpose. However, one of the drawbacks of these methods is the formation of secondary pollutants like sludge, moreover, phenols cannot be completely be removed in the wastewater. Recently, advanced oxidation processes (AOPs) like catalytic wet air oxidation (CWAO) has been used for removal of phenol from wastewaters by many researches [1,6,14,15]. In a typical CWAO method, phenol is oxidized by a solid catalyst under an activated O_2 species, normally at pressures of 10–50 bar and temperatures of 125–250 °C. With regards to economic point of view, it is still essential to develop operating conditions at reaction temperature and partial pressures as milder as promising in order to reduce capital cost of the process. Therefore, the development of efficient and durable catalysts with robust catalyst activity and high stability is very important.

To the best of our knowledge, very few studies have been reported to evaluate CWAO of phenol using pillared clay catalyst including the use of cutting edge method to reduce the synthesis costs. Due to its outstanding properties such as low cost, environmental friendly and high chemical stability pillared clay catalysts has been utilized in AOPs related to wastewater treatment. The Al/Cr-PILC, Al/Fe-PILC and Al/Zr-PILC pillared clay solid materials were prepared by ultrasonic method with direct addition of clay into the pillaring agents solution containing Fe, Cr and Al metals, thereafter, intercalated through ion exchange. Structural and textural properties of the catalysts were characterized by XRD, SEM-EDX, Zetasizer and BET surface area measurements and catalytic activity of the solids materials were assessed by CWAO of phenol in aqueous solutions. Clay based catalysts used in this study have a great advantage in the CWAO of removal of phenol in wastewater due to their low cost and they have demonstrated high chemical stability because the concentrations of metal oxides leached in the CWAO reaction do not exceed 1 mg/L. Moreover, the CWAO process takes place at milder operating conditions and shorter residence time, making the operating cost of CWAO process to be very low.

2. Experimental procedures

2.1. Materials and chemicals

The raw clay purchased from Cape Bentonite Mine, Republic of South Africa was used as received. Table 1 shows that the South African bentonite clay was mainly composed of montmorillonite; it had the chemical composition as follows, SiO_2 (62.2%), Al_2O_3 (18.6%), Fe_2O_3 (3.5%), TiO_2 (0.2%), CaO (0.8%), MgO (3.7%), K_2O (0.7%), Na_2O (2.4%). Its cation exchange capacity (CEC), was 0.786 meq/g, its basal spacing 1.24 nm and its BET specific surface area 61 m^2/g. All solutions and dispersions were synthesized using an ultrapure water from Milli-Q water purification unit. $Al(NO_3)_3{\cdot}9H_2O$, $Zr(NO_3)_4{\cdot}5H_2O$, $Cr(NO_3)_3{\cdot}9H_2O$, $Fe(NO_3)_3{\cdot}9H_2O$, HCl and NaOH were purchased from Sigma-Aldrich.

Table 1: South African bentonite clay physicochemical properties

Sample	BET (m^2/g)	Mg (%)	Na (%)	Si (%)	Al (%)	Ca (%)	Fe (%)	K (%)	Ti (%)	Si/Al	CEC (meq/g)	d_{001} (nm)
Bentonite	61	3.7	2.4	62.2	18.6	0.8	3.7	0.7	0.2	4.25	0.786	1.24

2.2. The pillaring solution synthesis

The Al-pillaring solutions were prepared by adding 0.4 M sodium hydroxide solution slowly into definite amount of 0.4 M $Al(NO_3)_3{\cdot}9H_2O$ solution at room temperature under vigorous stirring using a separating funnel at approximately 5 mL/min constant dropping rate in order to obtain a molar ratio $[OH^-]/[Al^{3+}]$ = 2:1. It was the same procedure with Cr, Fe and Zr-pillaring solution. The obtained Al, Cr, Fe and Zr-pillaring solutions were set aside for use during intercalation of bentonite clay and their SEM images are shown in Figure 1.

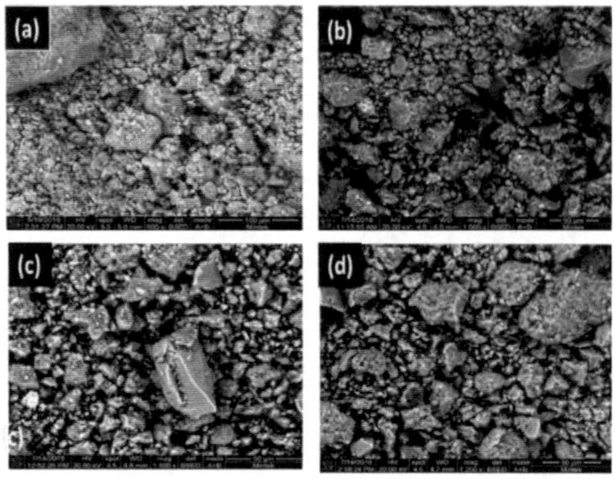

Figure 1: SEM images of (a) natural bentonite, (b) Al-PILC, (c) Cr-PILC, and (d) Fe-PILCs.

2.3. Synthesis of pillared clays

The pillared bentonite clay was prepared using ultrasonic treatment starting with dry clay material. In typical procedure, 0.01 M ratio of (Al + Cr, Fe or Zr)/g clay (25 mL /1 g) was produced by the direct addition of the dry bentonite clay into pillaring solutions and stirred vigorously for 30 minutes at 25 °C. Ion exchange was done by ultra-sonication of the suspensions for 10 min at 25 °C. Then, the solid was separated by centrifugation for 8 minutes and dried overnight at 120 °C and finally calcined for 2 h at 500 °C, at a 1 °C/min heating rate. These solids were designated as Al/Zr-PILC, Al/Cr-PILC and Al/Fe-PILC.

2.4. Characterization studies

All solids materials were characterized by X-ray diffraction (XRD), N_2 adsorption/desorption, Fourier transform infrared (FTIR), Thermogravimetric analysis (TGA) and zeta potential. X-ray diffraction patterns were obtained using a Brucker D8 Advance instrument employing Co-Kα emission of 40 kV using a LinxEye detector. The specific surface area, average pore diameters and pore volume of the raw bentonite and prepared solids were determined using N_2 adsorption/desorption at -196.15 °C, by the use of a Micromeritics (ASAP TriStar 3000) instrument. FTIR spectra were obtained by means of an Attenuated Total Reflectance (ATR) technique using a Varian 640-IR FTIR spectrometer (PIKE Technologies) at room temperature on the built-in diamond ATR crystal. Steady-state FTIR spectra were recorded over the range 400 to 4000 cm^{-1} over 200 scans. Thermogravimetry of the materials was performed using a Perkin–Elmer 6000 TG analyser. Zeta potential measurement was measured using Zetasizer Nano ZS (Malvern Instruments).

2.5. Phenol removal by catalytic wet air oxidation

The phenol oxidation experiments were performed in a autoclave semi-batch reactor (model 4531M) of 0.35 L capacity, stainless steel, equipped with an adjustable speed agitator, and a blanket which allows control of the reaction temperature and the gas phase addition was measured using needle valve that records the pressure input. The phenol removal experimental runs were stirred at 800 rpm, ensuring desirable mass transfer, which allow gas addition to the liquid phase and sampling of liquid. The reactor was charged with nitrogen gas at 10 bar for 5 minutes to create inert atmosphere inside the reactor and the vessel was evacuated again to take out any remaining nitrogen from the reactor system. After reaching preset reaction conditions the time was defined as zero time and volume of 5 mL solution was sampled at different time intervals during the reaction. An autoclave reactor was charged with a volume of 200 mL of 100 mg/L initial phenol concentration solutions and catalyst of 2 g/L to the reactor, partial oxygen of 5 bar and then 5 bar of pure nitrogen. All experimental runs were done under the following experimental conditions: phenol initial concentration = 1000 mg/L, partial pressure of oxygen = 10 bar, reaction temperature = 100 °C, reaction time = 120 min. All experimental runs were performed three times in order to check the reproducibility and evaluate the experimental error of the results. It was found that the standard deviation was less than 5%.

IOP Conf. Series: Materials Science and Engineering **655** (2019) 012048 doi:10.1088/1757-899X/655/1/012048

2.6. Analytical methods
Phenol concentration was measured by using the high performance liquid chromatography (HPLC) model LC-10AD (Shimadzu, Kyoto, Japan). The mobile phase was a mixture of methanol and water (v/v, 60:40) with 0.1% acetic acid aqueous solution added and with a flow rate of 1.0 mL min^{-1}. Total organic carbon (TOC) measurements were performed using TOC analyser by TOC-VCPN, Shimadzu, Japan.

3. Results and discussion
3.1. Phenol removal by catalytic wet air oxidation
In this series experiments, blank tests were conducted without a catalyst (WAO) and under raw bentonite clay (adsorption), and pillared clay catalysts modified with different metal oxides were investigated. Figure 2 shows the removals of phenol and TOC after 180 min of CWAO using different catalytic materials. No significant catalytic activity was observed during the blank tests experiments, showing the stability of phenol under our operating conditions in this study. These findings can be attributed high bentonite clay negative charge (as indicated in Figure 3) which repel the negatively charged phenol and also the absence of catalytic. In contrary, a maximum conversion of 100% phenol and very high removal of TOC was attained in 120 min of reaction. The catalytic activities of different catalytic materials were found to be in this; Al/Cr-PILC (1:1) < Al/Fe-PILC (3:1) < Al/Zr-PILC (3:1). The obtained results can be ascribed to the enhancement of surface electrical properties and new functional groups in the clay after pillaring [6]. Therefore, to study the relationship between pillared clay catalysts activity and catalysts structure and surface properties, solids materials were characterized.

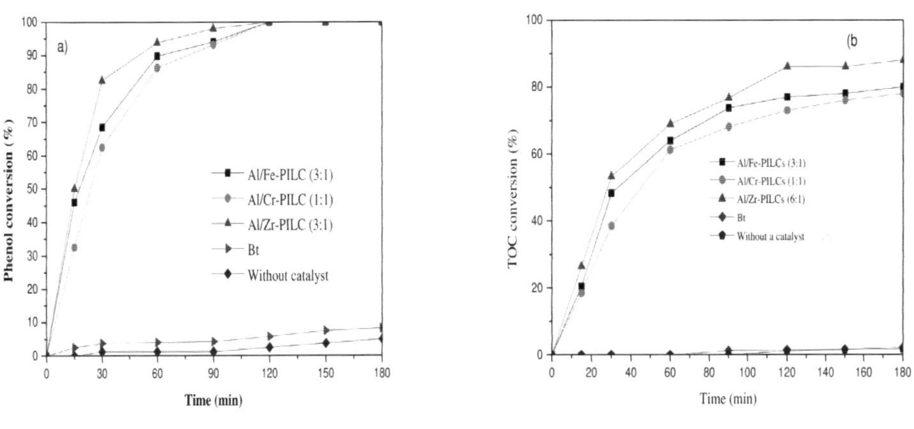

Figure 2: Phenol conversion (a) and TOC concentration removal (b) as a function of time for bentonite clay catalysts CWAO of phenol.

3.2. Characterization of catalysts
3.2.1. XRD
The XRD patterns of the samples were presented in Figure 3. It is observed that after intercalation with metal oxides pillars the basal spacing (d_{001}) peaks of bentonite clay shifted a little near small 2 theta values thus enhancing the basal spacing. The (d_{001}) was 1.24 nm for natural bentonite clay but increase to 1.61 nm, 1.50 nm and 1.92 nm for Al/Fe-PILC, Al/Cr-PILC and Al/Zr-PILC, respectively. Al/Cr-PILC and Al/Zr-PILC gives an intense and well defined (d001) diffraction peak whereas Al/Fe-PILC produce less intense diffraction peak, but noticeable peak. Therefore, the XRD data indicate that the pillared clay catalyst modified with Cr and Zr metal oxides have more ordered structure compared to Al/Fe-PILC. Main compounds composed of montmorillonite with impurities like muscovite, calcite and quartz. The obtained XRD patterns in this study were in concordance with similar studies reported in the literature [1,3,6]. The XRD patterns of the materials before and after pillaring were very similar, which suggests that the starting clay structure was not destroyed during the pillaring method.

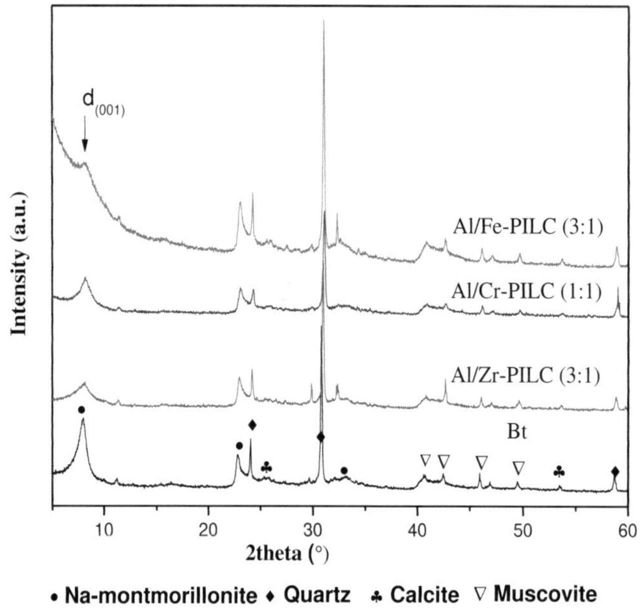

Figure 3: XRD patterns obtained for the natural and pillared bentonite clays materials.

3.2.2. The specific surface area

The specific surface area of pillared bentonite clay improved significantly after pillaring process, indicating that metal oxides were efficiently intercalated into the interlayer space of natural clay layers. Table 2 shows the specific surface area of pillared clay materials ranged from 150 to 230 m^2/g. The BET surface area of bentonite catalysts Zr and Cr metal oxides were higher than that of Fe metal oxide catalyst. Usually, the BET surface area and pore structure are significant factors of catalysts in a solid phase, since catalytic activity reported in this study is directly proportional to their specific surface area. Owing to more porous structures of pillared bentonite clay materials being formed, Al/Zr-PILC (230 m^2/g) and Al/Cr-PILC (200 m^2/g) display much higher BET surface area than Al/Fe-PILC (150 m^2/g).

Table 2: Textural characterization of prepared catalysts (specific surface area, pore volumes and pore diameter)

Sample	Metal ratio	S^1_{BET} (m^2/g)	V^2_{pore} (cm^3/g)	V^3_{mic} (cm^3/g)	d^4_{pore} (nm)
Bentonite	-	61	0.11	0.089	10.1
Al/Fe-PILC	3:1	150	0.29	0.1	8.1
Al/Cr-PILC	1:1	200	0.37	0.113	7.5
Al/Zr-PILC	3:1	230	0.38	0.115	7.1

[1] Specific surface area (S_{BET}); [2] Total pore volume at $p/p^0 \sim 0, 99$; [3] Micropore volume; [4] Pore diameter

IOP Conf. Series: Materials Science and Engineering **655** (2019) 012048 doi:10.1088/1757-899X/655/1/012048

3.2.3. SEM-EDX

SEM investigations of all samples revealed the presence of untied and randomly distributed closely packed porous materials (images not shown here). Thus, natural bentonite clay mechanical structure can be appreciated and was not destroyed by pillaring process supporting our reported XRD results. The elemental composition of the bentonite clay materials were determined by EDX analysis (Table 3). The EDX profiles confirm the presence of Fe, Al, Si, Mg, Na, Cr, Ca and K elements for all samples. EDX results, shows that Al, Fe and/ or Cr metal oxides contents enhanced with replacement of small interlayer polycations, such as Na and Ca due to pillaring by metal oxides pillars. The presence of impurities such as quartz and muscovite can be confirmed by the higher ratio of Si/Al [5].

Table 3: Physicochemical properties of the catalysts using EDX.

Materials	Al/Cr ratio	Mg (%)	Na (%)	Si (%)	Al (%)	Ca (%)	K (%)	Cr (%)	Fe (%)	Si/Al
Bentonite	-	2.29	3.73	65.09	15.31	2.09	2.39	0.00	9.10	4.25
Al/Fe-PILC	1:3	1.25	0.91	57.70	20.55	0.14	1.48	17.97	17.01	2.81
Al/Cr-PILC	1:1	1.22	0.97	59.01	19.02	0.21	1.61	13.58	4.38	3.10
Al/Zr-PILC	3:1	1.29	0.26	54.47	23.65	0.04	1.41	17.31	1.61	2.30

3.2.4. Zeta potential

The obtained results for zeta potential measurements are presented in Figure 4. The values of zeta potential for pure natural bentonite clay appear to be in agreement with those reported in the literature for a pH between 3 - 10 [16]. According to Baloyi et al., [6], the permanent negative charge of natural bentonite clay comes from isomorphous substitution of Si^{4+} by Al^{3+} and, also pH dependent charge from the hydroxyl groups at the edges of the clay sheets. As shown in Figure 4, at pH lower than the pH_{PZC}, all pillared clay catalysts are positively charged and at pH higher than the pH_{PZC}, it is negatively charged. These values may be associated with the formation of oxide pillars between starting bentonite clay sheets because of the hydration of Zr, Cr or Fe metal oxides intercalated into the intermellar space, during calcination.

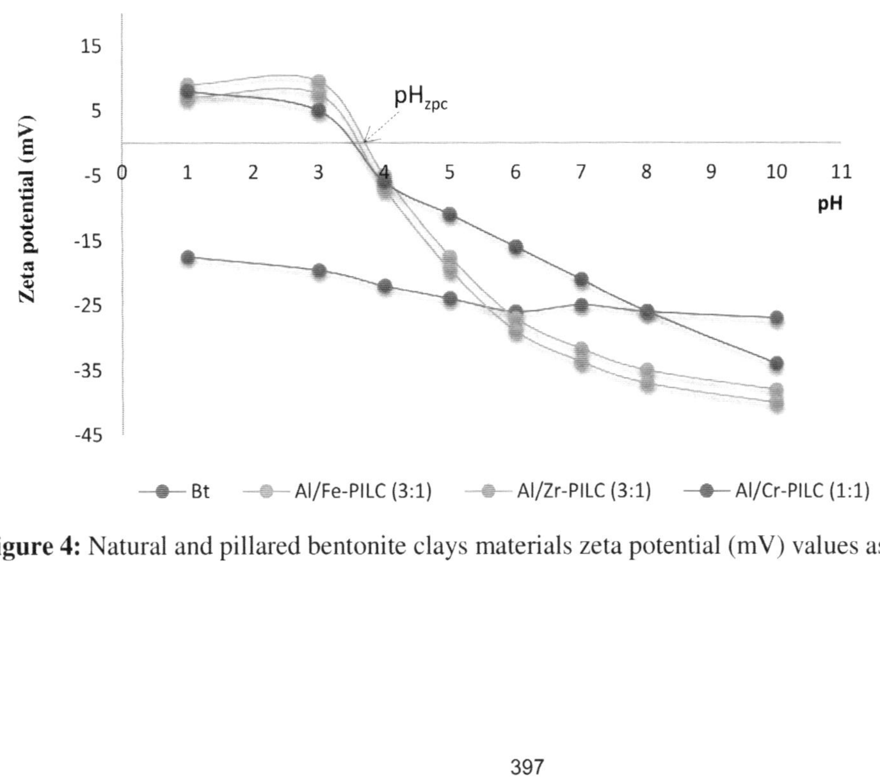

Figure 4: Natural and pillared bentonite clays materials zeta potential (mV) values as a function of pH.

3.3. Leaching test

Since the stability of a catalyst is significant in the evaluation of catalyst performance, six consecutive runs were performed with all pillared bentonite clay catalysts. After each run, the CWAO effluent was filtered using 0.45 µm size Whatmann filter paper and the residue was washed with distilled water followed by drying at 120 °C for 16 h in an oven. The amount of Fe^{3+}, Zr^{4+} and Cr^{3+} metal leached were detected, in order to measure the stability of pillared bentonite catalysts, and results are shown in Table 4. The leaching amount of all metal oxides were lower than 0.5 mg/L after six times catalysts reuse, which can be neglected and this showing a stronger interaction of the metal oxides with the pillared bentonite clay catalyst surfaces and the higher stability of the catalysts and its suitability for continuous operation.

Table 4: Stability, leaching of catalysts in the CWAO of phenol and removal of TOC under optimum reactions conditions.

No. of runs	**Leaching of Fe^{3+} ($Mg.L^{-1}$)**	**Leaching of Zr^{4+} ($Mg.L^{-1}$)**	**Leaching of Cr^{3+} ($Mg.L^{-1}$)**
First use	0.175	0.001	0.065
Re-use 1	0.198	0.003	0.096
Re-use 2	0.214	0.005	0.110
Re-use 3	0.221	0.009	0.118
Re-use 4	0.263	0.010	0.124
Re-use 5	0.275	0.162	0.160
Re-use 6	0.295	0.283	0.297

4. Conclusions

In conclusion, the pillared clay catalysts used showed an increase in CWAO catalytic activity and efficiency of the materials. Intercalation of polyoxocation in interlamellar region of bentonite clay enhanced phenol oxidation. The major drawback for application of these materials which is the cost of pillaring agents that modify the bentonite clay, was overcome by applying the ultrasound treatment method during the synthesis of pillared bentonite clay catalysts. Moreover, starting raw bentonite clay was added directly as received to the pillaring solution. Therefore reducing the periods of synthesis and volume of water utilized during the preparation method, which decreases the capital cost of process. Findings in the current study provides the advantages for pillared clay synthesis method to be escalated to industrial production scale.

Acknowledgement

The authors wish to acknowledge financial support from Mintek, Republic of South Africa through research Grant ADR 31904.

References

[1] Baloyi J, Ntho T and Moma J 2019 *J. Porous Mater.* **26** 3
[2] Baloyi J, Seadira T, Raphulu M and Ochieng A 2015 *Mater. Today Proc.* **27**
[3] Sassi H, Lafaye G, Ben Amor H, Gannouni A, Jeday M.R and Barbier J 2017 *Front. Environ. Sci.* Eng. **12** 2
[4] Potsi G, Ladavos A.K, Petrakis D, Douvalis A.P, Sanakis Y, Katsiotis M.S, Papavassiliou G, Alhassan S, Gournis D and Rudolf P 2018 *J. Colloid Interface Sci.* **510** 395
[5] Baloyi J, Ntho T and Moma J 2018 *RSC Adv.* **8**
[6] Baloyi J, Ntho T and Moma J 2018 *Catal. Letters.* **148**
[7] Tolosana-Moranchel A, Ovejero D, Barco B, Bahamonde A, Díaz E and Faraldos M 2019 *J. Environ. Chem. Eng.* **7** 103051

[8] Sarkar B, Rusmin R, Ugochukwu U.C, Mukhopadhyay R and Manjaiah K.M 2019 *Elsevier.* **113** 127
[9] Moma J, Baloyi J and T. Ntho 2018 *RSC Adv.* **8**
[10] Saini R and Kumar P 2016 *J. Environ. Chem. Eng.* **4** 673
[11] Jeong H, Lee J, Ju Y.M and Lee S.M 2019 *Renew. Energy.* **138** 971
[12] Hemmati M, Nazari N, Hemmati A and Shirazian S 2015 *J. Ind. Eng. Chem.* **21** 141
[13] Dehghani M.H, Mostofi M, Alimohammadi M, McKay G, Yetilmezsoy K, Albadarin A.B, Heibati B, AlGhouti M, Mubarak N.M and Sahu J.N 2016 *J. Ind. Eng. Chem.* **35** 63
[14] de los Monteros A.E, Lafaye G, Cervantes A, Del Angel G and Torres G 2015 *Catal. Today.* **258**
[15] Arena F, Di Chio R, Espro C, Palella A and Spadaro L 2017 *Springer International Publishing, Cham.* **349**
[16] Martinez J.M, Volzone C and Garrido L.B 2017 *Appl. Clay Sci.* **149** 20

Solar photocatalytic hydrogen production from glycerol reforming using ternary Cu/THS/Graphene

T W P Seadira[1*], S J Baloyi[1], C M Masuku[2,3], M S Scurrell[2]

[1] Advanced Materials Division, Catalysis Group, Mintek, Private Bag X3015, Randburg, 2125, South Africa.

[2] Department of Civil and Chemical Engineering, University of South Africa, Private Bag X6, FLORIDA, 1710, South Africa.

[3] Department of Chemical Engineering, Carnegie Mellon University, Pittsburgh, PA 15213, USA.

E-mail: tumelo.seadira@gmail.com

Abstract. A ternary Cu/THS/rGO photocatalyst was prepared using solvothermal method. Firstly, pure anatase TiO_2 hollow spheres (THS) were prepared via hydrothermal method using titanium butoxide, ethanol, ammonium sulphate, and urea via hydrothermal method. The copper (Cu) nanoparticles were subsequently loaded on the surface of the THS by wet impregnation. During the solvothermal process, the deposition and well dispersion of Cu/THS composites onto the graphene oxide surface. Subsequently, the reduction of graphene oxide to graphene was achieved. The morphological and structural properties of the prepared samples were characterized by Scanning Electron Microscope (SEM), UV-vis DRS, and photoluminescence (PL). The activities of the prepared catalysts were tested for hydrogen production via simultaneous photocatalytic water-splitting and glycerol reforming under visible light irradiation. The excellent photocatalytic activity of the Cu/THS/rGO catalyst was attributed the rGO which acts as both storage and transferor of electrons generated at the Cu and TiO_2 heterojunction, thus increasing the electro-hole pairs separation. Furthermore, the effects of reaction parameters photocatalytic hydrogen production were also studied.

1. Introduction

Photocatalysis has attracted a lot of attention as an alternative green and cost effective process for the production of clean-energy-source-hydrogen compared to the traditional high-energy-consuming processes such as steam reforming using fossil fuels, which leave a huge carbon footprint in the ecosystem [1]; [2]; [3]. However, the challenges associated with photocatalysis are the thermodynamic uphill of water splitting for hydrogen production, which results in low efficiency of hydrogen evolution [4]; and the development of highly active photocatalysts under solar irradiation [5]. It has been reported extensively that the thermodynamic uphill challenge can be addressed by the addition of sacrificial agents that are biomass derived feedstocks, that can be reformed easily and also act as the hole (h^+) scavengers of the semiconductor photocatalysts during photocatalysis; therefore, increasing the hydrogen evolution efficiency [6]; [7]; [8]. Among the number of biomass derived feedstocks studied for photocatalytic hydrogen evolution, many authors reported that methane provided the highest rates of hydrogen. However, due to the industrial demand which drives up the cost of methane, makes it an unideal sacrificial agent for low cost hydrogen production [9]. Therefore, given is production in large amounts as a by-product of the biodiesel industry, glycerol has been identified as an alternative sacrificial agent because of its low cost and abundance. Furthermore, the processing of glycerol can relieve the cost of growing storage attached to it [10]; [11].

It is widely reported that the TiO_2 based materials impregnated with various metal particles such as Pt [12], Au [13], Pd [14], Ni [15], Ag [16], and Cu [17] are impressively active for hydrogen production from the photocatalytic reforming of glycerol. However, most precious metals such as Pt, Pd, Au, are costly, and therefore limiting their utilization as co-catalysts. Therefore, research had been focused on developing nano-structured TiO_2 materials such as nanotubes [16], nanoflowers [18], and nano-hollow-spheres [18], in order to enhance the photocatalytic activity for hydrogen production using transition metals which are less costly. Furthermore, it has been reported that the incorporation of graphene sheets to these co-catalysts will further increase their photocatalytic activity [19]; [20]; [21].

In our recent studies, it has been demonstrated that the efficiency of photocatalytic hydrogen evolution activity can be enhanced by using anatase containing Cu/THS photocatalyst [22]. Furthermore, it was shown that the Cu loading on the surface of the THS particles played a crucial role on the activity of this composite photocatalyst. Moreover, it was shown that graphene also enhances the activity of the ternary Cu/THS/Graphene photocatalyst; again, it was found that the graphene content on the ternary Cu/THS/Graphene has a great effect on the activity on the ternary photocatalyst for solar-driven hydrogen evolution from glycerol reforming [23]. In the present study, we investigate the solar-driven photocatalytic reforming of glycerol for hydrogen production over ternary Cu/THS/Graphene catalysts; with great interest on the effect of process parameters such as initial glycerol concentration, pH, and photocatalyst dosage.

2. Methodology

The preparation of the photocatalysts and the catalytic H_2 production activity testing procedures are mentioned elsewhere [22]. The three photocatalysts prepared were THS, composite 0.5wt.%Cu/THS, and ternary 0.5wt.%Cu/THS/$Graphene_{4wt.\%}$. Both the composite and ternary photocatalysts were optimized from our previous studies, while studying the effect of Cu and graphene on the solar photocatalytic production of hydrogen activity from glycerol reforming [22]; [23]. The photocatalysts were characterized for SEM, XRD, UV-vis/DRS, and Photoluminescence (PL). The XRD analysis showed that all three catalysts contained 100% anatase THS particles; and no obvious peak of Cu or graphene was detected (results are not shown in this paper). In the next section, we discuss the SEM, UV-vis, and PL characterization of the three photocatalysts; we also discuss the solar photocatalytic hydrogen production activities of the three photocatalysts; as well as the effects of reaction parameters.

3. Results and Discussion

3.1. Characterization

3.1.1. SEM analysis

Figure 1 shows the SEM images of the prepared THS particles; composite 0.5wt.%Cu/THS; Elemental mapping of Cu nanoparticles dispersed on the surface of the THS particles; and the ternary 0.5wt.%Cu/THS/$Graphene_{4wt.\%}$. It can be observed from the picture that the THS were indeed achieved by the broken particle, which exposes its hollow chamber (insert). The main advantage of these THS is that they absorb light efficiently and produce multiple light reflections within the hollow chamber; and generate more electron/hole pairs [22]; [24]. The elemental mapping shows that the Cu nanoparticles are highly dispersed on the surface of the THS particles. This high dispersion of Cu nanoparticles allows the light photons to efficiently reach the surface of THS particles and photoexcite them in order to generate electron/hole pairs effeciently. The ternary Cu/THS/Graphene was successfully prepared as it can be seen from the picture, Figure 1(c). The Cu/THS composite particles are well dispersed and anchored on the inner and outer walls of the graphene sheets. The strong interaction between the Cu/THS composite particles and graphene sheets it attributed to the disconnection of the GO’s oxygen functional groups during calcination treatment, therefore resulting in the chemical bond between the Cu/THS composites and graphene sheets [25].

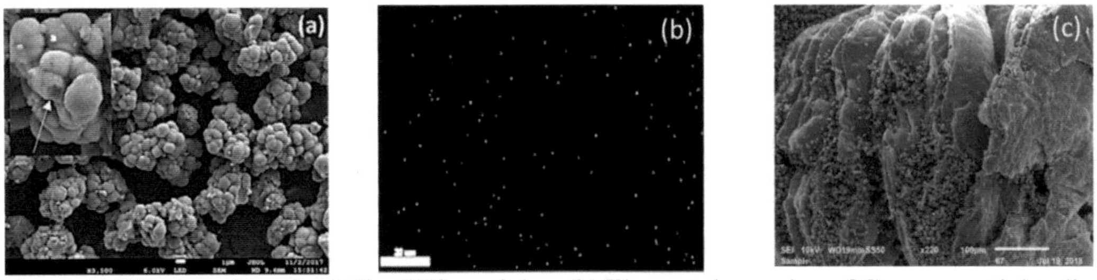

Figure 1. SEM imaging of: (a) The THS particles; (b) Elemental mapping of Cu nanoparticles dispersed on the surface of the THS particles; and (c) ternary 0.5wt.%Cu/THS/Graphene$_{4wt.\%}$.

3.1.2. UV-vis/DRS and Photoluminescence Analysis
The UV-vis/DRS was employed to study the optical properties of the prepared samples, Figure 2 (left). The DRS plots show that there was a red shift in the absorption band edge into the visible light region, which suggested that the band gap was narrowed. Furthermore, a decrease in bandgap was achieved with Cu nanoparticles loading and the graphene; which increased the absorption efficiency of the visible light [22]; [23]; [26]; [27]. The rate electron/hole pair recombination of the prepared samples was then studied using photoluminescence (PL) analysis.
The PL analyses of the prepared samples are depicted in Figure 2 (right). The PL plots showed that the rate of recombination of the THS particles decreased after they were impregnated with Cu nanoparticles; and was further decreased graphene was introduced to the Cu/THS composite framework as shown by the weakening of the peaks intensities. The suppression in recombination rate is attributed to the Cu nanoparticles, which trap and transport the photoinduced electrons to the graphene sheets, which act as electron storage for facilitation of proton reduction into hydrogen molecules; whilst leaving the holes on the THS surface vacant for glycerol oxidation [23]; [22]; [28] [29].

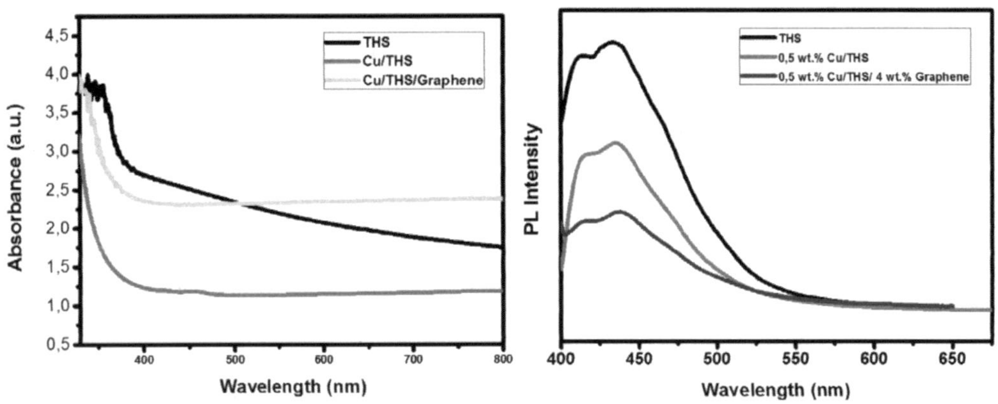

Figure 2. UV-vis/DRS analysis (left) and the photoluminescence analysis (right) of the prepared photocatalysts.

3.2. Photocatalytic Hydrogen production activity testing of the prepared photocatalysts
3.2.1. Effect of Cu and graphene loading
Figure 3 shows the photocatalytic hydrogen production activity of THS, composite Cu/THS, and ternary Cu/THS/Graphene photocatalysts. It can be observed that the activity of the THS particles increased exponentially after being loaded with Cu nanoparticles, and even increased further after the graphene was incorporated to its framework. The increase in the photocatalytic activity is attributed to high dispersion of Cu nanoparticles, which allows the light photons to reach the surface of the THS; the decrease in the bandgap and the suppression of the electron/hole recombination rate by both Cu nanoparticles and graphene sheets. The photocatalytic hydrogen evolution achieved for THS, Cu/THS, and Cu/THS/Graphene was 694 μmol.g_{cat}^{-1}, 25140 μmol.g_{cat}^{-1}, and 146824 μmol.g_{cat}^{-1}, respectively.

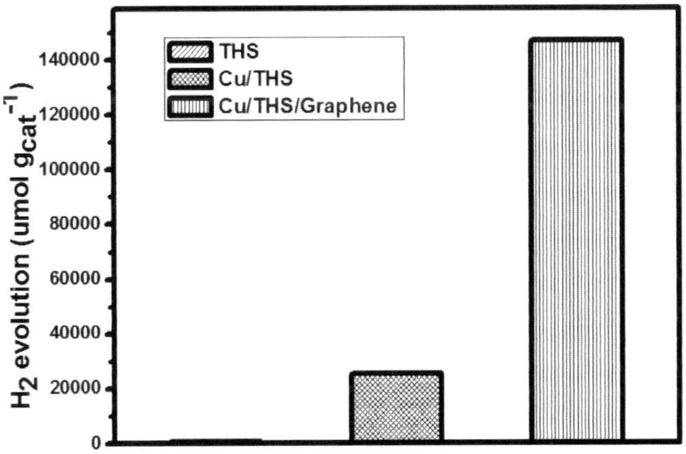

Figure 3. Photocatalytic hydrogen production activity of THS, composite Cu/THS, and ternary Cu/THS/Graphene photocatalysts.

3.2.2. Effect of initial glycerol concentration

Figure 4 shows the effect of glycerol concentration on the photocatalytic production of H_2 activity over 0.5wt.%Cu/THS/graphene$_{4wt.\%}$ ternary catalyst. It was observed from Figure 4(a) that the H_2 evolution increased with increasing glycerol concentration between 5 – 20 % (v/v), and levelled off when the glycerol concentration was increased to 25 % (v/v). The decrease in hydrogen evolution can be attributed to the presence of many glycerol molecules saturated on the active sites of the ternary photocatalyst; therefore, reducing the efficiency of electron/hole photogeneration, as well as the efficiency of glycerol oxidation [30]; [31]. Figure 4(b) depicts the initial hydrogen evolution rates (r_{H2}) as a function of initial glycerol concentration; which were determined from the slopes of fitting lines in Figure 4(a). It was observed that the rates of hydrogen evolution with variation of glycerol concentrations obeys the Langmuir-Hinshelwood mechanism [31]; [32]; [33]. According to the reaction mode, r_{H2} can be expressed as

$$r_{H2} = k_{H2}KCo/(1 + KCo) \quad (1)$$

where k_{H2} is the rate constant of hydrogen evolution, K is the adsorption constant of glycerol, and Co is the initial glycerol concentration. At low glycerol concentration (1 >> KCo), r_{H2} should be proportional to the glycerol concentration and therefore resulting in first-order with respect to glycerol with an apparent rate constant $k_{H2}K$. At high concentration (1 << KCo), r_{H2} should remain constant resulting in zero-order kinetics. The inversion of equation (1) results in a linear plot, $1/ r_{H2} = 1/ k_{H2} + 1/(k_{H2}KCo)$ as depicted in Figure 4(c); and it was observed that the calculated r_{H2} value fit well with the linear relationship and the k_{H2} and K were determined to be 0.051 mol.g^{-1}.h^{-1} and 0.162 L.mol^{-1}, respectively.

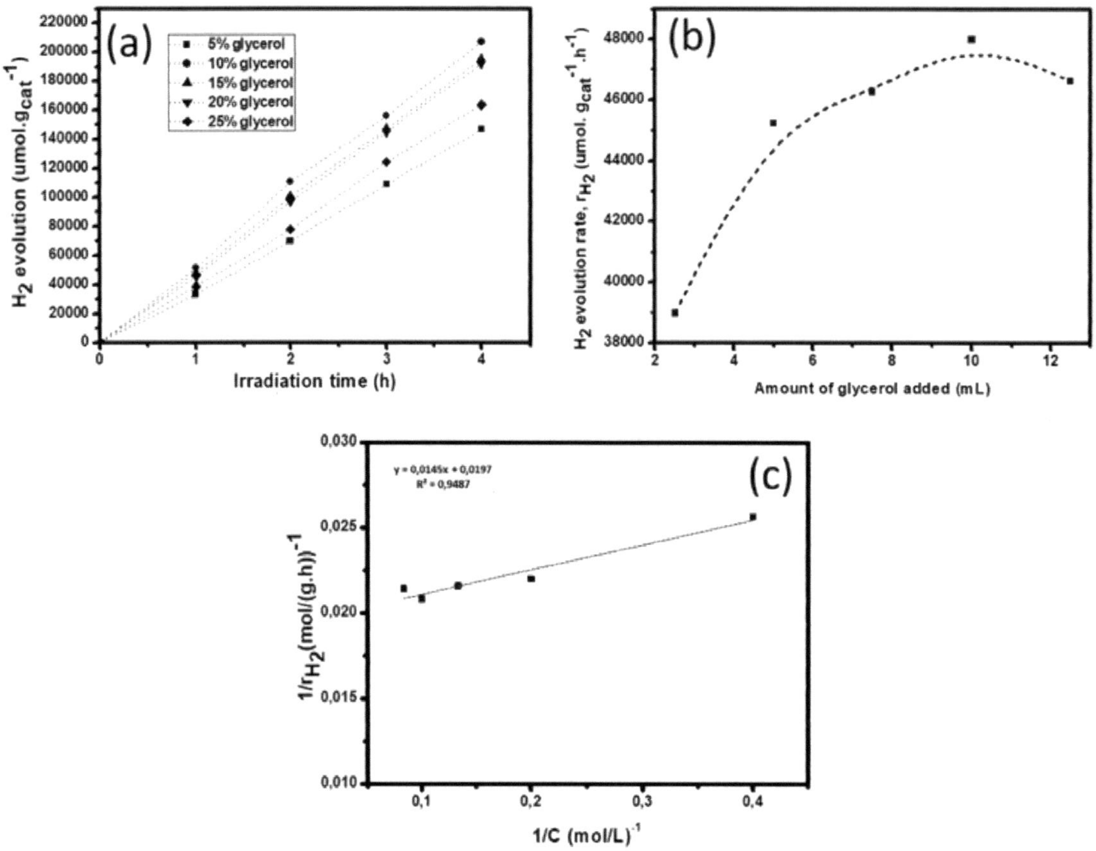

Figure 4. Effect of initial glycerol concentration on the photocatalytic hydrogen production activity of ternary 0.5wt.%Cu/THS/graphene$_{4wt.\%}$ catalyst, (a) hydrogen evolution, (b) rates of hydrogen evolution, (c) linear plots of r_{H2} and the k_{H2}.

3.2.3. Effect of pH

Solution pH is one of the most important parameter, which influences the photocatalytic hydrogen evolution, since it directly affects the surface charge of the photocatalyst and the sacrificial agent in the solution; resulting in a change of the electrostatic interaction between the photocatalyst surface and the sacrificial agent [34]; [35]. The influence of pH (2 - 10) on the photocatalytic hydrogen production from glycerol (10% v/v) reforming over a 0.5 wt.% Cu/THS/graphene$_{4wt.\%}$ photocatalyst is shown in Figure 5. It was observed that the hydrogen production increased with increasing pH from 2 to 6, and decreased drastically with increasing pH above 6. These results suggested that the photocatalytic hydrogen production over the 0.5 wt.% Cu/THS/graphene$_{4wt.\%}$ photocatalyst is most efficient under mild acidic conditions, with the best operating pH found to be 6.

The influence of the glycerol solution pH on the photocatalytic hydrogen evolution efficiency is complex. Therefore, it is important to understand the interaction between the glycerol molecules and active sites of the photocatalyst in order to put forward a plausible explanation. Since a ternary Cu/THS/graphene catalyst was employed in this study, an attempt to discuss the effect of pH on the individual components of the ternary catalyst will be made.

Firstly, it should be noted that the adsorption of glycerol onto the surface of the photocatalyst takes place through hydrogen bonding with the surface hydroxyl groups (TiOH) of the photocatalyst [36]; [37]. When the pH of the glycerol solution is extremely acidic (pH < 5), the surface charge of titania becomes positive; resulting in an electrostatic repulsion between the hydronium cations (H^+) present in the solution and the positively charged photocatalyst, which in turn reduces the adsorption efficiency of the hydronium cations; therefore, reducing the photocatalytic hydrogen evolution activity of the photocatalyst. This is attributed to the ionization state of the surface of the catalyst according to the reaction equations below:

$$Ti^{IV}\text{-}OH + H^+ \rightarrow Ti^{IV}\text{-}OH_2^+, \qquad pH < pH_{PZC} \qquad (2)$$
$$Ti^{IV}\text{-}OH + OH^- \rightarrow Ti^{IV}\text{-}O^- + H_2O \qquad pH > pH_{PZC} \qquad (3)$$

When the pH of the solution is high (pH > 7), the surface charge of titania is negative. Therefore, an electrostatic repulsion between a negatively charged titania and glycerol molecules occurs; retarding the glycerol adsorption on to the valence band holes for oxidation as well as the recombination of the photogenerated electron/hole pair. Furthermore, it is suggested that the photogenerated electron/hole pairs on the titania surface might recombine due to the negatively charged titania surface, which inhibits the migration of the photogenerated electrons [34]. It is widely reported that the point of zero charge (PZC) of titania occurs at pH of 5.8 – 6, [31]; [38]; [39]; [40]. Therefore, is it suggested that the efficient adsorption of glycerol molecules on to the active site of titania for enhanced oxidation can be achieved when the surface charge of titania is neutral; also the migration of photogenerated electrons to the conduction band is promoted, leading to an efficient photocatalytic activity (as shown on our results). It has been mentioned that the titania particles agglomerate under the acid conditions, resulting in less surface area exposed for glycerol adsorption and harvesting of photons; therefore, affecting the photocatalyst activity [41]. However, this agglomeration is not expected to occur with our ternary Cu/THS/graphene because the THS particles are well anchored on the surface of the graphene sheets. Furthermore, it is not expected that the THS particles will leach off the surface of the graphene. This is attributed to the surface charge of graphene, which is negative under both acidic and basic conditions; suggesting that the THS particles (which are positively charged under acidic conditions) will remain anchored on the surface of the graphene sheets [42]; [35]. Also, pH has a detrimental effect on the copper nanoparticles under extreme acidic conditions because under these conditions, the copper nanoparticles tend to leach off the surface of the titania into the solution [43]. Therefore, it can be concluded that at under extremely acidic conditions, the low photocatalyst activity is attributed to the electrostatic repulsion between THS nanoparticles and the glycerol molecules. While under the basic conditions, the low activity can be attributed to the light scattering of the THS particles because of the higher concentration of the OH^- present in the system; as well as the poor migration of the photogenerated electrons to the conduction band due to the change in the surface charge of the THS particles [34]; [44].

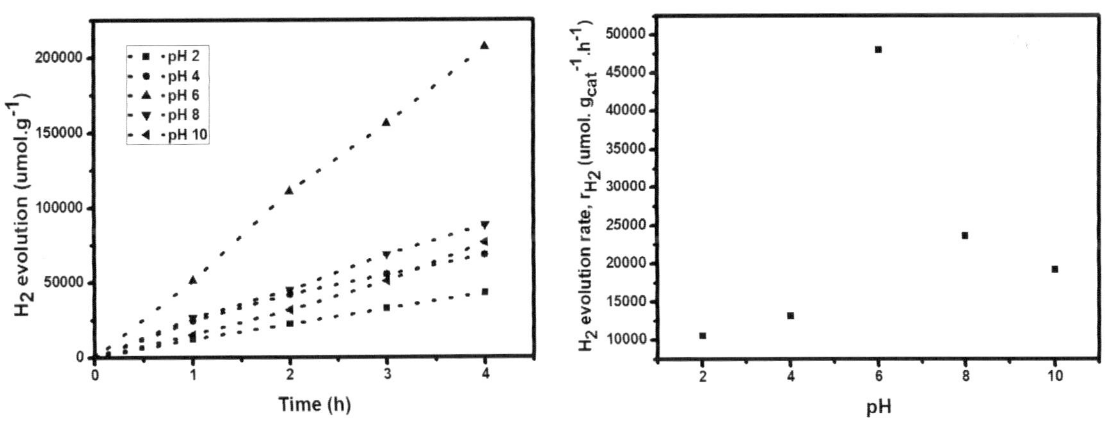

Figure 5. Effect of pH on the photocatalytic hydrogen production activity of ternary $0.5wt.\%Cu/THS/graphene_{4wt.\%}$ catalyst, (left) hydrogen evolution, (right) rates of hydrogen evolution.

3.2.4. Effect of catalyst dosage

The effect of catalyst dosage on the photocatalytic hydrogen activity of a ternary $0.5wt.\%Cu/THS/graphene_{4wt.\%}$ catalyst was studied by varying the catalyst dosage between 20 – 120 mg. Figure 6 illustrate the effect of catalyst dosage on the photacatalytic activity of a ternary $0.5wt.\%Cu/THS/graphene_{4wt.\%}$ photocatalyst for hydrogen evolution (left) as well as the rate of hydrogen evolution (right). It was observed that the hydrogen evolution increased with an increase in catalyst dosage, with the maximum evolution (and rate of evolution) achieved with the catalyst dosage of 80 mg (50 ml of glycerol solution, 10% v/v); and a decrease in hydrogen evolution was observed with a catalyst dosage higher than 80 mg. The decrease in photocatalytic hydrogen evolution due to an increase in catalyst dosage can be attributed to aggregation of the photocatalyst particles, which reduces the active surface for efficient photons resulting in light scattering; and therefore, hindering the bulk of photons from reaching the active sites of the catalyst surface for efficient photogenerated electron/hole pair [45]; [46]; [47]; [48].

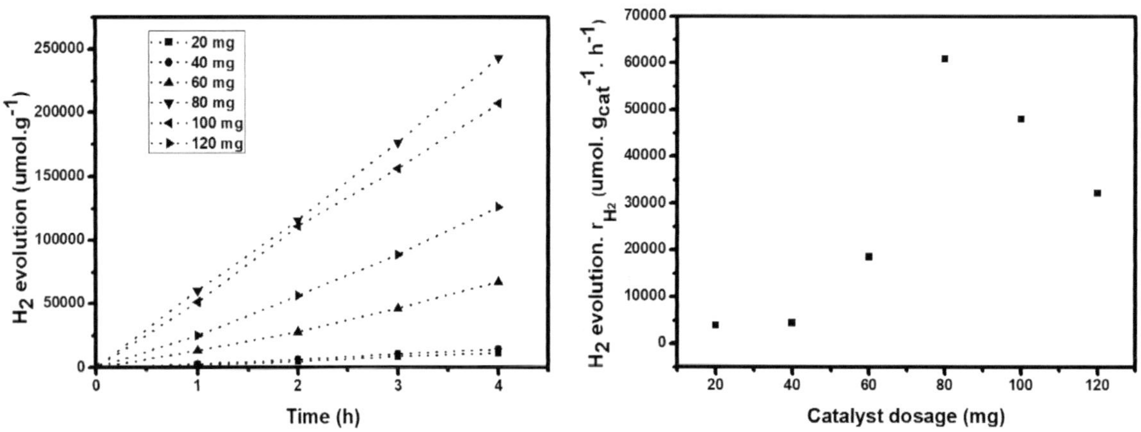

Figure 6. Effect of catalyst dosage on the photocatalytic hydrogen production activity of ternary 0.5wt.%Cu/THS/graphene$_{4wt.\%}$ catalyst, (a) hydrogen evolution, (b) rates of hydrogen evolution.

4. Conclusions

In conclusion, it has been found that the photocatalytic hydrogen evolution activity of a ternary 0.5wt.%Cu/THS/graphene$_{4wt.\%}$ photocatalyst is strongly influenced by initial glycerol concentration, pH, and photocatalyst dosage. An increase in initial glycerol concentration leads to a slightly linear increase in hydrogen production, and levels off when the concentration in increased above 10 % (v/v); which is consistent with the Langmuir-Hinshelwood mechanism. The photocatalytic hydrogen evolution activity of the ternary catalyst is dependent on the influence of the pH on the surface charge of catalyst and the glycerol solution. The maximum rate of hydrogen evolution was achieved when the glycerol solution was slightly acidic (pH 6), which was attributed to charge of the THS particles (in the ternary catalyst, responsible for the photocatalytic generation of electron/hole pairs) which is almost neutral; therefore, able to adsorb and oxidize the glycerol molecules; and subsequently reduce the H^+ into hydrogen molecules. Furthermore, the photogenerated electrons at the valence band can easily migrate to the conduction band, then to the surfaces of Cu nanoparticles and graphene sheets under slightly acidic conditions (pH 6). The photocatalytic hydrogen evolution activity increases with increasing photocatalyst dosage, and decreases when the photocatalyst is increased beyond the optimized catalyst dosage (80 mg/50ml of glycerol solution); because of the agglomeration of the catalyst particles, resulting in light scattering and the hindrance of efficient photons absorption for efficient photogeneration of electron/hole pairs.

Acknowledgements

We would like to acknowledge that this work has been supported in part by Mintek, South Africa, University of South Africa (Research Flagship CREATE – Catalytic Reaction Technology and the Transformation of Energy), and the National Research Foundation (NRF) of South Africa (Grant Number: 113652). The opinions, findings and conclusions/recommendations expressed in this publication are those of the authors, and the NRF accepts no liability whatsoever in this regard.
There are no competing interests to be declared.

References

[1] L. Gradisher, B. Dutcher and M. and Fan, "Catalytic hydrogen production from fossil fuels via the water gas shift reaction," *Applied Energy* , vol. 139, pp. 335-349, 2015.

[2] G. Corro, U. Pal and N. and Tellez, "Biodiesel production from Jatropha curcas crude oil using ZnO/SiO2 photocatalyst for free fatty acids esterification," *Applied Catalysis B: Environmental,* vol. 129, pp. 39-47, 2013.

[3] K. Shimura and H. and Yoshida, "Heterogeneous photocatalytic hydrogen production from water and biomass derivatives," *Energy & Environmental Science,* vol. 4, no. 7, pp. 2467-2481, 2011.

[4] A. Ismail and D. and Bahnemann, "Photochemical splitting of water for hydrogen production by photocatalysis: a review," *olar Energy Materials and Solar Cells,* vol. 128, pp. 85-101, 2014.

[5] L. Clarizia, D. Spasiano, I. Di Somma, R. Marotta, R. Andreozzi and D. and Dionysiou, "Copper modified-TiO2 catalysts for hydrogen generation through photoreforming of organics. A short review," *International Journal of Hydrogen Energy,* vol. 39, no. 30, pp. 16812-16831, 2014.

[6] V. Lanese, D. Spasiano, R. Marotta, I. Di Somma, L. Lisi, S. Cimino and R. and Andreozzi, "Hydrogen production by photoreforming of formic acid in aqueous copper/TiO2 suspensions under UV-simulated solar radiation at room temperature," *International Journal of Hydrogen Energy,* vol. 38, no. 23, pp. 9644-9654, 2013.

[7] J. Colmenares, A. Magdziarz, M. Aramendia, A. Marinas, J. Marinas, F. Urbano and J. and Navio, "Influence of the strong metal support interaction effect (SMSI) of Pt/TiO2 and Pd/TiO2 systems in the photocatalytic biohydrogen production from glucose solution," *Catalysis Communications,* vol. 16, no. 1, pp. 1-6, 2011.

[8] M. Languer, F. Scheffer, A. Feil, D. Baptista, P. Migowski, G. Machado, D. de Moraes, J. Dupont, S. Teixeira and D. and Weibel, "Photo-induced reforming of alcohols with improved hydrogen apparent quantum yield on TiO2 nanotubes loaded with ultra-small Pt nanoparticles," *International Journal of Hydrogen Energy,* vol. 38, no. 34, pp. 14440-14450, 2013.

[9] J. Highfield, M. Chen, P. Nguyen and Z. and Chen, "Mechanistic investigations of photo-driven processes over TiO 2 by in-situ DRIFTS-MS: part 1. Platinization and methanol reforming," *Energy & Environmental Science,* vol. 2, no. 9, pp. 991-1002, 2009.

[10] D. Sun, Y. Yamada, S. Sato and W. and Ueda, "Glycerol hydrogenolysis into useful C3 chemicals," *Applied Catalysis B: Environmental* , vol. 193, pp. 75-92, 2016.

[11] C. Zhou, J. Beltramini, Y. Fan and G. and Lu, "Chemoselective catalytic conversion of glycerol as a biorenewable source to valuable commodity chemicals," *Chemical Society Reviews,* vol. 37, no. 3, pp. 527-549, 2008.

[12] M. Stelmachowski, M. Marchwicka, E. Grabowska, M. Diak and A. and Zaleska, "The photocatalytic conversion of (biodiesel derived) glycerol to hydrogen-a short review and preliminary experimental results part 2: Photocatalytic conversion of glycerol to hydrogen in batch and semi-batch laboratory reactors," *Journal of Advanced Oxidation Technologies,* vol. 17, no. 2, pp. 179-186, 2004.

[13] W. Chen, A. Chan, Z. Al-Azri, A. Dosado, M. Nadeem, D. Sun-Waterhouse, H. Idriss and G. and Waterhouse, "Effect of TiO2 polymorph and alcohol sacrificial agent on the activity of Au/TiO2 photocatalysts for H2 production in alcohol–water mixtures," *Journal of catalysis,* vol. 329, pp. 499-513, 2015.

[14] F. López-Tenllado, J. Hidalgo-Carrillo, V. Montes, A. Marinas, F. Urbano, J. Marinas, L. Ilieva, T. Tabakova and F. and Reid, "A comparative study of hydrogen photocatalytic production from glycerol and propan-2-ol on M/TiO2 systems (M= Au, Pt, Pd)," *Catalysis Today,* vol. 280, pp. 58-64, 2017.

[15] F. Chong, E. Nurlaela and B. and Dutta, "Impact of glycerol as scavenger for solar hydrogen production from water," *Internation Journal of Energy and Environment* , pp. 19-26, 2014.

[16] D. Kumar, N. Reddy, M. Karthik, B. Neppolian, J. Madhavan and M. and Shankar, "Solar light sensitized p-Ag2O/n-TiO2 nanotubes heterojunction photocatalysts for enhanced hydrogen production in aqueous-glycerol solution," *Solar Energy Materials and Solar Cells,* vol. 154, pp. 78-87, 2016.

[17] R. Bashiri, N. Mohamed, C. Kait and S. and Sufian, "Hydrogen production from water photosplitting using Cu/TiO2 nanoparticles: Effect of hydrolysis rate and reaction medium," *International Journal of Hydrogen Energy,* vol. 40, no. 18, pp. 6021-6037, 2015.

[18] B. Wang, X. Lu, K. Lawrence, J. Xuan, M. Leung and H. and Guo, " Facile synthesis of TiO 2 hollow spheres composed of high percentage of reactive facets for enhanced photocatalytic activity," *CrystEngComm,* vol. 16, no. 43, pp. 10046-10055, 2014.

[19] H. Park, H. Ha, R. Ruoff and A. and Bard, "On the improvement of photoelectrochemical performance and finite element analysis of reduced graphene oxide–BiVO4 composite electrodes," . *Journal of Electroanalytical Chemistry,* vol. 716, pp. 8-15, 2014.

[20] P. Wang, N. Dimitrijevic, A. Chang, R. Schaller, Y. Liu, T. Rajh and E. and Rozhkova, "Photoinduced electron transfer pathways in hydrogen-evolving reduced graphene oxide-boosted hybrid nano-bio catalyst," *ACS Nano,* vol. 8, no. 8, pp. 7995-8002, 2014.

[21] X. Lv, S. Zhou, C. Zhang, H. Chang, Y. Chen and W. and Fu, "Synergetic effect of Cu and graphene as cocatalyst on TiO 2 for enhanced photocatalytic hydrogen evolution from solar water splitting," *Journal of Materials Chemistry,* vol. 22, no. 35, pp. 18542-18549, 2012.

[22] T. Seadira, G. Sadanandam, T. Ntho, C. Masuku and M. and Scurrell, "Preparation and characterization of metals supported on nanostructured TiO2 hollow spheres for production of hydrogen via photocatalytic reforming of glycerol," *Applied Catalysis B: Environmental,* vol. 222, pp. 133-145, 2018.

[23] T. Seadira, T. Ntho, C. Masuku and M. Scurrell, "Solar Photocatalytic Glycerol Reforming for Hydrogen over Ternary Cu/THS/Graphene Photocatalyst: Effect of Cu and Graphene loading," *Submitted for publication.*

[24] J. Xu, W. Dai, J. Li, Y. Cao, H. Li and K. and Fan, "Novel core-shell structured mesoporous titania microspheres: Preparation, characterization and excellent photocatalytic activity in phenol abatement," *Novel core-shell structured mesoporous titania microspheres: Preparation, characterization and excellent photocatalytic activity in phenol abatement,* vol. 195, no. 2 - 3, pp. 284-294, 2008.

[25] C. Nethravathi and M. and Rajamathi, "Chemically modified graphene sheets produced by the solvothermal reduction of colloidal dispersions of graphite oxide," *Carbon,* vol. 46, no. 14, pp. 1994-1998, 2008.

[26] B. Fang, A. Bonakdarpour, K. Reilly, Y. Xing, F. Taghipour and D. and Wilkinson, "Large-scale synthesis of TiO2 microspheres with hierarchical nanostructure for highly efficient photodriven reduction of CO2 to CH4," *ACS applied materials & interfaces,* vol. 6, no. 17, pp. 15488-15498, 2014.

[27] B. Fang, Y. Xing, A. Bonakdarpour, S. Zhang and D. and Wilkinson, "Hierarchical CuO–TiO2 hollow microspheres for highly efficient photodriven reduction of CO2 to CH4.," *ACS Sustainable Chemistry & Engineering,* vol. 3, no. 10, pp. 2381-2388, 2015.

[28] J. Liqiang, Y. Q. W. Baiqi, L. Shudan, J. Baojiang, Y. Libin, F. Wei, F. Honggang and S. and Jiazhong, "Review of photoluminescence performance of nano-sized semiconductor materials and its relationships with photocatalytic activity.," *Solar Energy Materials and Solar Cells,* vol. 90, no. 12, pp. 1773-1787, 2006.

[29] R. He and W. and He, " Ultrasonic assisted synthesis of TiO2–reduced graphene oxide nanocomposites with superior photovoltaic and photocatalytic activities," *Ceramics International,* vol. 42, no. 5, pp. 5766-5771, 2016.

[30] N. Strataki, V. Bekiari, D. Kondarides and P. and Lianos, "Hydrogen production by photocatalytic alcohol reforming employing highly efficient nanocrystalline titania films," *Applied Catalysis B: Environmental,* vol. 77, no. (1-2), pp. 184-189, 2007.

[31] S. Fujita, H. Kawamori, D. Honda, H. Yoshida and M. and Arai, "Photocatalytic hydrogen production from aqueous glycerol solution using NiO/TiO2 catalysts: Effects of preparation and reaction conditions," *Applied Catalysis B: Environmental,* vol. 181, pp. 818-824, 2016.

[32] P. Gomathisankar, D. Yamamoto, H. Katsumata, T. Suzuki and S. and Kaneco, "Photocatalytic hydrogen production with aid of simultaneous metal deposition using titanium dioxide from aqueous glucose solution," *International Journal of Hydrogen Energy,* vol. 38, no. 14, pp. 5517-5524, 2013.

[33] X. Jiang, X. Fu, L. Zhang, S. Meng and S. and Chen, "Photocatalytic reforming of glycerol for H 2 evolution on Pt/TiO 2: fundamental understanding the effect of co-catalyst Pt and the Pt deposition route," *Journal of Materials Chemistry A,* vol. 3, no. 5, pp. 2271-2282, 2015.

[34] T. Sreethawong, T. Puangpetch, S. Chavadej and S. and Yoshikawa, " Quantifying influence of operational parameters on photocatalytic H2 evolution over Pt-loaded nanocrystalline mesoporous TiO2 prepared by single-step sol–gel process with surfactant template," *Journal of power sources,* vol. 165, no. 2, pp. 861-869, 2007.

[35] E. Aboelfetoh, A. Elhelaly and A. and Gemeay, "Synergistic effect of Cu (II) in the one-pot synthesis of reduced graphene oxide (rGO/CuxO) nanohybrids as adsorbents for cationic and anionic dyes," *ournal of environmental chemical engineering,* vol. 6, no. 1, pp. 623-634, 2018.

[36] M. Estahbanati, M. Feilizadeh and M. and Iliuta, "Photocatalytic valorization of glycerol to hydrogen: Optimization of operating parameters by artificial neural network," *Applied Catalysis B: Environmental,* vol. 209, pp. 483-492, 2017.

[37] C. Kormann, D. Bahnemann and M. and Hoffmann, "Photolysis of chloroform and other organic molecules in aqueous titanium dioxide suspensions," *Environmental science & technology,* vol. 25, no. 3, pp. 494-500, 1991.

[38] M. Li, Y. Li, S. Peng, G. Lu and S. and Li, "Photocatalytic hydrogen generation using glycerol wastewater over Pt/TiO2," *Frontiers of Chemistry in China,* vol. 4, no. 1, pp. 32-38, 2009.

[39] K. Selvam, M. Muruganandham, I. Muthuvel and M. and Swaminathan, "The influence of inorganic oxidants and metal ions on semiconductor sensitized photodegradation of 4-fluorophenol," *Chemical Engineering Journal,* vol. 128, no. 1, pp. 51-57, 2007.

[40] L. Zang, P. Qu, J. Zhao, T. Shen and H. and Hidaka, "Photocatalytic bleaching of p-nitrosodimethylaniline in TiO2 aqueous suspensions: A kinetic treatment involving some primary events photoinduced on the particle surface," *Journal of Molecular Catalysis A: Chemical,* vol. 120, no. 1-3, pp. 235-245, 1997.

[41] M. Habibi, A. Hassanzadeh and S. and Mahdavi, "The effect of operational parameters on the photocatalytic degradation of three textile azo dyes in aqueous TiO2 suspensions," *Journal of Photochemistry and Photobiology A: Chemistry,* vol. 172, no. 1, pp. 89-96, 2005.

[42] H. Tan, F. Denny, M. Hermawan, R. Wong, R. Amal and Y. and Ng, "Reduced graphene oxide is not a universal promoter for photocatalytic activities of TiO2," *ournal of Materiomics,* vol. 3, no. 1, pp. 51-57, 2017.

[43] L. Clarizia, D. Spasiano, I. Di Somma, R. Marotta, R. Andreozzi and D. and Dionysiou, "Copper modified-TiO2 catalysts for hydrogen generation through photoreforming of organics. A short review," *International Journal of Hydrogen Energy,* vol. 39, no. 30, pp. 16812-16831, 2016.

[44] A. Mills, R. Davies and D. and Worsley, "Water purification by semiconductor photocatalysis," *Chemical Society Reviews,* vol. 22, no. 6, pp. 417-425, 1993.

[45] E. Melián, C. López, D. Santiago, R. Quesada-Cabrera, J. Méndez, J. Rodríguez and O. and Díaz, "Study of the photocatalytic activity of Pt-modified commercial TiO2 for hydrogen production

in the presence of common organic sacrificial agents," *Applied Catalysis A: General,* vol. 518, pp. 189-197, 2016.

[46] V. Sakkas, M. Islam, C. Stalikas and T. and Albanis, "Photocatalytic degradation using design of experiments: a review and example of the Congo red degradation," *Journal of hazardous materials,* vol. 175, no. 1-3, pp. 33-44, 2010.

[47] I. Konstantinou and T. and Albanis, "TiO2-assisted photocatalytic degradation of azo dyes in aqueous solution: kinetic and mechanistic investigations: a review," *Applied Catalysis B: Environmental,* vol. 49, no. 1, pp. 1-14, 2004.

[48] B. Neppolian, H. Choi, S. Sakthivel, B. Arabindoo and V. and Murugesan, "Solar/UV-induced photocatalytic degradation of three commercial textile dyes," *Journal of hazardous materials,* vol. 89, no. 2-3, pp. 303-317, 2002.

Kinetic study of phenol oxidation in a trickle bed reactor over Al/Zr-pillared clay catalyst

T J Makatsa[1,2*], S J Baloyi[2], T A Ntho[2] and C M Masuku[1]

[1] Department of Civil and Chemical Engineering, University of South Africa, Private Bag X6, Florida, 1710, South Africa

[2] Advanced Materials Division, Mintek, Private Bag X3015, Randburg, 2125, South Africa

E-mail: Corresponding. : tmakatsa@gmail.com

Abstract. Catalytic wet air oxidation of phenol was studied in a trickle bed reactor operated at 10 bar and temperatures between 120–160 °C over pillared clay wash-coated monolith catalyst. Different variables were studied including reaction temperature and liquid flow rate. The obtained results indicated that phenol conversion increase with an increase in temperature, whereas an increase in liquid flow rate has a negative effect on phenol conversion. Simple power law was used to model the kinetics of the process. When a simple power law model was used experimental results were predicted successfully with R^2 = 0.9988 at 160 °C and the activation energy was 42.289 kJ/mol.

1. Introduction

The world is moving away from the treatment of industrial wastewater by conventional processes like biological process due to the toxicity of phenol towards microorganisms. Emerging technologies like wet air oxidation (WAO) can be used to depollute wastewater that contains high concentrations of toxic organic pollutants. However, to completely mineralize phenol and its derivatives to carbon dioxide and water the process requires higher temperature and pressure which makes the process unfavourable due to high operating cost [1–3]. On the other hand, catalytic wet air oxidation (CWAO) is a more suitable process because of mild operating conditions due to catalyst incorporation resulting in lower activation energy. The reaction mechanism of phenol oxidation is influenced by reactor configuration and the type of catalyst used as well as reaction intermediates that can take place on the surface of the catalyst or in the homogeneous phase [4].

The use of a trickle bed reactor (TBR) in CWAO process is common due to the high ratio of catalyst to liquid, wide range of operating conditions (high to low pressure) and the reactor can be operated close to plug flow [5–7]. However, TBR has some drawbacks like high-pressure drop and liquid maldistribution, which can result in the formation of hot spots. On the other hand, monolithic catalyst can be used to reduce pressure drop and liquid channelling. Furthermore, scaling up of monolithic catalyst is not complicated since the conditions within individual channels remain the same [7].

In this work, the aim is to develop a kinetic model of phenol oxidation in a fixed bed monolithic reactor operated in a trickle flow mode over Al/Zr-pillared clay catalyst. Furthermore, the effect of operating conditions (temperature and liquid flow rate) on kinetic parameters will be investigated.

2. Materials and Method

2.1 Chemicals

Bentonite clay purchased from ECCA Holdings (Pty) Ltd., was used as a starting material during the pillaring process. The pillaring solution was prepared by mixing Al $(NO_3)_3 \cdot 9H_2O$, $ZrOCl_2 \cdot 8H_2O$, NaOH and monolith which was acid-treated using 37% H_2SO_4. Drierite was purchased from W.A Hammond Drierite Company LTD. Cordierite monolith was purchased from Ghophin Chemical. Synthetic wastewater was prepared by mixing phenol supplied by Sigma-Aldrich Chemical Co with high purity water. Anhydrous methanol used in this study was bought from Sigma-Aldrich Chemical Co.

2.2 Experimental Methods

2.2.1 Catalyst testing

The experiment was conducted in a stainless steel fixed bed reactor of 56 mm inside diameter, 9 mm wall thickness and length of 430 mm operated in a trickle flow region and the feed flow rates were kept constant (liquid flow = 10 mL/min, gas flow = 0.012 m/s) while reaction temperature and pressure were also kept constant (160 °C, 10 bar). The experiment continued for 3 h and a sample was withdrawn every 30 min and analyzed using High Performance Liquid Chromatography (HPLC). Two cordierite monoliths wash-coated with Al/Zr pillared clay catalyst (55 mm diameter and 50 mm height each) were packed inside the reactor. A top layer of (230 mm height) silica sand was used as a packing material and after catalyst packing another layer of silica was added (120 mm) at the bottom of the reactor. Drierite was used as a moisture trap to avoid liquid entering the gas chromatographer. Operating conditions of the reactor are listed in Table 1 for each experiment [8]. The performance of the catalyst was evaluated in terms of phenol conversion (x_{ph}) as defined by the following equation,

$$x_{ph} = \left(\frac{C_{ph}^{o} - C_{ph}}{C_{ph}^{o}}\right).100 \qquad (1)$$

where C_{ph}, is phenol concentration measured during the reaction and C_{ph}^{o} is initial phenol concentration

Table 1: Fixed bed monolithic reactor operating conditions for CWAO.

Kinetics experiment	
Temperature (°C)	(120-160)
Pressure (bar)	(10)
Liquid velocity (mL/min)	(10-40)
Gas velocity (m/s)	(0.012)
Phenol Conc. (mg/L)	1000

2.2.2 Analytical techniques

Phenol and intermediates were measured by Shimadzu HPLC with a UV detector at a wavelength of 210 nm. For stationary phase, waters spherisorb S5ODS2 (25 cm x 4.6 mm x 5 µm) column was used. The mobile phase of (65/35) % anhydrous methanol in water was used and the flow rate was adjusted to 1 mL/min with an injection volume of 0.5 µL [9]. An online gas chromatograph (GC) was connected to the reactor and used to quantify the CO_2 release. The liquid sample was taken every 30 min from the reactor to determine phenol conversion.

3. Results and Discussions

3.1 Catalyst activity test

Figure 1 shows phenol conversion with time on stream when Al/Zr pillared clay catalyst was used. The maximum conversion (100 %) was reached after 3 h and this increase can be attributed to an increase in the number of active sites available on the catalyst surface. Similar results were reported by Baloyi et al [10,11] when they studied the removal of phenol by CWAO using Al/Zr-PILCs.

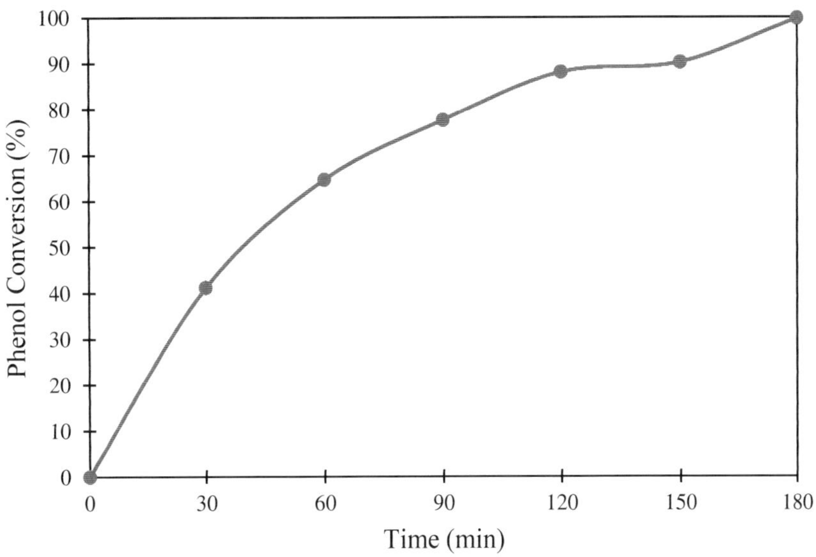

Figure 1: Phenol removal with time in a trickle bed reactor over Al/Zr-PILCs catalyst supported on a monolith (Experimental conditions: 160 °C, 10 bar, 0.012 m/s).

Generally, phenol is broken down to aromatics, carboxylic acid and CO_2 inside the reactor depending on the reaction pathway followed [10]. In this work, the amount of CO_2 released was measured by an online GC connected to the reactor and the results were plotted in Figure 2. As shown in Figure 2 the amount of CO_2 released increases with time and a large peak appeared within 3 h and immediately a sharp decrease was observed afterward signalling complete conversion of the pollutant.

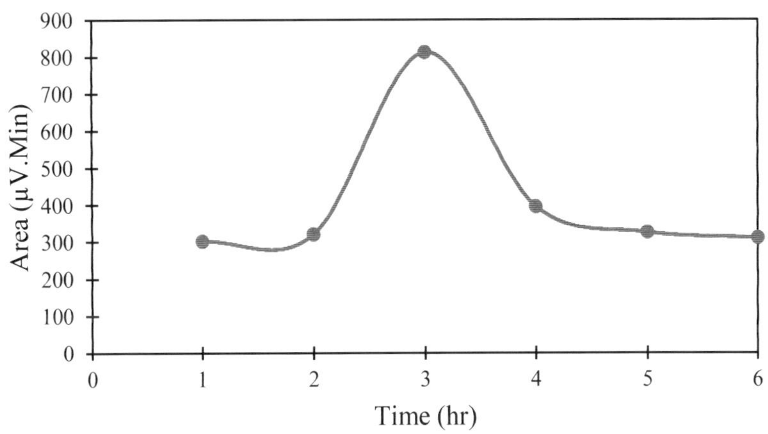

Figure 2: Amount of CO_2 released with time during phenol oxidation in CWAO process.

3.2 Influence of main operational parameters

Our preliminary investigation found 10 bar to be the optimum pressure and results are not shown in this paper. Figure 3 shows that at 120 °C and liquid flow rate of 10 mL/min the conversion of phenol is 86 %. Moreover, an increase in temperature to 140 °C or 160 °C while keeping liquid flow rate and pressure constant (10 mL/min, 10 bar) resulted in the high conversion of phenol 91 and 100%, respectively. This behaviour can be attributed to the fact that the reaction rate constant is a function of temperature and activation energy, according to the Arrhenius equation:

$$k = A.\exp(\frac{E_a}{RT}) \qquad (2)$$

where k is reaction rate constant, A is a pre-exponential factor, E_a is activation energy, R is ideal gas constant and T is the temperature. Furthermore, an increase in temperature results in a decrease in liquid viscosity which facilitates the transfer of reactants from bulk liquid to the surface of the catalyst [12]. The same observations were reported by Abid et al [13] when activated carbon catalyst was used. They reported that 64, 87 and 97% of phenol was converted at 120, 140 and 160 °C. Similarly, Mohammed [12] reported that 88% of phenol was converted at 120 °C and when the temperature was increased to 140 °C or 160 °C phenol conversion increased to 91 and 93%, respectively.

The effect of liquid flow rate on phenol conversion was also studied in the range (10, 20, 30 and 40 mL/min) while keeping other parameters constant (gas flow rate = 0.012 m/s and pressure = 10 bar). As shown in Figure 3 an increase in liquid flow rate has a negative impact on phenol conversion due to the fact that an increase in liquid flow rate reduces residence time resulting in less contact time between the phases. Furthermore, the high liquid flow rate increases film thickness and liquid holdup which decreases contact time between gas and liquid on the catalyst active sites resulting in high resistance to mass transfer. When the reaction temperature was kept constant at 120 °C while varying liquid flow rate between (10, 20, 30 and 40 mL/min), phenol conversion decreased as follow 86, 79, 69 and 64%, respectively. Similarly, phenol conversion decreased from 91 to 70% when the liquid flow rate was increased from 10 to 40 mL/min while keeping the temperature constant at 140 °C. Moreover, when the temperature was increased to 160 °C while varying liquid flow rate between 10 and 40 mL/min phenol conversion also decreased from 100 to 85%. Abid et al [13] reported similar behaviour, phenol conversion was 79% at 1.662 mL/min, while the conversion increased to 87% and 96% at 0.996 and 0.6 mL/min.

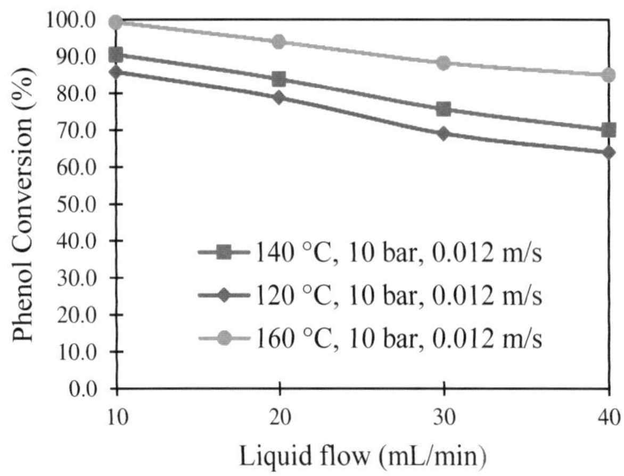

Figure 3: Phenol conversion with liquid flow rate and change in reaction temperature.

3.3 Kinetic model

The reaction kinetics is required to provide a complete interpretation of data obtained in a fixed bed monolithic reactor operated in a trickle flow mode during phenol oxidation. When a simple power law model was used to estimate kinetic parameters as in [4,13,14] equations 3 to 8 were employed.

$$-r_{ph} = k_{ob}.C_{ph}^{\alpha} \tag{3}$$

And k_{ob} can be expressed as follow,

$$K_{ob} = k_o.exp(\frac{-E_{ob}}{RT}).P_{O_2}^{\beta} \tag{4}$$

Since the reaction takes place in a liquid phase the above equation becomes,

$$K_{ob} = k_o.exp(\frac{-E_{ob}}{RT}).X_{O_2}^{\beta} \tag{5}$$

Assuming that the vapour phase behaves ideally [15], then the oxygen mole fraction (X_{O2}) was calculated using Henry's law [5]

$$P_{O_2} = X_{O_2}.\mathrm{H} \tag{6}$$

Equation 5 can be linearized by taking logarithm on both sides of the equation,

$$InK_{ob} = Ink_o + \beta InX_{O_2} - (\frac{-E_{ob}}{RT}) \tag{7}$$

Where α and β are reaction order with respect to phenol and oxygen, E_{ob} is activation energy, H is Henry's constant while k_o is the pre-exponential factor. By assuming ideal plug flow and first order with respect to phenol, equation 5 was integrated and rearranged to obtain K_{ob} as follow,

$$k_{ob} = -(\frac{1}{\tau})In(1 - x_{ph}) \tag{8}$$

K_{ob} was applied to the experimental data by plotting the graph of $In(1 - x_{ph})$ vs τ as in Figure 4 and a perfect fit was found at 160 °C with R^2 value equal to 0.9988 [4,13]. The activation energy was calculated from the slope of the line in Figure 5.

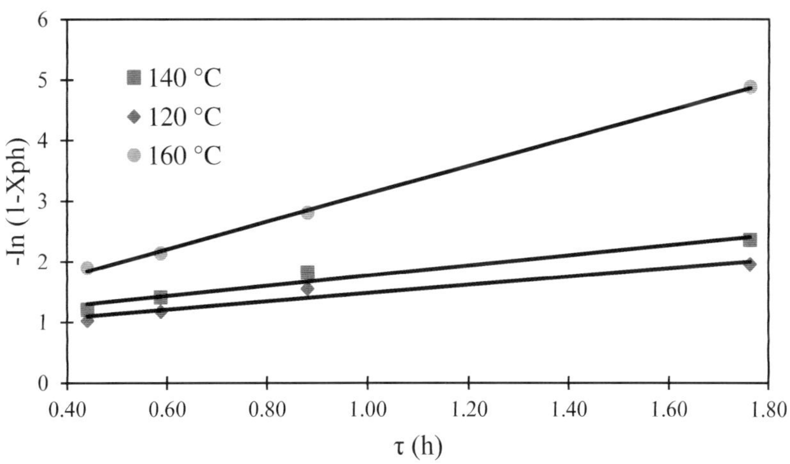

Figure 4: Graph of In(1-Xph) vs τ at 10 bar and temperatures ranging between 120, 140 and 160 °C.

$$Ink_{ob} = \frac{-5086.5}{T} + 12.425 \tag{9}$$

E_{ob} = 42.289 kJ/mol

A wide range of activation energies were reported in open literature as 21.306 kJ/mol [11], 85 kJ/mol [16], 29.299 kJ/mol [4], 35.4 kJ/mol [17] and 74.9 kJ/mol [3] using different catalysts and the results obtained in this study are within the range. These results provide a basis to develop a computational fluid dynamics model of this system which would be instrumental to identifying liquid maldistribution and preventing formation of hotspots [18].

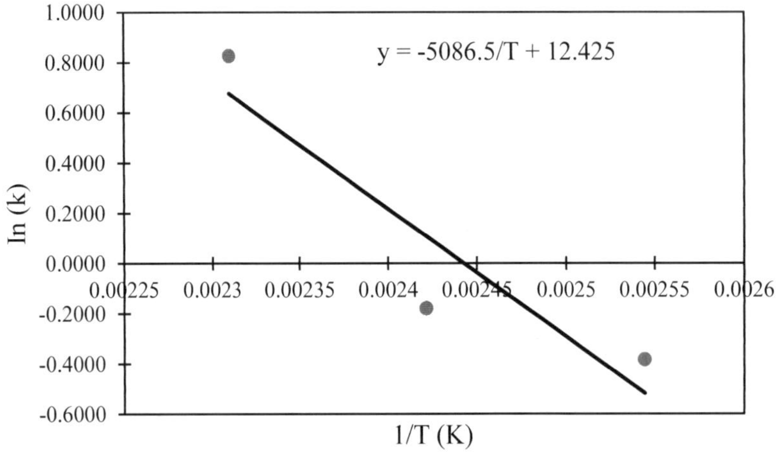

Figure 5: Plot of In(k) versus 1/T at a pressure of 10 bar and temperatures of 120, 140, and 160 °C.

4. Conclusions

Phenol oxidation was studied in a fixed bed monolithic reactor operated in a trickle flow mode over Al/Zr pillared clay catalyst. The rate of phenol degradation was improved by an increase in temperature whereas an increase in liquid flow rate showed an opposite trend. The complete conversion was reached after 3 h. A simple power law model was used to determine activation energy from a linear plot and found to be 42.289 kJ/mol.

Acknowledgments

The authors are grateful to the National Research Foundation (NRF)/Thuthuka NRF grant No. 113652 and Mineral Science Council of South Africa (Mintek) under CWO of Wastewater Project No. ADR 31904 for financial support.

References

[1] Zapico R R, Marín P, Díez F V and Ordóñez S 2017 Assessment of phenol wet oxidation on CuO/γ -Al2O3 catalysts: Competition between heterogeneous and leached-copper homogeneous reaction paths *J. Environ. Chem. Eng.* **5** 2570–8.

[2] Baloyi J, Ntho T and Moma J 2018 Synthesis and application of pillared clay heterogeneous catalysts for wastewater treatment: a review. *RSC Adv.* **8** 5197–211.

[3] Eftaxias A, Font J, Fortuny A, Giralt J, Fabregat A and Stüber F 2001 Kinetic modelling of catalytic wet air oxidation of phenol by simulated annealing *Appl. Catal. B Environ.* **33**175–90.

[4] Abid M F, Jasim F T and Ahmed L S 2014 Kinetic study of phenol removal from wastewater over a 0.5% Pt/γ-Al2O3 catalyst in a trickle bed reactor *Environ. Eng. Manag. J.* **13**1265–75.

[5] Mohammed W T 2014 Active carbon from date stones for phenol oxidation in trickle bed reactor, experimental and kinetic study *J. Eng.* **20** 169–89.

[6] Khadilkar M R 1998 *Performance studies of trickle bed reactors* (Washington University: Sever Institute of Technology).

[7] Cybulski A 2007 Catalytic wet air oxidation: Are monolithic catalysts and reactors feasible? *Ind. Eng. Chem. Res.* **46** 4007–33.

[8] Abid M F, Jasim F T and Ahmed L S 2014 Hydrodynamics and kinetics of phenols removal from industrial wastewater in a trickle bed reactor (Part I: Hydrodynamic Study) *Chem. Biochem. Eng. Q. J.* **28** 319–27.

[9] Guo J 2005 *Catalytic wet oxidation over pillared clay catalyst in packed bed reactors: Experiments and modeling* (Washington University in St. Louis).

[10] Baloyi J, Ntho T and Moma J 2018 Synthesis of highly active and stable Al/Zr pillared clay as catalyst for catalytic wet oxidation of phenol *J. Porous. Mater.* **26** 583–97.

[11] Moma J, Ntho T and Baloyi S J 2018 Influence of operational parameters and kinetic modelling of catalytic wet air oxidation of phenol by Al/Zr pillared clay catalyst *Iran. J. Chem. Chem. Eng.* In press.

[12] Mohammed A E, Jarullah A T, Gheni S A and Mujtaba I M 2016 Optimal design and operation of an industrial three phase reactor for the oxidation of phenol *Comput. Chem. Eng.* **94** 257–71.

[13] Abid M F, Alwan G M and Abdul-Ridha L A 2016 Study on catalytic wet air oxidation process for phenol degradation in synthetic wastewater using trickle bed reactor *Arab. J. Sci. Eng.* **41** 2659–70.

[14] Eftaxias A, Font J, Fortuny A, Fabregat A and Stüber F 2005 Kinetics of phenol oxidation in a trickle bed reactor over active carbon catalyst. *J. Chem. Technol. Biotechnol.* **80** 677–87.

[15] Eze P C and Masuku C M 2018 Vapour–liquid equilibrium prediction for synthesis gas conversion using artificial neural networks *SAJCE* **26** 80–5.

[16] Fortuny A, Bengoa C, Font J, Castells F and Fabregat A 1999 Water pollution abatement by catalytic wet air oxidation in a trickle bed reactor *Catal. Today* **53** 107–14.

[17] Wu Q, Hu X and Yue P L 2005 Kinetics study on heterogeneous catalytic wet air oxidation of phenol using copper/activated carbon catalyst *Int. J. Chem. React. Eng.* **3** A29.

[18] Kapfunde N, Masuku C M and Hildebrandt D 2018 Optimization of the thermal efficiency of a fixed-bed gasifier using computational fluid dynamics *Comput. Aided Chem. Eng.* **44** 1747–52.

AUTHOR INDEX

AUTHOR INDEX